METHODS IN MOLECULAR BIOLOGY

Series Editor
John M. Walker
School of Life and Medical Sciences,
University of Hertfordshire, Hatfield,
Hertfordshire AL10 9AB, UK

For further volumes:
http://www.springer.com/series/7651

Investigations of Early Nutrition Effects on Long-Term Health

Methods and Applications

Edited by

Paul C. Guest

Debden, Essex, UK

 Humana Press

Editor
Paul C. Guest
Debden, Essex, UK

ISSN 1064-3745   ISSN 1940-6029   (electronic)
Methods in Molecular Biology
ISBN 978-1-4939-8533-3   ISBN 978-1-4939-7614-0   (eBook)
https://doi.org/10.1007/978-1-4939-7614-0

Printed on acid-free paper

This Humana Press imprint is published by Springer Nature
The registered company is Springer Science+Business Media, LLC
The registered company address is: 233 Spring Street, New York, NY 10013, U.S.A.

Preface

Environmental factors such as suboptimal parental diet can lead to perturbations of in utero growth and development. In turn, this can have long-lasting negative effects on health in later life by leading to increased risk and earlier onset of diseases such as type 2 diabetes, obesity, cardiovascular conditions, cancer, and neurological disorders. In addition, future generations are at risk of inheriting these conditions through nongenetic mechanisms. Elucidating the molecular pathways involved in these processes may help in the design of prevention strategies and may also lead to the identification of biomarkers and potential drug targets for improved patient management.

This book contains several reviews in the field as well as the step-by-step use of targeted and global approaches within the areas of genomics, epigenetics, proteomics, transcriptomics, and metabolomics which aim to address this dilemma as well as help to pinpoint new treatment strategies. It also describes the generation of several models in the area and methods for assessing health as well as potential approaches for reversing or minimizing disease effects. The chapter contributions come from authors representing five out of the seven continents, including countries such as the UK, Germany, Switzerland, Italy, the USA, Mexico, Brazil, Chile, Australia, New Zealand, and Japan. This reflects the growing importance of this topic throughout the world.

The book will be of high interest to researchers, clinical scientists, physicians, as well as the major drug companies as it gives insights into the latest ideas and technologies enabling progress in this area. It will also be of high interest to both technical and bench scientists as it gives step-by-step instructions on how to carry out each of the protocols. Lastly, it will provide important information on disease mechanisms, as each method will be described in the context of specific disease or therapeutic areas.

Debden, Essex, UK   *Paul C. Guest*

Contents

Contributors

PHABLO ABREU • *Institute of Biomedical Sciences, University of Sao Paulo, Sao Paulo, Brazil; Department of Biochemistry, Institute of Chemistry, University of Sao Paulo, Sao Paulo, Brazil*

SONIA DE ASSIS • *Georgetown University Lombardi Comprehensive Cancer Center, Washington, DC, USA*

JULIAN AYER • *The Heart Centre for Children, The Children's Hospital at Westmead, The University of Sydney, Westmead, NSW, Australia*

BANNY SILVA BARBOSA • *Laboratório de Química Biológica, Departamento de Química Orgânica, Instituto de Química, Universidade Estadual de Campinas, Campinas, SP, Brazil*

A. BECKLEY • *MFB Fertility Inc., Boulder, CO, USA*

FRANK F. BIER • *Fraunhofer Institute for Cell Therapy and Immunology, Branch Bioanalytics and Bioprocesses (IZI-BB), Potsdam, Germany*

NIKITAS BISTOLAS • *Fraunhofer Institute for Cell Therapy and Immunology, Branch Bioanalytics and Bioprocesses (IZI-BB), Potsdam, Germany*

KIMBERLEY D. BRUCE • *Division of Endocrinology, Metabolism, and Diabetes, University of Colorado Anschutz Medical Campus, Aurora, CO, USA*

KEITH BURLING • *Core Biochemical Assay Laboratory, Cambridge University Hospitals NHS Foundation Trust, Cambridge, UK*

DIANA C. CASTRO-RODRÍGUEZ • *Reproductive Biology Department, Instituto Nacional de Ciencias Médicas y Nutrición Salvador Zubirán, Mexico City, Mexico*

TÁSSIA B. B. C. COSTA • *Laboratório de Química Biológica, Departamento de Química Orgânica, Instituto de Química, Universidade Estadual de Campinas, Campinas, SP, Brazil*

GUILHERME CRUZ • *Laboratório de Química Biológica, Departamento de Química Orgânica, Instituto de Química, Universidade Estadual de Campinas, Campinas, SP, Brazil*

RAQUEL SANTANA DA CRUZ • *Georgetown University Lombardi Comprehensive Cancer Center, Washington, DC, USA*

RUI CURI • *Department of Physiology and Biophysics, Institute of Biomedical Sciences, University of Sao Paulo, Sao Paulo, Brazil; Interdisciplinary Post-Graduate Program in Health Sciences, Cruzeiro do Sul University, Sao Paulo, Brazil*

RHIANNON DOWLA • *Discipline of Exercise and Sport Science, Faculty of Health Sciences and the Charles Perkins Centre, University of Sydney, Sydney, NSW, Australia*

CAMILE CASTILHO FONTELLES • *Department of Food and Experimental Nutrition, Food Research Center (FoRC), Faculty of Pharmaceutical Sciences, University of São Paulo, São Paulo, Brazil; Georgetown University Lombardi Comprehensive Cancer Center, Washington, DC, USA*

MARCO A. FORTES • *Department of Physiology and Biophysics, Institute of Biomedical Sciences, University of Sao Paulo, Sao Paulo, Brazil*

JONATHAN FREESTON • *Discipline of Exercise and Sport Science, Faculty of Health Sciences and the Charles Perkins Centre, University of Sydney, Sydney, NSW, Australia*

Francesca L. Guest • *Taunton and somerset NHS trust, Musgrove Park Hospital, Taunton, UK*

Paul C. Guest • *Laboratory of Neuroproteomics, Department of Biochemistry and Tissue Biology, Institute of Biology, University of Campinas, Campinas, Brazil*

Marjan Mosalman Haghighi • *Exercise Health and Performance Faculty Research Group, Faculty of Health Sciences, University of Sydney, Camperdown, NSW, Australia*

Ian Halsall • *Core Biochemical Assay Laboratory, Cambridge University Hospitals NHS Foundation Trust, Cambridge, UK*

Steve F. C. Hawkins • *Bioline Reagents Limited, Unit 16, The Edge Business Centre, London, UK*

Leena Hilakivi-Clarke • *Georgetown University Lombardi Comprehensive Cancer Center, Washington, DC, USA*

Sandro Massao Hirabara • *Institute of Biomedical Sciences, University of Sao Paulo, Sao Paulo, Brazil; Interdisciplinary Post-Graduate Program in Health Sciences, Cruzeiro do Sul University, Sao Paulo, Brazil*

Ulrich Höller • *Analytical Research Center, DSM Nutritional Products, Kaiseraugst, Switzerland*

Nozomi Itani • *Division of Women's Health, Women's Health Academic Centre, King's College London and King's Health Partners, London, UK*

Keiko Iwata • *Department of Biology, University of Padova, Padova, Italy; Research Center for Child Mental Development, University of Fukui, Fukui, Japan*

Karen R. Jonscher • *Department of Anesthesiology, University of Colorado Anschutz Medical Campus, Aurora, CO, USA*

Imran Y. Khan • *Division of Women's Health, Women's Health Academic Centre, King's College London and King's Health Partners, London, UK*

Ann-Katrin Kraeuter • *Laboratory of Psychiatric Neuroscience, Australian Institute of Tropical Health and Medicine, James Cook University, Townsville, QLD, Australia; Discipline of Biomedicine, College of Public Health, Medicine and Veterinary Sciences, James Cook University, Townsville, QLD, Australia*

Divya Krishnamurthy • *Department of Chemical Engineering and Biotechnology, University of Cambridge, Cambridge, UK*

Cecilia Laurisch • *Fraunhofer Institute for Cell Therapy and Immunology, Branch Bioanalytics and Bioprocesses (IZI-BB), Potsdam, Germany*

Dan Ma • *Department of Neurosciences, University of Cambridge, Cambridge, UK*

J. L. Martinez-Hurtado • *TUM Incubator, Technische Universität München, Munich, Germany*

Lucas Gelain Martins • *Laboratório de Química Biológica, Departamento de Química Orgánica, Instituto de Química, Universidade Estadual de Campinas, Campinas, SP, Brazil*

Gabriel Nasri Marzuca-Nassr • *Department of Internal Medicine, Faculty of Medicine, Universidad de La Frontera, Temuco, Chile*

P. R. Matías-García • *Institute of Medical Informatics, Biometry and Epidemiology (IBE), Ludwig-Maximilians-Universität München, Munich, Germany*

Phillippa A. Matthews • *Division of Women's Health, Women's Health Academic Centre, King's College London and King's Health Partners, London, UK*

Bridin Murnion • *Concord Repatriation and General Hospital, Sydney Local Health District, Sydney, NSW, Australia; Discpline of Addiction Medicine, Faculty of Medicine, University of Sydney, Sydney, NSW, Australia*

LESLIE MYATT • *Department of Obstetrics & Gynecology, Oregon Health & Science University, Portland, OR, USA; Bob and Charlee Moore Institute for Nutrition & Wellness, Oregon Health & Science University, Portland, OR, USA*

THOMAS PRATES ONG • *Department of Food and Experimental Nutrition, Food Research Center (FoRC), Faculty of Pharmaceutical Sciences, University of São Paulo, São Paulo, Brazil*

SUSAN E. OZANNE • *Metabolic Research Laboratories, University of Cambridge, Cambridge, UK; MRC Metabolic Diseases Unit, Wellcome Trust-MRC Institute of Metabolic Science, Addenbrooke's Hospital, Cambridge, UK*

DONALD B. PALMER • *Department of Comparative Biomedical Sciences, Royal Veterinary College, London, UK*

LUCAS CARMINATTI PANTALEÃO • *MRC Metabolic Diseases Unit, University of Cambridge Metabolic Research Laboratories, Wellcome Trust-MRC Institute of Metabolic Science, Addenbrooke's Hospital, Cambridge, UK*

HARALD PETER • *Fraunhofer Institute for Cell Therapy and Immunology, Branch Bioanalytics and Bioprocesses (IZI-BB), Potsdam, Germany*

CARLOS H. PINHEIRO • *Department of Physiology and Biophysics, Institute of Biomedical Sciences, University of Sao Paulo, Sao Paulo, Brazil*

LUCILLA POSTON • *Division of Women's Health, Women's Health Academic Centre, King's College London and King's Health Partners, London, UK*

HASSAN RAHMOUNE • *Department of Chemical Engineering and Biotechnology, University of Cambridge, Cambridge, UK*

DOUGLAS REES • *Division of Women's Health, Women's Health Academic Centre, King's College London and King's Health Partners, London, UK*

CLARE M. REYNOLDS • *Liggins Institute, University of Auckland, Auckland, New Zealand*

KIA ROBERTS • *Concord Repatriation and General Hospital, Sydney Local Health District, Sydney, NSW, Australia*

GUADALUPE L. RODRÍGUEZ-GONZÁLEZ • *Reproductive Biology Department, Instituto Nacional de Ciencias Médicas y Nutrición Salvador Zubirán, Mexico City, Mexico*

KIERON ROONEY • *Discipline of Exercise and Sport Science, Faculty of Health Sciences and the Charles Perkins Centre, University of Sydney, Sydney, NSW, Australia*

ZOLTÁN SARNYAI • *Laboratory of Psychiatric Neuroscience, Australian Institute of Tropical Health and Medicine, James Cook University, Townsville, QLD, Australia; Discipline of Biomedicine, College of Public Health, Medicine and Veterinary Sciences, James Cook University, Townsville, QLD, Australia*

MARIA V. M. SCERVINO • *Department of Physiology and Biophysics, Institute of Biomedical Sciences, University of Sao Paulo, Sao Paulo, Brazil*

M. SCHMIDMAYR • *Frauenklinik und Poliklinik, Technische Universität München, Munich, Germany*

SOEREN SCHUMACHER • *Fraunhofer Institute for Cell Therapy and Immunology, Branch Bioanalytics and Bioprocesses (IZI-BB), Potsdam, Germany*

V. SEIFERT-KLAUSS • *Frauenklinik und Poliklinik, Technische Universität München, Munich, Germany*

AMANDA N. SFERRUZZI-PERRI • *Department of Physiology, Development and Neuroscience, Centre for Trophoblast Research, University of Cambridge, Cambridge, UK*

LEONARDO R. SILVEIRA • *Department of Structural and Functional Biology, Obesity and Comorbidities Research Center, Institute of Biology, Unicamp, Campinas, Sao Paulo, Brazil*

LAURIE STEPHEN • *Ampersand Biosciences, Saranac Lake, NY, USA*

JANE L. TARRY-ADKINS • *MRC Metabolic Diseases Unit, University of Cambridge Metabolic Research Laboratories, Wellcome Trust-MRC Institute of Metabolic Science, Addenbrooke's Hospital, Cambridge, UK*

LJUBICA TASIC • *Laboratório de Química Biológica, Departamento de Química Orgânica, Instituto de Química, Universidade Estadual de Campinas, Campinas, SP, Brazil*

KEVIN TAYLOR • *Core Biochemical Assay Laboratory, Cambridge University Hospitals NHS Foundation Trust, Cambridge, UK*

PAUL D. TAYLOR • *Division of Women's Health, Women's Health Academic Centre, King's College London and King's Health Partners, London, UK*

KENT L. THORNBURG • *Bob and Charlee Moore Institute for Nutrition & Wellness, Oregon Health & Science University, Portland, OR, USA; Knight Cardiovascular Institute, Oregon Health & Science University, Portland, OR, USA*

DEREK TRAN • *Faculty of Health Sciences, University of Sydney, Sydney, NSW, Australia; Department of Clinical Medicine, Macquarie University, Sydney, NSW, Australia*

JOHANNES VEGT • *Appamedix, Innovations-Centrum CHIC, Berlin, Germany*

MARK H. VICKERS • *Liggins Institute, University of Auckland, Auckland, New Zealand*

KAIO F. VITZEL • *School of Health Sciences, College of Health, Massey University, Auckland, New Zealand*

ELENA ZAMBRANO • *Department of Reproductive Biology, Instituto Nacional de Ciencias Médicas y Nutrición Salvador Zubirán, Mexico City, Mexico*

Part I

Reviews

Chapter 1

Nutritional Programming Effects on Development of Metabolic Disorders in Later Life

Thomas Prates Ong and Paul C. Guest

Abstract

Developmental programming resulting from maternal malnutrition can lead to an increased risk of metabolic disorders such as obesity, insulin resistance, type 2 diabetes and cardiovascular disorders in the offspring in later life. Furthermore, many conditions linked with developmental programming are also known to be associated with the aging process. This review summarizes the available evidence about the molecular mechanisms underlying these effects, with the potential to identify novel areas of therapeutic intervention. This could also lead to the discovery of new treatment options for improved patient outcomes.

Key words Developmental programming, Metabolic disease, Diabetes, Obesity, Aging, Novel therapeutics

1 Introduction

The results of human epidemiological and animal model studies have led to the suggestion that imbalances in maternal nutrition during critical time windows of development can have long-term detrimental health consequences for the offspring [1–3]. These effects occur by developmental programming, and emerging evidence suggests that they are caused by epigenetic changes of chromatin structure and regulation of gene expression [4–6]. Such effects can also occur in later life and can influence the way in which an individual responds to environmental changes in a manner independent of DNA sequence [7, 8]. Epigenetic changes can also occur throughout life, leading to the potential for the prevention and treatment of diseases whether or not they are the result of developmental programming [9].

The molecular mechanisms underlying these effects are slowly being unraveled. The main findings indicate that nutritional imbalances during pregnancy can cause permanent changes to the physiology and metabolism of the offspring, which will lead to effects on

Paul C. Guest (ed.), *Investigations of Early Nutrition Effects on Long-Term Health: Methods and Applications*, Methods in Molecular Biology, vol. 1735, https://doi.org/10.1007/978-1-4939-7614-0_1, © Springer Science+Business Media, LLC 2018

their health with an increased risk of metabolic-related diseases in later life [2, 10]. This increased risk may be linked to an advanced aging phenotype [11]. Possible imbalances leading to these effects during pregnancy include both under- and overnutrition, as well as exposure to drugs or alcohol, and they can also occur due to maternal overweight, obesity, excess gestational weight, stress, or diabetes mellitus.

This chapter focuses on these possibilities as an explanation for the effects of developmental programming on long-term health. It also describes potential intervention strategies targeting the perturbed molecular pathways as a means of helping to achieve improved health outcomes for the affected individuals.

2   Developmental Programming

2.1   The Thrifty Phenotype Hypothesis

Approximately 25 years ago, Hales and Barker published influential papers describing associations between suboptimal growth in early life and increased risk in later life of impaired glucose tolerance [12], type 2 diabetes [13], metabolic syndrome, and cardiovascular disease [14]. They called this phenomenon the "thrifty phenotype hypothesis," which proposes that a fetus will undergo developmental programming changes in organ structure and adaption of metabolism to ensure immediate survival in a poor in utero environment. This may involve the preservation of specific vital organs such as the brain at the expense of other organs like the heart, pancreas, liver, kidney, and skeletal muscle. In utero growth restriction may also occur. However, if the postnatal experienced life is different than the predicted one, the programmed individual would be specially prone to develop these metabolic alteration in conditions of normal or excessive nutrition. Epidemiological investigations, such as the Uppsala, Sweden; Helsinki, Finland; and the Nurses' Health Study, USA, have identified correlations between low birth weight and development of cardiovascular disease, hypertension, type 2 diabetes, overweight, and obesity in later life (for review, see [15]).

2.2   Effects of Maternal Malnutrition

One of the most convincing epidemiological pieces of evidence for the thrifty phenotype hypothesis has come from follow-up studies of the Dutch Hunger Winter. The Dutch people experienced a severe famine due to a food blockade between November 1944 and May 1945, during World War II. One study found that offspring of mothers who were pregnant during the famine had a low body weight at birth and glucose intolerance in later life [16]. The timing of exposure to the famine was also important in the way this programming was manifested. For example, there was increased prevalence of coronary heart disease, atherogenic lipid profiles, and higher adiposity in offspring of mothers who were exposed during early gestation [17]. In contrast, offspring of mothers exposed to the famine in the middle of gestation had microalbuminuria and abnormal renal function in adulthood [18]. In addition, those exposed in late gestation had offspring with the highest risk of developing type 2 diabetes in later life [16]. These susceptibilities are likely to be due to the environmental effects and not the result of genetic programming as shown by studies in monozygotic twins of different birth weights. Such studies showed that the twin with the lowest birth weight was more likely to develop type 2 diabetes [19], glucose intolerance [20], impaired production of insulin, and insulin resistance [21, 22], compared to the normal birth weight twin. A poor maternal environment followed by an increased rate of *postnatal* growth can also have an impact on age-related disease risk and lead to effects such as poor glucose tolerance [23], insulin resistance [24], endothelial dysfunction [25], hypertension [26], cardiovascular disease [27], and nonalcoholic fatty liver disease [28]. In contrast, breastfeeding can induce slower postnatal growth [29] and reduce blood pressure [30], cholesterol [31], insulin resistance [32], and risk of childhood obesity [33], compared to cases of formula feeding.

2.3 Effects of Maternal Obesity

Obesity has reached epidemic proportions both in developing and developed countries. This is a major problem since it can lead to birth of children that are either larger or smaller than the average size from a normal weight pregnancy [34]. In addition, maternal obesity is often associated with gestational diabetes and hyperglycemia, which tends to give rise to children of a larger size. This occurs because maternal glucose crosses the placental barrier but insulin cannot. This means that the fetus attempts to regulate glucose homeostasis by increasing insulin production and release from its own pancreatic islet beta cells. This can result in a larger offspring size since insulin can serve as a growth factor. The Pima Indians are a well-known population with high levels of gestational diabetes as well as type 2 diabetes and a high prevalence of obesity. In this population, the association of birth weight with these conditions has been shown to be U-shaped, with the highest prevalence occurring in those offspring who had a low or high birth weight [35]. Larger offspring resulting from maternal diabetes or obesity have an increased risk of developing metabolic syndrome and are more prone to being overweight or obese [36, 37], with an increased risk of premature mortality from cardiovascular [38] and coronary heart disease [39].

3 Animal Models

Although epidemiological studies have led to new insights into the phenomena of developmental programming, a more complete understanding of the underlying mechanisms and potential intervention strategies requires the strategic use of animal models. The main reasons for this stem from the fact that it is difficult to control for the many confounding factors involved in human studies. In addition, the extensive human lifespan also makes most studies impractical. Furthermore, animal studies allow controlled interventions during specific developmental windows and the isolation of different tissues for metabolic and molecular analysis. Therefore, the following sections focus on animal models of suboptimal nutrition during development and the resulting effects on physiology and health in later life.

3.1 Maternal Protein Restriction

Maternal protein restriction has received widespread use as a model of suboptimal in utero nutrition. The earliest studies consisted of rats administered a maternal diet composed of 8% protein, compared to the control situation of 20% protein. These studies found that the low-protein maternal diet led to growth restriction of the offspring with an age-dependent loss of glucose tolerance and development of insulin resistance and diabetes later in life (by approximately 15 months of age) [40]. The maternal low-protein rat offspring showed abnormalities of specific molecules of the intracellular insulin signaling pathway in skeletal muscle tissue preceding the manifestation of diabetes, including alterations in protein kinase C-zeta, glucose transporter-4, and phosphatidylinositol 3-kinase p85-alpha [41]. As corroborating evidence, these same molecules were also disrupted in both skeletal muscle and adipose tissue in a study of low birth weight males at a young age [41, 42]. Other models of maternal protein restriction have identified physiological changes such as hypertension [43], renal and vascular dysfunction [44], and hepatic steatosis [45] in the offspring.

3.2 Maternal Protein Restriction Followed by Catch-Up Growth

Other studies have shown that the maternal low-protein diet followed by catch-up growth after birth can lead to deleterious effects on physiology and health. Investigations in which newborn pups of maternal low-protein rodents were cross-fostered to control-fed dams found that these offspring had reduced longevity compared to their control littermates and this effect could be exacerbated by a *postnatal* obesogenic diet [46, 47]. This was supported by the finding of an age-dependent development of fatty liver in offspring that were subjected to catch-up growth, and this was associated with increased transcription of genes associated with lipid accumulation [48]. Similarly, this model also displays insulin resistance in adipose tissue [49], hepatic fibrosis, and an increased inflammatory profile [50]. In contrast, other studies found that rodent offspring born to dams fed a control 20% protein diet, followed by cross-fostering to mothers fed with the low-protein diet, experienced a slower *postnatal* growth and increased longevity compared to those animals fed with the control diet throughout gestation and lactation [46, 47]. At the physiological and molecular levels, offspring of the same model also showed reduced aging of the thymus [51] and spleen [52], with improved insulin sensitivity [53] and protection of the kidneys [54]. These findings suggest that a slower postnatal growth can be beneficial to the health of offspring in later life. This hypothesis has been tested and confirmed in many other species such as yeast, flies, and worms in which caloric restriction has been found to increase longevity [55]. The effect may involve alterations in the nutrient-sensing network, including altered production of proteins such as growth hormone, insulin-like growth factor 1, and the mechanistic target of rapamycin (mTOR) [56].

Uteroplacental insufficiency is a common cause of growth restriction during pregnancy. One investigation modeled this by uterine artery ligation in rat dams and found that the offspring had restricted growth, with hepatic insulin resistance [57], and this progressed to diabetes [58]. Female rat offspring of mothers that had been subjected to bilateral uterine artery ligation had increased arterial stiffness [59] and minor renal dysfunction [60]. However, another group which tested the same model found that 12-month-old female offspring were hypertensive, but there was no evidence of glucose intolerance [61].

3.3 Maternal Calorie Restriction

Maternal caloric/nutrient restriction is a critical concern in developing countries, and animal models based on this trait have been developed to increase our understanding of the molecular mechanisms involved. One study tested a rat model involving maternal calorie restriction of 50% and found that the offspring had an accelerated age-dependent loss of glucose tolerance [58] and another study which used the same model found that the offspring were hypertensive with endothelial dysfunction [62, 63]. Another study which used nonhuman primates restricted calories to 70% of the control value and found changes in renal morphology and associated transcriptomic profiles in the offspring [64]. Furthermore, an ovine model which employed 40% maternal calorie restriction as well as *postnatal* catch-up growth found growth restriction in the offspring followed by development of obesity, elevated cortisol levels, as well as insulin and leptin resistance [65]. Finally, studies of a 30% maternal calorie-restricted rat model found hyperinsulinemia and hyperleptinemia in the offspring, and these conditions worsened by exposure of the offspring to a hypercaloric diet after weaning [66].

3.4 Maternal Obesity

Studies of animal models of maternal obesity have shown the development of multiple pathologies in the offspring in later life. Such models use a diet rich in simple sugars and saturated fats. The male offspring of these rats have a normal body weight from birth to 8 weeks of age, but after this time, they develop cardiac hypertrophy, cardiac dysfunction, and hyperinsulinemia, with increased phosphorylation of protein kinase B (AKT), extracellular signal-regulated kinase (ERK), and mTOR activation [67, 68]. Using the same maternal diet, female offspring from obese mothers became hyperphagic between 4 and 6 weeks of age and showed increased adiposity at 6 months of age [69]. Other studies of the same maternal obesity model have found that the offspring develop fatty livers along with disrupted lipid metabolism [70] and insulin resistance [71]. Nonhuman primate studies have also found hepatic lipotoxicity in offspring of mothers fed an obesogenic diet [72].

In the obesity models, studies have shown that rat dams fed a high-fat diet (consisting of 60% fat) transmitted the risk of obesity to the offspring [73]. Other studies modeled maternal obesity in dams through intragastric feeding, and these demonstrated the programming of adiposity and insulin resistance in offspring weaned onto a high-fat diet [74, 75]. A study which compared maternal obesity and diet has indicated that there may be differential effects on fetal growth during gestation [76], with the finding of reduced fetal weight in the maternal obesity model and abnormal placental growth in the cafeteria diet model. This indicates that further work is warranted on the differences between the various models of suboptimal nutrition and obesity. This could be important for future studies aiming at identifying novel intervention methods.

4   Mechanisms of Developmental Programming

4.1   Effects of Organ Structure and Function

Decrease of nutrient supply to an organ during a critical period of development can result in permanent alterations of structure and function. This has been demonstrated using animal models of intrauterine growth and protein restriction. These studies have found a reduction in pancreatic beta cell mass [77–79], a shift in liver function toward glucose production (as opposed to glucose utilization) [80–83], reduced skeletal muscle mass [84, 85], and adipocyte hypertrophy [86]. Furthermore, animal model studies have found that maternal protein restriction can negatively impact on lifespan [82]. This effect may be due to oxidative stress, inflammation, and altered hormonal profiles. Oxidative stress can lead to DNA damage, increased DNA methylation, and accelerated cellular aging [87–89]. In pancreatic islet cells, increased oxidative damage can lead to shorter telomeres and, therefore, advanced cellular aging, beta cell dysfunction, insulin resistance, and hyperglycemia in the offspring [90].

In one version of the high-fat diet model, female mice are fed a palatable obesogenic diet, comprised of 16% saturated fats and 30% simple sugars [91, 92]. These levels are similar to those found in the diets of obese humans [93]. Mice fed on this diet are normally heavier than control dams throughout the pregnancy, birth, and lactation periods, and this occurs due to increased adiposity. In addition, the obese dams have higher circulating levels of glucose, insulin, free fatty acids, and cholesterol compared to the controls. In the case of the offspring, there are normally no significant differences in body weight or percentage fat composition over the first 8 weeks although these parameters tend to be increased in older offspring [91, 92]. However, feeding behavior [94, 95], insulin resistance [91, 92, 96], lipid metabolism, and cardiac function are all affected negatively. Maternal obesity in rodents also appears to increase the preference of the offspring for fatty and sugary foods along with increased frequency and duration of feeding. Circulating insulin levels are elevated with a concomitant decrease in the levels of insulin signaling proteins in adipose tissue. This includes the insulin receptor, phosphoinositide 3-kinase, the serine-threonine protein kinases AKT1 and AKT2, and the insulin receptor substrate 1 protein. The negative effects on cardiovascular function are consistent with the findings of the Helsinki birth cohort study, which found a significant association between maternal obesity and cardiovascular disease in the offspring [97].

4.2 Epigenetics

The mechanisms of epigenetic programming include DNA methylation, histone modifications, and regulation by DNA-binding proteins and noncoding RNAs. A number of studies have now shown that maternal nutritional status can permanently alter the epigenome of the developing fetus, consequently impacting long-term health [98–103]. For example, in studies of human and rodent intrauterine growth restriction offspring, epigenetic changes in the expression of genes related to glucose homeostasis, insulin secretion, and pancreatic islet cell turnover have been observed [12, 104, 105]. Changes in microRNAs have also been identified in offspring of high-fat fed dams, including methyl-CpG-binding domain proteins [106].

Transgenerational effects of obesity and diet have also been observed as shown by studies which found that the maternal high-fat diet influenced body length and insulin sensitivity in mice at least as far as the third generation [107, 108].

5 Intervention Strategies

5.1 Antioxidants

As increased reactive oxygen species appears to be an underlying consequence of suboptimal maternal nutrition, some studies have focused on the use of antioxidants as potential therapeutic test compounds in models of developmental programming. These have included treatment with high concentrations of vitamins such as A, C, and E and minerals such as selenium to reduce adiposity and improve glucose tolerance and insulin signaling [109]. Another study showed that it was possible to prevent vascular dysfunction in the offspring in a maternal protein restriction model by antenatal treatment with the antioxidant lazaroid [110]. In addition, supplementation with 5 mg of melatonin increased umbilical artery blood flow in a nutrient restriction model in sheep [111]. As most signs of maternal malnutrition can only be observed after birth, four studies from the same group successfully tested postnatal supplementation of coenzyme Q10 (CoQ10) as a preventative of cardiac dysfunction [112, 113], accelerated aging, adipose tissue insulin signaling dysregulation and inflammation [49], and hepatic fibrosis and oxidative stress [50] using a rat model of maternal protein restriction followed by catch-up growth.

5.2 Exercise

The finding that insulin resistance is a common occurrence in offspring of maternally malnourished humans and animals suggests that intervention strategies aimed at increasing insulin sensitivity may be effective. In addition, lifestyle changes such as exercise may be beneficial. Exercise is known to improve insulin sensitivity, independent of adiposity levels [114]. The UK Pregnancies Better Eating and Activity Trial (UBPEAT) attempted to test this potential intervention strategy using a large cohort of obese pregnant women who were encouraged to include a mild exercise program in their routines showed [115]. This resulted in significantly reduced gestational weight gain [115]. Follow-up studies of this cohort will establish whether or not this has beneficial effects in the offspring. This possibility has also been tested in rodent models of maternal exercise during pregnancy, as well as in the offspring to determine whether or not exercise during the early postnatal period can help to improve offspring metabolic health [116]. In this study the exercise group was given continuous access to a running wheel in their home cage, and the sedentary group was provided with a locked wheel. Another study used a rat maternal obesity model to study the effects of exercise of dams before and throughout pregnancy [117]. Although, this showed no difference in offspring body weight, the exercise regime helped to reverse serum corticosterone levels. Exercise also helped to restore leptin and triglyceride levels, consistent with the findings that male offspring of exercised mothers had a leaner body mass and decreased fat mass compared to male control offspring [117].

5.3 Insulin-Sensitizing Agents

A clinical study tested whether or not treatment of obese pregnant women with the insulin-sensitizing drug metformin can improve maternal and offspring health [118], although the outcome may take several years considering that the offspring may have to reach a certain age before statistically meaningful differences can be seen.

6 Conclusions

An increased risk of developing disorders such as type 2 diabetes, obesity and cardiovascular disease may be linked with fetal growth under conditions of a suboptimal maternal diet or metabolic health (Fig. 1). The molecular mechanisms of this effect include

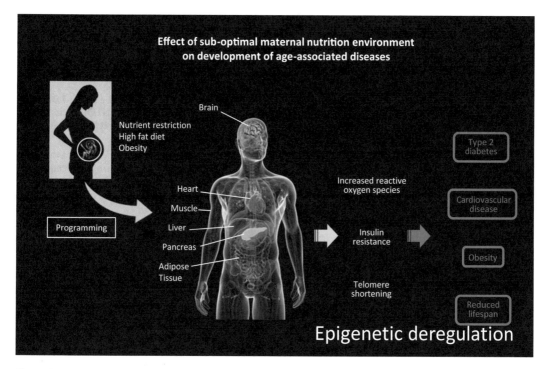

Fig. 1 Schematic diagram showing effects of maternal under- or overnutrition on development of metabolic diseases in the offspring in later life

epigenetic modifications, structural disturbances in some organs and tissues, generation of reactive oxygen species, insulin resistance, abnormal glucose metabolism, and accelerated aging. These effects have been identified in both epidemiological and animal model studies. Further studies of the molecular pathways underlying these phenomena of developmental programming could lead to development of new interventions suitable for use in human populations. This is critical considering the alarming effect that age-associated diseases linked to developmental programming can have at the individual and societal levels.

References

1. Kral JG, Biron S, Simard S, Hould FS, Lebel S, Marceau S et al (2006) Large maternal weight loss from obesity surgery prevents transmission of obesity to children who were followed for 2 to 18 years. Pediatrics 118: e1644–e1649

2. Waterland RA, Garza C (1999) Potential mechanisms of metabolic imprinting that lead to chronic disease. Am J Clin Nutr 69:179–197

3. Waterland RA (2005) Does nutrition during infancy and early childhood contribute to

later obesity via metabolic imprinting of epigenetic gene regulatory mechanisms? Nestle Nutrition workshop series. Paediatr Progr 56:157–171

4. Cao-Lei L, Massart R, Suderman MJ, Machnes Z, Elgbeili G, Laplante DP et al (2014) DNA methylation signatures triggered by prenatal maternal stress exposure to a natural disaster: Project Ice Storm. PLoS One 9:e107653

5. Gluckman PD, Hanson MA, Cooper C, Thornburg KL (2008) Effect of in utero and

early-life conditions on adult health and disease. N Engl J Med 359:61–73

6. Heijmans BT, Tobi EW, Stein AD, Putter H, Blauw GJ, Susser ES et al (2008) Persistent epigenetic differences associated with prenatal exposure to famine in humans. Proc Natl Acad Sci U S A 105:17046–17049

7. Milagro FI, Mansego ML, De Miguel C, Martinez JA (2013) Dietary factors, epigenetic modifications and obesity outcomes: progresses and perspectives. Mol Aspects Med 34:782–812

8. Jacobsen SC, Gillberg L, Bork-Jensen J, Ribel-Madsen R, Lara E, Calvanese V et al (2014) Young men with low birthweight exhibit decreased plasticity of genome-wide muscle DNA methylation by high-fat overfeeding. Diabetologia 57:1154–1158

9. Waterland RA (2014) Epigenetic mechanisms affecting regulation of energy balance: many questions, few answers. Annu Rev Nutr 34:337–355

10. Waterland RA, Michels KB (2007) Epigenetic epidemiology of the developmental origins hypothesis. Annu Rev Nutr 27:363–388

11. Duque-Guimarães D, Ozanne S (2017) Early nutrition and ageing: can we intervene? Biogerontology. https://doi.org/10.1007/s10522-017-9691-y

12. Hales CN, Barker DJP, Clark PMS, Cox LJ, Fall C, Osmond C et al (1991) Fetal and infant growth and impaired glucose tolerance at aged 64. BMJ 303:1019–1022

13. Hales CN, Barker DJ (1992) Type-2-(non-insulin-dependent) diabetes mellitus: the thrifty phenotype hypothesis. Diabetologia 35:595–601

14. Barker DJ, Hales CN, Fall CH, Osmond C, Phipps K, Clark PM (1993) Type 2(non-insulin-dependent) diabetes mellitus, hypertension, and hyperlipidaemia (syndrome X): relation to reduced fetal growth. Diabetologia 36:62–67

15. Sutton EF, Gilmore LA, Dunger DB, Heijmans BT, Hivert MF, Ling C et al (2016) Developmental programming: state-of-the-science and future directions-summary from a Pennington Biomedical symposium. Obesity (Silver Spring) 24:1018–1026

16. Ravelli AC, van der Meulen JH, Michels RP, Osmond C, Barker DJ, Hales CN et al (1998) Glucose tolerance in adults after prenatal exposure to famine. Lancet 351:173–177

17. Roseboom T, de Roijj S, Painter R (2006) The Dutch famine and its long-term consequences for adult health. Early Hum Dev 82:485–491

18. Painter RC, Roseboom TJ, van Montfrans GA, Bossuyt PMM, Krediet RT, Osmond C et al (2005) Microalbuminuria in adults after prenatal exposure to the Dutch famine. J Am Soc Nephrol 1:189–194

19. Poulsen P, Vaag AA, Kyvik KO, Moller Jensen D, Beck-Nielsen H (1997) Low birth weight is associated with NIDDM in discordant monozygotic and dizygotic twin pairs. Diabetologia 40:439–446

20. Grunnet L, Vielworth S, Vaag A, Poulsen P (2007) Birth weight is nongenetically associated with glucose intolerance in elderly twins, independent of adult obesity. J Intern Med 262:96–103

21. Poulsen P, Levin K, Beck-Nielsen H, Vaag A (2002) Age-dependent impact of zygosity and birth weight on insulin secretion and in twins. Diabetologia 45:1649–1657

22. Poulsen P, Vaag A (2006) The intrauterine environment as reflected by birth size and twin and zygosity status influences insulin action and intracellular glucose metabolism in an age- or time-dependent manner. Diabetes 55:1819–1825

23. Crowther NJ, Cameron N, Trusler J, Gray IP (1998) Association between poor glucose tolerance and rapid postnatal weight gain in seven-year-old children. Diabetologia 10:1163–1167

24. Ong KK, Petry CJ, Emmett PM, Sandhu MS, Kiess W, Hales CN et al (2004) Insulin sensitivity and secretion in normal children related to size at birth, postnatal growth and plasma-like growth factor-1 levels. Diabetologia 6:1064–1070

25. Touwslager RN, Houben AJ, Tan FE, Gielen M, Zeegers MP, Stehouwer CD et al (2015) Growth and endothelial function in the first two years of life. J Pediatr 166:666–667s1

26. Law CM, Shiell AW, Newsome CA, Syddall HE, Shinebourne EA, Fayers PM et al (2002) Fetal, infant and childhood growth and adult blood pressure: a longitudinal study from birth to 22 years of age. Circulation 105:1088–1092

27. Erikkson JG, Forsen T, Tuomilehto J, Winter PD, Osmond C, Barker DJP (2001) Catch-up growth in childhood and death from coronary heart disease: longitudinal study. BMJ 318:427–431

28. Faienza MF, Brunetti G, Ventura A, D'Aniello M, Pepe T, Giordano P et al (2013) Nonalcoholic fatty liver disease in prepubertal children born small for gestational age: influence of rapid catch up growth. Horm Res Paediatr 79:103–109

29. Fewtrell MS, Morley R, Abbott RA, Singhal A, Stephenson T, MacFadyen UM et al (2001) Catch-up growth in small for gestational-age infants: a randomized trial. Am J Clin Nutr 74:516–523

30. Martin RM, Gunnell D, Davey-Smith G (2005) Breast-feeding in infancy and blood-pressure in later life; systemic review and meta-analysis. Am J Epidemiol 161:15–26

31. Owen CG, Whincup PH, Odoki K, Gilg JA, Cook DG (2002) Infant feeding and blood cholesterol: a study in adolescents and a systematic review. Pediatrics 110:597–608

32. Ravelli ACJ, Van der Meulen JHP, Osmond C, Barker DJP, Bleker OP (2000) Infant feeding and adult glucose tolerance lipid profile, blood pressure and obesity. Arch Dis Child 82:248–252

33. Arenz S, Ruckerl R, Koletzko B, von Kries R (2004) Breastfeeding and childhood obesity: a systematic review. Int J Obes (Lond) 10:1247–1256

34. Djelanik AAAMJ, Kunst AE, van der Waal MF, Smit HA, Vrijkotte TGM (2011) Contribution of overweight and obesity to the occurrence of adverse pregnancy outcomes in a multi-ethic cohort: population attributive fractions for Amsterdam. Epidemiology 119:283–290

35. McCance DR, Pettitt DJ, Hanson RL, Jacobsson LT, Knowler WC, Bennett PH (1994) Birth weight and non-insulin dependent diabetes: thrifty genotype, thrifty phenotype, or surviving baby phenotype? BMJ 308:942–945

36. Daraki V, Georgiou V, Papavasiliou S, Chalkiadaki G, Karahaliou M, Koinaki S et al (2015) Metabolic profile in early pregnancy is associated with offspring adiposity at 4 years of age: the Rhea pregnancy cohort Crete, Greece. PLoS One 10:e0126327. https://doi.org/10.1371/journal.pone.0126327

37. Whitaker RC (2004) Predicting preschooler obesity at birth: the role of maternal obesity in early pregnancy. Pediatrics 114:e29–e36

38. Reynolds RM, Allan KM, Raja EA, Bhattacharya S, McNeill G, Hannaford PC et al (2013) Maternal obesity during pregnancy and premature mortality from cardiovascular event in adult offspring: follow-up of 1 323 275 person years. BMJ 347:f4539. https://doi.org/10.1136/bmj.f4539

39. Gaillard R (2015) Maternal obesity during pregnancy and cardiovascular disease and development in later life. Eur J Epidemiol 30:1141–1152

40. Petry CJ, Dorling MW, Pawlak DB, Ozanne SE, Hales CN (2001) Diabetes in old male offspring of rat dams fed a reduced protein diet. Int J Exp Diabetes Res 2:139–143

41. Ozanne SE, Jensen CB, Tingey KJ, Storgaard H, Madsbad S, Vaag A (2005) Low birth weight is associated with specific changes in muscle insulin-signalling protein expression. Diabetologia 48:547–552

42. Ozanne SE, Jensen CB, Tingey KJ, Martin-Gronert MS, Grunnet L, Brons C et al (2006) Decreased protein levels of key insulin signalling molecules in adipose tissue from young men with a low birthweight: potential link to increased diabetes? Diabetologia 49:2993–2999

43. Langley-Evans SC, Welham SJM, Jackson AA (1999) Fetal exposure to a maternal low protein diet impairs nephrogenesis and promotes hypertension in the rat. Life Sci 64:965–974

44. Black MJ, Lim K, Zimanyi MA, Sampson AK, Bubb KJ, Flower RL et al (2015) Accelerated age-related decline in renal and vascular function in female rats following early-life growth restriction. Am J Physiol Regul Intergr Comp Physiol 309:R1153–R1161

45. Kwon DH, Kang W, Nam YS, Lee MS, Lee IY, Kim HJ et al (2012) Dietary protein restriction leads to steatohepatitis and alters leptin/signal transducers and activators of transcription 3 signaling in lactating rats. J Nutr Biochem 23:791–799

46. Jennings BJ, Ozanne SE, Dorling MW, Hales CN (1999) Early growth determines longevity in male rats and may be related to telomere shortening in the kidney. FEBS Lett 448:4–8

47. Ozanne SE, Hales CN (2004) Lifespan: catch-up growth and obesity in male mice. Nature 427:411–412

48. Carr SK, Chen JH, Cooper WN, Constancia M, Yeo G, Ozanne SE (2014) Maternal diet amplifies the aging trajectory of Cidea in male mice and leads to the development of fatty liver. FASEB J 28:2191–2201

49. Tarry-Adkins JL, Fernandez-Twinn DS, Madsen R, Chen JH, Carpenter AMM, Hargreaves IPP et al (2015) Coenzyme Q10 prevents insulin signalling dysregulation and inflammation prior to development of insulin resistance in male offspring of a rat model of poor maternal nutrition and accelerated postnatal growth. Endocrinology 156:3528–3537

50. Tarry-Adkins JL, Fernandez-Twinn DS, Hargreaves IP, Neergheen V, Aiken CE, Martin-Gronert MS et al (2016) Coenzyme Q10 (CoQ) prevents hepatic fibrosis, inflammation and oxidative stress in a male rat model of poor maternal nutrition and accelerated postnatal growth. Am J Clin Nutr 103:579–588

51. Chen JH, Tarry-Adkins JL, Heppolette CAA, Palmer DB, Ozanne SE (2010) Early-life nutrition influences thymic growth in male mice that may be related to the regulation of longevity. Clin Sci 118:429–438

52. Heppolette CA, Chen JH, Carr SK, Palmer DB, Ozanne SE (2016) The effects of aging and maternal protein restriction during lactation on thymic involution and peripheral immunosenescence in adult mice. Oncotarget 7:6398–6409

53. Chen JH, Martin-Gronert MS, Tarry-Adkins JL, Ozanne SE (2009) Maternal protein restriction affects postnatal growth and the expression of key proteins involved in lifespan regulation in mice. PLoS One 4:e4950

54. Tarry-Adkins JL, Joles JA, Chen JH, Martin-Gronert MS, van der Giezen DM, Goldschmeding R (2007) Protein restriction in lactation confers nephroprotective effects in the male rat and is associated with increased antioxidant expression. Am J Physiol Regul Integr Comp Physiol 293:R1259–R1266

55. LeBourg E (2009) Hormesis, aging and longevity. Biochim Biophys Acta 1790:1030–1039

56. Fontana L, Partridge L, Longo VD (2010) Extending healthy life-span – from yeast to humans. Science 328:321–326

57. Vuguin P, Raab E, Liu B, Barzilai N, Simmons R (2004) Hepatic insulin resistance precedes the development of diabetes in a model of intrauterine growth restriction. Diabetes 53:2617–2622

58. Simmons RA, Templeton LJ, Gertz SJ (2001) Intrauterine growth retardation leads to development of type 2 diabetes in rats. Diabetes 50:2279–2286

59. Mazzuca MQ, Wlodek ME, Dragomir NM, Parkington HC, Tare M (2010) Uteroplacental insufficiency programs regional vascular dysfunction and alters arterial stiffness in female offspring. J Physiol 588:1997–2010

60. Moritz KM, Mazzuca MQ, Siebel AL, Milbus A, Arena D, Tare M et al (2009) Uteroplacental insufficiency causes a nephron deficit, modest renal insufficiency but no hypertension with ageing in female rats. J Physiol 587:2635–2646

61. Tran M, Young ME, Jefferies AJ, Hryciw DH, Ward MM, Fletcher EL et al (2015) Uteroplacental insufficiency leads to hypertension, but not glucose intolerance or impaired skeletal muscle mitochondrial biogenesis, in 12-month-old rats. Physiol Rep 3(9):pii: e12556. 10.14814/phy2.12556

62. Franco Mdo C, Arruda RM, Fortes ZB, de Olivera SF, Carvalho MH, Tostes RC et al (2002) Severe nutrient restriction in pregnant rats aggravates hypertension, altered vascular reactivity, and renal development in spontaneously hypertensive rat offspring. J Cardiovasc Pharmacol 39:369–377

63. Franco Mdo C, Dantas AP, Akamine EH, Kawamoto EM, Fortes ZB, Scavone C et al (2002) Enhanced oxidative stress as a potential mechanism underlying the programming of hypertension in utero. J Cardiovasc Pharmacol 40:501–509

64. Cox LA, Nijland MJ, Gilbert JS, Schlabritz-Loutsevitch NE, Hubbard GB, McDonald TJ et al (2006) Effect of 30 per cent maternal nutrient restriction from 0.16–0.5 gestation on fetal baboon kidney gene expression. J Physiol 572:67–85

65. Dallschaft NS, Alexandre-Gouabau MC, Gardner DS, Antignac JP, Keisler DH, Budge H et al (2015) Effect of pre-and postnatal growth and post-weaning activity on glucose metabolism in the offspring. J Endocrinol 224:171–182

66. Vickers MH, Reddy S, Ikenasio BA, Breier BH (2001) Dysregulation of the adipoinsular axis – a mechanism for the pathogenesis of hyperleptinemia and adipogenic diabetes induced by fetal programming. J Endocrinol 170:323–332

67. Fernandez-Twinn DS, Blackmore HL, Siggens L, Giussani DA, Cross CM, Foo R et al (2012) The programming of cardiac hypertrophy in the offspring by maternal obesity is associated with hyperinsulinemia, AKT, ERK, and mTOR activation. Endocrinology 153:5961–5971

68. Blackmore HL, Neo Y, Fernandez-Twinn DS, Tarry-Adkins JL, Giussani DA, Ozanne SE (2014) Maternal diet-induced obesity programmes cardiovascular dysfunction in adult male mouse offspring independent of current body weight. Endocrinology 155:3970–3980

69. Samuelsson AM, Matthews PA, Argenton M, Christie MR, McConnell JM, Jansen EHJM et al (2008) Diet-induced obesity in female mice leads to offspring hyperphagia, adiposity, hypertension, and insulin resistance: a novel murine model of developmental programming. Hypertension 51:383–392

70. Alfaradhi MZ, Fernandez-Twinn DS, Martin-Gronert MS, Musial B, Fowden A, Ozanne SE (2014) Oxidative stress and altered lipid homeostasis in the programming of offspring fatty liver by maternal obesity. Am J Physiol Regul Integr Comp Physiol 307:R26–R34

71. Oben JA, Mouralidarane A, Samuelsson AM, Matthews PJ, Morgan ML, McKee C et al (2010) Maternal obesity during pregnancy and lactation programs the development of offspring non-alcoholic fatty liver disease in mice. J Hepatol 52:913–920

72. McCurdy CE, Bishop JM, Williams SM, Grayson BE, Smith MS, Friedman JE et al (2009) Maternal high-fat diet triggers lipotoxicity in the fetal livers of nonhuman primates. J Clin Invest 119:323–335

73. White CL, Purpera MN, Morrison CD (2009) Maternal obesity is necessary for programming effect of high-fat diet on offspring. Am J Physiol Regul Integr Comp Physiol 296: R1464–R1472

74. Shankar K, Harrell A, Liu X, Gilchrist JM, Ronis MJJ, Badger TM (2008) Maternal obesity at conception programs obesity in the offspring. Am J Physiol Regul Integr Comp Physiol 294:R528–R538

75. Shankar K, Kang P, Harrell A, Zhong Y, Marecki JC, Ronis MJJ et al (2010) Maternal overweight programs insulin and adiponectin signaling in the offspring. Endocrinology 151:2577–2589

76. Akyol A, Langley-Evans SC, McMullen S (2009) Obesity induced by cafeteria feeding and pregnancy outcome in the rat. Br J Nutr 102:1601–1610

77. Snoeck A, Remacle C, Reusens B, Hoet JJ (1990) Effect of a low protein diet during pregnancy on the fetal rat endocrine pancreas. Biol Neonate 57:107–118

78. Styrud J, Eriksson UJ, Grill V, Swenne I (2005) Experimental intrauterine growth retardation in the rat causes a reduction of pancreatic B-cell mass, which persists into adulthood. Biol Neonate 88:122–128

79. Berney DM, Desai M, Palmer DJ, Greenwald S, Brown A, Hales CN et al (1997) The effects of maternal protein deprivation on the fetal rat pancreas: major structural changes and their recuperation. J Pathol 183:109–115

80. Desai M, Crowther NJ, Ozanne SE, Lucas A, Hales CN (1995) Adult glucose and lipid metabolism may be programmed during fetal life. Biochem Soc Trans 23:331–335

81. Burns SP, Desai M, Cohen RD, Hales CN, Iles RA, Germain JP et al (1997) Gluconeogenesis, glucose handling, and structural changes in livers of the adult offspring of rats partially deprived of protein during pregnancy and lactation. J Clin Invest 100:1768–1774

82. Hales CN, Desai M, Ozanne SE, Crowther NJ (1996) Fishing in the stream of diabetes: from measuring insulin to the control of fetal organogenesis. Biochem Soc Trans 24:341–350

83. Desai M, Byrne CD, Zhang J, Petry CJ, Lucas A, Hales CN (1997) Programming of hepatic insulin-sensitive enzymes in offspring of rat dams fed a protein-restricted diet. Am J Physiol 272:G1083–G1090

84. Sayer AA, Syddall HE, Dennison EM, Gilbody HJ, Duggleby SL, Cooper C et al (2004) Birth weight, weight at 1 y of age, and body composition in older men: findings from the Hertfordshire Cohort Study. Am J Clin Nutr 80:199–203

85. Gale CR, Martyn CN, Kellingray S, Eastell R, Cooper C (2001) Intrauterine programming of adult body composition. J Clin Endocrinol Metab 86:267–272

86. Cettour-Rose P, Samec S, Russell AP, Summermatter S, Mainieri D, Carrillo-Theander C et al (2005) Redistribution of glucose from skeletal muscle to adipose tissue during catch-up fat: a link between catch-up growth and later metabolic syndrome. Diabetes 54:751–756

87. Thompson LP, Al-Hasan Y (2012) Impact of oxidative stress in fetal programming. J Pregnancy 2012:582748. https://doi.org/10.1155/2012/582748

88. Richter T, von Zglinicki T (2007) A continuous correlation between oxidative stress and telomere shortening in fibroblasts. Exp Gerontol 42:1039–1042

89. Kawanishi S, Oikawa S (2004) Mechanism of telomere shortening by oxidative stress. Ann N Y Acad Sci 1019:278–284

90. Tarry-Adkins JL, Chen JH, Jones RH, Smith NH, Ozanne SE (2010) Poor maternal nutrition leads to alterations in oxidative stress, antioxidant defense capacity, and markers of fibrosis in rat islets: potential underlying mechanisms for development of the diabetic phenotype in later life. FASEB J 24:2762–2771

91. Bayol SA, Simbi BH, Stickland NC (2005) A maternal cafeteria diet during gestation and lactation promotes adiposity and impairs skeletal muscle development and metabolism in rat offspring at weaning. J Physiol 567:951–961

92. Beck B, Burlet A, Nicolas JP, Burlet C (1990) Hyperphagia in obesity is associated with a central peptidergic dysregulation in rats. J Nutr 120:806–811

93. Boney CM, Verma A, Tucker R, Vohr BR (2005) Metabolic syndrome in childhood: association with birth weight, maternal

obesity, and gestational diabetes mellitus. Pediatrics 115:e290–e296

94. Bruce KD, Cagampang FR, Argenton M, Zhang J, Ethirajan PL, Burdge GC et al (2009) Maternal high-fat feeding primes steatohepatitis in adult mice offspring, involving mitochondrial dysfunction and altered lipogenesis gene expression. Hepatology 50:1796–1808

95. Buckley AJ, Keserü B, Briody J, Thompson M, Ozanne SE, Thompson CH (2005) Altered body composition and metabolism in the male offspring of high fat-fed rats. Metab Clin Exp 54:500–507

96. Campfield LA, Smith FJ, Guisez Y, Devos R, Burn P (1995) Recombinant mouse OB protein: evidence for a peripheral signal linking adiposity and central neural networks. Science 269:546–549

97. Chen JH, Cottrell EC, Ozanne SE (2010) Early growth and ageing. Nestlé Nutr Workshop Ser Paediatr Program 65:41–50

98. Tobi EW, Goeman JJ, Monajemi R, Gu H, Putter H, Zhang Y et al (2014) DNA methylation signatures link prenatal famine exposure to growth and metabolism. Nat Commun 5:5592. https://doi.org/10.1038/ncomms6592

99. Waterland RA, Jirtle RL (2003) Transposable elements: targets for early nutritional effects on epigenetic gene regulation. Mol Cell Biol 23:5293–5300

100. Cooper WN, Khulan B, Owens S, Elks CE, Seidel V, Prentice AM et al (2012) DNA methylation profiling at imprinted loci after periconceptional micronutrient supplementation in humans: results of a pilot randomized controlled trial. FASEB J 26:1782–1790

101. Lillycrop KA, Slater-Jefferies JL, Hanson MA, Godfrey KM, Jackson AA, Burdge GC (2007) Induction of altered epigenetic regulation of the hepatic glucocorticoid receptor in the offspring of rats fed a protein-restricted diet during pregnancy suggests that reduced DNA methyltransferase-1 expression is involved in impaired DNA methylation and changes in histone modifications. Br J Nutr 97:1064–1073

102. Lillycrop KA, Phillips ES, Jackson AA, Hanson MA, Burdge GC (2005) Dietary protein restriction of pregnant rats induces and folic acid supplementation prevents epigenetic modification of hepatic gene expression in the offspring. J Nutr 135:1382–1386

103. Dominguez-Salas P, Moore SE, Baker MS, Bergen AW, Cox SE, Dyer RA et al (2014) Maternal nutrition at conception modulates DNA methylation of human metastable epialleles. Nat Commun 5:3746. https://doi.org/10.1038/ncomms4746

104. Thompson RF, Fazzari MJ, Niu H, Barzilai N, Simmons RA, Greally JM (2010) Experimental intrauterine growth restriction induces alterations in DNA methylation and gene expression in pancreatic islets of rats. J Biol Chem 285:15111–15118

105. Quilter CR, Cooper WN, Cliffe KM, Skinner BM, Prentice PM, Nelson L et al (2014) Impact on offspring methylation patterns of maternal gestational diabetes mellitus and intrauterine growth restraint suggest common genes and pathways linked to subsequent type 2 diabetes risk. FASEB J 28:4868–4879

106. Zhang J, Zhang F, Didelot X, Bruce KD, Cagampang FR, Vatish M et al (2009) Maternal high fat diet during pregnancy and lactation alters hepatic expression of insulin like growth factor-2 and key microRNAs in the adult offspring. BMC Genomics 10:478. https://doi.org/10.1186/1471-2164-10-478

107. Dunn GA, Bale TL (2009) Maternal high-fat diet promotes body length increases and insulin insensitivity in second-generation mice. Endocrinology 150:4999–5009

108. Dunn GA, Bale TL (2011) Maternal high-fat diet effects on third-generation female body size via the paternal lineage. Endocrinology 152:2228–2236

109. Sen S, Simmons RA (2010) Maternal antioxidant supplementation prevents adiposity in Western diet fed rats. Diabetes 59:3058–3065

110. Cambonie G, Comte B, Yzydorczyk C, Ntimbane T, Germaine N, Le NL et al (2007) Antenatal antioxidant prevents adult hypertension, vascular dysfunction, and microvascular rarefaction associated with in utero exposure to a low-protein diet. Am J Physiol Regul Integr Comp Physiol 292:R1236–R1245

111. Shukla P, Lemley CO, Dubey N, Meyer AM, O'Rourke ST, Vonnahme KA (2014) Effect of maternal nutrient restriction and melatonin supplementation from mid to late gestation on vascular reactivity of maternal and fetal placental arteries. Placenta 35:461–466

112. Tarry-Adkins JL, Blackmore HL, Martin-Gronert MS, Fernandez-Twinn DS, McConnell JM, Hargreaves IP et al (2013) Coenzyme Q10 prevents accelerated cardiac aging in a rat model of poor maternal nutrition and accelerated postnatal growth. Mol Metab 2:480–490

113. Tarry-Adkins JL, Fernandez-Twinn DS, Chen JH, Hargreaves IP, Martin-Gronert

MS, McConnell JM et al (2014) Nutritional programming of coenzyme Q: potential for prevention and intervention? FASEB J 28:5398–5405

114. Duncan GE, Perri MG, Theriaque DW, Hutson AD, Eckel RH, Stacpoole PW (2003) Exercise training, without weight loss, increases insulin sensitivity and postheparin plasma lipase activity in previously sedentary adults. Diabetes Care 26:557–562

115. Poston L, Bell R, Croker H, Flynn AC, Godfrey KM, Goff L et al (2015) Effect of a behavioural intervention in obese pregnant women (the UPBEAT study): a multicentre, randomised controlled trial. Lancet Diabetes Endocrinol 3:767–777

116. Raipuria M, Bahari H, Morris MJ (2015) Effects of maternal diet and exercise during pregnancy on glucose metabolism in skeletal muscle and fat of weanling rats. PLoS One 10: e0120980

117. Vega CC, Reyes-Castro LA, Bautista CJ, Larrea F, Nathanielsz PW, Zambrano E (2015) Exercise in obese female rats has beneficial effects on maternal and male and female offspring metabolism. Int J Obes (Lond) 39:712–729

118. Chiswick CA, Reynolds RM, Denison FC, Drake AJ, Forbes S, Newby DE et al (2016) Does metformin reduce excess birthweight in offspring of obese pregnant women? A randomised controlled trial of efficacy, exploration of mechanisms and evaluation of other pregnancy complications. NIHR Journals Library, Southampton (UK). https://www.ncbi.nlm.nih.gov/pubmed/27606384

Chapter 2

Effects of Prenatal Nutrition and the Role of the Placenta in Health and Disease

Leslie Myatt and Kent L. Thornburg

Abstract

Epidemiologic studies identified the linkage between exposures to stresses, including the type and plane of nutrition in utero with development of disease in later life. Given the critical roles of the placenta in mediating transport of nutrients between the mother and fetus and regulation of maternal metabolism, recent attention has focused on the role of the placenta in mediating the effect of altered nutritional exposures on the development of disease in later life. In this chapter we describe the mechanisms of nutrient transport in the placenta, the influence of placental metabolism on this, and how placental energetics influence placental function in response to a variety of stressors. Further the recent "recognition" that the placenta itself has a sex which affects its function may begin to help elucidate the mechanisms underlying the well-known dimorphism in development of disease in adult life.

Key words Placenta, Fetal programming, Nutrition, Energy, Pregnancy

1 Developmental Programming of Adult Disease

It is now well accepted that stresses encountered during the reproductive interval spanning oocyte maturation through infancy lead to vulnerability for lethal diseases in later life. The discovery of this fact arose from a surge of epidemiological studies beginning with a now-classic study showing that mortality from ischemic heart disease among English residents in Hertfordshire was inversely related to their birthweight [1, 2]. The relationship was similar for men and women and encompassed the entire birthweight range [3]. This general relationship has also been found in studies in other countries [4–8]. The Barker team found that the geographic distribution of high mortality rates of neonates overlapped with the mortality rates of adults with ischemic heart disease in the North of England in the early twentieth century [1]. The finding was subject to different theoretical interpretations. The relationship could have been considered completely coincidental, or it could have been argued that babies and adults in the North were exposed to a

Paul C. Guest (ed.), *Investigations of Early Nutrition Effects on Long-Term Health: Methods and Applications*, Methods in Molecular Biology, vol. 1735, https://doi.org/10.1007/978-1-4939-7614-0_2, © Springer Science+Business Media, LLC 2018

common chemical or biological toxin that affected vulnerable adults and neonates. However, the Barker team suggested neither of those reasons to explain the relationship between early and later life mortality rates. Rather, they proposed that babies weakened by poor nutrition and growth before birth were more likely to die as infants, whereas those who survived infancy remained in a weakened state and were more likely to die of heart disease as adults. This led to the idea that low birthweight babies are more likely to live shorter lives than those who are born with a more substantial body mass.

However, that theory could not be proven without data. Barker was fortunate to locate records showing weight at birth, weight at 1 year, and cause of death in a large population in Hertfordshire, UK. The fact that the records existed was due to the personal mission of one community nurse, Ethel Margaret Burnside, who died without knowing that her records would bring new powerful insight to the field of medicine decades later. She was the health visitor and "inspector of midwives." She organized a host of trained nurses to provide assistance for women in childbirth and give health advice to women on how to keep their infants healthy after birth [9]. Her data provided the perfect opportunity to test whether or not an inadequate flow of nutrients before birth precedes the risk for cardiovascular death in later life. Thus, lower birthweight or weight at 1 year was associated with a higher risk for later cardiovascular disease. In the intervening years, it has become evident that, in some populations, the relationship between cardiovascular disease and birthweight is more "U" shaped than linear across the birthweight scale. It appears now that the risks for chronic disease also increase with birthweight at the high end of the birthweight scale [10]. This is not entirely surprising because many "macrosomic" babies in the West are born to mothers who eat a high-fat diet and have poor glucose control and the offspring are known to accumulate fat before birth.

The Barker team wondered whether or not the low birthweight effect would also apply to other chronic diseases. Further studies on populations ranging from the Dutch Hunger Winter of 1944 [11] to Swedish cohorts [12] and the highly productive Helsinki Birth Cohort [13] showed that birthweight predicts other chronic diseases. As of today, there are a host of chronic conditions and diseases that are known to be related to birthweight. These include hypertension [2, 14], type 2 diabetes [15], obesity [16], ischemic heart disease [17], osteoporosis [18], neurological function [19], and others. The understanding of the relationship between early life growth and later life disease would have been more straightforward had birthweight been the only, or even the most powerful, predictor of later life consequences. However, as the new layers of discovery followed studies across the globe, it became evident that many factors reflecting the life of the mother, the father, the grandparents, the anthropomorphic features of the neonate, as well as the environmental history of the offspring were important influences on the development of the offspring and risk for disease after puberty.

2   The Placenta and Developmental Programming

There is a general and highly significant association between birthweight and placental weight in any given large population because placental weight tends to rise with fetal weight. The placental weight to birthweight ratio is often called "placental efficiency" to indicate the amount of fetal tissue that is generated for a given placental mass [20]. The idea is that the more fetal tissue that accumulates per gram of placental tissue, the more efficient the placenta must have been. While many people have challenged the use of the word "efficiency," because it does not account for the nutrient exchange area, it remains in common use. It was again Barker's team that showed that the relationship between placental weight and birthweight predicts cardiovascular disease in a "U"--shaped relationship. A placenta whose weight was about 19% of the birthweight had the lowest risk for ischemic heart disease. Placentas that were heavier or lighter compared to the offspring birthweight were more highly associated with the disease.

This relationship is particularly interesting because it is well documented that boys have more efficient placentas than girls. Boys grow faster over the course of gestation than girls and stimulate their mothers to consume more calories per day compared to girls [21]. However the more rapid growth strategy of boys puts them at higher risk for fetal demise compared to girls [22] under conditions of adversity. In the Chinese Great Leap Forward study, the birth ratio dropped during the famine of 1959–1961, with boys dying more often in the womb [23]. Thus, while babies tend to have placentas in proportion to their birthweight, those that fall out of that optimal proportion are at risk for chronic disease in later life.

As mentioned above several maternal and paternal factors predict disease in offspring. Among those, maternal body phenotype and placental size and shape are particularly important. The features of the placenta that have been independently associated with adult disease conditions include placental weight, placental weight as a ratio of birthweight, placental length, placental width, the difference between length and width, number of cotyledons (lobules), and umbilical cord length (Table 1). Each is associated with a specific condition.

Placental size and shape predict cancers as well as cardiovascular disease [24]. Three cancers have been reported to be associated with placental phenotypic features in the Helsinki Birth Cohort: lung cancer, colorectal cancer, and Hodgkin's leukemia. Lung

Table 1
Placental phenotypes associated with adult diseases

Disease	Phenotype	Reference
Sudden cardiac death	Placental thinness	[32]
Heart failure	Small placental area	[33]
Coronary heart disease	Small or large placenta	[31]
Hypertension	Number of lobes/cotyledons	[173]
Diabetes	Low placental weight, large placenta/birthweight ratio	[30, 174]
Asthma	Placental length	[175]
Lung cancer	Large or small placenta	[25]
Hodgkin's lymphoma	Short placental length	[26]
Colorectal cancer	Placental ovality	[28]

cancer was found to be independently associated with a short mother and a newborn that was short for its weight [25]. In addition, the association of lung cancer with placental phenotype was "U" shaped. Thus, the disease was associated with a surface area of the delivered placenta at the extremes. While the study revealed three placental phenotypes that showed predictive associations, in the most significant, a 100 cm^2 increase in placental surface area had a hazard ratio of 2.31 (1.45–3.69, $p < 0.001$) for acquiring lung cancer. The authors speculated that deficits in the transport of specific nutrients by the placenta resulted in a relatively stunted newborn which offered stress to pulmonary progenitor cells.

Hodgkin's lymphoma is a cancer affecting the lymphocytes that are key cells in the immune system. The disease is characterized by abnormal growth of these cells contained in the lymphatic system that once transformed may spread to other areas of the body. People with the disease lose their normal ability to resist infections. In the Helsinki Birth Cohort, people who contracted the disease had a prolonged gestational period, and they also had a placenta that was shorter than average [26]. After correcting for the length of gestation, the hazard ratio was 0.70 (0.53–0.92; $p = 0.01$) for every increase in the length of the surface of the placenta. The authors offered speculation that the cancer was related to two separate events: one which caused stress to immune progenitors and another related to a nutrition deficit which resulted in growth suppression of the placenta.

Most colorectal cancers are adenocarcinomas that originate from epithelial cells of the colorectal mucosa [27]. In the USA, about 5% of the population acquires the disease in their lifetime. The relationship was studied in the Helsinki Birth Cohort of

20,000 people born during 1924–1944 [28]. Among the 275 cases studied in 2013, the hazard ratios for colorectal cancer increased as the placental surface became longer and more oval. Among people born with a placenta that had a difference between the length and breadth of the surface exceeding 6 cm, the hazard ratio was 2.3 (95% CI 1.2–4.7), compared to people in whom there was no difference. Colorectal cancer was not related to other placental dimensions or to weight or to body size at birth.

The cardiovascular system is among the most vulnerable of organs to be programmed during fetal development and during infancy. One major underlying factor is the metabolic syndrome which has been studied in humans and in animal models [29]. However, because the syndrome is defined by several independent features, it is more difficult to identify specific biological changes that occur during development and lead to the disease. One study, however, showed an odds ratio as high as 18 for the common metabolic syndrome in men from Hertfordshire with a birthweight as low as 5 lb compared to the 9 lb category [30]. The metabolic syndrome is increasingly common in Western countries and predisposes to cardiovascular disease. The size and shape of the placenta are also related to cardiovascular disease. However, there have been no studies that have determined how the size and shape of the placenta at term predict the syndrome itself.

Thus far, the placental size and shape, in relation to maternal body type, are associated with at least six cardiac conditions (see Table 1). Among the studies in the Helsinki Birth Cohort (1934–1944), coronary heart disease was associated with three maternal placental combinations in addition to a low ponderal index (birthweight/length^3) [31]. In offspring born to short primiparous mothers, the hazard ratio for acquiring the disease was 1.14 (95% confidence interval 1.08–1.21, $p < 0.0001$) for each cm increase in the difference between the length and width of the delivered placental surface. The next two groups were separated by maternal body mass index (BMI) and not height alone. In mothers who were tall and had a BMI above the median, the hazard ratio was 1.25 (1.10–1.42, $p = 0.0007$) per 40 cm^2 decrease in the surface area. However, in tall mothers whose BMI was below the median, the hazard ratio was 1.07 (1.02–1.13, $p = 0.01$) per 1% increase in placental weight/birthweight ratio. In other words, it was the placental efficiency that was the predictor of later disease. The reason that maternal body size is important is that it has a large influence on the provision of nutrients for the conceptus and is thus linked to the development and function of the placenta, for which shape and size are apparent markers. The authors suggested that variations in the processes that regulate normal placental development lead to fetal malnutrition. These include implantation and spiral artery invasion, growth of the chorionic surface, and compensatory expansion of the chorionic surface under conditions of nutritional stress.

In a similar vein, sudden cardiac death was found to be associated with a thin placenta [32]. Heart failure was associated with a small surface area of the placenta delivered by mothers who were of short stature [33]. These findings imply that growth of the placenta was compromised from midgestation onward. Heart failure was also associated with a particular pattern of rapid growth in early childhood, a condition also known to be associated with insulin resistance in adults [34].

2.1 How Does the Placenta Mediate the Effect of Nutrition on Health and Disease?

The placenta is the interface between mother and fetus and acts as the immune barrier preventing rejection of the semi-allogeneic fetus, transports nutrients and gases to the fetus, and removes waste products from the fetus. Importantly the placenta also produces a range of peptide and steroid hormones that control maternal metabolism during gestation to ensure the supply of nutrients to the fetus but also may directly and/or indirectly regulate placental and fetal growth and metabolism. This has earned the placenta the name the "director of pregnancy" [35]. It is important to realize that the placenta itself grows, develops, and changes its function across gestation and grows exponentially in the third trimester. The placenta itself has considerable metabolic activity both to support its own growth and development, to power transport mechanisms, and to support the considerable synthesis of peptide and steroid hormones. Indeed it is estimated that half of the glucose taken up by the placenta is used to support placental metabolism [36]. Therefore alterations in placental metabolism can directly impact nutrient availability to the fetus in several ways.

It is recognized that placental growth and development occur in a carefully controlled sequence across gestation. Any insult or disruption in that developmental sequence will therefore affect placental growth and function which then impacts the fetus, which depends solely on the placenta for support [35]. Indeed alterations in placental morphology and function are well described in pregnancies with adverse outcomes, including preeclampsia, IUGR, obesity, diabetes, preterm birth, etc. While such pregnancies can give rise to inappropriate fetal outcome recognized at birth (e.g., growth restriction), evidence is accumulating of the long-term adverse outcomes seen in adult life, such as programming of cardiovascular, metabolic disease or cancer. What is now becoming obvious are additional programming outcomes in offspring including neurodevelopmental and behavioral abnormalities, such as depression [37], ADHD [38], anxiety [39], as well as neurodegenerative diseases including Parkinson's disease [40] and Alzheimer's disease [41]. Such diseases are not obvious at birth but also appear to have in utero antecedents linked to placental function. Fetal programming occurs when the normal pattern of fetal growth and development is disrupted leading to changes in function of fetal organs that are associated with development of disease in adult life. Hence placental function is tightly related to fetal programming. The disruptions to placental function can occur as a result of a range of exposures and can be manifested in many ways. Exposures that may alter placental function include endocrine disruptors and other environmental toxins, physical and physiologic stressors, differences and changes in the level and type of nutrition, micronutrient levels, and adverse intrauterine environments created by maternal medical conditions including obesity, diabetes, and hypertension. Many of these exposures are thought to lead to epigenetic changes in the placenta that may impact function. In addition the role of inflammation, oxidative and nitrative stress, and redox status in altering placental function is an active area of research.

Effects on placental structure and function that may result from the insults listed above include changes in vasculogenesis and vascular reactivity and in trophoblast surface area; changes in expression of glucose, amino acid, and fatty acid transporters; changes in hormone production; and changes in cellular metabolism. These will be examined in relation to the insults described above (Fig. 1). While there have been a large number of studies of the human placenta, these are necessarily descriptive and conducted mainly at term when tissue is available. They also have limited ability to address cause and effect due to the range of confounders present in human studies that cannot be easily controlled for. Correspondingly, animal models have been adopted to investigate mechanism. The majority of these studies have been carried out in mice, but primates, with their similar placental structure to humans, have also

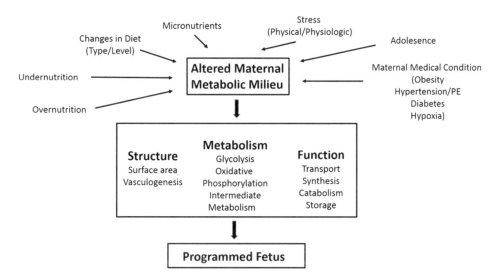

Fig. 1 Impact of stresses that can alter maternal nutrient type and availability on placental structure/function and hence programming of the fetus

been widely used employing under- and overnutrition strategies, alterations in dietary composition, and other procedures to mimic the human situation. We will describe the manipulations used in different species, the comparison with the human situation, and the effect on various placental pathways that are thought to be relevant to placental function and ultimately programming.

2.2 Placental Energetics

The tremendous synthetic activity and transport capacity of the trophoblast require an equally robust energy production in the form of ATP from mitochondria. Mitochondria produce ATP but are also a major source of production of reactive oxygen species (ROS). Indeed, 2–3% of oxygen is converted to superoxide in mitochondria. Not only are mitochondria a major source of ROS but elevated ROS also damages mitochondria, leading to a dangerous feedback loop. Increased production of ROS and disruption of the electron transport chain have been demonstrated multiple times in the setting of preeclampsia [42], the maternal hypertensive disorder postulated to have a primary etiology that lies in inadequate trophoblast invasion and a relative placental ischemia. Decreased placental mitochondrial respiration and expression of mitochondrial complexes are also found with preeclampsia [42] and are associated with increased expression of miR-210 which targets iron-sulfur cluster units (ISCU) in the mitochondrial electron transport chain. We have also found that the chronic inflammatory milieu of maternal obesity is associated with significant reduction in trophoblast mitochondrial respiration [43] and a significant increase in miR-210. This increase in miR-210 was activated by NFkB1 but only in the female placenta suggesting a sexual dimorphism of this effect. Obesity predisposes to the development of gestational diabetes (GDM), and we found that type A2 GDM, which requires treatment of women with medication, was associated with a further significant decrease in trophoblast respiration and reduction in expression of components of the electron transport chain compared to controls matched for adiposity. Interestingly in this case, the placenta was larger and displayed an increase in expression of the glucose transport GLUT 1 and in glycolysis which may be a compensatory response to support the fetus [44]. The switch between oxidative phosphorylation and glycolysis was mediated by miR-143. In addition to the inflammatory milieu of preeclampsia and obesity, other environmental manipulations can affect mitochondrial respiration. Maternal caloric restriction by 30% reduces placental and fetal weights and placental ATP levels in the rat in the face of increased oxygen consumption, biogenesis, and bioenergetic efficiency [45]. This was related to differential expression of several placental proteins, including mitochondrial proteins. Together this suggests that disruption of placental bioenergetics is seen with adverse conditions and may occur in a sexually dimorphic manner that may relate to fetal sex-specific growth and development and developmental outcomes. We will explore this below in the context of alterations in placental processes including transport of nutrients.

3   Sexual Dimorphism in Placental Function and Pregnancy Outcomes

A male fetus is more at risk of poor pregnancy outcomes than a female fetus in association with complications such as placental insufficiency, preeclampsia, infection, IUGR, and preterm delivery [46]. An increasing number of placental functions and placental disorders are now being shown to be sexually dimorphic. However the underlying mechanism(s) underlying the dimorphism remain to be elucidated. Microarray analysis of placental tissue [47–49] or isolated placental cell types [50] have shown a distinct sexual dimorphism in gene expression in the human placenta. In particular immune genes were reported to be expressed at higher levels although this was in the female placenta compared to male in one study [47] with the opposite situation reported in another [50]. Gene expression in the placenta also responds to maternal inflammatory status in a sex-dependent manner [51]. For example, expression of 59 genes was changed within the placenta of women with asthma versus no asthma in cases of those with a female fetus compared to only 6 genes changed in those with a male fetus [52]. Changes in diet provide distinctive signatures of sexually dimorphic genes in the murine placenta with expression generally being higher in female than male placentas [53]. The male placenta has higher TLR4 expression and a greater production of TNFα in response to lipopolysaccharide than the female placenta, which may underlie the propensity to preterm birth in males [54].

4   Placental Nutrient Transport Mechanisms

The human placental is hemomonochorial, with maternal blood in direct contact with the syncytiotrophoblast layer of the placenta. At term maternal blood and fetal blood are separated by the syncytio-trophoblast and the underlying fetal vascular endothelium. The vast majority of transport studies have focused on mechanisms within the syncytiotrophoblast which has two polarized membranes: the maternal-facing microvillous membrane (MVM) and the basal membrane (BM) facing the fetal blood vessels. Transport between maternal and fetal circulation is regulated by the combination of various factors including uterine and umbilical blood flows, the concentration gradient of the nutrients and trophoblast surface area, thickness, expression of transporters, and metabolism of sub-strates by the placenta itself. Blood flows regulate the transfer of membrane-permeable molecules including oxygen and carbon dioxide, whereas the transfer of large less permeable compounds depends on membrane transport proteins. Substrates and transporters that have been the most investigated include those for glucose, amino acids, and fatty acids.

Glucose transport across the placenta is via facilitated diffusion mediated by a number of specific glucose transporters (GLUT) found in both the basal and microvillous membranes [55]. Several members of the GLUT family are found in the placenta [56], but GLUT 1 is the major isoform expressed and the principal transporter [55]. Amino acid transporters are also found on both syncytiotrophoblast membranes [57]. These transporters are classified by substrate specificity (referred to as system) sequence homology or physiologic function, reviewed by Gaccioli and Lager 2016 [58]. In the human placenta, 20 distinct amino acid transporters have been identified, being variously distributed in microvillous and basal membranes [57]. Transfer of fatty acids across the placenta involves a sequence of steps. Fatty acids mainly circulate in the blood as triglycerides that have to be hydrolyzed at the syncytiotrophoblast surface by lipoprotein lipase [59]. The fatty acids are then taken up into trophoblast and then effluxed into the fetal circulation by fatty acid transport proteins (FATPs) of which five isoforms are expressed in the human placenta [60] or by CD36, a fatty acid translocase [61], both found in microvillous and basal membranes. In addition to transport across the placenta, fatty acids can also be metabolized in placenta and stored in various structures including lipid droplets. Fatty-acid-binding proteins (FABPs) deliver the fatty acids to the intracellular compartments. Four distinct FABP isoforms are found in the human placenta [62].

5   Effects of Changes in Maternal Nutrient Supply to the Placenta

The supply of nutrients to the placenta can be altered in several ways. Inadequate development of the uterine blood supply to the placenta, which would limit supply of flow-limited substrates (e.g., oxygen) and substrates that cross by facilitated diffusion (e.g., glucose), can lead to intrauterine growth restriction. This may simply be caused by the overall reduced amount of substrate (nutrient) reaching the placenta and fetus or by responses in the placenta to preserve its own function or adapt to the reduced maternal supply and preserve both placental and fetal growth and development [35] (Fig. 2). Maternal undernutrition will also alter both the supply and the composition of nutrients reaching the placenta and fetus, resulting in a programming insult. An often quoted example of this is the Dutch Hunger Winter of 1944 [63, 64]. Limitations in food supply in developing countries and even in developed countries where individuals and families face "food insecurity" [65] mean that maternal undernutrition still remains a common

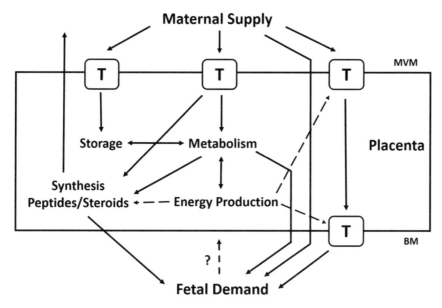

Fig. 2 Mechanisms whereby the placenta can utilize substrates supplied by the mother to meet fetal demand or to support placental function per se

problem today. In addition physiologic changes of pregnancy occurring in adolescent pregnancy where there is a competition between ongoing maternal growth and development and fetal growth and development [66] and where adolescents may have inadequate or marginal nutrition during pregnancy or with hyperemesis in pregnancy [67] will give rise to a relative undersupply of nutrients to the placenta and fetus. In contrast increased maternal nutrient availability can be considered to occur in the setting of diabetes where maternal glucose, amino acids, and fatty acids can be increased. Type 1, type 2, and GDM affect pregnancy, and the incidence of GDM (currently 2–10% of pregnancies in the USA) is steadily increasing with subsequent programming of offspring for development of type 2 diabetes. The effects of alterations in maternal nutrition on placental transport in a range of animal models, rat, mouse, primate, and ovine, and in human studies have been comprehensively reviewed [68]. Pregnant animals of several species including rodents, rabbits, pigs, sheep, and nonhuman primates have been given calorie- and/or protein-insufficient diets or high-energy (Western) diets and their offspring carefully studied [29]. These diet regimens have led to offspring with cardiovascular disease, hypertension, obesity, type 2 diabetes, and all general attributes of the metabolic syndrome in people. These diets may also lead to impaired cognitive, behavioral, and equivalent psychiatric states. While there are differences in offspring outcome between so-called underfeeding and overfeeding pregnant animals as models for developmental programming, what is remarkable is that the two forms of maternal stress often have common outcomes. Both lead to elevated risks for hypertension, insulin resistance, obesity, and cardiovascular dysfunction [69, 70].

5.1 Maternal Undernutrition

A series of ovine studies revealed that the approach used to model the events of adolescent pregnancy gave differing outcomes. When overfeeding of adolescent sheep was used to promote rapid maternal growth and give a growth-restricted hypoglycemic fetus, reduced placental trophectoderm and placental angiogenic factors were found at midgestation [71] with reduced placental mass but normal weight-specific glucose transport found at term, suggesting in this model that size of the placenta and not nutrient metabolism and transfer capacity regulates fetal growth. In contrast underfeeding of the adult ovine dam gave a lower transplacental glucose gradient and a slowing of fetal growth in late gestation [72]. Redmer et al. showed that nutrient restriction of the adult dam and overnutrition of adolescent dams both suppressed placental cell proliferation and vascularity and expression of placental angiogenic factors and receptors suggesting vascular growth, and hence nutrient supply are affected by maternal nutrition in pregnancy [73]. Human studies are limited to cross-sectional observation on delivered placentas, usually at term. Teenagers have an increased incidence of SGA babies, but no major differences in placental weight or composition were found between growing and non-growing teenagers, whereas birthweight to placental weight ratio, a crude measure of placental efficiency, was higher in growing versus non-growing teenagers [66]. In humans, SGA pregnancies have reduced placental amino acid transport. Teenagers with either AGA or SGA babies have lower placental system A amino acid transporter activity than adults, and the activity was higher in growing versus non-growing teenagers [74]. However placental mRNA expression of system A amino acid transporters was lower in teenagers than adults but with no difference between growing and non-growing teenagers. Therefore it appears that teenagers have reduced placental amino acid transport which may relate to their susceptibility to SGA babies, but growing teenagers can adapt to this by increasing activity but not expression of system A transporters [74]. Overall these studies illustrate the complex responsive and adaptive relationship between maternal nutrient supply and the placenta that influences fetal growth and development and potential programming for adult disease.

5.2 Maternal Diabetes

Women with type 1 diabetes often enter pregnancy with poor glycemic control and are at risk for a variety of adverse fetal and neonatal outcomes including IUGR. Type 1 diabetics often have vascular problems that may be reflected by inadequate development of the uterine and maternal vasculature during early pregnancy that can lead to compromised blood flow and nutrient delivery to the placenta and to their more frequent development of preeclampsia. In contrast women who develop GDM later in gestation commonly have a higher incidence of LGA babies, so-called fetal overgrowth [75], perhaps related to increased nutrient abundance. The different type and duration of nutrient "environments" that the placenta is exposed to between type 1 and 2 diabetes and GDM will obviously alter placental responses and fetal programming events. This may underlie the lack of consistency of data on the effect of diabetes on placental glucose, amino acid, and fatty acid transporters (reviewed in [68]). Whereas basal membrane glucose transport and GLUT 1 expression [76] and MVM GLUT 9 protein expression [77] are increased in type 1 diabetes, glucose transport across the in vitro perfused placenta was decreased in untreated GDM [78]. An early study of GDM pregnancies versus controls found expression and activity of placental GLUT 1 were unaltered [79]. However, when controlling for treatment modality and maternal BMI in nonobese women, those with insulin-controlled GDM had higher GLUT 1 mRNA and protein but lower GLUT4 mRNA and protein when compared to nonobese diet-controlled GDM and healthy controls [80]. Hence, there are further interactive effects of obesity and treatment modality.

Data on placental amino acid transport are similarly inconsistent. Activity of MVM system A was found to be increased in type 1 diabetics regardless of fetal overgrowth by Jansson et al. [81], whereas MVM system A activity was reduced and system L unaltered in type 1 diabetics with fetal overgrowth [82]. In GDM placenta Jansson et al. reported leucine transport was increased [81], whereas Dicke and Henderson found no differences in transport of neutral amino acids [83]. Similarly, data on fatty acid transport across the placenta with diabetes is inconsistent but also sparse. With fetal overgrowth the activity of LPL is increased in type 1 diabetes and the expression of FABP1 protein increased in both type 1 and GDM [84], whereas obese women with GDM show increased FABP4 [85] and endothelial lipase activity [86]. Microarray profiling revealed a 2.4-fold upregulation of FABP4 and a 3.4-fold downregulation of LPL expression in placenta of type 1 diabetes compared to control [87].

5.3 Maternal Obesity   Obesity in pregnant women may affect placental function and hence fetal growth and development in several ways. The metabolic adjustments of normal pregnancy are characterized by increased fat mass, insulin resistance, low-grade inflammation [88], and mild hyperlipidemia [89, 90], where phospholipids, total LDL and HDL cholesterol, and triglycerides are all increased. The metabolic changes become exacerbated with pre-gravid obesity [91, 92]. Obese pregnant women are characterized by high levels of free fatty acids (FFA) but in addition have higher circulating levels of leptin, TNFα, IL-1, IL-6, and IL-8 and reduced levels of adiponectin [93]. This altered cytokine and adipokine milieu gives rise to chronic low-grade inflammation which is associated with oxidative stress. With maternal obesity, the placenta is exposed to an altered nutrient composition available both to support its own metabolism and for transport to the fetus. Secondly and perhaps independently, the chronic inflammatory milieu may alter placental function, either directly via actions on placental transport mechanisms or indirectly via increased oxidative stress. Among the inflammatory molecules, IL-6 [94], leptin [95], IGF-1 [96], and insulin [97] acting via MVM surface receptors stimulate where as adiponectin [98] inhibits amino acid transporter activity in the placenta. This may be a mechanism whereby the maternal inflammatory state can alter placental function to then give rise to the fetal programming effect. As we have pointed out above, maternal obesity is known to be associated with increased placental oxidative stress and reduced mitochondrial respiration which provides energy to support placental function [43].

Studies of the effect of obesity on placental transporters are as disparate as those studying the effect of diabetes. Obesity does not appear to alter GLUT 1 mRNA or protein levels but is associated with a significant reduction in GLUT4 mRNA in human placenta [80]. However, in a high-fat diet mouse model of obesity, increased placental GLUT 1 was observed in association with increased fetal weight [99]. When examining placental amino acid transporters, decreased expression and activity of SNAT4 but no change in SNAT1 and SNAT2 expression were seen in obese compared to lean women [100]. However no differences in placental system A and system L transporters were found in overweight/obese women compared to lean women although MVM system A correlated with birthweight but not maternal BMI [101], and SNAT2 but not SNAT 1 or SNAT 4 correlated with birthweight and maternal BMI.

Although maternal hyperglycemia has long been associated with increased fetal growth [102], maternal triglycerides also contribute to fetal growth with aberrant fetal growth seen with GDM despite good glucose control. Indeed in multivariate analysis, increased birthweight positively correlates only with hypertriglyceridemia in women with GDM [103, 104]. Prior studies showing the effects of maternal obesity on fatty acid (FA) transport and metabolism via fatty acid oxidation (FAO) are limited, and a small number of human and animal studies have shown that obesity impacts placental expression of FA transporters and enzymes involved in FAO. Again human studies show conflicting results for CD36 transport protein expression with increasing maternal adiposity [105], but with the majority of studies not considering fetal gender [106]. Other studies have demonstrated decreased human placental SLC27A4, FABP1, and FABP3 with increasing adiposity [105] and associations of maternal obesity with decreased expression of the oxidation enzyme LCAD [107] and increased expression of

FATP4 protein [108]. We have recently examined expression of FA transport proteins and FAO enzymes in placentas of lean and obese women and the influence of fetal gender. We found that the placenta is susceptible to maternal adiposity in a sex-specific manner. When evaluating for obesity alone, SLC27A4 expression was decreased with increasing BMI, as shown in a similar study [105], but we found increased expression of fatty-acid-binding protein FABP3, decreased CD36, and FAO enzyme VLCAD with increasing BMI. These findings suggest alterations in uptake (CD36 and SLC72A4) as well as intracellular transport of fatty acids with obesity. This could be a mechanism by which the placenta attempts to prevent or limit transport into the fetal circulation in the face of the increased supply of FAs in the maternal circulation. A recent study has also concluded that extramitochondrial (peroxisomal) FAO and lipid storage were enhanced in the placenta of obese women perhaps as a mechanism to limit excess fat delivery [109]. While no other placental FA oxidation enzymes were affected by obesity, we found very long-chain acyl-dehydrogenase (VLCAD) expression was decreased suggesting possible decreased oxidation of very long-chain fatty acids with obesity. This observed alteration of placental VLCAD and the potential ensuing increase in very long-chain FAs reaching the offspring of obese women (due to lack of placental metabolism) may be a factor that contributes to increased risk of cardiovascular disease later in life and perhaps fatty liver disease (hepatic steatosis).

In our study, sexual dimorphism is evident in placental expression of LPL and enzymes involved in FAO. The male placenta appears to respond to adverse intrauterine conditions by altering gene and protein expression so that continued fetal growth occurs in a less than optimal state, while the female adopts a more conservative strategy to ensure survival under adverse conditions. This, however, results in diminished reserve and results in worse neonatal outcomes for males [22]. As seen in mothers with chronic asthma, their male fetuses grow normally and are larger than females. However, if the asthma worsens acutely, the males have increased incidence of growth restriction, preterm delivery, and stillbirth [46]. A similar growth strategy has also been described with preeclampsia, with males maintaining normal growth [110]. Overall, these data again suggest that the placenta has a central role in regulating the amount of nutrient reaching the fetus, thereby mediating a programming effect.

6   Effect of Diet on Placental Gene Expression

The rise in obesity and diabetes has been attributed in part to the availability of cheap, high-calorie food of low nutrient quality. The quality of diet therefore is thought to be partially responsible for the cycle of obesity, diabetes, and metabolic syndrome via fetal programming. As outlined in this chapter, diet may influence programming via direct effects as substrates for fetal growth and development but also indirectly via altering placental function. There is evidence that diet per se can alter overall placental gene expression and expression of specific genes (e.g., those encoding transporters). Studies of dietary composition in humans are of necessity associative. Therefore, animal models have been adopted to assess the influence of diet on global placental gene expression and expression of transporters in the context of fetal programming using murine, ovine, and primates with a variety of dietary manipulations, such as altering caloric value and composition of lipids, protein, and carbohydrate. Reduction of protein content by 50% but maintenance of calories resulted in upregulation of genes involved in negative regulation of cell growth and metabolism (the p53 pathway, apoptosis) and in epigenesis, but decreased expression of genes involved in nucleotide metabolism. Subsequent studies have also examined sexual dimorphism of effect. An in-depth study [53] comparing the effect of low-fat and very high-fat versus control diets in the mouse showed that diet affected placental gene expression in a sexually dimorphic manner. Each diet resulted in a distinct profile of sexually dimorphic genes with expression being higher in the female placenta. In the mouse a high-fat diet was found to cause global demethylation and deregulation of clusters of imprinted genes important in the control of many cellular, metabolic, and physiological functions in the placenta but only in the case of female fetuses [111, 112]. Imprinted genes in the mouse placenta have been shown to be susceptible to reduced methylation in association with caloric restriction [113]. More recent studies in the rabbit [114] and baboon [115] have shown that a high-fat diet/high-cholesterol diet or maternal nutrient restriction respectively gave sex-dependent changes in placental gene expression.

In general terms, alteration of the diet in any of these ways leads to a programming effect but with varying effects on expression of specific transporters. For example, in mice with an ad libitum high-fat diet, increased placental glucose and amino acid transport accompanied by increased GLUT 1 and SNAT2 expression were seen [99]. However, a high-sugar, high-fat diet resulted in decreased placental weight, increased glucose and amino acid transport, and increased SNAT2 and GLUT3 expression [116], and when the amount of fiber and fat in the diet were altered, placental expression of GLUT 1, GLUT3, and SNAT4 were found [117]. Some sexually dimorphic effects have been described (e.g., with a high-fat, high-sucrose, so-called cafeteria diet) such as increased SNAT2 in male placentas and increased SNAT4 in female placentas [118]. It is clear that not only caloric value but composition of diet alters placental transport mechanisms although there is a complex interaction leading to the programming effect.

7   Diet and Epigenetic Effects in Placenta

It is well known that diet alters DNA methylation and hence gene expression, but individual nutrients or components of the diet can achieve this (reviewed in [119]). Close regulation of epigenetic changes in placenta is essential especially in the early phase of gestation where global DNA demethylation in the zygote is seen but may subsequently be influenced by the maternal metabolic environment. Chromatin-modifying enzymes including DNA methyltransferases sense and respond to alterations to the nutritional environment through their effects on intermediary metabolites [120]. Studies of individuals exposed to famine during the Dutch Hunger Winter [121, 122] have shown that nutritional exposures in utero can induce epigenetic changes in offspring. In later life, the epigenome appears to be capable of responding to changes in nutrients including deficiencies in methyl donors [123], folic acid supplementation [124], and fat [125] and caloric restriction [126]. As of yet, there is little data on the effects of nutrients on DNA methylation in placenta beyond the murine studies reported above [53, 111, 112, 127]. In placentas of women exposed to GDM, Bouchard et al. found decreased DNA methylation of leptin and adiponectin genes with increasing blood glucose levels [128, 129] and correlation of leptin gene expression with its DNA methylation. A candidate gene approach studying metabolic programming found decreased methylation of the maternally imprinted gene *MEST* which is essential for fetal and placental growth in placentas from GDM mothers [130]. We have reported on the differences in DNA methylation and hydroxymethylation, particularly of genes on chromosome 17 and 19 corresponding to pregnancy-specific glycoproteins and chorionic somatomammotropin, in placentas of obese compared to lean women, suggesting that dietary-induced epigenetic changes in placenta can in turn regulate maternal metabolism.

7.1   One-Carbon Metabolism and Epigenetic Programming

DNA and histone methylation are linked to the one-carbon cycle in which folate, vitamin B12, methionine, choline, and betaine play essential roles [131]. In this pathway, S-adenosyl-methionine (SAM) is used as a methyl donor cofactor of DNA and histone methyltransferases for transfer to DNA or histone macromolecules. In rodents methyl donor supplements can influence the degree of DNA methylation of specific genes during pregnancy [132] and may prevent the offspring metabolic phenotype induced by maternal dietary protein restriction [133]. Another product of one-carbon metabolism, S-adenosylhomocysteine, is a product inhibitor of methyltransferases. These micronutrients may act on the placental epigenome, and the human genome encodes 208 methyltransferases, including 70 histone and four DNA methyltransferases [134]. Biotin, niacin, and pantothenic acid, which are also water-soluble B vitamins, may have a role in epigenetic regulation as they are involved in histone modification [135]. Further, the activity of DNA methyltransferases and histone acetyltransferases can be altered by bioactive food components including genistein, polyphenols, tea catechin, resveratrol, butyrate, and curcumin (reviewed in [119, 136]). Thus nutrients may influence placental epigenetics at several levels, affecting placental function and programming.

8   Imprinted Genes

Imprinted genes are expressed exclusively or preferentially on one of the two parental chromosomes but are repressed on the other [137]. Approximately 120 imprinted genes have been described as being either paternally or maternally expressed in humans and mice. Most imprinted genes are expressed in both placenta and embryo with a subset of imprinted genes being expressed predominantly in the placenta, and some genes are more widely expressed but imprinted selectively in the placenta and not in the embryo [138, 139]. Imprinted genes can regulate the ability of placenta to support embryonic development to transfer nutrients [140] and regulate placentation, embryonic growth, and energy homeostasis [141]. Paternally expressed genes (e.g., *IGF2*) generally increase resource transfer from the mother to the fetus, while maternally expressed genes (e.g., *GRB10*) limit this transfer in the womb to ensure the maternal well-being [142]. IGF2 regulates placental growth and as such its transport capacity and has been shown to mediate the placental response to maternal undernutrition in a mouse model [143]. Maternal and paternal metabolic environments such as those seen with famine [122] or obesity [144] can alter methylation of imprinted genes and methylation of imprinted genes such as *IGF2* at birth are associated with development of metabolic disorders [145]. Imprinted genes in mouse placenta have been shown to be susceptible to reduced methylation in association with caloric restriction [113].

9   Morphologic/Structural Changes in Response to Nutrition and Relation to Function

Placental development occurs continuously in a complex tightly regulated manner across the entirety of gestation. Components that are critical to placental transport mechanisms include development of both the uteroplacental and fetoplacental vasculatures to deliver and abstract nutrients from the placenta, the size of the trophoblast exchange area, and thickness of the placental barrier. Nutritional cues (e.g., type and amount of nutrients) can affect development at all stages of gestation. For example, this includes the depth of trophoblast invasion and adaptation of the uteroplacental circulation in the first trimester and the exponential increase in trophoblast surface area in the third trimester. The limited numbers of human studies are typically cross-sectional from tissues obtained at delivery usually at term. In early studies the hyperglycemic milieu of diabetes was associated with an increase in the volume of parenchyma and in villous surface area [146], whereas a more recent study that may reflect tighter glycemic control reported no difference in villous and capillary surface area, membrane thickness, or morphometric diffusing capacity between type 1 diabetic and control placentas [147]. IUGR has been reported to be associated with a reduced trophoblast surface area [148]. Recently differences in depth of trophoblast invasion and in placental surface area were found between placentas of male and female fetuses in Saudi Arabia [149]. Animal models have been widely employed in this research. In the guinea pig, maternal food restriction was shown to result in a reduction of exchange surface area and an increase in barrier thickness of the placenta [150], whereas in the rat, high fat, high salt, or a combination of the two decreased the size of the placenta in males compared to females [151], and in mice, a high-fat diet that gave maternal obesity increased proliferation and thickness of the placental labyrinth (exchange area) in early gestation but resulted in increased inflammation and macrophage accumulation particularly in male placentas in late gestation [152]. While this evidence supports a role for maternal nutrition in altering placental structure, more studies are needed to define the relationship of structure to function and ultimately programming.

10   Inflammation/Oxidative Stress/Redox Status

In pregnancies associated with adverse outcomes (e.g., preeclampsia, obesity, GDM, IUGR), there is abundant evidence for increased inflammation and oxidative stress in the placenta [153–156], the severity of which occurs in a sexually dimorphic manner. The altered maternal nutrition and metabolic milieu seen in these conditions may cause the inflammation and oxidative stress to impact placental function and indirectly result in programming. There are reports of increased maternal amino acid [157] and fatty acid [158] levels and increased cord blood amino acids [157] and altered cord blood fatty acid [159] concentrations with preeclampsia. Placentas from preeclamptic women also have reduced expression of transporter for the amino acid taurine [160], whereas they have no difference in mRNA expression and activity of system A transporters [161, 162] and an increase in arginine transport [163]. Expression of placental fatty acid transporters FATP 1 and FATP4 is also reduced in placentas from preeclamptic pregnancies [164]. It is however uncertain which aspect(s) of the placental environment, blood flow, oxidative stress, and inflammation in a pregnancy with preeclampsia alters expression of the various transporters.

While oxidative stress is implicated in various disease states, ROS also act as signaling molecules in a range of cellular pathways (reviewed in [165]) which are tightly regulated. Transcription factors responsive to inflammatory cytokines [166] and redox (oxidative stress) [167] may mediate some of the effects. Thioredoxin and glutathione, produced by their respective reductases, are the two major factors responsible for maintaining low redox potential and high free sulfhydryl levels in cells [168]. The expression and activity of thioredoxin reductase and glutathione reductase are sensitive to oxidative stress, and we have recently shown induction of both enzymes in the placenta presumably, in response to oxidative stress, in the placenta of obese women [169]. Interestingly, induction was significantly greater in the placenta of males than females, and this is potentially a reflection of the greater degree of inflammation/oxidative stress seen in the male placenta. Both of these enzymes are selenoproteins with selenium in their active site, and variations in the soil content of selenium have been linked to the incidence of conditions such as preeclampsia [170]. Also, obese children [171] and morbidly obese females [172] have significant depletion of selenium. Therefore, as dietary content of selenium or other antioxidants may impact oxidative stress, placental function and ultimately fetal programming supplementation with trace elements and antioxidants (e.g., vitamin C and E) are promising therapeutic avenues.

11   Conclusions

In this chapter we have described the antecedents of the burgeoning field now known as fetal programming or the developmental origins of health and disease and the epidemiologic work that lead to the placenta being identified as a major influence in this phenomenon. Given that the majority of human studies have necessarily been observational and conducted at term where the changing responses throughout gestation cannot be studied, many animal models employing a range of dietary strategies to mimic the insult or insults that give rise to programming have been used. However few have studied the mechanism(s) at the placental level that mediate the effect. Here we have outlined the roles and function of the placenta and the various mechanisms at the morphologic, structural, and functional levels, whereby the alterations in prenatal and pregnancy nutrition shown to be associated with fetal programming may act via the placenta. The sexual dimorphism in fetal programming responses has long been recognized, and more recently sexual dimorphism in placental function, including in response to diet, has been recognized and investigated although the underlying mechanisms remain to be elucidated. Our advancing knowledge offers up the potential for interventions at the placental level to prevent or ameliorate the programming effects and alleviate a tremendous burden of disease.

References

1. Barker DJ, Osmond C (1986) Infant mortality, childhood nutrition, and ischaemic heart disease in England and Wales. Lancet 1 (8489):1077–1081

2. Barker DJ, Osmond C, Golding J, Kuh D, Wadsworth ME (1989) Growth in utero, blood pressure in childhood and adult life, and mortality from cardiovascular disease. BMJ 298(6673):564–567

3. Barker DJ, Winter PD, Osmond C, Margetts B, Simmonds SJ (1989) Weight in infancy and death from ischaemic heart disease. Lancet 2(8663):577–580

4. Barker DJ, Eriksson JG, Forsen T, Osmond C (2002) Fetal origins of adult disease: strength of effects and biological basis. Int J Epidemiol 31(6):1235–1239

5. Leon DA, Lithell HO, Vagero D, Koupilova I, Mohsen R, Berglund L et al (1998) Reduced fetal growth rate and increased risk of death from ischaemic heart disease: cohort study of 15 000 Swedish men and women born 1915-29. BMJ 317(7153):241–245

6. Stein CE, Fall CH, Kumaran K, Osmond C, Cox V, Barker DJ (1996) Fetal growth and coronary heart disease in south India. Lancet 348(9037):1269–1273

7. Rich-Edwards JW, Stampfer MJ, Manson JE, Rosner B, Hankinson SE, Colditz GA et al (1976) Birth weight and risk of cardiovascular disease in a cohort of women followed up since 1976. BMJ 315(7105):396–400

8. Fan Z, Zhang ZX, Li Y, Wang Z, Xu T, Gong X et al (2010) Relationship between birth size and coronary heart disease in China. Ann Med 42(8):596–602

9. Barker D (2003) The midwife, the coincidence, and the hypothesis. BMJ 327 (7429):1428–1430

10. Kapral N, Miller SE, Scharf RJ, Gurka MJ, DeBoer MD (2017) Associations between birthweight and overweight and obesity in school-age children. Pediatr Obes. https://doi.org/10.1111/ijpo.12227

11. Roseboom TJ, Painter RC, de Rooij SR, van Abeelen AF, Veenendaal MV, Osmond C et al (2011) Effects of famine on placental size and efficiency. Placenta 32(5):395–399

12. Zoller B, Sundquist J, Sundquist K, Crump C (2015) Perinatal risk factors for premature ischaemic heart disease in a Swedish national cohort. BMJ Open 5(6):e007308. https://doi.org/10.1136/bmjopen-2014-007308

13. Barker DJ, Osmond C, Kajantie E, Eriksson JG (2009) Growth and chronic disease: findings in the Helsinki Birth Cohort. Ann Hum Biol 36(5):445–458

14. Bagby SP (2007) Maternal nutrition, low nephron number, and hypertension in later life: pathways of nutritional programming. J Nutr 137(4):1066–1072

15. Hales CN, Barker DJ (2013) Type 2 (non-insulin-dependent) diabetes mellitus: the thrifty phenotype hypothesis. Int J Epidemiol 42 (5):1215–1222

16. Ravelli AC, van Der Meulen JH, Osmond C, Barker DJ, Bleker OP (1999) Obesity at the age of 50 y in men and women exposed to famine prenatally. Am J Clin Nutr 70 (5):811–816

17. Barker DJ (1995) Fetal origins of coronary heart disease. BMJ 311(6998):171–174

18. Cooper C, Walker-Bone K, Arden N, Dennison E (2000) Novel insights into the pathogenesis of osteoporosis: the role of intrauterine programming. Rheumatology (Oxford) 39(12):1312–1315

19. Buss C, Entringer S, Wadhwa PD (2012) Fetal programming of brain development: intrauterine stress and susceptibility to psychopathology. Sci Signal 5(245):pt7. https://doi.org/10.1126/scisignal.2003406

20. Wilson ME, Ford SP (2001) Comparative aspects of placental efficiency. Reprod Suppl 58:223–232

21. Tamimi RM, Lagiou P, Mucci LA, Hsieh CC, Adami HO, Trichopoulos D (2003) Average energy intake among pregnant women

carrying a boy compared with a girl. BMJ 326 (7401):1245–1246

22. Eriksson JG, Kajantie E, Osmond C, Thornburg K, Barker DJ (2010) Boys live dangerously in the womb. Am J Hum Biol 22(3):330–335

23. Song S (2014) Malnutrition, sex ratio, and selection: a study based on the great leap forward famine. Hum Nat 25(4):580–595

24. Barker DJ, Thornburg KL (2013) The obstetric origins of health for a lifetime. Clin Obstet Gynecol 56(3):511–519

25. Barker DJ, Thornburg KL, Osmond C, Kajantie E, Eriksson JG (2010) The prenatal origins of lung cancer. II. The placenta. Am J Hum Biol 22(4):512–516

26. Barker DJ, Osmond C, Thornburg KL, Kajantie E, Eriksson JG (2013) The intrauterine origins of Hodgkin's lymphoma. Cancer Epidemiol 37(3):321–323

27. Hamilton S, Bosman FT, Boffetta P, Ilyas M, Morreau H, Nakamura S-I et al (2010) Carcinoma of the colon and rectum. In: Bosman FT, Carneiro F, Hruban RH, Theise ND (eds) WHO classification of tumours of the digestive system, WHO/IARC classification of tumours, vol 3, 4th edn. IARC Press, Lyon, pp 134–146. ISBN-10: 9283224329

28. Barker DJ, Osmond C, Thornburg KL, Kajantie E, Eriksson JG (2013) The shape of the placental surface at birth and colorectal cancer in later life. Am J Hum Biol 25 (4):566–568

29. McMillen IC, Robinson JS (2005) Developmental origins of the metabolic syndrome: prediction, plasticity, and programming. Physiol Rev 85(2):571–633

30. Hales CN, Barker DJ (2001) The thrifty phenotype hypothesis. Br Med Bull 60:5–20

31. Eriksson JG, Kajantie E, Thornburg KL, Osmond C, Barker DJ (2011) Mother's body size and placental size predict coronary heart disease in men. Eur Heart J 32(18):2297–2303

32. Barker DJ, Larsen G, Osmond C, Thornburg KL, Kajantie E, Eriksson JG (2012) The placental origins of sudden cardiac death. Int J Epidemiol 41(5):1394–1399

33. Barker DJ, Gelow J, Thornburg K, Osmond C, Kajantie E, Eriksson JG (2010) The early origins of chronic heart failure: impaired placental growth and initiation of insulin resistance in childhood. Eur J Heart Fail 12(8):819–825

34. Barker DJ, Osmond C, Forsen TJ, Kajantie E, Eriksson JG (2005) Trajectories of growth among children who have coronary events as adults. N Engl J Med 353(17):1802–1809

35. Myatt L (2006) Placental adaptive responses and fetal programming. J Physiol 572 (Pt 1):25–30

36. Hauguel S, Desmaizieres V, Challier JC (1986) Glucose uptake, utilization, and transfer by the human placenta as functions of maternal glucose concentration. Pediatr Res 20(3):269–273

37. Rofey DL, Kolko RP, Iosif AM, Silk JS, Bost JE, Feng W et al (2009) A longitudinal study of childhood depression and anxiety in relation to weight gain. Child Psychiatry Hum Dev 40(4):517–526

38. Waring ME, Lapane KL (2008) Overweight in children and adolescents in relation to attention-deficit/hyperactivity disorder: results from a national sample. Pediatrics 122 (1):e1–e6. https://doi.org/10.1542/peds. 2007-1955

39. Pasinetti GM, Eberstein JA (2008) Metabolic syndrome and the role of dietary lifestyles in Alzheimer's disease. J Neurochem 106 (4):1503–1514

40. Peterson LJ, Flood PM (2012) Oxidative stress and microglial cells in Parkinson's disease. Mediators Inflamm 2012:401264. https://doi.org/10.1155/2012/401264

41. Priyadarshini M, Kamal MA, Greig NH, Reale M, Abuzenadah AM, Chaudhary AG et al (2012) Alzheimer's disease and type 2 diabetes: exploring the association to obesity and tyrosine hydroxylase. CNS Neurol Disord Drug Targets 11(4):482–489

42. Muralimanoharan S, Maloyan A, Mele J, Guo C, Myatt LG, Myatt L (2012) MIR-210 modulates mitochondrial respiration in placenta with preeclampsia. Placenta 33(10):816–823

43. Mele J, Muralimanoharan S, Maloyan A, Myatt L (2014) Impaired mitochondrial function in human placenta with increased maternal adiposity. Am J Physiol Endocrinol Metab 307(5):E419–E425

44. Muralimanoharan S, Maloyan A, Myatt L (2016) Mitochondrial function and glucose metabolism in the placenta with gestational diabetes mellitus: role of miR-143. Clin Sci (Lond) 130(11):931–941

45. Mayeur S, Lancel S, Theys N, Lukaszewski MA, Duban-Deweer S, Bastide B et al (2013) Maternal calorie restriction modulates placental mitochondrial biogenesis and bioenergetic efficiency: putative involvement in fetoplacental growth defects in rats. Am J Physiol Endocrinol Metab 304(1):E14–E22

46. Clifton VL (2010) Review: sex and the human placenta: mediating differential strategies of

fetal growth and survival. Placenta 31(Suppl): S33–S39

47. Sood R, Zehnder JL, Druzin ML, Brown PO (2006) Gene expression patterns in human placenta. Proc Natl Acad Sci U S A 103 (14):5478–5483

48. Buckberry S, Bianco-Miotto T, Bent SJ, Dekker GA, Roberts CT (2014) Integrative transcriptome meta-analysis reveals widespread sex-biased gene expression at the human fetal-maternal interface. Mol Hum Reprod 20(8):810–819

49. Sedlmeier EM, Brunner S, Much D, Pagel P, Ulbrich SE, Meyer HH et al (2014) Human placental transcriptome shows sexually dimorphic gene expression and responsiveness to maternal dietary n-3 long-chain polyunsaturated fatty acid intervention during pregnancy. BMC Genomics 15:941. https://doi.org/10.1186/1471-2164-15-941

50. Cvitic S, Longtine MS, Hackl H, Wagner K, Nelson MD, Desoye G et al (2013) The human placental sexome differs between trophoblast epithelium and villous vessel endothelium. PLoS One 8(10):e79233. https://doi.org/10.1371/journal.pone.0079233

51. Scott NM, Hodyl NA, Murphy VE, Osei-Kumah A, Wyper H, Hodgson DM et al (2009) Placental cytokine expression covaries with maternal asthma severity and fetal sex. J Immunol 182(3):1411–1420

52. Osei-Kumah A, Smith R, Jurisica I, Caniggia I, Clifton VL (2011) Sex-specific differences in placental global gene expression in pregnancies complicated by asthma. Placenta 32(8):570–578

53. Mao J, Zhang X, Sieli PT, Falduto MT, Torres KE, Rosenfeld CS (2010) Contrasting effects of different maternal diets on sexually dimorphic gene expression in the murine placenta. Proc Natl Acad Sci U S A 107 (12):5557–5562

54. Yeganegi M, Leung CG, Martins A, Kim SO, Reid G, Challis JR et al (2011) Lactobacillus rhamnosus GR-1 stimulates colony-stimulating factor 3 (granulocyte) (CSF3) output in placental trophoblast cells in a fetal sex-dependent manner. Biol Reprod 84(1):18–25

55. Baumann MU, Deborde S, Illsley NP (2002) Placental glucose transfer and fetal growth. Endocrine 19(1):13–22

56. Lager S, Powell TL (2012) Regulation of nutrient transport across the placenta. J Pregnancy 2012:179827. https://doi.org/10.1155/2012/179827

57. Cleal JK, Lewis RM (2008) The mechanisms and regulation of placental amino acid transport to the human foetus. J Neuroendocrinol 20(4):419–426

58. Gaccioli F, Lager S (2016) Placental nutrient transport and intrauterine growth restriction. Front Physiol 7:40. https://doi.org/10.3389/fphys.2016.00040

59. Herrera E, Ortega-Senovilla H (2014) Lipid metabolism during pregnancy and its implications for fetal growth. Curr Pharm Biotechnol 15(1):24–31

60. Schaiff WT, Bildirici I, Cheong M, Chern PL, Nelson DM, Sadovsky Y (2005) Peroxisome proliferator-activated receptor-gamma and retinoid X receptor signaling regulate fatty acid uptake by primary human placental trophoblasts. J Clin Endocrinol Metab 90 (7):4267–4275

61. Campbell FM, Bush PG, Veerkamp JH, Dutta-Roy AK (1998) Detection and cellular localization of plasma membrane-associated and cytoplasmic fatty acid-binding proteins in human placenta. Placenta 19(5-6):409–415

62. Biron-Shental T, Schaiff WT, Ratajczak CK, Bildirici I, Nelson DM, Sadovsky Y (2007) Hypoxia regulates the expression of fatty acid-binding proteins in primary term human trophoblasts. Am J Obstet Gynecol 197(5):516.e1–516.e6

63. Stein Z, Susser M (1975) The Dutch famine, 1944-1945, and the reproductive process. II. Interrelations of caloric rations and six indices at birth. Pediatr Res 9(2):76–83

64. Stein Z, Susser M (1975) The Dutch famine, 1944-1945, and the reproductive process. I. Effects on six indices at birth. Pediatr Res 9(2):70–76

65. Dowler EA, O'Connor D (2012) Rights-based approaches to addressing food poverty and food insecurity in Ireland and UK. Soc Sci Med 74(1):44–51

66. Hayward CE, Greenwood SL, Sibley CP, Baker PN, Jones RL (2011) Effect of young maternal age and skeletal growth on placental growth and development. Placenta 32 (12):990–998

67. Snell LH, Haughey BP, Buck G, Marecki MA (1998) Metabolic crisis: hyperemesis gravidarum. J Perinat Neonatal Nurs 12(2):26–37

68. Gaccioli F, Lager S, Powell TL, Jansson T (2013) Placental transport in response to altered maternal nutrition. J Dev Orig Health Dis 4(2):101–115

69. Armitage JA, Taylor PD, Poston L (2005) Experimental models of developmental programming: consequences of exposure to an energy rich diet during development. J Physiol 565(Pt 1):3–8

70. Brawley L, Poston L, Hanson MA (2003) Mechanisms underlying the programming of small artery dysfunction: review of the model using low protein diet in pregnancy in the rat. Arch Physiol Biochem 111(1):23–35

71. Wallace JM, Aitken RP, Milne JS, Hay WW Jr (2004) Nutritionally mediated placental growth restriction in the growing adolescent: consequences for the fetus. Biol Reprod 71 (4):1055–1062

72. Wallace JM, Luther JS, Milne JS, Aitken RP, Redmer DA, Reynolds LP et al (2006) Nutritional modulation of adolescent pregnancy outcome -- a review. Placenta 27(Suppl A): S61–S68

73. Redmer DA, Wallace JM, Reynolds LP (2004) Effect of nutrient intake during pregnancy on fetal and placental growth and vascular development. Domest Anim Endocrinol 27(3):199–217

74. Hayward CE, Greenwood SL, Sibley CP, Baker PN, Challis JR, Jones RL (2012) Effect of maternal age and growth on placental nutrient transport: potential mechanisms for teenagers' predisposition to small-for-gestational-age birth? Am J Physiol Endocrinol Metab 302(2):E233–E242

75. Group HSCR, Metzger BE, Lowe LP, Dyer AR, Trimble ER, Chaovarindr U et al (2002) Hyperglycemia and adverse pregnancy outcomes. N Engl J Med 358(19):1991–2002

76. Jansson T, Wennergren M, Powell TL (1999) Placental glucose transport and GLUT 1 expression in insulin-dependent diabetes. Am J Obstet Gynecol 180(1 Pt 1):163–168

77. Bibee KP, Illsley NP, Moley KH (2011) Asymmetric syncytial expression of GLUT9 splice variants in human term placenta and alterations in diabetic pregnancies. Reprod Sci 18(1):20–27

78. Osmond DT, Nolan CJ, King RG, Brennecke SP, Gude NM (2000) Effects of gestational diabetes on human placental glucose uptake, transfer, and utilisation. Diabetologia 43 (5):576–582

79. Jansson T, Ekstrand Y, Wennergren M, Powell TL (2001) Placental glucose transport in gestational diabetes mellitus. Am J Obstet Gynecol 184(2):111–116

80. Colomiere M, Permezel M, Riley C, Desoye G, Lappas M (2009) Defective insulin signaling in placenta from pregnancies complicated by gestational diabetes mellitus. Eur J Endocrinol 160(4):567–578

81. Jansson T, Ekstrand Y, Bjorn C, Wennergren M, Powell TL (2002) Alterations in the activity of placental amino acid transporters in pregnancies complicated by diabetes. Diabetes 51(7):2214–2219

82. Kuruvilla AG, D'Souza SW, Glazier JD, Mahendran D, Maresh MJ, Sibley CP (1994) Altered activity of the system A amino acid transporter in microvillous membrane vesicles from placentas of macrosomic babies born to diabetic women. J Clin Invest 94(2):689–695

83. Dicke JM, Henderson GI (1988) Placental amino acid uptake in normal and complicated pregnancies. Am J Med Sci 295(3):223–227

84. Magnusson AL, Waterman IJ, Wennergren M, Jansson T, Powell TL (2004) Triglyceride hydrolase activities and expression of fatty acid binding proteins in the human placenta in pregnancies complicated by intrauterine growth restriction and diabetes. J Clin Endocrinol Metab 89 (9):4607–4614

85. Scifres CM, Chen B, Nelson DM, Sadovsky Y (2011) Fatty acid binding protein 4 regulates intracellular lipid accumulation in human trophoblasts. J Clin Endocrinol Metab 96(7): E1083–E1091

86. Gauster M, Hiden U, van Poppel M, Frank S, Wadsack C, Hauguel-de Mouzon S et al (2011) Dysregulation of placental endothelial lipase in obese women with gestational diabetes mellitus. Diabetes 60(10):2457–2464

87. Radaelli T, Lepercq J, Varastehpour A, Basu S, Catalano PM, Hauguel-De Mouzon S (2009) Differential regulation of genes for fetoplacental lipid pathways in pregnancy with gestational and type 1 diabetes mellitus. Am J Obstet Gynecol 201(2):209.e1–209e10

88. Stewart FM, Freeman DJ, Ramsay JE, Greer IA, Caslake M, Ferrell WR (2007) Longitudinal assessment of maternal endothelial function and markers of inflammation and placental function throughout pregnancy in lean and obese mothers. J Clin Endocrinol Metab 92(3):969–975

89. Ghio A, Bertolotto A, Resi V, Volpe L, Di Cianni G (2011) Triglyceride metabolism in pregnancy. Adv Clin Chem 55:133–153

90. Mazurkiewicz JC, Watts GF, Warburton FG, Slavin BM, Lowy C, Koukkou E (1994) Serum lipids, lipoproteins and apolipoproteins in pregnant non-diabetic patients. J Clin Endocrinol Metab 47(8):728–731

91. Okereke NC, Huston-Presley L, Amini SB, Kalhan S, Catalano PM (2004) Longitudinal changes in energy expenditure and body composition in obese women with normal and impaired glucose tolerance. Am J Physiol Endocrinol Metab 287(3):E472–E479

92. Catalano PM, Huston L, Amini SB, Kalhan SC (1999) Longitudinal changes in glucose metabolism during pregnancy in obese women with normal glucose tolerance and gestational diabetes mellitus. Am J Obstet Gynecol 180(4):903–916

93. Aviram A, Hod M, Yogev Y (2011) Maternal obesity: implications for pregnancy outcome and long-term risks-a link to maternal nutrition. Int J Gynaecol Obstet 115(Suppl 1): S6–10

94. Jones HN, Jansson T, Powell TL (2009) IL-6 stimulates system A amino acid transporter activity in trophoblast cells through STAT3 and increased expression of SNAT2. Am J Physiol Cell Physiol 297(5):C1228–C1235

95. Jansson N, Greenwood SL, Johansson BR, Powell TL, Jansson T (2003) Leptin stimulates the activity of the system A amino acid transporter in human placental villous fragments. J Clin Endocrinol Metab 88 (3):1205–1211

96. Johansson M, Karlsson L, Wennergren M, Jansson T, Powell TL (2003) Activity and protein expression of Na+/K+ ATPase are reduced in microvillous syncytiotrophoblast plasma membranes isolated from pregnancies complicated by intrauterine growth restriction. J Clin Endocrinol Metab 88(6):2831–2837

97. Karl PI, Alpy KL, Fisher SE (1992) Amino acid transport by the cultured human placental trophoblast: effect of insulin on AIB transport. Am J Physiol 262(4 Pt 1):C834–C839

98. Jones HN, Jansson T, Powell TL (2010) Full-length adiponectin attenuates insulin signaling and inhibits insulin-stimulated amino Acid transport in human primary trophoblast cells. Diabetes 59(5):1161–1170

99. Jones HN, Woollett LA, Barbour N, Prasad PD, Powell TL, Jansson T (2009) High-fat diet before and during pregnancy causes marked up-regulation of placental nutrient transport and fetal overgrowth in C57/BL6 mice. FASEB J 23(1):271–278

100. Farley DM, Choi J, Dudley DJ, Li C, Jenkins SL, Myatt L et al (2010) Placental amino acid transport and placental leptin resistance in pregnancies complicated by maternal obesity. Placenta 31(8):718–724

101. Jansson N, Rosario FJ, Gaccioli F, Lager S, Jones HN, Roos S et al (2013) Activation of placental mTOR signaling and amino acid transporters in obese women giving birth to large babies. J Clin Endocrinol Metab 98 (1):105–113

102. Metzger BE, Persson B, Lowe LP, Dyer AR, Cruickshank JK, Deerochanawong C et al (2010) Hyperglycemia and adverse pregnancy outcome study: neonatal glycemia. Pediatrics 126(6):e1545–e1552

103. Schaefer-Graf UM, Graf K, Kulbacka I, Kjos SL, Dudenhausen J, Vetter K et al (2008) Maternal lipids as strong determinants of fetal environment and growth in pregnancies with gestational diabetes mellitus. Diabetes Care 31(9):1858–1863

104. Whyte K, Kelly H, O'Dwyer V, Gibbs M, O'Higgins A, Turner MJ (2013) Offspring birth weight and maternal fasting lipids in women screened for gestational diabetes mellitus (GDM). Eur J Obstet Gynecol Reprod Biol 170(1):67–70

105. Dube E, Gravel A, Martin C, Desparois G, Moussa I, Ethier-Chiasson M et al (2012) Modulation of fatty acid transport and metabolism by maternal obesity in the human full-term placenta. Biol Reprod 87(1):14. 1-11

106. Brass E, Hanson E, O'Tierney-Ginn PF (2013) Placental oleic acid uptake is lower in male offspring of obese women. Placenta 34 (6):503–509

107. Borengasser SJ, Lau F, Kang P, Blackburn ML, Ronis MJ, Badger TM et al (2011) Maternal obesity during gestation impairs fatty acid oxidation and mitochondrial SIRT3 expression in rat offspring at weaning. PLoS One 6(8):e24068. https://doi.org/10. 1371/journal.pone.0024068

108. Zhu MJ, Ma Y, Long NM, Du M, Ford SP (2010) Maternal obesity markedly increases placental fatty acid transporter expression and fetal blood triglycerides at midgestation in the ewe. Am J Physiol Regul Integr Comp Physiol 299(5):R1224–R1231

109. Calabuig-Navarro V, Haghiac M, Minium J, Glazebrook P, Ranasinghe GC, Hoppel C et al (2017) Effect of maternal obesity on placental lipid metabolism. Endocrinology. https://doi.org/10.1210/en.2017-00152

110. Stark MJ, Clifton VL, Wright IM (2009) Neonates born to mothers with preeclampsia exhibit sex-specific alterations in microvascular function. Pediatr Res 65(3):292–295

111. Gheorghe CP, Goyal R, Holweger JD, Longo LD (2009) Placental gene expression responses to maternal protein restriction in the mouse. Placenta 30(5):411–417

112. Gabory A, Ferry L, Fajardy I, Jouneau L, Gothie JD, Vige A et al (2012) Maternal diets trigger sex-specific divergent trajectories of gene expression and epigenetic systems in mouse placenta. PLoS One 7(11):e47986. https://doi.org/10.1371/journal.pone. 0047986

113. Chen PY, Ganguly A, Rubbi L, Orozco LD, Morselli M, Ashraf D et al (2013) Intrauterine calorie restriction affects placental DNA methylation and gene expression. Physiol Genomics 45(14):565–576

114. Tarrade A, Rousseau-Ralliard D, Aubriere MC, Peynot N, Dahirel M, Bertrand-Michel J et al (2013) Sexual dimorphism of the feto-placental phenotype in response to a high fat and control maternal diets in a rabbit model. PLoS One 8(12):e83458. https://doi.org/10.1371/journal.pone.0083458

115. Cox LA, Li C, Glenn JP, Lange K, Spradling KD, Nathanielsz PW et al (2013) Expression of the placental transcriptome in maternal nutrient reduction in baboons is dependent on fetal sex. J Nutr 143(11):1698–1708

116. Sferruzzi-Perri AN, Vaughan OR, Haro M, Cooper WN, Musial B, Charalambous M et al (2013) An obesogenic diet during mouse pregnancy modifies maternal nutrient partitioning and the fetal growth trajectory. FASEB J 27(10):3928–3937

117. Lin Y, Zhuo Y, Fang ZF, Che LQ, Wu D (2012) Effect of maternal dietary energy types on placenta nutrient transporter gene expressions and intrauterine fetal growth in rats. Nutrition 28(10):1037–1043

118. King V, Hibbert N, Seckl JR, Norman JE, Drake AJ (2013) The effects of an obesogenic diet during pregnancy on fetal growth and placental gene expression are gestation dependent. Placenta 34(11):1087–1090

119. Choi SW, Friso S (2010) Epigenetics: a new bridge between nutrition and health. Adv Nutr 1(1):8–16

120. Gut P, Verdin E (2013) The nexus of chromatin regulation and intermediary metabolism. Nature 502(7472):489–498

121. Heijmans BT, Tobi EW, Lumey LH, Slagboom PE (2009) The epigenome: archive of the prenatal environment. Epigenetics 4 (8):526–531

122. Tobi EW, Lumey LH, Talens RP, Kremer D, Putter H, Stein AD et al (2009) DNA methylation differences after exposure to prenatal famine are common and timing- and sex-specific. Hum Mol Genet 18 (21):4046–4053

123. Waterland RA, Lin JR, Smith CA, Jirtle RL (2006) Post-weaning diet affects genomic imprinting at the insulin-like growth factor 2 (Igf2) locus. Hum Mol Genet 15 (5):705–716

124. Keyes MK, Jang H, Mason JB, Liu Z, Crott JW, Smith DE et al (2007) Older age and dietary folate are determinants of genomic

and p16-specific DNA methylation in mouse colon. J Nutr 137:1713–1717

125. Hoile SP, Irvine NA, Kelsall CJ, Sibbons C, Feunteun A, Collister A et al (2013) Maternal fat intake in rats alters 20:4n-6 and 22:6n-3 status and the epigenetic regulation of Fads2 in offspring liver. J Nutr Biochem 24 (7):1213–1220

126. Hass BS, Hart RW, MH L, Lyn-Cook BD (1993) Effects of caloric restriction in animals on cellular function, oncogene expression, and DNA methylation in vitro. Mutat Res 295(4-6):281–289

127. Gallou-Kabani C, Gabory A, Tost J, Karimi M, Mayeur S, Lesage J et al (2010) Sex- and diet-specific changes of imprinted gene expression and DNA methylation in mouse placenta under a high-fat diet. PLoS One 5(12):e14398. https://doi.org/10.1371/journal.pone.0014398

128. Bouchard L, Hivert MF, Guay SP, St-Pierre J, Perron P, Brisson D (2012) Placental adiponectin gene DNA methylation levels are associated with mothers' blood glucose concentration. Diabetes 61(5):1272–1280

129. Bouchard L, Thibault S, Guay SP, Santure M, Monpetit A, St-Pierre J et al (2010) Leptin gene epigenetic adaptation to impaired glucose metabolism during pregnancy. Diabetes Care 33(11):2436–2441

130. El Hajj N, Pliushch G, Schneider E, Dittrich M, Muller T, Korenkov M et al (2013) Metabolic programming of MEST DNA methylation by intrauterine exposure to gestational diabetes mellitus. Diabetes 62 (4):1320–1328

131. Scott JM, Weir DG (1998) Folic acid, homocysteine and one-carbon metabolism: a review of the essential biochemistry. J Cardiovasc Risk 5(4):223–227

132. Waterland RA, Jirtle RL (2003) Transposable elements: targets for early nutritional effects on epigenetic gene regulation. Mol Cell Biol 23(15):5293–5300

133. Lillycrop KA, Phillips ES, Jackson AA, Hanson MA, Burdge GC (2005) Dietary protein restriction of pregnant rats induces and folic acid supplementation prevents epigenetic modification of hepatic gene expression in the offspring. J Nutr 135(6):1382–1386

134. Petrossian TC, Clarke SG (2011) Uncovering the human methyltransferasome. Mol Cell Proteomics 10(1):M110.000976. https://doi.org/10.1074/mcp.M110.000976

135. Kirkland JB (2009) Niacin status impacts chromatin structure. J Nutr 139 (12):2397–2401

136. Bacalini MG, Friso S, Olivieri F, Pirazzini C, Giuliani C, Capri M et al (2014) Present and future of anti-ageing epigenetic diets. Mech Ageing Dev 136-137:101–115

137. Reik W, Walter J (2001) Genomic imprinting: parental influence on the genome. Nat Rev Genet 2(1):21–32

138. Coan PM, Burton GJ, Ferguson-Smith AC (2005) Imprinted genes in the placenta--a review. Placenta 26(Suppl A):S10–S20

139. Monk D, Arnaud P, Frost J, Hills FA, Stanier P, Feil R et al (2009) Reciprocal imprinting of human GRB10 in placental trophoblast and brain: evolutionary conservation of reversed allelic expression. Hum Mol Genet 18(16):3066–3074

140. Constancia M, Kelsey G, Reik W (2004) Resourceful imprinting. Nature 432 (7013):53–57

141. Cleaton MA, Edwards CA, Ferguson-Smith AC (2014) Phenotypic outcomes of imprinted gene models in mice: elucidation of pre- and postnatal functions of imprinted genes. Annu Rev Genomics Hum Genet 15:93–126

142. Reik W, Constancia M, Fowden A, Anderson N, Dean W, Ferguson-Smith A et al (2003) Regulation of supply and demand for maternal nutrients in mammals by imprinted genes. J Physiol 547(Pt 1):35–44

143. Sferruzzi-Perri AN, Vaughan OR, Coan PM, Suciu MC, Darbyshire R, Constancia M et al (2011) Placental-specific Igf2 deficiency alters developmental adaptations to undernutrition in mice. Endocrinology 152(8):3202–3212

144. Soubry A, Murphy SK, Wang F, Huang Z, Vidal AC, Fuemmeler BF et al (2013) Newborns of obese parents have altered DNA methylation patterns at imprinted genes. Int J Obes (Lond) 39(4):650–657

145. Perkins E, Murphy SK, Murtha AP, Schildkraut J, Jirtle RL, Demark-Wahnefried W et al (2012) Insulin-like growth factor 2/H19 methylation at birth and risk of overweight and obesity in children. J Pediatr 161 (1):31–39

146. Boyd PA, Scott A, Keeling JW (1986) Quantitative structural studies on placentas from pregnancies complicated by diabetes mellitus. Br J Obstet Gynaecol 93(1):31–35

147. Nelson SM, Coan PM, Burton GJ, Lindsay RS (2009) Placental structure in type 1 diabetes: relation to fetal insulin, leptin, and IGF-I. Diabetes 58(11):2634–2641

148. Mayhew TM, Manwani R, Ohadike C, Wijesekara J, Baker PN (2007) The placenta in pre-eclampsia and intrauterine growth restriction: studies on exchange surface areas, diffusion distances and villous membrane diffusive conductances. Placenta 28 (2-3):233–238

149. Alwasel SH, Harrath AH, Aldahmash WM, Abotalib Z, Nyengaard JR, Osmond C et al (2014) Sex differences in regional specialisation across the placental surface. Placenta 35 (6):365–369

150. Roberts CT, Sohlstrom A, Kind KL, Earl RA, Khong TY, Robinson JS et al (2001) Maternal food restriction reduces the exchange surface area and increases the barrier thickness of the placenta in the guinea-pig. Placenta 22 (2-3):177–185

151. Reynolds CM, Vickers MH, Harrison CJ, Segovia SA, Gray C (2015) Maternal high fat and/or salt consumption induces sex-specific inflammatory and nutrient transport in the rat placenta. Physiol Rep 3(5):pii: e12399. 10.14814/phy2.12399

152. Kim DW, Young SL, Grattan DR, Jasoni CL (2014) Obesity during pregnancy disrupts placental morphology, cell proliferation, and inflammation in a sex-specific manner across gestation in the mouse. Biol Reprod 90 (6):130. https://doi.org/10.1095/biolreprod.113.117259

153. Muralimanoharan S, Maloyan A, Myatt L (2013) Evidence of sexual dimorphism in the placental function with severe preeclampsia. Placenta 34(12):1183–1189

154. Coughlan MT, Vervaart PP, Permezel M, Georgiou HM, Rice GE (2004) Altered placental oxidative stress status in gestational diabetes mellitus. Placenta 25(1):78–84

155. Myatt L, Cui X (2004) Oxidative stress in the placenta. Histochem Cell Biol 122 (4):369–382

156. Roberts VH, Smith J, McLea SA, Heizer AB, Richardson JL, Myatt L (2009) Effect of increasing maternal body mass index on oxidative and nitrative stress in the human placenta. Placenta 30(2):169–175

157. Evans RW, Powers RW, Ness RB, Cropcho LJ, Daftary AR, Harger GF et al (2003) Maternal and fetal amino acid concentrations and fetal outcomes during pre-eclampsia. Reproduction 125(6):785–790

158. Alvino G, Cozzi V, Radaelli T, Ortega H, Herrera E, Cetin I (2008) Maternal and fetal fatty acid profile in normal and intrauterine growth restriction pregnancies with and without preeclampsia. Pediatr Res 64(6):615–620

159. Roy S, Dhobale M, Dangat K, Mehendale S, Wagh G, Lalwani S et al (2014) Differential levels of long chain polyunsaturated fatty

acids in women with preeclampsia delivering male and female babies. Prostaglandins Leukot Essent Fatty Acids 91(5):227–232

160. Desforges M, Ditchfield A, Hirst CR, Pegorie C, Martyn-Smith K, Sibley CP et al (2013) Reduced placental taurine transporter (TauT) activity in pregnancies complicated by pre-eclampsia and maternal obesity. Adv Exp Med Biol 776:81–91

161. Malina A, Daftary A, Crombleholme W, Markovic N, Roberts JM (2005) Placental system A transporter mRNA is not different in preeclampsia, normal pregnancy, or pregnancies with small-for-gestational-age infants. Hypertens Pregnancy 24(1):65–74

162. Shibata E, Hubel CA, Powers RW, von Versen-Hoeynck F, Gammill H, Rajakumar A et al (2008) Placental system A amino acid transport is reduced in pregnancies with small for gestational age (SGA) infants but not in preeclampsia with SGA infants. Placenta 29 (10):879–882

163. Speake PF, Glazier JD, Ayuk PT, Reade M, Sibley CP, D'Souza SW (2003) L-Arginine transport across the basal plasma membrane of the syncytiotrophoblast of the human placenta from normal and preeclamptic pregnancies. J Clin Endocrinol Metab 88 (9):4287–4292

164. Wadhwani N, Patil V, Pisal H, Joshi A, Mehendale S, Gupte S et al (2014) Altered maternal proportions of long chain polyunsaturated fatty acids and their transport leads to disturbed fetal stores in preeclampsia. Prostaglandins Leukot Essent Fatty Acids 91 (1-2):21–30

165. Ray PD, Huang BW, Tsuji Y (2012) Reactive oxygen species (ROS) homeostasis and redox regulation in cellular signaling. Cell Signal 24 (5):981–990

166. Muralimanoharan S, Guo C, Myatt L, Maloyan A (2015) Sexual dimorphism in miR-210 expression and mitochondrial dysfunction in the placenta with maternal obesity. Int J Obes (Lond) 39(8):1274–1281

167. Sen CK, Packer L (1996) Antioxidant and redox regulation of gene transcription. FASEB J 10(7):709–720

168. Arner ES, Holmgren A (2000) Physiological functions of thioredoxin and thioredoxin reductase. Eur J Biochem 267 (20):6102–6109

169. Evans L, Myatt L (2017) Sexual dimorphism in the effect of maternal obesity on antioxidant defense mechanisms in the human placenta. Placenta 51:64–69

170. Vanderlelie J, Perkins AV (2011) Selenium and preeclampsia: a global perspective. Pregnancy Hypertens 1(3-4):213–224

171. Blazewicz A, Klatka M, Astel A, Korona-Glowniak I, Dolliver W, Szwerc W et al (2015) Serum and urinary selenium levels in obese children: a cross-sectional study. J Trace Elem Med Biol 29:116–122

172. Alasfar F, Ben-Nakhi M, Khoursheed M, Kehinde EO, Alsaleh M (2011) Selenium is significantly depleted among morbidly obese female patients seeking bariatric surgery. Obes Surg 21(11):1710–1713

173. Barker D, Osmond C, Grant S, Thornburg KL, Cooper C, Ring S et al (2013) Maternal cotyledons at birth predict blood pressure in childhood. Placenta 34(8):672–675

174. Barker DJ, Hales CN, Fall CH, Osmond C, Phipps K, Clark PM (1993) Type 2 (non-insulin-dependent) diabetes mellitus, hypertension and hyperlipidaemia (syndrome X): relation to reduced fetal growth. Diabetologia 36(1):62–67

175. Barker DJ, Osmond C, Forsen TJ, Thornburg KL, Kajantie E, Eriksson JG (2013) Foetal and childhood growth and asthma in adult life. Acta Paediatr 102(7):732–738

Chapter 3

Developmental Origins of Stress and Psychiatric Disorders

Francesca L. Guest and Paul C. Guest

Abstract

Over the last few decades, evidence has emerged that the pathogenesis of psychiatric disorders such as schizophrenia can involve perturbations of the hypothalamic-pituitary-adrenal (HPA) axis and other neuroendocrine systems. Variations in the manifestation of these effects could be related to differences in clinical symptoms between affected individuals and to differences in treatment response. Such effects can also arise from the complex interaction between genes and environmental factors. Here, we review the effects of maternal stress on abnormalities in HPA axis regulation and the development of psychiatric disorders such as schizophrenia. Studies in this area may prove critical for increasing our understanding of the multidimensional nature of mental disorders and could lead to the development of improved diagnostics and novel therapeutic approaches for treating individuals who suffer from these conditions.

Key words Nutritional deprivation, Maternal stress, Psychiatric disorders, Schizophrenia, HPA axis dysfunction, Diagnosis, Biomarkers

1 Introduction

In response to stress, biochemicals such as adrenaline and glucocorticoids are released into the bloodstream, which leads to a chain reaction of molecular events inside the body. The effect of this stress-induced molecular cascade has also been investigated in mothers during pregnancy. In a seminal work, David Barker proposed the "fetal programming hypothesis" which described the role of the intrauterine environment in the development of a fetus [1]. The hypothesis states that any deviation from the ideal intrauterine environment may result in lasting effects on long-term health [1, 2]. This idea can be extrapolated to the brain and neuroendocrine organs which would mean that such effects may increase the risk of the offspring developing neuroendocrine disorders such as schizophrenia, major depression, autism spectrum disorders, Alzheimer's disease, and Parkinson's disease.

Paul C. Guest (ed.), *Investigations of Early Nutrition Effects on Long-Term Health: Methods and Applications*, Methods in Molecular Biology, vol. 1735, https://doi.org/10.1007/978-1-4939-7614-0_3, © Springer Science+Business Media, LLC 2018

A number of studies have now documented correlations between maternal stress and neuroendocrine abnormalities in the offspring [3–6]. In this scenario, hormones released into the bloodstream of the mother in response to stressful stimuli can also have direct effects on the brain, other organs, and systems of the developing fetus. This is due to the fact that any deviation from the normal concentration of neuroendocrine hormones and other bioactive molecules can produce abnormalities in development, such as in the synaptic connections within and across different regions of the brain [7]. In turn, this can lead to the development of psychiatric and neurodegenerative conditions in the offspring in later life, which may be linked to disruptive effects on the hypothalamic-pituitary-adrenal (HPA) axis and other organs of the diffuse neuroendocrine system.

This chapter discusses the existing literature on the effects of prenatal stress on development and how this can be disruptive to the psychiatric and overall health of the offspring in the long term. Since most of the current literature on this subject derives from studies of animal models, we will attempt to extrapolate the findings to the human situation, as possible. We also suggest the likely effects of prenatal stress that can lead to psychiatric disorders in humans and discuss the effects that this has on the HPA axis, other neuroendocrine systems, and brain structure and function.

2   Forms of Stress

Stress is an event that is either real or perceived which can disturb an individual's homeostasis [8]. For example, emotional stress can be caused by the perception of conditions such as important deadlines, crucial examinations, overbearing bosses, and important decisions [9]. Interestingly, different people can respond in different ways or have different thresholds to the same stressful situation [10]. Intense emotional stresses occur in close relationships, such as between family members or partners [11–13]. Other types of stressors include those of a mechanical or physical nature, such as impact injuries or trauma, and biochemical stressors which can affect the physiology through alterations in molecular signaling. The latter could be brought about by ingestion of a harmful substance, deprivation of a substance that the body needs, or other deviations from the norm, such as experiencing the effects of high altitude, lack of a protein or an essential amino acid in the diet, or enduring extremes of temperature. The degree to which a specific stress affects a person depends on that own perception of whether or not they can cope with it. Those who cannot cope may experience depression or anxiety, whereas some may feel spurred on by stress to increase their performance [14, 15]. The latter category may include those who have a higher stress threshold.

3   The Stress Response in Humans

In reaction to a stressful situation, the body can respond by activation of the HPA axis. This is an important circuit that prepares the body for a "fight or flight" response to a potentially dangerous situation [16, 17]. There are three main phases of the stress response that can be initiated by a stressful situation. The first phase is the production of noradrenaline and adrenaline which are released rapidly from the adrenal medulla into the bloodstream from where they can affect multiple bodily actions. As examples, both of these molecules can act to increase the heart rate, constrict blood vessels, dilate the pupils, and open the respiratory airways. All of these would act together to enhance the fight or flight response. In addition insulin secretion is inhibited, and glycogenolysis is stimulated in the muscles and liver, leading to a higher concentration of glucose in the bloodstream to support the sudden bursts of energy that may be required [16, 18]. In the next phase, corticotrophin-releasing factor (CRF) is released from the hypothalamus and stimulates release of adrenocorticotropic hormone (ACTH) from the anterior pituitary into the bloodstream. ACTH circulates through the blood and encounters the adrenal glands, stimulating release of glucocorticoids such as cortisol from the adrenal cortex. Cortisol produces similar responses as noradrenaline and adrenaline, by acting to increase glucose levels for enhanced energy. Finally, the increased levels of cortisol in the blood block the further release of ACTH and CRF via a negative feedback loop [8].

4   Stress Programming

Chronic maternal stress through HPA axis activation has been linked with morphological changes in the developing brain of the fetus, although the exact mechanisms underlying these effects have not been elucidated [19]. It is well known that during pregnancy a mother body undergoes both physical and biochemical changes including a change in the production of key hormones and growth factors. For example, cortisol levels are normally found to be elevated in the bloodstream of a pregnant mother [20, 21]. Increased levels of this hormone are essential for fetal growth, although persistently high levels can lead to permanent detrimental changes [22]. In addition, the increased levels of circulating adrenaline in prenatal stress can lead to restriction of blood flow to the placenta and potentially decreased levels of nutrients and oxygen that reach the fetus [19]. Other effects include increased levels of estrogen, placental lactogen, and hypothalamic CRH [23]. Anatomical studies have shown that stress-induced increases in maternal CRH levels can result in both macroscopic and microscopic changes in the hippocampi of the offspring [24–27].

5 Psychological Disorders Linked to Stress

A variety of psychological conditions have linked to prenatal stress. For example, studies carried out during the 1990s showed that exposure of rhesus monkeys to loud noises during pregnancy led to an increased incidence of attention deficit hyperactivity disorder (ADHD) in the offspring [28, 29], and the offspring of mothers exposed to long-term stress showed greater responses to stress themselves along with disturbed social behavior [30, 31]. Likewise, human studies have also reported a greater prevalence of ADHD in the offspring of stressed mothers [32, 33]. A Danish study of human participants showed that the incidence of ADHD was increased by 72% in offspring of mothers who experienced bereavements during pregnancy [34].

A number of reports have also suggested an increased incidence of schizophrenia in the offspring of mothers who were stressed during pregnancy. For examples, studies have reported an increased incidence of schizophrenia in the offspring of mothers who had experienced the loss of a family member [35] or disasters such as a tornado [36] during pregnancy. The German army invasion of the Netherlands in May 1940 resulted in nationwide stress, and the offspring of mothers who had been exposed to this stress during their pregnancies showed a greater incidence of schizophrenia [37]. Similar effects have also been reported to occur following a lack of essential nutrients in the diet. Such a situation occurred again in World War II during the winter of 1944–1945 when a Nazi blockade led to food shortages for approximately 40,000 individuals in the Netherlands (now referred to as "The Dutch Hunger Winter") [38–41]. This led to a twofold increase in the incidence of schizophrenia in the offspring, with the largest effects observed for those who had been conceived during the peak of the famine. Similar findings resulted from studies of the Chinese Famine of 1958–1961, which found that individuals conceived at the peak of the famine showed an increased risk of schizophrenia and other conditions [42].

Other psychiatric disorders have also been linked to maternal stress. One study found a greater than twofold increase in the incidence of depression in the offspring of mothers who were exposed to a severe earthquake compared to nonexposed mothers [43]. When the US state of Louisiana was hit by a series of hurricanes and tropical storms, one report showed that the incidence of autism was seven times higher in the offspring of mothers in the high-exposure group, compared to those in the low-exposure group [44]. Other studies found increased incidences of autism [45], as well as cognitive, behavioral, and emotional issues [46–50], in the offspring of mothers who experienced a natural disaster or other stressful situations.

Studies involving deprivation of nutrients such as protein during pregnancy in rats have shown that this can have long-term effects on brain structure and function in the offspring. This includes morphological changes in the hippocampus as well as differences in the binding of neurotransmitters such as glutamate and dopamine to their receptors [51]. Such effects may be due to changes in maternal metabolism of lipids, which are an essential component of all cellular membranes, including those in the brain [52]. The behavioral abnormalities experienced by the offspring following maternal protein restriction appear during early adulthood and include observations such as decreased pre-pulse inhibition of the startle response and hyperlocomotion, which are tests that are traditionally used as a measure of schizophrenia-like behavior [53, 54]. A pre-pulse is a low-level stimulus given shortly before a full version (the pulse) of the stimulus. In normal individuals, the brain will naturally temper a startle response when a stimulus is preceded by a pre-pulse. However, individuals with schizophrenia often show an untempered startle response of the same magnitude [55]. Prenatal stress has also been linked to increased depression which may be manifested through abnormalities in the HPA axis [56]. Rat studies have shown that the offspring of mothers who were restrained in the last week of pregnancy are more likely to self-administer the drugs cocaine and amphetamine, compared to the offspring of non-restrained mothers [57]. This indicates that antenatal stress may increase the propensity toward drug addiction, most likely through effects on the neuronal circuitry.

6   Effects of Antenatal Stress on Brain Structure

Certain areas of the brain have been shown to be affected by prenatal stress both at the macroscopic and microscopic levels. These include the amygdala, cerebellum, cerebral cortex, corpus callosum, hippocampus, hypothalamus, and midbrain [58–74]. As these regions separately and in a various combinations regulate behavior, it is not surprising that structural changes in these could lead to the behavioral deficits seen in some psychiatric disorders. For example, changes in the size of the corpus callosum and the composition of specific cell types within this brain region have been linked with ADHD, autism, and schizophrenia [75–77]. The hippocampus is a region of the brain that is well known to be involved in memory formation and learning [78]. In line with this, several studies have now reported that prenatal stress decreases memory consolidation and learning performance [79–81]. This may be linked to defects in synaptogenesis and neurogenesis since one study showed that the number of hippocampal granule neurons was reduced in adult rats from mothers exposed to stress during pregnancy [79]. Interestingly, these effects could be counteracted by handling of the offspring in the neonatal period.

7   Timing and Severity of the Stress

Rodent and human studies have established that the timing of the stressor during pregnancy is important in cases of long-term effects on brain function in the offspring [82, 83]. A study using rats found a 64% increase in the production of the stress-related hormone corticosterone after handling pregnant rats during the last week of gestation and placing them in unfamiliar cages [84]. The timing and severity of a stressor which affects the human fetus is harder to determine due to ethical reasons, although natural disasters can be used to provide estimates for this purpose since such events can provide data on a large cohort of people affected by an identical stress. This allows retrospective studies of females who were pregnant during the disaster, as well as follow-up studies of the offspring.

Analyses of the cases concerning the Dutch Hunger Winter and the Chinese Famine described above, both indicate that the highest incidence of schizophrenia occurred for offspring who had been conceived at the peak of the famine [38–42, 85]. A study termed "Project Ice Storm" specifically examined the timing of the effects of antenatal stress on the offspring. In this study, more than three million people were exposed to extreme cold due to power outages in Quebec, Canada [46, 86]. The results showed an abnormality in fingerprint formation in offspring of mothers exposed to the storm during weeks 14–22 of pregnancy. Fingerprint formation was used in this analysis due to its overlap with formation of the hippocampus [87, 88]. Therefore, it is possible that changes in the hippocampus may be seen in the same individuals. The project is still ongoing and a study in 2008 found that Project Ice Storm offspring had lower cognitive and language abilities at age 5½ [49].

8   Advantages and Disadvantages of Stress

The hippocampal dentate gyrus region is more active during acute stress. This is the area of the brain mainly associated with learning, memory, and formation of new neurons and new synapses. As this is part of the mechanism of how memories are made, it is supposed that this process enables individuals to remember the details associated with the stress and enables them to avoid similar situations in the future [89]. In contrast, chronic stress appears to decrease the formation of new memories. For example, one study showed that chronically stressed rats performed poorer on a spatial maze test compared to non-stressed control rats [90]. Similar findings have been replicated on many occasions, linking stress to behavioral deficits along with alterations in neurogenesis and new synapse formation [91]. Likewise, human studies have shown that many years of stressful living can lead to increased incidences of depression and other psychiatric disorders [92, 93].

9   Effects of Stress on Insulin Resistance and the HPA Axis

Maternal stress during pregnancy has been shown to predict low birth weight and preterm delivery in humans [25]. Studies have shown that this is linked with metabolic dysfunctions such as decreased glucose tolerance and increased insulin resistance in the offspring [94]. One study used an oral glucose tolerance test to investigate young adults whose mothers had experienced stressors during pregnancy, such as relationship conflicts, death of a family member or close friend, financial difficulties, and car accidents, as compared to controls from stress-free pregnancies [95]. Although there was no significant difference in glucose levels between the groups, the antenatal stress group had significantly higher insulin levels at the 120 min time point of the test. Another study by the same authors showed that adults whose mothers were stressed during pregnancy had higher cortisol levels during the Trier Social Stress Test (TSST) and a decreased cortisol response after ACTH stimulation, suggesting possible HPA axis dysregulation [96]. The effect on the HPA axis has been confirmed by many separate investigations [97]. In line with these findings, antenatal stress has been found to cause long-term changes in feeding behavior, glucose metabolism, and insulin signaling, as seen in type 2 diabetes [98]. Another study found that administration of stress hormones to sheep in the early stages of pregnancy resulted in impaired glucose tolerance and hyperinsulinemia in the offspring [99]. A link between HPA axis and neuroendocrine dysfunction and psychiatric disorders has been shown by studies that have identified higher levels of CRH, arginine vasopressin (AVP), ACTH, cortisol, insulin, and other pancreatic islet hormones in studies of individuals suffering from these conditions [100, 101].

10   Conclusions and Future Prospects

It is clear that the environmental trigger stress in utero is not the only catalyst for development of psychiatric disease. It is likely that there are also genetic components that act in unison to trigger the disease in accordance with the two-hit hypothesis (Fig. 1). Maternal

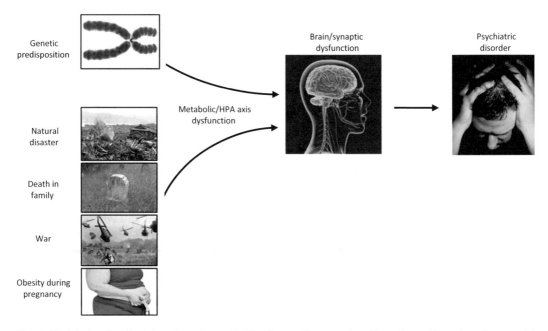

Fig. 1 Model showing the triggering of psychiatric disease by a combination of genetic and environmental factors

stress has been linked with many psychological and behavioral problems in the offspring, such as schizophrenia, ADHD, autism, and major depression. In particular, structural and functional changes are seen in areas of the brain that are responsible for the control of behavior. The mechanisms by which these changes occur most likely involve the maternal and fetal HPA axis, leading to alterations of the intrauterine environment with lasting effects on fetal brain development. There is increasing evidence for metabolic and hormonal components in conditions such as schizophrenia and other psychiatric disorders, which may be linked to antenatal stress. Abnormalities in glucose metabolism, insulin action, and HPA axis function appear to be present in the earliest stages of these disorders and may therefore provide the basis for identification and development of molecular biomarker tests for psychiatric conditions. The use of such tests could result in earlier and more accurate diagnosis allowing the use of personalized medicine approaches and more effective treatments. In addition, this could lead to identification of novel drug targets based on metabolic and hormonal pathways for the development of novel treatment approaches.

References

1. Barker DJ, Winter PD, Osmond C, Margetts B, Simmonds SJ (1989) Weight in infancy and death from ischaemic heart disease. Lancet 2:577–580

2. Hales CN, Barker DJ (1992) Type 2 (non-insulin-dependent) diabetes mellitus: the thrifty phenotype hypothesis. Diabetologia 35:595–601

3. Bale TL (2015) Epigenetic and transgenerational reprogramming of brain development. Nat Rev Neurosci 16:332–344

4. Wood CE, Walker CD (2015) Fetal and neonatal HPA axis. Compr Physiol 6:33–62

5. Scheinost D, Sinha R, Cross SN, Kwon SH, Sze G, Constable RT et al (2017) Does prenatal stress alter the developing connectome? Pediatr Res 81:214–226

6. Edlow AG (2017) Maternal obesity and neurodevelopmental and psychiatric disorders in offspring. Prenat Diagn 37:95–110

7. Mychasiuk R, Gibb R, Kolb B (2012) Prenatal stress alters dendritic morphology and synaptic connectivity in the prefrontal cortex and hippocampus of developing offspring. Synapse 66:308–314

8. Lovejoy DA (2005) Stress, arousal and homeostatic challenge. In: Neuroendocrinology, an integrated approach. John Wiley & Sons Ltd, Chichester, pp 243–256. ISBN-10: 0470844329

9. Perry BD, Pollard R (1998) Homeostasis, stress, trauma, and adaptation. A neurodevelopmental view of childhood trauma. Child Adolesc Psychiatr Clin N Am 7:33–51. viii

10. Tyssen R, Dolatowski FC, Røvik JO, Thorkildsen RF, Ekeberg O, Hem E et al (2007) Personality traits and types predict medical school stress: a six-year longitudinal and nationwide study. Med Educ 41:781–787

11. Stuber ML (1996) Psychiatric sequelae in seriously ill children and their families. Psychiatr Clin North Am 19:481–493

12. Hickman RL Jr, Douglas SL (2010) Impact of chronic critical illness on the psychological outcomes of family members. AACN Adv Crit Care 21:80–89

13. Falconier MK, Nussbeck F, Bodenmann G, Schneider H, Bradbury T (2015) Stress from daily hassles in couples: its effects on intradyadic stress, relationship satisfaction, and physical and psychological well-being. J Marital Fam Ther 41:221–235

14. Simmons BL, Nelson DL (2001) Eustress at work: the relationship between hope and health in hospital nurses. Health Care Manage Rev 26:7–18

15. Harvey SB, Modini M, Joyce S, Milligan-Saville JS, Tan L, Mykletun A et al (2017) Can work make you mentally ill? A systematic meta-review of work-related risk factors for common mental health problems. Occup Environ Med 74:301. https://doi.org/10.1136/oemed-2016-104015. pii: oemed-2016-104015

16. Arun CP (2007) Fight or flight, forbearance and fortitude: the spectrum of actions of the catecholamines and their cousins. Ann N Y Acad Sci 1018:137–140

17. Rohleder N, Kirschbaum C (2007) Effects of nutrition on neuro-endocrine stress responses. Curr Opin Clin Nutr Metab Care 10:504–510

18. Halter JB, Beard JC, Porte D Jr (1984) Islet function and stress hyperglycemia: plasma glucose and epinephrine interaction. Am J Physiol 247:E47–E52

19. Charil A, Laplante DP, Vaillancourt C, King S (2010) Prenatal stress and brain development. Brain Res Rev 65:56–79

20. Ng PC (2011) Effect of stress on the hypothalamic-pituitary-adrenal axis in the fetus and newborn. J Pediatr 158(2 Suppl):e41–e43

21. Newby EA, Myers DA, Ducsay CA (2015) Fetal endocrine and metabolic adaptations to hypoxia: the role of the hypothalamic-pituitary-adrenal axis. Am J Physiol Endocrinol Metab 309:E429–E439

22. Seckl JR, Holmes MC (2007) Mechanisms of disease: glucocorticoids, their placental metabolism and fetal 'programming' of adult pathophysiology. Nat Clin Pract Endocrinol Metab 3:479–488

23. Wadhwa PD, Porto M, Chicz-DeMet A, Sandman CA (1988) Maternal CRH levels in early third trimester predict length of gestation in human pregnancy. Am J Obstet Gynecol 179:1079–1085

24. Sandman CA, Wadhwa PD, Glynn L, Chicz-Demet A, Porto M, Garite TJ (1999) Corticotrophin-releasing hormone and fetal responses in human pregnancy. Ann N Y Acad Sci 897:66–75

25. Wadhwa PD, Sandman CA, Garite TJ (2001) The neurobiology of stress in human pregnancy: implications for prematurity and development of the fetal central nervous system. Prog Brain Res 133:131–142

26. Kastin AJ, Akerstrom V (2002) Differential interactions of urocortin/corticotropin-releasing hormone peptides with the blood-brain barrier. Neuroendocrinology 75:367–374

27. Viltart O, Vanbesien-Mailliot CC (2007) Impact of prenatal stress on neuroendocrine programming. ScientificWorldJournal 7:1493–1537

28. Schneider ML (1992) Delayed object permanence development in prenatally stressed rhesus monkey infants (Macaca mulatta). Occup Ther J Res 12:96–110

29. Schneider ML, Coe CL (1993) Repeated social stress during pregnancy impairs neuromotor development in the primate infant. J Dev Behav Pediatr 14:81–87

30. Clarke AS, Schneider ML (1993) Prenatal stress has long-term effects on behavioral responses to stress in juvenile rhesus monkeys. Dev Psychobiol 26:293–304

31. Clarke AS, Wittwer DJ, Abbott DH, Schneider ML (1994) Long-term effects of prenatal stress on HPA axis activity in juvenile rhesus monkeys. Dev Psychobiol 27:257–269

32. Linnet KM, Dalsgaard S, Obel C, Wisborg K, Henriksen TB, Rodriguez A et al (2003) Maternal lifestyle factors in pregnancy risk of attention deficit hyperactivity disorder and associated behaviors: review of the current evidence. Am J Psychiatry 160:1028–1040

33. Grizenko N, Shayan YR, Polotskaia A, Ter-Stepanian M, Joober R (2008) Relation of maternal stress during pregnancy to symptom severity and response to treatment in children with ADHD. J Psychiatry Neurosci 33:10–16

34. Li J, Olsen J, Vestergaard M, Obel C (2010) Attention-deficit/hyperactivity disorder in the offspring following prenatal maternal bereavement: a nationwide follow-up study in Denmark. Eur Child Adolesc Psychiatry 19:747–753

35. Huttenen MO, Niskanen P (1978) Prenatal loss of father and psychiatric disorders. Arch Gen Psychiatry 35:429–431

36. Clarke MC, Harley M, Cannon M (2006) The role of obstetric events in schizophrenia. Schizophr Bull 32:3–8

37. Van Os J, Selten JP (1998) Prenatal exposure to maternal stress and subsequent schizophrenia. The invasion of The Netherlands. Br J Psychiatry 172:324–326

38. Susser E, Neugebauer R, Hoek HW, Brown AS, Lin S, Labovitz D et al (1996) Schizophrenia after prenatal famine. Further evidence. Arch Gen Psychiatry 53:25–31

39. Hoek HW, Brown AS, Susser E (1998) The Dutch famine and schizophrenia spectrum disorders. Soc Psychiatry Psychiatr Epidemiol 33:373–379

40. Brown AS, Susser ES (2008) Prenatal nutritional deficiency and risk of adult schizophrenia. Schizophr Bull 34:1054–1063

41. Kahn HS, Graff M, Stein AD, Lumey LH (2009) A fingerprint marker from early gestation associated with diabetes in middle age: the Dutch Hunger Winter Families Study. Int J Epidemiol 38:101–109

42. St Clair D, Xu M, Wang P, Yu Y, Fang Y, Zhang F et al (2005) Rates of adult schizophrenia following prenatal exposure to the Chinese famine of 1959-1961. JAMA 294:557–562

43. Watson JB, Mednick SA, Huttunen M, Wang X (1999) Prenatal teratogens and the development of adult mental illness. Dev Psychopathol 11:457–466

44. Kinney DK, Miller AM, Crowley DJ, Huang E, Gerber E (2008) Autism prevalence following prenatal exposure to hurricanes and tropical storms in Louisiana. J Autism Dev Disord 38:481–488

45. Beversdorf DQ, Manning SE, Hillier A, Anderson SL, Nordgren RE, Walters SE et al (2005) Timing of prenatal stressors and autism. J Autism Dev Disord 35:47147–47148

46. King S, Laplante DP (2005) The effects of prenatal maternal stress on children's cognitive development: Project Ice Storm. Stress 8:35–45

47. King S, Mancini-Marie A, Brunet A, Walker E, Meaney MJ, Laplante DP (2009) Prenatal maternal stress from a natural disaster predicts dermatoglyphic asymmetry in humans. Dev Psychopathol 21:343–353

48. Laplante DP, Barr RG, Brunet A, Galbaud du Fort G, Meaney ML, Saucier JF et al (2004) Stress during pregnancy affects general intellectual and language functioning in human toddlers. Pediatr Res 56:400–410

49. Laplante DP, Brunet A, Schmitz N, Ciampi A, King S (2008) Project Ice Storm: prenatal maternal stress affects cognitive and linguistic functioning in 5 1/2-year-old children. J Am Acad Child Adolesc Psychiatry 47:1063–1072

50. Talge NM, Neal C, Glover V (2007) Antenatal maternal stress and long-term effects on child neurodevelopment: how and why? J Child Psychol Psychiatry 48:245–261

51. Cripps RL, Martin-Gronert MS, Archer ZA, Hales CN, Mercer JG, Ozanne SE (2009) Programming of hypothalamic neuropeptide gene expression in rats by maternal dietary protein content during pregnancy and lactation. Clin Sci (Lond) 117:85–93

52. Torres N, Bautista CJ, Tovar AR, Ordáz G, Rodríguez-Cruz M, Ortiz V et al (2010) Protein restriction during pregnancy affects maternal liver lipid metabolism and fetal brain lipid composition in the rat. Am J Physiol Endocrinol Metab 298:E270–E277

53. Palmer AA, Printz DJ, Butler PD, Dulawa SC, Printz MP (2004) Prenatal protein

deprivation in rats induces changes in prepulse inhibition and NMDA receptor binding. Brain Res 996:193–201

54. Palmer AA, Brown AS, Keegan D, Siska LD, Susser E, Rotrosen J et al (2008) Prenatal protein deprivation alters dopamine-mediated behaviors and dopaminergic and glutamatergic receptor binding. Brain Res 1237:62–74

55. Csomor PA, Yee BK, Feldon J, Theodoridou A, Studerus E, Vollenweider VX (2009) Impaired prepulse inhibition and prepulse-elicited reactivity but intact reflex circuit excitability in unmedicated schizophrenia patients: a comparison with healthy subjects and medicated schizophrenia patients. Schizophr Bull 35:244–255

56. Weinstock M (2008) The long-term behavioural consequences of prenatal stress. Neurosci Biobehav Rev 32:1073–1086

57. Deminière JM, Piazza PV, Guegan G, Abrous N, Maccari S, Le Moal M et al (1992) Increased locomotor response to novelty and propensity to intravenous amphetamine self-administration in adult offspring of stressed mothers. Brain Res 586:135–139

58. Uno H, Tarara R, Else J, Sulemen M, Sapolsky RM (1989) Hippocampal damage associated with prolonged and fatal stress in primates. J Neurosci 9:1705–1711

59. Uno H, Lohmiller L, Thieme C, Kemnitz JW, Engle MJ, Roecker EB et al (1990) Brain damage induced by prenatal exposure to dexamethasone in fetal rhesus macaques, 1. Hippocampus. Dev Brain Res 53:157–167

60. Kerchner M, Ward IL (1992) SDN-MPOA volume in male rats is decreased by prenatal stress, but is not related to ejaculatory behavior. Brain Res 581:244–251

61. Uno H, Eisele S, Sakai A, Shelton S, Baker E, DeJesus O et al (1994) Neurotoxicity of glucocorticoids in the primate brain. Horm Behav 28:336–348

62. Anderson DK, Rhees RW, Fleming DE (1995) Effects of prenatal stress on differentiation of the sexually dimorphic nucleus of the preoptic area (SDN-POA) of the rat brain. Brain Res 332:113–118

63. Poland RE, Cloak C, Lutchmansingh PJ, McCracken JT, Chang L, Ernst T (1999) Brain N-acetyl aspartate concentrations measured by H MRS are reduced in adult male rats subjected to perinatal stress: preliminary observations and hypothetical implications for neurodevelopmental disorders. J Psychiatr Res 33:41–51

64. Schmitz C, Rhodes ME, Bludau M, Kaplan S, Ong P, Ueffing I et al (2002) Depression: reduced number of granule cells in the hippocampus of female, but not male, rats due to prenatal restraint stress. Mol Psychiatry 7:810–813

65. Coe CL, Lulbach GR, Schneider M (2002) Prenatal disturbance alters the size of the corpus callosum in young monkeys. Dev Psychobiol 41:178–185

66. Coe CL, Kramer M, Czéh B, Gould E, Reeves AJ, Kirschbaum C et al (2003) Prenatal stress diminishes neurogenesis in the dentate gyrus of juvenile rhesus monkeys. Biol Psychiatry 54:1025–1034

67. Salm AK, Pavelko M, Krouse EM, Webster W, Kraszpulski M, Birkle DL (2004) Lateral amygdaloid nucleus expansion in adult rats is associated with exposure to prenatal stress. Brain Res Dev Brain Res 148:159–167

68. Kraszpulski M, Dickerson PA, Salm AK (2006) Prenatal stress affects the developmental trajectory of the rat amygdala. Stress 9:85–95

69. Kawamura T, Chen J, Takahashi T, Ichitani Y, Nakahara D (2006) Prenatal stress suppresses cell proliferation in the early developing brain. Neuroreport 17:1515–1518

70. Barros VG, Duhalde-Vega M, Caltana L, Brusco A, Antonelli MC (2006) Astrocyte-neuron vulnerability to prenatal stress in the adult rat brain. J Neurosci Res 83:787–800

71. Ulupinar E, Yucel F (2005) Prenatal stress reduces interneuronal connectivity in the rat cerebellar granular layer. Neurotoxicol Teratol 27:475–484

72. Ulupinar E, Yucel F, Ortug G (2006) The effects of prenatal stress on the Purkinje cell neurogenesis. Neurotoxicol Teratol 28:86–94

73. Lee YA, Goto Y (2013) The effects of prenatal and postnatal environmental interaction: prenatal environmental adaptation hypothesis. J Physiol Paris 107:483–492

74. Gillies GE, Virdee K, Pienaar I, Al-Zaid F, Dalley JW (2016) Enduring, sexually dimorphic impact of in utero exposure to elevated levels of glucocorticoids on midbrain dopaminergic populations. Brain Sci 7(1):pii: E5. https://doi.org/10.3390/brainsci7010005

75. Egaas B, Courchesne E, Saitoh O (1995) Reduced size of corpus callosum in autism. Arch Neurol 52:794–801

76. Innocenti GM, Ansermet F, Parnas J (2003) Schizophrenia, neurodevelopment and corpus callosum. Mol Psychiatry 8:261–274

77. Seidman LJ, Valera EM, Makris N (2005) Structural brain imaging of attention-deficit/hyperactivity disorder. Biol Psychiatry 57:1263–1272

78. Guest PC (2017) Biomarkers and mental illness: it's not all in the mind. copernicus, 1st

edn. Göttingen, Germany. ISBN-10: 3319460870

79. Lemaire V, Koehl M, Le Moal M, Abrous DN (2000) Prenatal stress produces learning deficits associated with an inhibition of neurogenesis in the hippocampus. Proc Natl Acad Sci U S A 97:11032–11037

80. Li H, Li X, Jia N, Cai Q, Bai Z, Chen R et al (2008) NF-kappaB regulates prenatal stress-induced cognitive impairment in offspring rats. Behav Neurosci 122:331–339

81. Gonzalez-Perez O, Gutiérrez-Smith Y, Guzmán-Muñiz J, Moy-López NA (2011) Intrauterine stress impairs spatial learning in the progeny of Wistar rats. Rev Invest Clin 63:279–286

82. Li N, Wang Y, Zhao X, Gao Y, Song M, Yu L et al (2015) Long-term effect of early-life stress from earthquake exposure on working memory in adulthood. Neuropsychiatr Dis Treat 11:2959–2965

83. Bennett GA, Palliser HK, Walker D, Hirst J (2016) Severity and timing: how prenatal stress exposure affects glial developmental, emotional behavioural and plasma neurosteroid responses in guinea pig offspring. Psychoneuroendocrinology 70:47–57

84. Ward HE, Johnson EA, Salm AK, Birkle DL (2000) Effects of prenatal stress on defensive withdrawal behavior and corticotropin releasing factor systems in rat brain. Physiol Behav 70:359–366

85. Bygren LO (2013) Intergenerational health responses to adverse and enriched environments. Annu Rev Public Health 34:49–60

86. Liu GT, Dancause KN, Elgbeili G, Laplante DP, King S (2016) Disaster-related prenatal maternal stress explains increasing amounts of variance in body composition through childhood and adolescence: Project Ice Storm. Environ Res 150:1–7

87. Bayer SA, Altman J, Russo RJ, Zhang X (1993) Timetables of neurogenesis in the human brain based on experimentally determined patterns in the rat. Neurotoxicology 14:83–144

88. Van Oel CJ, Baare WF, Hulshoff Pol HE, Haag J, Balazs J, Dingemans A et al (2001) Differentiating between low and high susceptibility to schizophrenia in twins: the significance of dermatoglyphic indices in relation to other determinants of brain development. Schizophr Res 52:181–193

89. Izquierdo I, Furini CR, Myskiw JC (2016) Fear memory. Physiol Rev 96:695–750

90. Conrad CD, Galea LAM, Kuroda Y, McEwen BS (1996) Chronic stress impairs rat spatial memory on the Y maze, and this effect is blocked by tianeptine treatment. Behav Neurosci 110:1321–1334

91. Fenoglio KA, Brunson KL, Baram TZ (2006) Hippocampal neuroplasticity induced by early-life stress: functional and molecular aspects. Front Neuroendocrinol 27:180–192

92. Marin MF, Lord C, Andrews J, Juster RP, Sindi S, Arsenault-Lapierre G et al (2011) Chronic stress, cognitive functioning and mental health. Neurobiol Learn Mem 96:583–595

93. Slavich GM (2016) Life stress and health: a review of conceptual issues and recent findings. Teach Psychol 43:346–355

94. Newsome CA, Shiell AW, Fall CH, Phillips DI, Shier R, Law CM (2003) Is birth weight related to later glucose and insulin metabolism? A systematic review. Diabet Med 20:339–348

95. Entringer S, Wüst S, Kumsta R, Layes IM, Nelson EL, Hellhammer DH et al (2008) Prenatal psychosocial stress exposure is associated with insulin resistance in young adults. Am J Obstet Gynecol 199:498.e1–498.e7

96. Entringer S, Kumsta R, Hellhammer DH, Wadhwa PD, Wüst S (2009) Prenatal exposure to maternal psychosocial stress and HPA axis regulation in young adults. Horm Behav 55:292–298

97. Silberman DM, Acosta GB, Zorrilla Zubilete MA (2016) Long-term effects of early life stress exposure: role of epigenetic mechanisms. Pharmacol Res 109:64–73

98. Lesage J, Del-Favero F, Leonhardt M, Louvart H, Maccari S, Vieau D et al (2004) Prenatal stress induces intrauterine growth restriction and programmes glucose intolerance and feeding behaviour disturbances in the aged rat. J Endocrinol 81:291–296

99. De Blasio MJ, Dodic M, Jefferies AJ, Moritz KM, Wintour EM, Owens JA (2007) Maternal exposure to dexamethasone or cortisol in early pregnancy differentially alters insulin secretion and glucose homeostasis in adult male sheep offspring. Am J Physiol Endocrinol Metab 293:E75–E82

100. Guest PC, Martins-de-Souza D, Vanattou-Saifoudine N, Harris LW, Bahn S (2011) Abnormalities in metabolism and hypothalamic-pituitary-adrenal axis function in schizophrenia. Int Rev Neurobiol 101:145–168

101. Steiner J, Guest PC, Rahmoune H, Martins-de-Souza D (2017) The application of multiplex biomarker techniques for improved stratification and treatment of schizophrenia patients. Methods Mol Biol 1546:19–35

Chapter 4

Proteomic Studies of Psychiatric Disorders

Paul C. Guest

Abstract

Many diseases result from programming effects in utero. This chapter describes recent advances in proteomic studies which have improved our understanding of the underlying pathophysiological pathways in the major psychiatric disorders, resulting in the development of potential novel biomarker tests. Such tests should be based on measurement of blood-based proteins given the ease of accessibility of this medium and the known connections between the periphery and the central nervous system. Most importantly, emerging biomarker tests should be developed on lab-on-a-chip and other handheld devices to enable point-of-care use. This should help to identify individuals with psychiatric disorders much sooner than ever before, which will allow more rapid treatment options for the best possible patient outcomes.

Key words Proteomics, 2D gel electrophoresis, Mass spectrometry, Multiplex immunoassay, Lab-on-a-chip, Psychiatric disorders, Neurodegenerative disorders

1 Introduction

According to World Health Organization (WHO) statistics, the major psychiatric diseases of major depressive disorder (MDD), schizophrenia, and bipolar disorder affect an estimated 430 million people worldwide [1]. Furthermore, these disorders are ranked among the highest as a leading cause of disease burden in terms of disability-adjusted life-years (DALYs) [2] and monetary costs [3]. Although these diseases have been recognized as distinct entities for more than 100 years, the diagnostic procedures have remained at a standstill and are still based on symptom manifestation. Furthermore, diagnosis still relies on communications between the patient and a clinician or psychiatrist. This communication normally consists of an interview and may employ the Diagnostic and Statistical Manual of Mental Disorders (DSM) [4] or the International Classification of Diseases (ICD-10) [5] categorizations as guidelines. However, these are only used to define symptoms, and they make no grounds toward increasing our understanding of the underlying pathophysiological processes.

Paul C. Guest (ed.), *Investigations of Early Nutrition Effects on Long-Term Health: Methods and Applications*, Methods in Molecular Biology, vol. 1735, https://doi.org/10.1007/978-1-4939-7614-0_4, © Springer Science+Business Media, LLC 2018

Another problem is that classification of a person as having a particular psychiatric disorder is not straightforward given the overlap of symptoms across multiple diagnoses [6–8]. This has led to systematic attempts by scientists and clinicians to identify specific biomarker tests that can be used to predict the onset of these diseases, improve diagnostic accuracy, monitor disease progression and treatment response, and inform treatment options. For ease of use in the doctor's office or applications in clinical studies, it is important that such tests are developed for use in blood serum or plasma [9]. Blood-based biomarkers are relatively more accessible than other bio-samples and uncomplicated to standardize due to the simplicity and low costs of the procedure.

The application of biomarker tests to accurately classify patients according to the type or subtype of a psychiatric disorder will contribute to reduce the interval of untreated disease and improve responses due to the fact that patients have been administered the best possible treatment at the earliest possible phase of the disease. Early and accurate treatment is critical since there is a direct correlation between longer periods without treatment and worse patient outcomes [10]. By initiating earlier more effective treatments, we should see a reduction in visits of the patients to hospitals, community groups, and crisis centers. This will have the knock-on effect of decreasing the financial burden associated with these conditions, which totaled more than 60 billion dollars per year in the 1990s, in the USA alone [11]. Most importantly for the patients, an earlier and more effective intervention may help to reduce symptom severity.

Biomarkers are physical characteristics that can be analyzed or measured in biological samples as readouts of physiological status [12]. Since changes in physiology can be dynamic, these are likely to be reflected by rapid alterations in the levels of many proteins that feed into different biological pathways. As such, most biomarker scientists now consider that proteomic methods are likely to be the most informative about physiological status [13]. It is also becoming more accepted by researchers and clinicians that brain conditions such as psychiatric disorders can be investigated by looking in peripheral tissues as well as the blood [14]. This makes sense considering the two-way communication system between the brain and the periphery, and much of this occurs by fluctuations in blood-borne molecules.

The development of proteomic-based biomarker panels that correspond to clinical data will advance healthcare for psychiatric illnesses significantly. This will be facilitated by the incorporation of such tests into the clinical decision-making process as user-friendly, cost-effective point-of-care devices. This chapter describes the current and future proteomic methods which have been employed in studies of psychiatric disorders, as well as the challenges of developing and implementing novel tests for these conditions.

2   Major Depression, Schizophrenia, and Bipolar Disorder

There are a number of differences in the way the major psychiatric disorders of depression, schizophrenia, and bipolar disorder are manifested which could be reflected by differences in the underlying pathophysiology and proteomic profiles.

Depression is the most common of these disorders and accounts for some 350 million cases worldwide, with more females affected than males [1, 15–17]. This condition is characterized by sadness, loss of interest or pleasure, feelings of guilt, low self-worth, disturbed sleep patterns, abnormal appetite, excessive tiredness, and poor concentration. People affected by this disorder may also suffer from physical complaints with no tangible cause. In chronic or recurrent cases, depression can significantly impair function at school, at work, and in society in general and may lead to suicide in the most severe cases. People suffering with mild or moderate depression may be treated by cognitive behavioral therapy or psychotherapy, and those with more severe forms are mostly given antidepressants. These drugs lead to increased levels of neurotransmitters such as serotonin, dopamine, and norepinephrine [18, 19].

Schizophrenia affects approximately 21 million people worldwide with an earlier onset seen in males compared to females, although there is no gender difference in the numbers of individuals affected [1, 20]. The disease is characterized by psychosis and disturbances in thinking, perception, emotion, language, and behavior. Common psychotic experiences include hallucinations and delusions, and affected individuals can also suffer from negative symptoms, such as those seen in depression, as well as cognitive deficits [21]. As with depression, these symptoms make it difficult for the sufferers to function in society. Treatment with antipsychotics can be effective at reducing the symptoms, and these drugs appear to work by reducing the levels of the neurotransmitter dopamine, although other neurotransmitter systems are also affected such as those involving serotonin which ultimately regulate the dopaminergic system [22, 23].

Bipolar disorder affects around 60 million people worldwide [1]. This disorder often presents as major depression with later manifestation as bipolar disorder, which typically consists of both mania and depression episodes separated by periods of normal moods [24]. Due to the initial presentation, the disease is often misdiagnosed as major depression since the symptoms can be identical. The mania phase can involve elevated or irritable moods, overactivity, excessive speech, inflated self-esteem, and decreased need for sleep. Effective treatments are available for the treatment of the acute mania phase of the disease through use of mood stabilizers such as lithium and anticonvulsants including valproate and lamotrigine [25, 26].

3   Proteomic Biomarker Platforms

Several proteomic platforms are in current use for studies of psychiatric disorders. These include two-dimensional gel electrophoresis for studies of brain and other tissues and antibody-based approaches for analysis of body fluids, such as blood serum and plasma. These basic platforms and the principles behind them are described below.

3.1 Two-Dimensional Gel Electrophoresis

Two-dimensional gel electrophoresis (2DGE) provides a means of measuring changes in intact proteins. This includes changes in abundance and effects on posttranslational modifications such as proteolysis, phosphorylation, or glycosylation, which are all more difficult to obtain using most mass spectrometry-based methods (described in the following section) [27]. In 2DGE, sample extracts are added to an immobilized gel strip for electrophoretic focusing in the first dimension according to the isoelectric point of each protein. In the second dimension step, the proteins on the strip are subjected to sodium dodecyl-sulfate (SDS) polyacrylamide gel electrophoresis (PAGE) according to their apparent molecular weights (small proteins migrate faster than the larger ones). Finally, the resulting 2D spots in the gels can be visualized using specific protein stains such as Coomassie Blue or Sypro Ruby and quantitated using an imaging software. This enables the simultaneous display of hundreds or thousands of protein spots on one gel. In 1997, a variation of this method was developed which provided a way of analyzing two or three proteomes on a single gel, which circumvented irreproducibility issues due to comparing samples on different gels [28]. This method was called 2D difference gel electrophoresis (DIGE). In a 2D-DIGE experiment, up to three proteomes can be run on a single gel by covalently labeling the proteins prior to electrophoresis with dyes that fluoresce at distinct wavelengths [29]. Thus, imaging of the gel at the wavelengths specific for each CyDye produces distinct images of each proteome after electrophoresis. These images can be overlaid and analyzed for viewing any proteins which are present at different levels (Fig. 1). These proteins can be identified by a number of mass spectrometry-based platforms. Although the technique allows the study of proteins in many tissue types, there are some problems with analysis of blood serum or plasma samples due to the wide concentration range of proteins in these body fluids [30, 31]. This creates a problem such that abundant proteins like albumin appears as a series of large poorly resolved spots on the gels, obscuring other proteins present at much lower levels.

3.2 Mass Spectrometry

Mass spectrometry approaches emerged near the end of the Human Genome Project as a sensitive method for detection and quantitation of proteomic arker candidates [32]. The characteristic

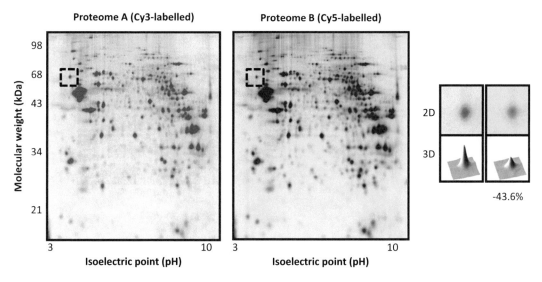

Fig. 1 2D-DIGE analysis showing 2D and 3D Cy3 and Cy5 images of two different mouse brain proteomes. In this example, there is a decrease of 43% in the level of the selected spot as shown in the 2D and 3D images on the right side of the gels

stage of this approach is that the proteins in biological samples are cleaved with a proteolytic enzyme such as trypsin prior to analysis. This is essential as most intact proteins are too large and complex to be analyzed directly in most mass spectrometry instruments. After proteolysis, the resulting peptides are subjected to separation by liquid chromatography based on physiochemical properties so they can be analyzed in the mass spectrometer in a manageable stream. As the peptides enter the mass spectrometer, they are ionized by electro-spray or other approaches, which convert the peptides to a charged plasma state. This allows electromagnetic acceleration toward a detector with a speed that is inversely proportional to their molecular weight (specifically, the mass over charge ratios). Thus, small proteins fly faster than larger ones. The amount of each peptide (and the corresponding protein) present can be determined essentially through the number of strikes at the detector per unit of time. At the same time, the sequence of each peptide can be determined by bombardment with a gas such as nitrogen, which results in fragmentation of the peptides (Fig. 2). This process is called collision-induced dissociation. The mass of each fragment can then be used to determine amino acid sequences which are used to search an appropriate database for identification of the parent protein.

Many mass spectrometry-based techniques such as matrix-assisted laser desorption/ionization time of flight (MALDI-TOF) mass spectrometry are used in proteomic studies [33]. This method is typically used for peptide fingerprint analysis in combination with 2DGE approaches for identification of the differentially expressed

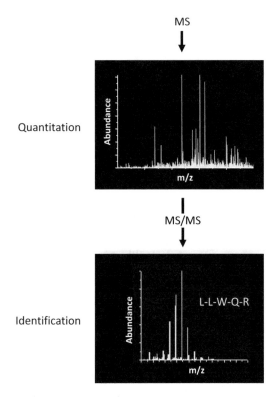

Fig. 2 Tandem liquid chromatography mass spectrometry (LC-MS/MS) analysis. The figure on the top shows a typical MS spectrum which provides quantitative information about each peptide ion. The image on the bottom shows a typical MS/MS spectrum (produced by collision-induced dissociation) which leads to amino acid sequence information for identification purposes

proteins. Another method called selective reaction monitoring (SRM) mass spectrometry is based on selecting and quantifying a specific set of peptides on a target list of proteins identified through a standard liquid chromatography tandem mass spectrometry experiment [34]. This approach leverages the ion-filtering capabilities of tandem quadrupole-based mass spectrometry instruments for selection of precursor and fragment ions produced by collision-induced dissociation. Due to the rapid cycling capability, numerous precursor-fragment ion pairs can be monitored simultaneously [35]. Another approach is called isobaric tagging for relative and absolute quantitation (iTRAQ) mass spectrometry. In the iTRAQ 4-plex approach, samples are proteolyzed and labeled with up to four different mass tags containing reporter, balance, and reactive groups [36, 37]. After this the samples are mixed together and analyzed simultaneously in the same mass spectrometry instrument. Since each tag has the same mass/change (m/z) ratio of 145 Da using a combination of ^{13}C, ^{15}N, and ^{18}O in the reporter (m/z 114–117) and balance groups (28–31 Da), the labeled peptides from each proteome appear as single peak in a mass spectrometry (MS1) spectrum. However, selection of the precursor ion for collision-induced dissociation leads to a tandem mass spectrometry (MS2) spectrum with reporter ion peak intensities at 114, 115, 116, and 117 m/z, which are directly related to the peptide abundance in the different proteomes. A significant advantage of mass spectrometry-based methods over 2DGE approaches is the capability of analyzing more difficult types of proteins, such as those in the extreme basic range, low molecular weight peptides, or high molecular weight proteins. However, a major disadvantage is the loss of intact protein information since the proteins are proteolyzed before analysis.

3.3 Multiplex Immunoassay

The blood contains hundreds of bioactive and regulatory proteins including inflammation factors, hormones, growth factors, transport proteins, and components of the clotting cascade. However, some of these proteins are present at exceeding low concentrations and cannot be measured by conventional 2DGE or mass spectrometry approaches [30, 31]. This means that biomarker measurement systems should have sufficient sensitivity and dynamic range capability of measuring proteins in serum and plasma samples. One way of achieving this is through the use of antibody-based approaches, such as multiplex immunoassay [38, 39]. These assays are constructed and carried out in a similar manner as a standard "sandwich" immunoassay, which was developed almost 50 years ago [40]. In the first stage, microspheres are loaded with dyes in the red and infrared range (700–800 nm) at different ratios to give a unique fluorescent signature for each sphere. In the next step, specific capture antibodies are covalently attached to the surface of each fluorescently distinct sphere. For example, an antibody against growth hormone could be linked to a 710 nm sphere, and one for insulin could be bound to a sphere that is detectable at 720 nm, and so on (Fig. 3). After this, the antibody-sphere

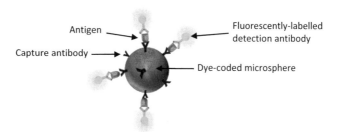

Antigen

Fluorescently-labelled detection antibody

Capture antibody

Dye-coded microsphere

Fig. 3 Schematic diagram of a dye-coded microsphere, capture antibody, targeted antigen, and detection antibody in the sandwich format. Multiplexing is achieved using mixtures of microsphere-capture antibody complexes with proteome extracts

conjugates are combined to form the multiplex. In an actual analysis, a serum or plasma sample is added to the multiplex spheres, and the targeted molecules are captured by the corresponding antibodies. The next step is to wash away the unbound material. After this, fluorescently labeled detection antibodies specific for the same target molecules are added, and each of these binds in a sandwich configuration. Again, the unbound material is washed away, and the samples are streamed through a reader in a flow-cytometric manner such as the Luminex® instrument [41]. Within the reader, the microspheres are simultaneously analyzed by one laser for identification and another for quantification of the bound protein. This is possible since the lasers identify which proteins are present using the unique fluorescent signature of each microsphere and quantity can be measured by the intensity of the fluorescent signals associated with the tags on the secondary antibodies.

3.4 Lab-on-a-Chip

There are some disadvantages regarding current multiplex immunoassay protocols that hinder the broad use of this approach in routine analysis, such as in the analysis of clinical samples. These include relatively long experimental times and the necessity of complex laboratory-based procedures, as well as the need for skilled operational personnel. Lab-on-a-chip-based approaches can be used to overcome these limitations as they provide a rapid and automated analysis which integrates most of the laboratory-based steps. One such approach is the integrated in vitro diagnostic platform, developed by the Fraunhofer Institute in Germany [42, 43]. The platform can be used to run a singleplex or multiplex immunoassay to deliver a biomarker "score" from a single drop of blood or a serum or plasma sample in less than 15 min. This is achieved using a combination of antibody microarrays with microfluidics to automate and increase the speed of multiplex proteomic assays. The platform consists of a credit card-sized microfluidic cartridge that contains all of the relevant elements necessary for attachment of antibodies as well as reservoirs for all of the reagents, integrated pumping systems, and an optical transducer to allow integrated sensing (Fig. 4) [44, 45]. Once the blood drop is applied, the card is inserted into a small base unit, which contains the necessary electronics to control the cartridge, a readout system to analyze and display the results, and a touch screen to control the assay and monitor the results.

These devices can be used directly in a doctor's office or a clinic, with the result being obtained within the time frame of a short visit. In addition, these devices can contain a universal serial bus (USB) for transmission of the data to a smartphone or other handheld devices. This is important as new point-of-care methods should incorporate mobile communication and internet capabilities to facilitate formatting of the resulting data for ease of presentation and interpretation by the users. It is anticipated that this may lead to

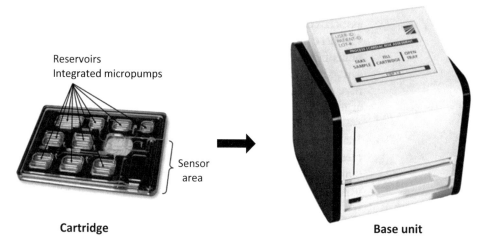

Reservoirs
Integrated micropumps

Sensor area

Cartridge

Base unit

Fig. 4 Lab-on-a-chip cartridge and base unit

the emergence of smartphone-readable bio-signatures for each individual, consisting of proteomic scores, physical readings, and medical histories, to support personalized medicine approaches [9]. The testing of smartphone apps has already begun in some fields of medicine with some success. Clinical trials incorporating smartphone-assisted interventions showed improvements in many outcome measures, compared to non-assisted control cases [46–48], and multiplex smartphone hybrid assays are already in development in medical areas such as viral detection [49–51] and for measurement of prostate-specific antigen levels [52].

4   Mass Spectrometry Analysis of Brain, Neuroendocrine, and Peripheral Tissues

This section describes MS-based proteomic studies of the brain, pituitary, and peripheral tissues which have led to the discovery of potential biomarkers. The main diseases focused on are depression, schizophrenia, and bipolar disorder considering their high ranking in terms of disease burden and associated costs on the healthcare systems, as well as on society in general [2, 3].

4.1 Brain Analysis

Given that psychiatric disorders are manifested through dysfunctions of the brain, it is not surprising that most 2DGE and mass spectrometry profiling studies have focused on analysis of this tissue. One of the first liquid chromatography mass spectrometry profiling studies examined postmortem dorsolateral prefrontal cortex (DLPFC) tissue from schizophrenia patients and controls [53]. This resulted in identification of significant differences in the levels of 30 out of approximately 500 identifiable proteins, and these were involved mainly in synaptic function (e.g., myelin-associated glycoprotein precursor, myristoylated alanine-rich C-kinase substrate, neural cell adhesion molecule 1, neurofascin, neurofilament medium polypeptide, neuromodulin, synapsin-2, syntaxin-1A) and energy/metabolism (e.g., glutamate dehydrogenase 1, glyceraldehyde-3-phosphate dehydrogenase, protein-arginine deiminase type-2, pyruvate dehydrogenase) pathways.

In 2012, the same technique was also applied in a similar analysis of DLPFC tissue from major depressive disorder patients, resulting in distinct proteome fingerprints between patients and controls [54]. The differentially expressed proteins were mainly involved in energy metabolism and synaptic function, consistent with the above studies of this tissue, and changes were also found in the histidine triad nucleotide-binding protein 1 (HINT1), which has been implicated in mood and behavior regulation [55–57]. This same study also found distinct proteome profiles in samples from major depression patients with and without psychosis, with the former showing an overlap with proteomic changes seen in the analysis of the same brain region from schizophrenia patients. This suggested that it may be possible to distinguish different subtypes of depression based on mass spectrometry-based proteomic profiling analyses. Another study carried out a comparative phosphoproteome analysis of postmortem DLPFC from major depression patient and control donors using tandem mass spectrometry profiling [58]. This led to identification of more than 800 nonredundant proteins containing phosphorylated sequences. Interestingly, 90 of these proteins showed differential levels of phosphorylation in samples from depressed patients compared to controls, and most of the proteins were associated with synaptic transmission and cellular architecture.

A more recent study investigated postmortem prefrontal cortex tissue of patients with schizophrenia, bipolar disorder, and major depression with and without psychotic features, compared to that from healthy controls using a combined liquid chromatography tandem mass spectrometry and SRM mass spectrometry approach [59]. In silico analyses of the resulting biological annotations showed that there were effects on common pathways across all of these disorders, associated mainly with presynaptic glutamatergic neurotransmission and energy metabolism. However, these appeared to be opposite in that schizophrenia patients showed changes linked to a hypoglutamatergic state and lower energy metabolism, whereas bipolar disorder and depressed patients showed increased activation of these same pathways. The same group used a targeted SRM mass spectrometry analysis of 56 proteins previously implicated in the aetiology of major psychiatric disorders using the same samples as above [60]. This confirmed previous findings of changes in Wnt-signaling and glutamate receptor abundance predominantly in bipolar disorder and alterations in energy metabolism in all diseases. Calcium signaling was affected mostly in the schizophrenia and in the depressed patients with psychosis. In addition there were changes in four oligodendrocyte-specific proteins (myelin oligodendrocyte glyco-protein, myelin basic protein, myelin proteolipid protein, 2′,3′-cyclic-nucleotide 3′-phosphodiesterase) in bipolar disorder and to a lesser extent in the schizophrenia and depressed patients with psychosis. Finally, the study showed that ankyrin 3 was specifically altered in the depressed patients with psychosis and septin 5 was altered in schizophrenia.

4.2 Neuroendocrine Tissue Analysis

Given the known dysfunction of the hypothalamic-pituitary-adre-nal (HPA) axis in psychiatric disorders [61], two mass spectrometry studies have been carried out investigating postmortem pituitaries from patients with schizophrenia [52], bipolar disorder [63], and major depression [63]. Krishnamurthy and colleagues identified altered protein levels in pituitaries from schizophrenia patients compared to controls including changes in several hormones (e.g., agouti-related protein, arginine vasopressin, growth hor-mone, prolactin, pro-opiomelanocortin, and the apolipoproteins A1, A2, C3, and H) [62]. Stelzhammer et al. found perturbations of the hormones galanin and pro-opiomelanocortin in pituitaries from bipolar disorder patients along with effects on proteins involved in gene transcription, the stress response, lipid metabo-lism, and growth pathways [63]. In contrast, pituitaries from the depressed patients had significantly decreased levels of the prohormone-converting enzymes carboxypeptidase E and prolyl-oligopeptidase convertase in addition to changes in proteins with intracellular transport and cytoskeletal roles. Given that the pitui-tary secretes hormones and other bioactive molecules directly into the bloodstream [64], many of these molecules may have some use as potential proteomic biomarkers in studies of serum or plasma in clinical studies of living patients [14].

4.3 Peripheral Tissues

Several studies have now been carried out using peripheral cells taken from psychiatric patients as potential cellular models of these diseases. Peripheral blood mononuclear cells (PBMCs) have been used in a number of such studies since these cells contain many of the same neurotransmitter receptor systems as in brain tissues [65–67]. Herberth et al. carried out mass spectrometry analysis of PBMCs isolated from first-onset schizophrenia patients and con-trols which revealed differences in the levels of 18 proteins [68]. Eight of these proteins were associated with glycolysis, con-sistent with the findings of alterations in this pathway in brain tissues from patients with psychiatric disorders, as described above. The authors suggested that the combination of the glyco-lytic changes and the PBMC model could be of diagnostic and prognostic value as well as a useful tool for drug discovery. The same group carried out a separate mass spectrometry analysis of PBMCs from bipolar disorder patients and identified changes that were associated mostly with cytoskeletal and stress response pathways [69]. These findings suggested that there may be some differences in the effects on PBMCs between schizophrenia and bipolar disorder patients although further studies involving a direct comparison by mass spectrometry profiling are required to test this possibility. Another mass spectrometry profiling study found significant differences in the levels of 16 proteins in skin fibroblasts from schizophrenia patients compared to controls [70]. The proteins were associated mostly with cell division and growth pathways. These findings were validated at the functional level by confirmation of decreased proliferation of cells from the schizophrenia patients in a time course analysis. Similar abnormalities have been reported to occur in analyses of post mortem brain samples from schizophrenia patients [71–73].

4.4   Serum

Mass spectrometry-based analyses have been used on only a few occasions to investigate proteomic differences in serum or plasma from patients with psychiatric disorders due the fact that only the most abundant proteins are typically detected [30, 31]. Levin and colleagues used a liquid chromatography tandem mass spectrometry approach and found that several proteins were present at significantly lower levels in first-onset schizophrenia patients compared to controls, including the lipid transport proteins apolipoproteins A1, A2, A4, C1, and D, as well as the inflammation-related proteins CD5L, immunoglobulin m, clotting factor XIII B, and serotransferrin [74]. Since these are some of the most abundant proteins in serum, the authors suggested that future analyses should focus on the lower abundance proteins in serum using depletion methods. In line with this, Stelzhammer et al. carried out a mass spectrometry analysis of depleted serum from first-onset major depression patients [75]. This showed that the levels of ceruloplasmin (CP) and haptoglobin-related protein (HP) were altered between patients and controls and four proteins (complement C4-B, CP, HP and plasminogen) were correlated with symptom severity. Taken together, these findings suggested that inflammation-related proteins may be involved in the early stages of major depression.

5   2DGE Analyses of Brain and Peripheral Tissues

The sections below describe both 2DGE and 2D-DIGE approaches in proteomic analyses of brain and peripheral tissues from patients with psychiatric disorders.

5.1   Brain Analysis

An early study used a 2DGE approach for display and mass spectrometry sequencing to identify proteins with altered levels in postmortem frontal cortex tissues from individuals with schizophrenia, bipolar disorder, and major depressive disorder, compared to controls [76]. This showed that the levels of carbonic anhydrase 1 and fructose biphosphate aldolase C were increased, and those of glial fibrillary acidic protein, dihydropyrimidinase-related protein 2, and ubiquinone cytochrome c reductase core protein 1 were decreased in one or more of these diseases compared to controls. In a later study, Pennington and coworkers characterized protein-level differences in postmortem DLPFC tissue from schizophrenia and bipolar disorder patients using 2DGE [77]. This analysis showed that 11 and 48 protein spots were altered in schizophrenia and bipolar disorder patients, respectively, and seven spots were changed in both diseases compared to controls. Their follow-up use of mass spectrometry analysis led to identification of 15 altered proteins in schizophrenia and 51 altered proteins in bipolar disorder. In silico analysis showed that the functional groups most affected included changes in synaptic proteins in schizophrenia and metabolic or mitochondrial-associated proteins in bipolar disorder. The same group carried out a 2D-DIGE approach for analysis of the same samples and found that 70 protein spots were significantly altered between disease and control subjects and 46 of these were subsequently identified by mass spectrometry [78]. They confirmed the changes in three of these proteins (neurofilament light, amphiphysin II, and Rab-GDP-α) using immunoassay. Another group carried out a 2D-DIGE investigation of postmortem human hippocampus from individuals with schizophrenia and bipolar disorder, compared to controls [79]. This led to identification of 58 and 70 altered proteins in schizophrenia and bipolar disorder, respectively, using MS. In silico analysis of the data using the Ingenuity Pathway Analysis software (www.ingenuity. com/) revealed prominent effects on 14-3-3 and aryl hydrocarbon receptor signaling in schizophrenia and gluconeogenesis/glycolysis in bipolar disorder, and both disorders showed changes in proteins involved in the oxidative stress response, mitochondrial function, and protein endocytosis, trafficking, degradation, and ubiquitination. Taken together, these findings confirmed that proteomic changes occur in synaptic function and energy metabolism pathways in psychiatric disorders.

5.2 Analysis of Peripheral Tissues

Huang and colleagues showed that the levels of apolipoprotein A1 were decreased in postmortem brain and liver tissues from schizophrenia patients compared to controls using Western blot and 2DGE analysis [80]. This suggested that central nervous system changes may be reflected in the periphery. Kazuno et al. showed that approximately 200 protein spots were present at differential levels in lymphoblastoid cells obtained from monozygotic twins who were discordant for bipolar disorder using 2D-DIGE [81]. They identified some of these spots by mass spectrometry as proteins involved in cell death and glycolysis, and they used

Western blot analysis to confirm that the glycolytic enzyme phosphoglycerate mutase 1 was significantly increased in bipolar disorder compared to control co-twins. Another research group found that eight proteins were present at altered levels in blood platelets from patients with major depression compared to controls using a 2D-DIGE approach [82]. Their findings included higher levels of protein disulfide-isomerase A3 (PDIA3) and F-actin-capping protein subunit beta and lower levels of fibrinogen beta chain, fibrinogen gamma chain, retinoic acid receptor beta, glutathione peroxidase 1, SH3 domain-containing protein 19, and T-complex protein 1 subunit beta in patients with major depression compared to healthy controls. Again, these findings support the case that some central nervous system changes are recapitulated in the periphery of patients with psychiatric disorders. This indicates that these diseases can be investigated by analyses of peripheral tissues and body fluids using living patients.

6   Clinical Biomarker Profiling Studies

For decades, most psychiatrists, clinicians, and research scientists have looked at mental illnesses as brain disorders. However, advances in other areas of science such as endocrinology and immunology have now opened the door to the likelihood that these diseases can also be manifested through various dysfunctions of the periphery as well as by somatic conditions and physical traumas [13]. For example, fluctuations of cytokine levels in the periphery can induce changes in levels of central nervous system neurotransmitters involved in regulation of mood and behavior [83]. Likewise, inflammatory conditions such as asthma have been linked to anxiety disorders and depression [84]; some kidney and liver diseases show increased incidences of dementia, psychosis, and various anxiety and adjustment disorders [85, 86]; and diabetes has been linked with a number of psychiatric conditions [87]. Another example is the fight-or-flight reflex which kicks a feedback loop in motion between the brain and the periphery, regulating the response to a threatening situation [88]. This circuit begins with the perception of danger by the brain, which triggers secretion of corticotrophin-releasing factor (CRF). In turn, CRF stimulates release of adrenocorticotrophic hormone (ACTH) from the pituitary into the peripheral circulation. ACTH then circulates throughout the body and acts on the adrenal glands to produce and release the hormone cortisol, which acts to increase blood pressure and glucose levels in preparation for the fight-or-flight response. Finally, the rise in cortisol exerts a negative feedback effect on receptors in the brain and pituitary to switch off the response. Other well-known whole body-brain circuits which can be detected by changes in blood molecules include immune system dysfunctions as well as metabolic disorders and insulin resistance [89]. Interestingly, several studies have now shown that impairments in HPA axis function and insulin signaling can contribute to inflammation, neurological dysfunctions, and memory disturbances [90, 91]. The following sections describe serum and plasma profiling studies which have been carried out in attempts to identify biomarkers for major depression, schizophrenia, and bipolar disorder.

6.1 Major Depression

Investigations have shown that circulating ACTH and cortisol levels are increased in major depression patients with melancholia compared to the levels in depressed patients without melancholia [92, 93]. In addition, several studies have confirmed HPA axis hyperactivity in patients with major depression [94–96]. Many patients with depression also develop glucose intolerance accompanied by hyperinsulinemia. One study carried out a glucose tolerance test of depressed patients which showed that insulin efficacy was lower in patients before treatment compared to control subjects and antidepressant treatment resulted in a significant improvement [97]. This has led to hypotheses that peripheral insulin resistance results in poor metabolism in neuronal cells, which in turn leads to perturbations in central nervous system functions, such as mood, emotion, and cognition. Another study showed an association between the levels of leptin with depressed mood and sleep [98]. Other studies have shown that the levels of another lipid metabolism-related hormone, adiponectin, are also reduced in major depressive disorder patients. These findings are consistent with the results of studies which have shown altered levels of adiponectin [99, 100] and circulating lipids such as cholesterol in major depression patients [101, 102]. Studies have found that other hormones including thyroid-stimulating hormone (TSH) and free thyroxine 4 levels may also be correlated with symptom severity [103].

A number of studies have also found altered levels of growth factors like brain-derived neurotrophic factor (BDNF) in patients with major depression which may be correlated with symptom severity, although these findings have not been consistent [104–107]. Other growth factor effects include vascular endothelial growth factor (VEGF) and fibroblast growth factor 2 (FGF-2), which both have angiogenic and neurogenic properties [108]. Furthermore, increased levels of angiotensin-converting enzyme activity have been found in serum from depressed patients in association with higher diastolic blood pressure [109]. Another showed that the serum amyloid peptide ratio of Ab_{40}/Ab_{42} was significantly higher in patients with major depression compared to controls, suggesting that the proteolytic conversion of the amyloid precursor polypeptide may be altered in depression as seen on Alzheimer's disease [110].

Potentially the most widely reported finding in depression is that of altered inflammation-related markers in serum and plasma. A number of studies have used either single analyte or multiplex immunoassay profiling and found changes in the levels of both pro-inflammatory and anti-inflammatory cytokines in depressed patients [93, 111–115]. Other studies have found increased levels of several members of the acute phase response protein family [103, 116], including serum amyloid polypeptide and C-reactive protein [117]. A recent study found that elevated serum levels of interleukin-5 are associated with increased risk of depression [118], and another found higher levels of eotaxin and lower levels of monocyte chemotactic protein-1 and RANTES (regulated-on-activation normal T cell expressed and secreted) in depressed females with suicidal thoughts [119]. Finally, other studies have described the presence of autoantibodies in depressed patients such as glutamic acid decarboxylase (GAD)-2, thyroid peroxidase, and the gastric H+/K+ ATPase (ATP4B) [120]. Some biomarker profiling studies have also found increased production of reactive oxygen species in depressed patients as shown by the finding of decreased levels of antioxidant molecules including superoxide dismutase and glutathione peroxidase and higher levels of the lipid peroxidation biomarker, malondialdehyde [121].

A multiplex immunoassay profiling study by Stelzhammer and colleagues attempted to identify serum molecular differences and, in first onset, antidepressant-naïve major depression patients compared to controls [75]. This led to identification of several molecules that distinguished patients from controls, and/or which correlated with symptom severity, including angiotensin-converting enzyme, BDNF, complement C4-B, cortisol, ferritin, growth hormone, interleukin-16, macrophage migration inhibitory factor, serotransferrin, and superoxide dismutase. The nature of these changes confirmed increased that pro-inflammatory and oxidative stress response effects were occurring, followed by a hyperactivation of the HPA axis in the acute stages of major depression, along with dysregulation in growth factor pathways. In 2016, Ramsey et al. found 28 gender-specific serum biomarkers of major depression in a logistic regression analysis [122]. This included mostly male-specific elevated levels of proteins involved in the immune response such as C-reactive protein, trefoil factor 3, -cystatin-C, fetuin-A, β2-microglobulin, CD5L, FASLG receptor, and tumor necrosis factor receptor 2. The authors suggested that this gender specificity may at least partly explain why females are more susceptible to depression compared to males.

In 2013, a 9-plex serum biomarker test for alpha1 antitrypsin, apolipoprotein-CIII, BDNF, cortisol, epidermal growth factor, myeloperoxidase, prolactin, resistin, and soluble tumor necrosis factor-α receptor type II (designated MDDScore) was used to distinguish major depressive disorder patients from controls with a sensitivity of 92% and a specificity of 81% [123]. The same group achieved similar results in 2015 through a validation study using a larger subset of patients and controls [124]. Although this demonstrates the utility of using multiplex algorithms to aid diagnosis, further research is essential before such a test can be implemented in the clinical setting. One aspect that should be a major focus is the usefulness of this test in distinguishing individuals with major depression from those with other psychiatric diseases. In particular, it would be important to determine if this test could separate major depression from bipolar disorder patients in the depressive phase, given the high rate of misdiagnosis. Two recent studies attempted to identify patients at risk of a developing a depressive disorder in samples from individuals suffering from either social anxiety disorder [125] or panic disorder [126], made available through the Netherlands Study for Depression and Anxiety (NESDA). Logistic regression analysis resulted in identification of four serum proteins (AXL receptor tyrosine kinase, vascular cell adhesion molecule 1, vitronectin, and collagen IV) and four clinical variables (Inventory of Depressive Symptomatology, Beck Anxiety Inventory somatic subscale, depressive disorder lifetime diagnosis, and body mass index) which yielded an area under the curve (AUC) of 0.86 for identification of social anxiety disorder patients who later developed a depressive disorder [125]. In the case of the panic disorder patients, logistic regression analysis identified an optimal predictive panel comprised of tetranectin and creatine kinase MB combined with the clinical factors of gender and scores from the Inventory of Depressive Symptomatology rating scale which gave an AUC of 0.87 for identifying those patients who developed a depressive disorder within 2 years [126]. In both studies, the combination of both molecular analytes and clinical information resulted in greater performance than was achieved with either one of these components alone.

6.2 Schizophrenia

A number of studies have now found impaired fasting glucose tolerance, insulin resistance, and hyperinsulinemia in some first episode [127, 128] and chronic [129, 130] antipsychotic-free schizophrenia patients. This was confirmed in a separate study using a hyperinsulinemic clamp method [131], although one study found no differences in the plasma levels of glucose, insulin, or connecting C-peptide in a cohort of antipsychotic-free patients compared with healthy controls under non-fasting conditions [129]. In addition, leptin levels were found to be elevated in antipsychotic-free female schizophrenia patients [132]. In line with the changes in metabolic hormones, elevated levels of HPA axis hormones such as ACTH [133] and cortisol have been found in first or recent onset patients [134]. Interestingly, cortisol levels appear to be correlated with negative symptoms in chronic schizophrenia patients [135]. Antipsychotic treatments may further aggravate metabolism dysfunctions in schizophrenia as shown by medication-induced changes in biomarkers such as prolactin [136], leptin [137], ghrelin [137], and lipids [138, 139]. A study carried out in 2014 found that a large group of acute schizophrenia patients ($n = 180$) could be classified into two groups based on serum biomarker patterns compared to controls [140]. One group had marked differences in the levels of hormones and growth factors such as insulin, leptin, prolactin, resistin, testosterone, platelet-derived growth factor, and angiotensinogen, and the other had predominant changes in the levels of inflammation-related molecules, including macrophage migration inhibitory factor and interleukins 1RA, 8, 16, and 18. These findings are in line with studies from the same group which showed that some but not all first-onset antipsychotic-free schizophrenia patients have increased levels of proinsulin, split proinsulin, C-peptide, and insulin [141–143].

A systematic meta-analysis carried out in 2008 of 2298 schizophrenia patients compared to 1858 healthy controls found increased levels of soluble interleukin-2R, interleukin-1RA, and interleukin-6 [144], and the latter two were changed in first-onset, antipsychotic-free patients. Another meta-analysis carried out in 2011 found elevated levels of interleukin-1β, interleukin-6, and transforming growth factor-β in first-onset and relapsed patients, and all of these proteins were normalized by antipsychotic treatment suggesting that they may be state biomarkers [145]. The same study also found that soluble interleukin-2R, interleukin-12, and tumor necrosis factor-α were increased in patients but did not respond to treatment, suggesting that these may be trait biomarkers. A meta-analysis that focused on antipsychotic drug effects in 762 patients found that the levels of interleukin-1β and interferon-γ were decreased, and the levels of soluble interleukin-2R and interleukin-12 were increased by treatment [146]. A review carried out in 2011 confirmed that the most frequently reported serum biomarker candidates for schizophrenia are proteins associated with inflammation pathways [147]. The same research group performed a meta-analysis of 180 first-onset antipsychotic-free schizophrenia patients and 350 controls, matched for age, gender, smoking, and body mass index [148]. They found increased levels of interleukin-1RA, interleukin-10, and interleukin-15, which indicated both pro- and anti-inflammatory changes were occurring in patients at first onset. This study showed that only interleukin-1RA and interleukin-10 were decreased with treatment and the latter change was correlated with decreased negative and cognitive symptoms. Interestingly, another review carried out in 2014 provided evidence for a link between immune and inflammatory changes in schizophrenia with effects on cognitive performance [149]. A meta-study of 570 first-onset antipsychotic-free schizophrenia patients and 683 controls found elevated levels of interleukin-1β, soluble interleukin-2R, interleukin-6, and tumor necrosis factor-α [150]. The variations in these studies could be due to differences in the patient populations, specific demographic variables, as well as the biomarker platforms used. Consistent with the inflammation-related changes described above, two studies found that 30% of the schizophrenia patients tested had elevated levels of the acute phase response protein C-reactive protein [151, 152]. Earlier studies showed that high levels of this protein are associated with a more severe psychopathology in some patients with schizophrenia [153]. Other inflammation-related proteins found to be increased in patients include proteins of the complement cascade such as complements C3 and C4 [154, 155]. Another meta-analysis carried out in 2013 found a disruption in antioxidant status in schizophrenia patients which improved with antipsychotic treatment [147].

Several studies have now been carried out aimed at identification of a multiplex biomarker signature for schizophrenia. A multiplex immunoassay study performed in 2010 found that combined measurement of BDNF, epidermal growth factor, and several chemokines could be used to distinguish schizophrenia patients from healthy controls and from major depression patients [156]. The same year another group used the same platform and identified a 51-plex signature comprised of inflammation factors, hormones, growth factors, stress response, and transport proteins that could discriminate schizophrenia patients from controls with a sensitivity of 83% and a specificity of 83% [157]. This was commercialized as a multiplex kit called Veripsych which cost 2500 US dollars per test to perform. Unfortunately a later validation study showed that this test had a high false-positive rate in the analysis of 25 schizophrenia patients and 50 healthy controls (specificity = 34%), and it was later withdrawn from the market [158]. Another study showed that a signature of 34 molecules could separate schizophrenia patients from controls with good sensitivity and specificity, and a subset of these markers (betacellulin, carcinoembryonic antigen, glutathione S-transferase, haptoglobin, intercellular adhesion molecule 1, -interleukin-17, luteinizing hormone, and thrombospondin-1) were altered in more than 20% of the patients [159].

Going one step further, two recent studies showed that it may be possible to identify individuals at risk of developing schizophrenia several months to years before the actual clinical presentation. The first of these was carried out by researchers as part of the North American Prodrome Longitudinal Study project [160]. This investigation measured blood plasma analytes using multiplex immunoassay which led to identification of 15 molecular tests that could be used to separate 32 persons with high-risk symptoms who later presented with schizophrenia from 35 similar subjects who did not develop a psychiatric disorder over a 2-year follow-up study. In the second investigation, Chan and colleagues carried out a three-step procedure in order to develop and validate their test

[161]. The first step involved analyses of serum samples from 127 first-onset schizophrenia patients and 204 controls which led to identification of a panel of 26 discriminative biomarkers. Next, they validated this biomarker panel using an additional 93 patients and 88 controls, and this gave an accuracy of 97% for detection of schizophrenia. In the final step, they tested the predictive performance of this panel for identifying at-risk individuals up to 2 years before the onset of schizophrenia using samples from 445 at-risk individuals. By also including symptom scores into a complex algorithm, they were able to identify those individuals who were later diagnosed with schizophrenia with 90% accuracy. Although these results are promising, considerable further work is required in order to validate and implement these approaches. It should also be noted that there are disadvantages regarding current multiplex immuno-assay protocols that hinder the broad use of this approach in routine point-of-care analysis [43]. This includes the relatively long experimental times, the necessity of complex laboratory-based procedures, and the need for skilled operators.

6.3 Bipolar Disorder

Bipolar disorder is one of the most difficult psychiatric disorders to investigate considering its presentation, heterogeneity, and alternating swings of distinct mood states. As with the other psychiatric disorders described above, a number of hormones associated with the HPA axis and other neuroendocrine systems have been found to be altered in bipolar disorder patients. The levels of ACTH, cortisol, luteinizing hormone, and prolactin have all been found to be increased in bipolar disorder patients experiencing manic episodes [162, 163]. In addition, changes in insulin sensitivity have been implicated in bipolar disorder as found in one study which showed that bipolar disorder females had higher levels of fasting insulin and glucose, combined with a higher body mass index compared to control females [164]. Another study found reduced insulin sensitivity and lower adiponectin levels in males with bipolar disorder or major depressive disorder compared to controls [165]. The impaired insulin signaling in bipolar disorder suggests that general metabolism is disturbed, consistent with the finding that unmedicated bipolar disorder patients have decreased levels of apolipoprotein E [166] and apolipoprotein A1 [167], which could be normalized by lithium treatment. One potential biomarker which has been identified most frequently in bipolar disorder is BDNF. A meta-analysis carried out in 2011 [168] described the finding of consistently decreased serum levels of BDNF in bipolar disorder patients during both manic and depressive episodes. Furthermore, this study also suggests that there is a negative correlation between BDNF levels and symptom severity in bipolar disorder patients. Another study found reduced levels of plasma nerve growth factor in bipolar patients compared to controls, which again was negatively correlated with symptom severity [169].

Alterations in circulating interleukin-2R levels have been reported for all mood states in symptomatic and euthymic bipolar disorder patients [170–173]. In addition, interleukin-6R [174] and interleukin-8 [175] levels have been found to be increased in bipolar disorder patients experiencing a mania phase. Other inflammation-related proteins which have been associated with bipolar disorder include other cytokines. Berk and colleagues have described an inflammatory cascade in bipolar disorder, which begins by changes in interleukin-1β and tumor necrosis factor-α-mediated activation of NFkB [176], followed by activation of interleukin-6, interferon-8, and interferon-γ [177], and increased levels of haptoglobin and C-reactive protein [178]. As another aspect of inflammation during the mania phase of bipolar disorder, the complement proteins C3, C4, C6, and B have all been found to be increased [179]. Stress may also be involved as the levels of pro-inflammatory cytokines, such as interleukin-1 and interleukin-6, correlate with HPA axis hyperactivity in patients [180]. Other studies have suggested impairment of antioxidant pathways in bipolar disorder, as seen by the finding of lower levels of total glutathione [181]. Oxidative stress appears to be mood state independent and has been shown to be constantly elevated in bipolar disorder patients [182, 183]. In turn, the antioxidant defense system has been found to be increased during mania and depressive episodes, as shown by altered levels of superoxide dismutase [184, 185].

A major challenge in biomarker studies of bipolar disorder is to develop tests which can be used to distinguish this condition from major depressive disorder and to develop tests for predicting conversion from one phase to another. One way of addressing this would be through the analysis of samples from patients in different mood states. A multiplex immunoassay study carried out in 2015 found differential plasma levels of the hormones C-peptide, progesterone, and insulin and the inflammatory protein cancer antigen 125 in both mania and mixed mood states of the disease [186]. However, the hormone peptide YY and the growth factor trafficking protein sortilin were changed only in mania, and the inflammation factors haptoglobin, chemokine CC4, and matrix metalloproteinase 7 were altered only in mixed mood patients. A meta-analysis of eight case-control studies identified a panel comprised of 20 distinguishing analytes [187]. The researchers used this panel to test serum samples from 249 patients with established bipolar disorder, 122 pre-diagnostic bipolar disorder patients, 75 - pre-diagnostic schizophrenia patients, 90 first-onset major depression patients, and 371 controls. The panel yielded excellent predictive performance with an AUC > 0.90 for identification of bipolar disorder patients compared to controls, good predictive performance (AUC = 0.84) for differentiating 12 misdiagnosed bipolar disorder patients from major depression patients, and an AUC of 0.79 for separating the pre-diagnostic bipolar disorder patients from the controls. Although these findings are promising, considerable further work is required before reliable tests can be implicated at the clinical or point-of-care levels.

7   Future Perspectives and Conclusions

Developments over recent years have resulted in formation of new concepts in psychiatry that involves the whole body in the onset, course, and treatment of mental disorders. This leads to the possibility of monitoring at least some of these aspects by looking for proteomic changes in the blood [9, 14, 89]. In the future, it is likely that increased biomarker testing by psychiatrists and clinicians will lead to the concept of "bio-profiles" that reflect the ongoing physiological changes occurring in healthy or mentally ill individuals. At present, tests for schizophrenia and depression have been developed on multiplex immunoassay formats, although these still require extensive validation testing and then translation to clinically useful and inexpensive platforms to enhance their use in the clinic and point-of-care situations. The existing methods leave considerable room for improvement and are financially costly due to the required medium to large format instrumentation, the need for skilled operators, and longer processing times. Several ways of overcoming these issues have emerged recently, including increasing use of portable mass spectrometry devices [188], lab-on-a-chip cards [42–44], and smartphone apps [189]. Some point-of-care testing devices already consist of disposable strips within small cassettes such as the widely available home pregnancy tests [190]. The cassette provides a space for application of the sample and visualization of the results that are usually interpreted visually or by using a low-cost reader. Coupling the readouts of these devices to smartphone apps would represent a significant step forward in patient self-monitoring. This also suits current goals of global companies such as Apple and Google which have shown increasing interests in this market via connection of test scores with app readouts using smart software. As for merging multiplex assay results with smartphone technology, this is already being done. Such assays have been developed on smartphone-based devices using 3D printed optomechanical interfaces, and the data are transmitted to a database for analysis and the results returned to the user in under a minute [49–52]. This mobile platform has already proven successful in clinical settings through rapid analysis of multiplex assays for mumps, measles, and herpes simplex I and II virus immunoglobulins. In each case, these use smartphone camera optics function for accumulation and transmission of the readouts [9]. It is conceivable that other lab-on-a-chip and smartphone-based systems will be developed for other diseases such as mental disorders in the near future. This should help to improve patient outcomes by placing the right patients on the right treatments at the earliest possible phase, in line with personalized medicine objectives.

References

1. http://www.who.int/mediacentre/factsheets/fs396/en/

2. Kassebaum NJ, Arora M, Barber RM, Bhutta ZA, Brown J, Carter A et al (2016) Global, regional, and national disability-adjusted life-years (DALYs) for 315 diseases and injuries and healthy life expectancy (HALE), 1990-2015: a systematic analysis for the global burden of disease study 2015. Lancet 388 (10053):1603–1658

3. http://www.who.int/nmh/publications/ncd_report_full_en.pdf

4. American Psychiatric Association (2013) Diagnostic and statistical manual of mental disorders, fifth edition (DSM-5), 5th edn American Psychiatric Publishing Arlington, VA. ISBN-10: 0890425558

5. World Health Organization. ICD-10: the ICD-10 classification of mental and behavioural disorders: clinical descriptions and diagnostic guidelines. World Health Organisation Geneva. ISBN-10: 9241544228

6. Möller HJ (2003) Bipolar disorder and schizophrenia: distinct illnesses or a continuum? J Clin Psychiatry 64(Suppl 6):23–27

7. Domschke K (2013) Clinical and molecular genetics of psychotic depression. Schizophr Bull 39:766–775

8. Lee J, Rizzo S, Altshuler L, Glahn DC, Miklowitz DJ, Sugar CA et al (2016) Deconstructing bipolar disorder and schizophrenia: a cross-diagnostic cluster analysis of cognitive phenotypes. J Affect Disord 209:71–79

9. Guest PC (2017) Multiplex biomarker approaches to enable point-of-care testing and personalized medicine. Methods Mol Biol 1546:311–315

10. Penttilä M, Jääskeläinen E, Hirvonen N, Isohanni M, Miettunen J (2014) Duration of untreated psychosis as predictor of long-term outcome in schizophrenia: systematic review and meta-analysis. Br J Psychiatry 205:88–94

11. Jablensky A (2000) Epidemiology of schizophrenia: the global burden of disease and disability. Eur Arch Psychiatry Clin Neurosci 250:274–285

12. Atkinson AJ, Colburn WA, DeGruttola VG, DeMets DL, Downing GJ, Hoth DF et al (2001) Biomarkers and surrogate endpoints: Preferred definitions and conceptual framework. Clin Pharmacol Ther 69:89–95

13. Guest PC, Guest FL, Martins-de Souza D (2015) Making sense of blood-based proteomics and metabolomics in psychiatric research. Int J Neuropsychopharmacol pii: pyv138. https://doi.org/10.1093/ijnp/pyv138

14. Guest FL, Guest PC, Martins-de-Souza D (2016) The emergence of point-of-care blood-based biomarker testing for psychiatric disorders: enabling personalized medicine. Biomark Med 10:431–443

15. Weissman MM, Bland RC, Canino GJ, Faravelli C, Greenwald S, Hwu HG et al (1996) Cross-national epidemiology of major depression and bipolar disorder. JAMA 276:293–299

16. Kessler R (2003) Epidemiology of women and depression. J Affect Disord 74:5–13

17. Bromet E, Andrade LH, Hwang I, Sampson NA, Alonso J, de Girolamo G et al (2011) Cross-national epidemiology of DSM-IV major depressive episode. BMC Med 9:90. https://doi.org/10.1186/1741-7015-9-90

18. Fuller RW, Wong DT (1985) Effects of antidepressants on uptake and receptor systems in the brain. Prog Neuro-Psychopharmacol Biol Psychiatry 9:485–490

19. Richelson E (1990) Antidepressants and brain neurochemistry. Mayo Clin Proc 65:1227–1236

20. Leung A, Chue P (2000) Sex differences in schizophrenia, a review of the literature. Acta Psychiatr Scand 101:3–38

21. van Os J, Kapur S (2009) Schizophrenia. Lancet 374:635–645

22. Meltzer HY (1991) The mechanism of action of novel antipsychotic drugs. Schizophr Bull 17:263–287

23. Kapur S, Remington G (1996) Serotonin-dopamine interaction and its relevance to schizophrenia. Am J Psychiatry 153:466–476

24. Cerimele JM, Chwastiak LA, Chan YF, Harrison DA, Unützer J (2013) The presentation, recognition and management of bipolar depression in primary care. J Gen Intern Med 28:1648–1656

25. Gelenberg AJ, Hopkins HS (1993) Report on efficacy of treatments for bipolar disorder. Psychopharmacol Bull 29:447–456

26. Vieta E, Valentí M (2013) Pharmacological management of bipolar depression: acute treatment, maintenance, and prophylaxis. CNS Drugs 27:515–529

27. Oliveira BM, Coorssen JR, Martins-de-Souza D (2014) 2DE: the phoenix of proteomics. J Proteome 104:140–150

28. Unlü M, Morgan ME, Minden JS (1997) Difference gel electrophoresis: a single gel method for detecting changes in protein extracts. Electrophoresis 18:2071–2077

29. Knowles MR, Cervino S, Skynner HA, Hunt SP, de Felipe C, Salim K et al (2003) Multiplex proteomic analysis by two-dimensional differential in-gel electrophoresis. Proteomics 3:1162–1171

30. Anderson NL, Anderson NG (2002) The human plasma proteome: history, character, and diagnostic prospects. Mol Cell Proteomics 1:845–867

31. Simpson KL, Whetton AD, Dive C (2009) Quantitative mass spectrometry-based techniques for clinical use: biomarker identification and quantification. J Chromatogr B Analyt Technol Biomed Life Sci 877:1240–1249

32. Link AJ, Eng J, Schieltz DM, Carmack E, Mize GJ, Morris DR et al (1999) Direct analysis of protein complexes using mass spectrometry. Nat Biotechnol 17:676–682

33. Yip TT, Hutchens TW (1992) Mapping and sequence-specific identification of phosphopeptides in unfractionated protein digest mixtures by matrix-assisted laser desorption/ionization time-of-flight mass spectrometry. FEBS Lett 308:149–153

34. Mallick P, Schirle M, Chen SS, Flory MR, Lee H, Martin D et al (2007) Computational prediction of proteotypic peptides for quantitative proteomics. Nat Biotechnol 25:125–131

35. Faça VM (2017) Selective reaction monitoring for quantitation of cellular proteins. Methods Mol Biol 1546:213–221

36. DeSouza L, Diehl G, Rodrigues MJ, Guo J, Romaschin AD, Colgan TJ et al (2005) Search for cancer markers from endometrial tissues using differentially labeled tags iTRAQ and cICAT with multidimensional liquid chromatography and tandem mass spectrometry. J Proteome Res 4:377–386

37. Núñez EV, Domont GB, Nogueira FC (2017) iTRAQ-based shotgun proteomics approach for relative protein quantification. Methods Mol Biol 1546:267–274

38. Fulton RJ, RL MD, Smith PL, Kienker LJ, Kettman JR Jr (1997) Advanced multiplexed analysis with the FlowMetrix system. Clin Chem 43:1749–1756

39. Stephen L (2017) Multiplex immunoassay profiling. Methods Mol Biol 1546:169–176

40. Salmon SE, Mackey G, Fudenberg HH (1969) "Sandwich" solid phase radioimmunoassay for the quantitative determination of human immunoglobulins. J Immunol 103:129–137

41. Vignali DA (2000) Multiplexed particle-based flow cytometric assays. J Immunol Methods 243:243–255

42. Schumacher S, Nestler J, Otto T, Wegener M, Ehrentreich-Förster E, Michel D et al (2012) Highly-integrated lab-on-chip system for point-of-care multiparameter analysis. Lab Chip 12:464–473

43. Peter H, Wienke J, Bier FF (2017) Lab-on-a-chip multiplex assays. Methods Mol Biol 1546:283–294

44. Schumacher S, Ludecke C, Ehrentreich-Förster E, Bier FF (2013) Platform technologies for molecular diagnostics near the patient's bedside. Adv Biochem Eng Biotechnol 133:75–87

45. Streit P, Nestler J, Shaporin A, Schulze R, Gessner T (2016) Thermal design of integrated heating for lab-on-a-chip systems. Proceedings of the 17th international conference on thermal, mechanical and multiphysics simulation and experiments in microelectronics and microsystems (EuroSimE), 18–20 Apr, pp 1–6

46. Klasnja P, Pratt W (2012) Healthcare in the pocket: mapping the space of mobile-phone health interventions. J Biomed Inform 45:184–198

47. Ventola CL (2014) Mobile devices and apps for health care professionals: uses and benefits. P T 39:356–364

48. Krishna S, Boren SA, Balas EA (2009) Healthcare via cell phones: a systematic review. Telemed J E Health 15:231–240

49. Liao SC, Peng J, Mauk MG, Awasthi S, Song J, Friedman H (2016) Smart cup: a minimally-instrumented, smartphone-based point-of-care molecular diagnostic device. Sens Actuators B Chem 229:232–238

50. Yeo SJ, Choi K, Cuc BT, Hong NN, Bao DT, Ngoc NM et al (2016) Smartphone-based fluorescent diagnostic system for highly pathogenic H5N1 viruses. Theranostics 6:231–242

51. Guo T, Patnaik R, Kuhlmann K, Rai AJ, Sia SK (2015) Smartphone dongle for simultaneous measurement of hemoglobin concentration and detection of HIV antibodies. Lab Chip 15:3514–3520

52. Barbosa AI, Gehlot P, Sidapra K, Edwards AD, Reis NM (2015) Portable smartphone quantitation of prostate specific antigen (PSA) in a fluoropolymer microfluidic device. Biosens Bioelectron 70:5–14

53. Chan MK, Tsang TM, Harris LW, Guest PC, Holmes E, Bahn S (2011) Evidence for disease and antipsychotic medication effects in post-mortem brain from schizophrenia patients. Mol Psychiatry 16:1189–1202

54. Martins-de-Souza D, Guest PC, Harris LW, Vanattou-Saifoudine N, Webster MJ, Rahmoune H et al (2012) Identification of proteomic signatures associated with depression and psychotic depression in post-mortem brains from major depression patients. Transl Psychiatry 2:e87. https://doi.org/10.1038/tp.2012.13

55. Barbier E, Zapata A, Oh E, Liu Q, Zhu F, Undie A et al (2007) Supersensitivity to amphetamine in protein kinase-C interacting protein/HINT1 knockout mice. Neuropsychopharmacology 32:1774–1782

56. Barbier E, Wang JB (2009) Anti-depressant and anxiolytic like behaviors in PKCI/HINT1 knockout mice associated with elevated plasma corticosterone level. BMC Neurosci 10:132. https://doi.org/10.1186/1471-2202-10-132

57. Varadarajulu J, Lebar M, Krishnamoorthy G, Habelt S, Lu J, Bernard Weinstein I et al (2011) Increased anxiety-related behaviour in Hint1 knockout mice. Behav Brain Res 220:305–311

58. Martins-de-Souza D, Guest PC, Vanattou-Saifoudine N, Rahmoune H, Bahn S (2012) Phosphoproteomic differences in major depressive disorder postmortem brains indicate effects on synaptic function. Eur Arch Psychiatry Clin Neurosci 262:657–666

59. Gottschalk M, Wesseling H, Guest PC, Bahn S (2014) Proteomic enrichment analysis of psychotic and affective disorders reveals common signatures in presynaptic glutamatergic signaling and energy metabolism. Int J Neuropsychopharmacol 18:pii: pyu019. https://doi.org/10.1093/ijnp/pyu019

60. Wesseling H, Gottschalk MG, Bahn S (2014) Targeted multiplexed selected reaction monitoring analysis evaluates protein expression changes of molecular risk factors for major psychiatric disorders. Int J Neuropsychopharmacol 18(1):pii: pyu015. https://doi.org/10.1093/ijnp/pyu015

61. Guest PC, Martins-de-Souza D, Vanattou-Saifoudine N, Harris LW, Bahn S (2011) Abnormalities in metabolism and hypothalamic-pituitary-adrenal axis function in schizophrenia. Int Rev Neurobiol 101:145–168

62. Krishnamurthy D, Harris LW, Levin Y, Koutroukides TA, Rahmoune H, Pietsch S et al (2012) Metabolic, hormonal and stress-related molecular changes in post-mortem pituitary glands from schizophrenia subjects. World J Biol Psychiatry 14:478–489

63. Stelzhammer V, Alsaif M, Chan MK, Rahmoune H, Steeb H, Guest PC et al (2015) Distinct proteomic profiles in post-mortem pituitary glands from bipolar disorder and major depressive disorder patients. J Psychiatr Res 60:40–48

64. Dreifuss JJ (1975) A review on neurosecretory granules: their contents and mechanisms of release. Ann N Y Acad Sci 248:184–201

65. Gladkevich A, Kauffman HF, Korf J (2004) Lymphocytes as a neural probe: potential for studying psychiatric disorders. Prog Neuro-Psychopharmacol Biol Psychiatry 28:559–576

66. Torres KCL, Souza BR, Miranda DM, Nicolato R, Neves FS, Barros AGA et al (2009) The leukocytes expressing DARPP-32 are reduced in patients with schizophrenia and bipolar disorder. Progress Neuropsychopharmacol Biol Psychiatry 33:214–219

67. Rollins B, Martin MV, Morgan L, Vawter MP (2010) Analysis of whole genome biomarker expression in blood and brain. Am J Med Genet B Neuropsychiatr Genet 153B:919–936

68. Herberth M, Koethe D, Cheng TM, Krzyszton ND, Schoeffmann S, Guest PC et al (2011) Impaired glycolytic response in peripheral blood mononuclear cells of first-onset antipsychotic-naive schizophrenia patients. Mol Psychiatry 16:848–859

69. Herberth M, Koethe D, Levin Y, Schwarz E, Krzyszton ND, Schoeffmann S et al (2011) Peripheral profiling analysis for bipolar disorder reveals markers associated with reduced cell survival. Proteomics 11:94–105

70. Wang L, Lockstone HE, Guest PC, Levin Y, Palotás A, Pietsch S et al (2010) Expression

profiling of fibroblasts identifies cell cycle abnormalities in schizophrenia. J Proteome Res 9:521–527

71. Jarskog LF, Glantz LA, Gilmore JH, Lieberman JA (2005) Apoptotic mechanisms in the pathophysiology of schizophrenia. Prog Neuro-Psychopharmacol Biol Psychiatry 29:846–858

72. Reif A, Fritzen S, Finger M, Strobel A, Lauer M, Schmitt A et al (2006) Neural stem cell proliferation is decreased in schizophrenia, but not in depression. Mol Psychiatry 11:514–522

73. Katsel P, Davis KL, Li C, Tan W, Greenstein E, Kleiner Hoffman LB et al (2008) Abnormal indices of cell cycle activity in schizophrenia and their potential association with oligodendrocytes. Neuropsychopharmacology 33:2993–3009

74. Levin Y, Wang L, Schwarz E, Koethe D, Leweke FM, Bahn S (2010) Global proteomic profiling reveals altered proteomic signature in schizophrenia serum. Mol Psychiatry 15:1088–1100

75. Stelzhammer V, Haenisch F, Chan MK, Cooper JD, Steiner J, Steeb H et al (2014) Proteomic changes in serum of first onset, antidepressant drug-naïve major depression patients. Int J Neuropsychopharmacol 17:1599–1608

76. Johnston-Wilson NL, Sims CD, Hofmann JP, Anderson L, Shore AD, Torrey EF et al (2000) Disease-specific alterations in frontal cortex brain proteins in schizophrenia, bipolar disorder, and major depressive disorder. The Stanley Neuropathology Consortium. Mol Psychiatry 5:142–149

77. Pennington K, Beasley CL, Dicker P, Fagan A, English J, Pariante CM et al (2008) Prominent synaptic and metabolic abnormalities revealed by proteomic analysis of the dorsolateral prefrontal cortex in schizophrenia and bipolar disorder. Mol Psychiatry 13:1102–1117

78. English JA, Dicker P, Föcking M, Dunn MJ, Cotter DR (2009) 2-D DIGE analysis implicates cytoskeletal abnormalities in psychiatric disease. Proteomics 9:3368–3382

79. Schubert KO, Föcking M, Cotter DR (2015) Proteomic pathway analysis of the hippocampus in schizophrenia and bipolar affective disorder implicates 14-3-3 signaling, aryl hydrocarbon receptor signaling, and glucose metabolism: potential roles in GABAergic interneuron pathology. Schizophr Res 167:64–72

80. Huang JT, Wang L, Prabakaran S, Wengenroth M, Lockstone HE, Koethe D et al (2008) Independent protein-profiling studies show a decrease in apolipoprotein A1 levels in schizophrenia CSF, brain and peripheral tissues. Mol Psychiatry 13:1118–1128

81. Kazuno AA, Ohtawa K, Otsuki K, Usui M, Sugawara H, Okazaki Y et al (2013) Proteomic analysis of lymphoblastoid cells derived from monozygotic twins discordant for bipolar disorder: a preliminary study. PLoS One 8 (2):e53855. https://doi.org/10.1371/journal.pone.0053855

82. Huang TL, Sung ML, Chen TY (2014) 2D-DIGE proteome analysis on the platelet proteins of patients with major depression. Proteome Sci 12(1). https://doi.org/10.1186/1477-5956-12-1

83. Miller AH, Maletic V, Raison CL (2009) Inflammation and its discontents: the role of cytokines in the pathophysiology of major depression. Biol Psychiatry 65:732–741

84. Goodwin RD, Bandiera FC, Steinberg D, Ortega AN, Feldman JM (2012) Asthma and mental health among youth: etiology, current knowledge and future directions. Expert Rev Respir Med 6:397–406

85. Collis I, Lloyd G (1992) Psychiatric aspects of liver disease. Br J Psychiatry 161:12–22

86. Makara-Studzińska M, Ksiazek P, Koślak A, Załuska W, Ksiazek A (2011) Prevalence of depressive disorders in patients with end-stage renal failure. Psychiatr Pol 45:187–195

87. Duda-Sobczak A, Wierusz-Wysocka B (2011) Diabetes mellitus and psychiatric diseases. Psychiatr Pol 45:589–598

88. Laborit H (1976) On the mechanism of activation of the hypothalamo-pituitary-adrenal reaction to changes in the environment (the 'alarm reaction'). Resuscitation 5:19–30

89. Guest PC, Chan MK, Gottschalk MG, Bahn S (2014) The use of proteomic biomarkers for improved diagnosis and stratification of schizophrenia patients. Biomark Med 8:15–27

90. Reagan LP (2007) Insulin signaling effects on memory and mood. Curr Opin Pharmacol 7:633–637

91. Grizzanti J, Lee HG, Camins A, Pallas M, Casadesus G (2016) The therapeutic potential of metabolic hormones in the treatment of age-related cognitive decline and Alzheimer disease. Nutr Res 8:pii:S0271-5317(16) 30275-5. https://doi.org/10.1016/j.nutres.2016.11.002

92. Wong ML, Kling MA, Munson PJ, Listwak S, Licinio J, Prolo P et al (2000) Pronounced and sustained central hypernoradrenergic function in major depression with melancholic features: relation to hypercortisolism and corticotropin-releasing hormone. Proc Natl Acad Sci U S A 97:325–330

93. Kaestner F, Hettich M, Peters M, Sibrowski W, Hetzel G, Ponath G et al (2005) Different activation patterns of proinflammatory cytokines in melancholic and non-melancholic major depression are associated with HPA axis activity. J Affect Disord 87:305–311

94. Jokinen J, Nordstrom P (2009) HPA axis hyperactivity and attempted suicide in young adult mood disorder inpatients. J Affect Disord 116:117–120

95. Karlović D, Serretti A, Vrkić N, Martinac M, Marčinko D (2012) Serum concentrations of CRP, IL-6, TNF-α and cortisol in major depressive disorder with melancholic or atypical features. Psychiatry Res 198:74–80

96. Matsuzaka H, Maeshima H, Kida S, Kurita H, Shimano T, Nakano Y et al (2013) Gender differences in serum testosterone and cortisol in patients with major depressive disorder compared with controls. Int J Psychiatry Med 46:203–221

97. Okamura F, Tashiro A, Utumi A, Imai T, Suchi T, Tamura D et al (2000) Insulin resistance in patients with depression and its changes during the clinical course of depression: minimal model analysis. Metabolism 49:1255–1260

98. Häfner S, Baumert J, Emeny RT, Lacruz ME, Thorand B, Herder C et al (2012) Sleep disturbances and depressed mood: a harmful combination associated with increased leptin levels in women with normal weight. Biol Psychol 89:163–169

99. Lehto SM, Huotari A, Niskanen L, Tolmunen T, Koivumaa-Honkanen H, Honkalampi K et al (2010) Serum adiponectin and resistin levels in major depressive disorder. Acta Psychiatr Scand 121:209–215

100. Diniz BS, Teixeira AL, Campos AC, Miranda AS, Rocha NP, Talib LL et al (2012) Reduced serum levels of adiponectin in elderly patients with major depression. J Psychiatr Res 46:1081–1085

101. Olusi SO, Fido AA (1996) Serum lipid concentrations in patients with major depressive disorder. Biol Psychiatry 40:1128–1131

102. Sevincok L, Buyukozturk A, Dereboy F (2001) Serum lipid concentrations in patients with comorbid generalized anxiety disorder

and major depressive disorder. Can J Psychiatr 46:68–71

103. Maes M, Meltzer HY, Cosyns P, Suy E, Schotte C (1993) An evaluation of basal hypothalamic-pituitary-thyroid axis function in depression: results of a large-scaled and controlled study. Psychoneuroendocrinology 18:607–620

104. Shimizu E, Hashimoto K, Okamura N, Koike K, Komatsu N, Kumakiri C et al (2003) Alterations of serum levels of brain-derived neurotrophic factor (BDNF) in depressed patients with or without antidepressants. Biol Psychiatry 54:70–75

105. Karege F, Bondolfi G, Gervasoni N, Schwald M, Aubry JM, Bertschy G (2005) Low brain-derived neurotrophic factor (BDNF) levels in serum of depressed patients probably results from lowered platelet BDNF release unrelated to platelet reactivity. Biol Psychiatry 57:1068–1072

106. Deveci A, Aydemir O, Taskin O, Taneli F, Esen-Danaci A (2007) Serum brain-derived neurotrophic factor levels in conversion disorder: comparative study with depression. Psychiatry Clin Neurosci 61:571–573

107. Jevtović S, Karlović D, Mihaljević-Peleš A, Serić V, Vrkić N, Jakšić N (2011) Serum brain-derived neurotrophic factor (BDNF): the severity and symptomatic dimensions of depression. Psychiatr Danub 23:363–369

108. Kahl KG, Bens S, Ziegler K, Rudolf S, Kordon A, Dibbelt L et al (2009) Angiogenic factors in patients with current major depressive disorder comorbid with borderline personality disorder. Psychoneuroendocrinology 34:353–357

109. Firouzabadi N, Shafiei M, Bahramali E, Ebrahimi SA, Bakhshandeh H, Tajik N (2012) Association of angiotensin-converting enzyme (ACE) gene polymorphism with elevated serum ACE activity and major depression in an Iranian population. Psychiatry Res 200:336–342

110. Baba H, Nakano Y, Maeshima H, Satomura E, Kita Y, Suzuki T et al (2012) Metabolism of amyloid-beta protein may be affected in depression. J Clin Psychiatry 73:115–120

111. Maes M, Bosmans E, De Jongh R, Kenis G, Vandoolaeghe E, Neels H (1997) Increased serum IL-6 and IL-1 receptor antagonist concentrations in major depression and treatment resistant depression. Cytokine 9:853–858

112. Mikova O, Yakimova R, Bosmans E, Kenis G, Maes M (2001) Increased serum tumor necrosis factor alpha concentrations in major

depression and multiple sclerosis. Eur Neuropsychopharmacol 11:203–208

113. Kim YK, Suh IB, Kim H, Han CS, Lim CS, Choi SH et al (2002) The plasma levels of interleukin-12 in schizophrenia, major depression, and bipolar mania: effects of psychotropic drugs. Mol Psychiatry 7:1107–1114

114. Penninx BW, Kritchevsky SB, Yaffe K, Newman AB, Simonsick EM, Rubin S et al (2003) Inflammatory markers and depressed mood in older persons: results from the Health. Aging body composition study. Biol Psychiatry 54:566–572

115. Simon NM, McNamara K, Chow CW, Maser RS, Papakostas GI, Pollack MH et al (2008) A detailed examination of cytokine abnormalities in major depressive disorder. Eur Neuropsychopharmacol 18:230–233

116. Maes M, Bosmans E, Meltzer HY (1995) Immunoendocrine aspects of major depression. Relationships between plasma interleukin-6 and soluble interleukin-2-receptor, prolactin and cortisol. Eur Arch Psychiatry Clin Neurosci 245:172–178

117. Kling MA, Alesci S, Csako G, Costello R, Luckenbaugh DA, Bonne O et al (2007) Sustained low-grade pro-inflammatory state in unmedicated, remitted women with major depressive disorder as evidenced by elevated serum levels of the acute phase proteins C-reactive protein and serum amyloid A. Biol Psychiatry 62:309–313

118. Elomaa AP, Niskanen L, Herzig KH, Viinamäki H, Hintikka J, Koivumaa-Honkanen H et al (2012) Elevated levels of serum IL-5 are associated with an increased likelihood of major depressive disorder. BMC Psychiatry 12:2. https://doi.org/10.1186/1471-244X-12-2

119. Grassi-Oliveira R, Brieztke E, Teixeira A, Pezzi JC, Zanini M, Lopes RP et al (2012) Peripheral chemokine levels in women with recurrent major depression with suicidal ideation. Rev Bras Psiquiatr 34:71–75

120. Ching KH, Burbelo PD, Carlson PJ, Drevets WC, Iadarola MJ (2010) High levels of Anti-GAD65 and Anti-Ro52 autoantibodies in a patient with major depressive disorder showing psychomotor disturbance. J Neuroimmunol 222:87–89

121. Stefanescu C, Ciobica A (2012) The relevance of oxidative stress status in first episode and recurrent depression. J Affect Disord 143:34–38

122. Ramsey JM, Cooper JD, Bot M, Guest PC, Lamers F, Weickert CS et al (2016) Sex differences in serum markers of major depressive disorder in the netherlands study of depression and anxiety (NESDA). PLoS One 11:e0156624. https://doi.org/10.1371/journal.pone.0156624

123. Papakostas GI, Shelton RC, Kinrys G, Henry ME, Bakow BR, Lipkin SH et al (2013) Assessment of a multi-assay, serum-based biological diagnostic test for major depressive disorder: a pilot and replication study. Mol Psychiatry 18:332–339

124. Bilello JA, Thurmond LM, Smith KM, Pi B, Rubin R, Wright SM et al (2015) MDDScore: confirmation of a blood test to aid in the diagnosis of major depressive disorder. J Clin Psychiatry 76:e199–e206

125. Gottschalk MG, Cooper JD, Chan MK, Bot M, Penninx BW, Bahn S (2015) Discovery of serum biomarkers predicting development of a subsequent depressive episode in social anxiety disorder. Brain Behav Immun 48:123–131

126. Gottschalk MG, Cooper JD, Chan MK, Bot M, Penninx BW, Bahn S (2016) Serum biomarkers predictive of depressive episodes in panic disorder. J Psychiatr Res 73:53–62

127. Ryan MC, Collins P, Thakore JH (2003) Impaired fasting glucose tolerance in first-episode, drug-naive patients with schizophrenia. Am J Psychiatry 160:284–489

128. Spelman LM, Walsh PI, Sharifi N, Collins P, Thakore JH (2007) Impaired glucose tolerance in first-episode drug-naive patients with schizophrenia. Diabet Med 24:481–445

129. Arranz B, Rosel P, Ramírez N, Dueñas R, Fernández P, Sanchez JM et al (2004) Insulin resistance and increased leptin concentrations in noncompliant schizophrenia patients but not in antipsychotic-naive first-episode schizophrenia patients. J Clin Psychiatry 65:1335–1342

130. Cohn TA, Remington G, Zipursky RB, Azad A, Connolly P, Wolever TM (2006) Insulin resistance and adiponectin levels in drug-free patients with schizophrenia: a preliminary report. Can J Psychiatr 51:382–386

131. van Nimwegen LJ, Storosum JG, Blumer RM, Allick G, Venema HW, de Haan L et al (2008) Hepatic insulin resistance in antipsychotic naive schizophrenic patients: stable isotope studies of glucose metabolism. J Clin Endocrinol Metab 93:572–577

132. Wang HC, Yang YK, Chen PS, Lee IH, Yeh TL, Lu RB (2007) Increased plasma leptin in antipsychotic-naive females with schizophrenia, but not in males. Neuropsychobiology 56:213–215

133. Ryan MC, Sharifi N, Condren R, Thakore JH (2004) Evidence of basal pituitary-adrenal overactivity in first episode, drug naive patients with schizophrenia. Psychoneuroendocrinology 29:1065–1070

134. Guest PC, Schwarz E, Krishnamurthy D, Harris LW, Leweke FM, Rothermundt M et al (2011) Altered levels of circulating insulin and other neuroendocrine hormones associated with the onset of schizophrenia. Psychoneuroendocrinology 36:1092–1096

135. Zhang XY, Zhou DF, Cao LY, Wu GY, Shen YC (2005) Cortisol and cytokines in chronic and treatment-resistant patients with schizophrenia: association with psychopathology and response to antipsychotics. Neuropsychopharmacology 30:1532–1538

136. Haddad PM, Wieck A (2004) Antipsychotic-induced hyperprolactinaemia: mechanisms, clinical features and management. Drugs 64:2291–2314

137. Jin H, Meyer JM, Mudaliar S, Jeste DV (2008) Impact of atypical antipsychotic therapy on leptin, ghrelin, and adiponectin. Schizophr Res 100:70–85

138. Wirshing DA, Boyd JA, Meng LR, Ballon JS, Marder SR, Wirshing WC (2002) The effects of novel antipsychotics on glucose and lipid levels. J Clin Psychiatry 63:856–865

139. Huang TL, Chen JF (2005) Serum lipid profiles and schizophrenia: effects of conventional or atypical antipsychotic drugs in Taiwan. Schizophr Res 80:55–59

140. Schwarz E, van Beveren NJ, Ramsey J, Leweke FM, Rothermundt M, Bogerts B et al (2014) Identification of subgroups of schizophrenia patients with changes in either immune or growth factor and hormonal pathways. Schizophr Bull 40:787–795

141. Steiner J, Walter M, Guest P, Myint AM, Schiltz K, Panteli B et al (2010) Elevated S100B levels in schizophrenia are associated with insulin resistance. Mol Psychiatry 15:3–4

142. Guest PC, Wang L, Harris LW, Burling K, Levin Y, Ernst A et al (2010) Increased levels of circulating insulin-related peptides in first-onset, antipsychotic naive schizophrenia patients. Mol Psychiatry 15:118–119

143. Harris LW, Guest PC, Wayland MT, Umrania Y, Krishnamurthy D, Rahmoune H et al (2013) Schizophrenia: metabolic aspects of aetiology, diagnosis and future treatment strategies. Psychoneuroendocrinology 38:752–766

144. Potvin S, Stip E, Sepehry AA, Gendron A, Bah R, Kouassi E (2008) Inflammatory cytokine alterations in schizophrenia: a systematic quantitative review. Biol Psychiatry 63:801–808

145. Miller BJ, Buckley P, Seabolt W, Mellor A, Kirkpatrick B (2011) Meta-analysis of cytokine alterations in schizophrenia: clinical status and antipsychotic effects. Biol Psychiatry 70:663–671

146. Tourjman V, Kouassi É, Koué MÈ, Rocchetti M, Fortin-Fournier S, Fusar-Poli P et al (2013) Antipsychotics' effects on blood levels of cytokines in schizophrenia: a meta-analysis. Schizophr Res 151:43–47

147. Chan MK, Guest PC, Levin Y, Umrania Y, Schwarz E, Bahn S et al (2011) Converging evidence of blood-based biomarkers for schizophrenia. Rev Neurobiol 101:95–144

148. De Witte L, Tomasik J, Schwarz E, Guest PC, Rahmoune H, Kahn RS et al (2014) Cytokine alterations in first-episode schizophrenia patients before and after antipsychotic treatment. Schizophr Res 154:23–29

149. Ribeiro-Santos A, Lucio Teixeira A, Salgado JV (2014) Evidence for an immune role on cognition in schizophrenia: a systematic review. Curr Neuropharmacol 12:273–280

150. Upthegrove R, Manzanares-Teson N, Barnes NM (2014) Cytokine function in medication-naive first episode psychosis: a systematic review and meta-analysis. Schizophr Res 155:101–108

151. Dickerson F, Stallings C, Origoni A, Vaughan C, Khushalani S, Yang S et al (2013) C-reactive protein is elevated in schizophrenia. Schizophr Res 143:198–202

152. Miller BJ, Culpepper N, Rapaport MH (2014) C-Reactive protein levels in schizophrenia: a review and meta-analysis. Clin Schizophr Relat Psychoses 7:223–230

153. Fan X, Pristach C, Liu EY, Freudenreich O, Henderson DC, Goff DC (2007) Elevated serum levels of C-reactive protein are associated with more severe psychopathology in a subgroup of patients with schizophrenia. Psychiatry Res 149:267–271

154. Wong CT, Tsoi WF, Saha N (1996) Acute phase proteins in male Chinese schizophrenic patients in Singapore. Schizophr Res 22:165–171

155. Maes M, Delange J, Ranjan R, Meltzer HY, Desnyder R, Cooremans W (1997) Acute phase proteins in schizophrenia, mania and major depression: modulation by psychotropic drugs. Psychiatry Res 66:1–11

156. Domenici E, Willé DR, Tozzi F, Prokopenko I, Miller S, McKeown A et al (2010) Plasma protein biomarkers for depression and schizophrenia by multi analyte

profiling of case-control collections. PLoS One 5:e9166. https://doi.org/10.1371/journal.pone.0009166

157. Schwarz E, Izmailov R, Spain M, Barnes A, Mapes JP, Guest PC et al (2010) Validation of a blood-based laboratory test to aid in the confirmation of a diagnosis of schizophrenia. Biomark Insights 5:39–47

158. Wehler CA, Preskorn SH (2016) High false-positive rate of a putative biomarker test to aid in the diagnosis of schizophrenia. J Clin Psychiatry 77:e451–e456

159. Schwarz E, Guest PC, Rahmoune H, Harris LW, Wang L, Leweke FM et al (2012) Identification of a biological signature for schizophrenia in serum. Mol Psychiatry 17:494–502

160. Perkins DO, Jeffries CD, Addington J, Bearden CE, Cadenhead KS, Cannon TD et al (2014) Towards a psychosis risk blood diagnostic for persons experiencing high-risk symptoms: preliminary results from the NAPLS project. Schizophr Bull 41:419–428

161. Chan MK, Krebs MO, Cox D, Guest PC, Yolken RH, Rahmoune H et al (2015) Development of a blood-based molecular biomarker test for identification of schizophrenia before disease onset. Transl Psychiatry 5:e601. https://doi.org/10.1038/tp.2015.91

162. Whalley LJ, Christie JE, Bennie J, Dick H, Blackburn IM, Blackwood D et al (1085) Selective increase in plasma luteinising hormone concentrations in drug free young men with mania. Br Med J (Clin Res Ed) 290:99–102

163. Schmider J, Lammers CH, Gotthardt U, Dettling M, Holsboer F, Heuser IJ (1995) Combined dexamethasone/corticotropin-releasing hormone test in acute and remitted manic patients, in acute depression, and in normal controls: I. Biol Psychiatry 38:797–802

164. Rasgon NL, Kenna HA, Reynolds-May MF, Stemmle PG, Vemuri M, Marsh W et al (2010) Metabolic dysfunction in women with bipolar disorder: the potential influence of family history of type 2 diabetes mellitus. Bipolar Disord 12:504–513

165. Hung YJ, Hsieh CH, Chen YJ, Pei D, Kuo SW, Shen DC et al (2007) Insulin sensitivity, proinflammatory markers and adiponectin in young males with different subtypes of depressive disorder. Clin Endocrinol 67:784–789

166. Dean B, Digney A, Sundram S, Thomas E, Scarr E (2008) Plasma apolipoprotein E is decreased in schizophrenia spectrum and bipolar disorder. Psychiatry Res 158:75–78

167. Sussulini A, Dihazi H, Banzato CE, Arruda MA, Stühmer W, Ehrenreich H et al (2011) Apolipoprotein A-I as a candidate serum marker for the response to lithium treatment in bipolar disorder. Proteomics 11:261–269

168. Fernandes BS, Gama CS, Ceresér KM, Yatham LN, Fries GR, Colpo G et al (2011) Brain-derived neurotrophic factor as a state-marker of mood episodes in bipolar disorders: a systematic review and meta-regression analysis. J Psychiatr Res 45:995–1004

169. Barbosa IG, Huguet RB, Neves FS, Reis HJ, Bauer ME, Janka Z et al (2011) Impaired nerve growth factor homeostasis in patients with bipolar disorder. World J Biol Psychiatry 12:228–232

170. Tsai S-YY, Chen KP, Yang YY, Chen CC, Lee JC, Singh VK et al (1999) Activation of indices of cell-mediated immunity in bipolar mania. Biol Psychiatry 45:989–994

171. Tsai S-YY, Yang YY, Kuo CJ, Chen CC, Leu SJ (2001) Effects of symptomatic severity on elevation of plasma soluble interleukin-2-receptor in bipolar mania. J Affect Disord 64:185–193

172. Tsai S-YY, Lee HC, Chen CC, Lee CH (2003) Plasma levels of soluble transferrin receptors and Clara cell protein (CC16) during bipolar mania and subsequent remission. J Psychiatr Res 37:229–235

173. Breunis MN, Kupka RW, Nolen WA, Suppes T, Denicoff KD, Leverich GS et al (2003) High numbers of circulating activated T cells and raised levels of serum IL-2 receptor in bipolar disorder. Biol Psychiatry 53:157–165

174. Maes M, Bosmans E, Calabrese J, Smith R, Meltzer HY (1995) Interleukin-2 and interleukin-6 in schizophrenia and mania: effects of neuroleptics and mood stabilizers. J Psychiatr Res 29:141–152

175. O'Brien SM, Scully P, Scott LV, Dinan TG (2006) Cytokine profiles in bipolar affective disorder: focus on acutely ill patients. J Affect Disord 90:263–267

176. Berk M, Kapczinski F, Andreazza AC, Dean OM, Giorlando F, Maes M et al (2011) Pathways underlying neuroprogression in bipolar disorder: focus on inflammation, oxidative stress and neurotrophic factors. Neurosci Biobehav Rev 35:804–817

177. Gordon S, Martinez FO (2010) Alternative activation of macrophages: mechanism and functions. Immunity 32:593–604

178. Maes M, Scharpé S, Meltzer HY, Bosmans E, Suy E, Calabrese J et al (1993) Relationships between interleukin-6 activity, acute phase proteins, and function of the hypothalamic-pituitary-adrenal axis in severe depression. Psychiatry Res 49:11–27

179. Wadee AA, Kuschke RH, Wood LA, Berk M, Ichim L, Maes M (2002) Serological observations in patients suffering from acute manic episodes. Hum Psychopharmacol 17:175–179

180. Padmos RC, Hillegers MH, Knijff EM, Vonk R, Bouvy A, Staal FJ et al (2008) A discriminating messenger RNA signature for bipolar disorder formed by an aberrant expression of inflammatory genes in monocytes. Arch Gen Psychiatry 65:395–407

181. Rosa AR, Singh N, Whitaker E, de Brito M, Lewis AM, Vieta E et al (2014) Altered plasma glutathione levels in bipolar disorder indicates higher oxidative stress; a possible risk factor for illness onset despite normal brain-derived neurotrophic factor (BDNF) levels. Psychol Med 44:2409–2418

182. Andreazza AC, Cassini C, Rosa AR, Leite MC, de Almeida LM, Nardin P et al (2007) Serum S100B and antioxidant enzymes in bipolar patients. J Psychiatr Res 41:523–529

183. Machado-Vieira R, Andreazza AC, Viale CI, Zanatto V, Cereser V Jr, da Silva Vargas R et al (2007) Oxidative stress parameters in unmedicated and treated bipolar subjects during initial manic episode: a possible role for lithium antioxidant effects. Neurosci Lett 421:33–36

184. Kunz M, Gama CS, Andreazza AC, Salvador M, Cereser KM, Gomes FA et al (2008) Elevated serum superoxide dismutase and thiobarbituric acid reactive substances in different phases of bipolar disorder and in schizophrenia. Prog Neuro-Psychopharmacol Biol Psychiatry 32:1677–1681

185. Mao Y, Ge X, Frank CL, Madison JM, Koehler AN, Doud MK et al (2009) Disrupted in schizophrenia 1 regulates neuronal progenitor proliferation via modulation of GSK3beta/beta-catenin signaling. Cell 136:1017–1031

186. Haenisch F, Alsaif M, Guest PC, Rahmoune H, Yolken RH, Dickerson F et al (2015) Multiplex immunoassay analysis of plasma shows differences in biomarkers related to manic or mixed mood states in bipolar disorder patients. J Affect Disord 185:12–16

187. Haenisch F, Cooper JD, Reif A, Kittel-Schneider S, Steiner J, Leweke FM et al (2016) Towards a blood-based diagnostic panel for bipolar disorder. Brain Behav Immun 52:49–57

188. Leary PE, Dobson GS, Reffner JA (2016) Development and applications of portable gas chromatography-mass spectrometry for emergency responders, the military, and law-enforcement organizations. Appl Spectrosc 70:888–896

189. Berg B, Cortazar B, Tseng D, Ozkan H, Feng S, Wei Q et al (2015) Cellphone-based hand-held micro-plate reader for point-of-care testing of enzyme-linked immunosorbent assays. ACS Nano 9:7857–7866

190. Johnson S, Cushion M, Bond S, Godbert S, Pike J (2015) Comparison of analytical sensitivity and women's interpretation of home pregnancy tests. Clin Chem Lab Med 53:391–402

Chapter 5

Developmental Origins of Breast Cancer: A Paternal Perspective

Camile Castilho Fontelles, Raquel Santana da Cruz, Leena Hilakivi-Clarke, Sonia de Assis, and Thomas Prates Ong

Abstract

The developmental origins of breast cancer have been considered predominantly from a maternal perspective. Although accumulating evidence suggests a paternal programming effect on metabolic diseases, the potential impact of fathers' experiences on their daughters' breast cancer risk has received less attention. In this chapter, we focus on the developmental origins of breast cancer and examine the emerging evidence for a role of fathers' experiences.

Key words Breast cancer, Paternal programming, Preconceptional diet, Female offspring

1 Introduction

Breast cancer is the most commonly diagnosed cancer and the second most frequent cause of cancer death after lung cancer in women worldwide [1]. This major global public health problem is the second most common cancer overall with nearly 1.7 million cases and the fifth cause of global cancer mortality with approximately 522,000 deaths per year [2]. Nearly 70% of all breast cancer deaths occur in developing countries, which could be attributed to increased life expectancy and to increased obesity and adoption of a westernized sedentary lifestyle [3]. Environmental exposures such as diet, during early life, can further modulate later susceptibility to the disease, indicating that breast cancer prevention efforts should be initiated decades prior to breast cancer diagnosis [4].

Windows of susceptibility to breast cancer comprise specific periods during a woman's lifetime in which certain environmental exposures can alter mammary gland development and program increased disease risk in adulthood [5]. Mammary gland development is a lifelong process that starts during the fetal stage and is only completed many decades afterward with a first full-term

Paul C. Guest (ed.), *Investigations of Early Nutrition Effects on Long-Term Health: Methods and Applications*, Methods in Molecular Biology, vol. 1735, https://doi.org/10.1007/978-1-4939-7614-0_5, © Springer Science+Business Media, LLC 2018

pregnancy [6]. Intense remodeling of the mammary gland during in utero life and puberty, including high plasticity with continuous cell proliferation, differentiation, and death, makes the mammary gland particularly susceptible to environment-induced deregulation [7].

Studies investigating the developmental origins of breast cancer have focused on maternal experiences during gestation and lactation [8]. Maternal hormone levels and toxicant and nutritional exposures have been shown to alter susceptibility to breast cancer in the female offspring [9–12]. Although accumulating evidence suggest a paternal programming effect on metabolic diseases, the potential impact of fathers' experiences on their daughters' breast cancer risk has received less attention. Thus, in this chapter, we focus on the developmental origins of breast cancer and examine the emerging evidence for a role of fathers' experiences.

2   Maternal Experiences During Gestation and Lactation and Breast Cancer Programming in Daughters

In the early 1990s, Dr. David Barker proposed the hypothesis that metabolic diseases such as cardiovascular disease, diabetes, and obesity have a fetal origin [13]. Around the same time, Dr. Dimitrios Trichopoulos also hypothesized that breast cancer would be initiated in utero due to increased levels of maternal hormones including estrogen [14]. Since then, epidemiological, clinical, and experimental studies have provided evidence to support this hypothesis [15–19].

Increased birth weight, a proxy of fetal development that can reflect increased estrogen exposure, is associated with increased pre- and postmenopausal breast cancer risk [8, 20]. Other in utero experiences such as dizygotic twinning and maternal pre-eclampsia have also been associated with later increased and decreased breast cancer risk, respectively [21, 22]. Recent proof of concept of an in utero origin of the disease has been provided by the *quasi*-experiment where pregnant women took the synthetic estrogen diethylstilbestrol (DES) to avoid miscarriages. Their daughters presented almost a twofold higher breast cancer incidence in adult life [23]. Further evidence of maternal exposures and daughter's breast cancer comes from a recent human study showing that high maternal organochlorine levels (DDT) during pregnancy increased daughters' breast cancer risk by 3.7-fold [24], indicating that early life exposures to these chemicals increases breast cancer risk. Studies conducted in rodents to investigate the consequences of in utero exposures to DES revealed that the amplified breast cancer risk could be associated to impairment of mammary gland development, such as increased number of terminal end buds (TEBs), which are characterized as sites of cancer initiation, as well as modulation of epigenetic mechanisms such as DNA hypermethylation and modifications in histones and miRNA expression [25].

Maternal nutrition is a key factor for proper fetal growth and development of different systems including the mammary gland itself [26]. Animal models have been useful to elucidate the influence of maternal dietary interventions during key developmental stages on their female offspring susceptibility to mammary carcinogenesis. Poor maternal nutritional patterns could program breast cancer through alterations of in utero hormonal levels, which could trigger permanent epigenetic changes in the fetus, possibly at terminal end buds (TEBs) [5]. In rats higher birth weight, induced by a maternal high-fat diet, was associated with higher number of TEBs, increased number of proliferating cells and denser epithelial tree in the mammary gland, and increased susceptibility to breast cancer in rats [27]. Also in rats, in utero protein restriction and ensuing low birth weight increased breast cancer risk [28], particularly when offspring received a high-calorie diet in adulthood [29]. Maternal consumption of high-fat diet based on vegetal sources (soybean and cottonseed oil) was associated with higher leptin levels and increased breast cancer incidence in the female offspring of rats [30]. On the other hand, maternal consumption of a lard-based high-fat diet during gestation or gestation and lactation decreased rat female offspring susceptibility to mammary carcinogenesis [10]. These protective effects were accompanied by alterations in mammary gland lipidome and transcriptome [31].

Taken together, these data show that in utero and neonatal life represent key developmental stages that can impact a woman's later susceptibility to breast cancer. Thus, opportunities exist to establish early disease preventive strategies focusing on improving mothers' health and nutritional patterns.

3   Paternal Programming of Chronic Diseases

3.1 Metabolic Diseases

Due to the intrinsic relationship between mother and fetus, the contribution of maternal experiences, including diet, on offspring health and disease susceptibility has been more intensely investigated than possible paternal contribution. However, growing evidence suggests that past paternal experiences, including obesity, exercise, and exposures to toxicants and dietary factors, also impact offspring's' health and metabolic disease outcomes [32–35].

Spermatogenesis comprises dynamic chromatin remodeling mediated by alterations in epigenetic processes at the level of DNA methylation and histone posttranslational modifications [36]. At the primordial germ cell stage in early embryogenesis, an almost complete DNA-wide hypomethylation takes place in prospermatogonia to remove somatic epigenetic patterns and subsequently introduce sex-specific marks including gene imprinting [37]. During spermiogenesis, DNA-bound histones are replaced with protamines in the sperm head, which facilitates high-density packing. However, this replacement is not complete, with some specific histone modifications being transmitted to the zygote, suggesting that this mechanism could be involved in the inheritance of chromosomal architecture [38]. Despite the marked demethylation of the parental genomes after fertilization, in addition to imprinted genes, some other DNA regions such as intracisternal A particle elements maintain their methylation status [39, 40]. Thus, it has been proposed that epigenetics could represent a nongenetic inheritance mechanism through the male germ line [41] in controlling gene expression patterns during embryogenesis [42].

Because of the plasticity of the epigenome during sperm development, paternal lifestyle and environmental exposures could alter epigenetic processes including genomic imprinting and histone to protamine substitutions. This would have not only short-term impact (i.e., infertility and altered embryo development) but also long-term impact on offsprings' disease onset in adulthood [43]. In addition, because epigenetic remodeling occurs already in the primordial germ cells of the fetus, inadequate in utero environments through maternal exposures not only program F1 male offspring but also its descendants [44]. Thus, it has been proposed that paternal programming of metabolic diseases could potentially be mediated by inheritance of epigenetic changes in germ cells induced by environmental factors during certain windows of susceptibility in a man's lifetime: the intrauterine period, prepuberty, and the reproductive and zygotic phases [34]. Although the terms inter- and transgenerational are used interchangeably, they represent distinct concepts. From a paternal perspective, intergenerational effects comprise those in F1 offspring when a father's germ cells are affected either during fetal or adult life, while transgenerational effects comprise those taking place even in the absence of paternal germ cell exposure to the deregulating factor (F2 and beyond) [45]. Longitudinal epidemiological studies provide evidence for a transgenerational effect of male experiences. Data from the Överkalik cohort in northern Sweden shows that paternal grandfather's food abundance during the slow growth period in mid-childhood was associated to increased mortality in their grandsons [46, 47]. Human observations further show that other paternal factors including smoking and betel nut consumption are linked to descendants' metabolic phenotypes [48].

One of the first experimental studies to show an influence of paternal metabolic condition on offspring health was provided by Linn et al. [49]. The study found that diabetic male mice reared offspring with alterations in insulin metabolism. It was further observed that male mice that were fasted before mating reared offspring with decreased serum glucose, corticosterone, and insulin-like growth factor 1 levels [50]. More recently, the publications by Ng et al. [51] and Carone et al. [52] further increased awareness of the importance of fathers' experiences for their children health. Ng et al. [51] observed that the female offspring of male rats that consumed a high-fat diet presented in adulthood altered insulin secretion and glucose tolerance impairment, although their body weights and adiposity were not altered. Gene expression analysis of the female offsprings' pancreatic islets revealed alterations in a significant number of genes, suggesting epigenetic as a potential underlying mechanism of these paternal metabolic programming effects [51]. Diet-induced obesity in male mice impaired embryo preimplantation development and implantation [53]. In addition to overnutrition, paternal suboptimal nutrition was shown to impact offspring health. In a mice study conducted by Carone et al. [52], male consumption of a low-protein diet during preconception altered the expression of genes associated with the upregulation of fatty acid and cholesterol pathways in the offsprings' livers. Epigenetic analysis identified altered methylation of a putative enhancer of PPARα gene in the offspring.

The impact of maternal dietary habits during gestation on transgenerational programming through the male germ line has been evaluated. Female mouse fed with a high-fat diet induced in F1 and F2 offspring (male and females) increased body length and insulin resistance, but it was determined that the transmission to F3 offspring occurred only via paternal lineage [54]. In contrast, maternal low-protein diet during gestation induced increased blood pressure and reduced number of nephrons in F1 generation (males and females) that was intergenerationally transmitted to F2 via both maternal and paternal lines [55].

Evidence of sperm epigenetic disturbances induced by paternal experiences during the entire life course has been found. Male mice that were exposed to excessive levels of folic acid from the in utero period until sexual maturation had alterations in germ cell development and imprinted gene methylation that were associated with adverse effects in their offspring, such as increased neonatal mortality and variation in imprinted gene methylation [56]. Wei et al. [57] showed that a paternal prediabetes condition transgenerationally programmed alterations in glucose and insulin metabolism up to F2 generation in mice. Of notice, there was an overlap of genes identified having altered methylation patterns in the sperm of prediabetic fathers and pancreatic islets of their offspring, providing evidence for the inheritance of acquired traits through male germ line epigenetic marks [57].

A growing number of studies also provide evidence that male in utero suboptimal conditions can alter the sperm methylome in adult life. In one of them [58], pregnant mice were submitted to nutrition restriction from day 12.5 to 18.5, a period that comprises epigenetic remodeling in male primordial germ cells. In this model of intergenerational nutritional programming, glucose metabolic dysregulation is transmitted from F1 to F2 through the male germ line. Sperm analysis of the in utero undernourished adult F1 males showed 111 hypomethylated regions that were, however, not retained in F2 somatic tissues [58]. Nevertheless, these data show that nutritional inadequacies starting in fetal life can have long-lasting effects on the sperm epigenome profile and possibly mediate metabolic programming effects. A study by Martinez et al. [59] provided evidence that epigenetic alterations induced in the male germ line due to adverse nutritional conditions can be transmitted to the next generation. In this study, both male F1 and F2 generations of mice that underwent caloric restriction during the last week of gestation displayed hyperglycemia and glucose intolerance. Importantly, altered epigenetic pattern of the lipid metabolism-associated transcription factor liver X receptor was observed in F1 sperm and F2 somatic tissue [59].

Most of the studies suggesting a role of epigenetic deregulation as a mediator of paternal programming effects have been done using animal models. There are a growing number of studies in humans investigating fathers' experiences on offspring epigenetic marks. Soubry et al. [60] provided the first epidemiological evidence that fathers' preconceptional obesity was associated with hypomethylation of the imprinted insulin-like growth factor 2 (*IGF2*) gene. They suggested that obesity-associated hormone imbalances could have mediated incomplete methylation of this gene during spermatogenesis [60]. Influences of paternal obesity on hypomethylation of imprinted genes in newborns were further shown for mesoderm-specific transcript, paternally expressed gene 3 (*MEST*), and neuronatin (*NNAT*) [61]. In another study, obese men presented several altered sperm parameters, including cell morphology and function [62], decreased mitochondrial activity, and increased DNA fragmentation [63]. In addition, sperm of obese men contains lower levels of methylation at different imprinted genes, such as paternally expressed gene 10, necdin (*NDNL2*), and small nuclear ribonucleoprotein polypeptide N (*SNRPN*), which code growth effectors associated with embryonic and fetal growth [64]. Further, another recent study showed that obesity-associated epigenetic abnormalities in sperm can be reversed by weight loss [65]. It was also observed that supplementation of men with high (5 mg) but not low (400 µg) dose of folic acid altered sperm DNA methylation patterns [66, 67]. Because several other dietary compounds, including vitamin A, selenium, and polyphenols, have shown to modulate DNA methylation and histone posttranslational modifications [68], human studies on sperm epigenetics need to also consider these nutrients and bioactive compounds.

In addition to DNA methylation and histone modifications, it is now acknowledged that small noncoding RNAs including miR-NAs also have key roles during spermatogenesis [37]. It has been proposed that miRNAs in spermatozoa that are delivered to the oocyte at fertilization could influence embryonic development [69]. Early-life traumatic stress elicited altered behavioral responses in male mice that were transmitted to their male offspring [70]. Sperm small noncoding RNA analysis revealed that stress upregulated several miRNAs including miR375, which has been implicated in stress response. Importantly microinjection of RNAs purified from sperm from traumatized males into fertilized eggs elicited the same behavioral responses previously observed in the offspring [70]. Thus, alterations in sperm miRNA could represent a mechanism of transmitting paternal "experience" to offspring [71] and be a key factor involved in paternal metabolic programming effects [72]. Obesity-induced sperm miRNA alterations in mice were associated with transmission of obesity and insulin resistance through two subsequent generations [73]. Another recent class of small noncoding RNAs, tRNA fragments, has also been shown to play a role in male germ line-mediated epigenetic inheritance [74, 75]. It is noteworthy that specific paternal interventions, such as balanced diet and exercise in obese mice, normalized obesity-induced aberrant epigenetic marks in sperm cells and enhanced female offspring health outcomes [76]. This suggest that by adopting a healthier lifestyle, obese men could still reverse epigenetic alterations in their germ cells and improve the health of their descendants.

3.2 Breast Cancer

Although the number of studies investigating paternal metabolic programming effects has increased in the last years, there is very limited information in the literature on the relevance of fathers' experiences on their daughters' breast cancer risk. In humans, increased paternal age [77] and education [78] are associated with increased breast cancer risk in the offspring. To our knowledge, our laboratories provided first experimental evidence of an influence of paternal nutritional experiences during preconception period on the female offspring's susceptibility to breast carcinogenesis [79–81].

In a first study in a mouse model [79], we investigated the impact of diet-induced paternal overweight before and during conception on their daughters' mammary gland development and breast cancer risk. We found that overweight males mated with lean females led to increased birth weight, delayed mammary gland development, and increased susceptibility to 7,12-dimethylbenz [a]anthracene (DMBA)-induced breast carcinogenesis in the female offspring. In addition, we observed that paternal overweight was associated with similar miRNA alterations in the father's sperm and in daughters' mammary tissue. Genes associated with hypoxia signaling pathway, regulated by some of these miRNAs, had altered expression both in mammary tissue and mammary tumors of overweight fathers' offspring [79].

In a second study done using a rat model [80], we compared the influence of paternal consumption of animal-based and plant-based high-fat diets during preconception period on their daughters' mammary gland and DMBA-induced mammary tumor development. Opposing effects by these high-fat diets were observed, with paternal consumption of lard-based high-fat diet rich in saturated fatty acids increasing and the corn oil-based diet rich in n-6 polyunsaturated fatty acids decreasing breast cancer risk in the female offspring. These effects by the lard-based high-fat diet were associated with altered mammary gland development, increased mammary cell proliferation, and decreased apoptosis in the female offspring. These paternal programming effects were further associated with altered miRNA profiles in fathers' sperm and their daughters' mammary glands, particularly those targeting genes regulating epithelial-to-mesenchymal transition-related protein expression in this tissue [80]. Thus, our studies in mice and rats indicate that in addition to male obesity, the fatty acid profile of the diet is also a principal factor influencing paternal programming of the daughter's breast cancer risk.

We have further shown in rats that paternal micronutrient status also influences female offspring breast cancer risk [81]. Consumption of a selenium-deficient diet during preconception by fathers altered mammary gland development and increased DMBA-induced mammary carcinogenesis in female offspring. On the other hand, paternal selenium supplementation did not alter any of these processes in female offspring. This indicate that instead of supplementing males with high levels of the micronutrients, assuring intake at the daily recommended levels to prevent selenium deficiency would represent a potential way to prevent breast cancer in the daughters [81].

Taken together, our data [79–81] provide support to the hypothesis that a woman's risk of developing breast cancer can be set early in development and through the male germ line. For this new research field to expand, it would be important that these results are confirmed in other breast cancer models and that other paternal environmental factors such as toxicants and additional nutritional exposures be incorporated in new breast cancer programming studies. The findings that paternal interventions with a mix of green extract, vitamins, and antioxidants including lycopene reversed paternal malnutrition metabolic programming in the offspring [82] suggest that interventions with bioactive compounds in functional foods could represent a promising approach to reverse diet-induced paternal breast cancer programming. Furthermore, because interactions between mothers' and fathers' nutrition on their offspring health have been suggested [83], it would be also important to test this possibility in breast cancer model systems since same dietary interventions (high-fat diets) have resulted opposite outcomes in daughters depending whether intervention was through the maternal or paternal line [10, 80]. There is also need to elucidate the exact mechanisms mediating the observed paternal programming effects and whether these could be transgenerationally transmitted. In addition to miRNAs, sperm epigenetic disturbances at DNA methylation and histone modifications represent key candidates. Finally, a major challenge would be to confirm in human populations the relevance of fathers' experiences on their daughters' breast cancer risk. Although it will take time to generate data from cohort studies, feasible initial approaches could be to associate paternal health parameters and diet intake before conception with perinatal factors postulated to increase later risk of breast cancer [81], such as birth weight and umbilical cord hormone and stem cell levels [84–86].

4   Conclusions

Breast cancer is currently a major global public health problem. Despite large investment in improving treatments, mortality rates are still high, especially in developing countries. This indicates that novel preventive strategies should be developed. A woman's early life represents a valuable developmental stage at which to start risk reduction interventions. In addition to improving mothers' health and nutrition during gestation and lactation, emerging evidence indicates that applying this to future fathers could also improve health and decrease breast cancer risk in their daughters.

Acknowledgments

C. C. F. was a recipient of a PhD scholarship from the Brazilian National Council for Scientific and Technological Development (CNPq; Proc. 153478/2012-8). T. P. O. is the recipient of a researcher fellowship from CNPq (Proc.307910/2016-4) and is supported by grants from CNPq (Proc. 448501/2014-7), the Food Research Center (FoRC), and the São Paulo State Research Funding Agency (Proc.2013/07914-8).

References

1. Oeffinger KC, Fontham ET, Etzioni R, Herzig A, Michaelson JS, Shih YC et al (2015) Breast cancer screening for women at average risk. JAMA 314:1599–1614

2. Ferlay J, Soerjomataram I, Dikshit R, Eser S, Mathers C, Rebelo M et al (2015) Cancer incidence and mortality worldwide: sources, methods and major patterns in GLOBOCAN 2012. Int J Cancer 136:E359–E386

3. Lee BL, Liedke PE, Barrios CH, Simon SD, Finkelstein DM, Goss PE (2012) Breast cancer in Brazil: present status and future goals. Lancet Oncol 13:e95–102

4. Colditz GA, Bohlke K, Berkey CS (2014) Breast cancer risk accumulation starts early: prevention must also. Breast Cancer Res Treat 145:567–579

5. Hilakivi-Clarke L (2007) Nutritional modulation of terminal end buds: its relevance to breast cancer prevention. Curr Cancer Drug Targets 7:465–474

6. Oakes SR, Gallego-Ortega D, Ormandy CJ (2014) The mammary cellular hierarchy and breast cancer. Cell Mol Life Sci 71:4301–4324

7. Russo J (2015) Significance of rat mammary tumors for human risk assessment. Toxicol Pathol 43:145–170

8. Hilakivi-Clarke L, de Assis S (2006) Fetal origins of breast cancer. Trends Endocrinol Metab 17:340–348

9. de Assis S, Wang M, Jin L, Bouker KB, Hilakivi-Clarke LA (2013) Exposure to excess estradiol or leptin during pregnancy increases mammary cancer risk and prevents parity-induced protective genomic changes in rats. Cancer Prev Res 6:1194–1211

10. de Oliveira Andrade F, Fontelles CC, Rosim MP, de Oliveira TF, de Melo Loureiro AP, Mancini-Filho J et al (2014) Exposure to lard-based high-fat diet during fetal and lactation periods modifies breast cancer susceptibility in adulthood in rats. J Nutr Biochem 25:613–622

11. Soto AM, Brisken C, Schaeberle C, Sonnenschein C (2013) Does cancer start in the womb? Altered mammary gland development and predisposition to breast cancer due to in utero exposure to endocrine disruptors. J Mammary Gland Biol Neoplasia 18:199–208

12. Osborne G, Rudel R, Schwarzman M (2015) Evaluating chemical effects on mammary gland development: a critical need in disease prevention. Reprod Toxicol 54:148–155

13. Barker DJ, Gluckman PD, Godfrey KM, Harding JE, Owens JA, Robinson JS (1993) Fetal nutrition and cardiovascular disease in adult life. Lancet 341:938–941

14. Trichopoulos D (1990) Is breast cancer initiated in utero? Epidemiology 1:95–96

15. Hilakivi-Clarke L, Clarke R, Onojafe I, Raygada M, Cho E, Lippman M (1997) A maternal diet high in n-6 polyunsaturated fats alters mammary gland development, puberty onset, and breast cancer risk among female rat offspring. Proc Natl Acad Sci U S A 94:9372–9377

16. Hilakivi-Clarke L, Clarke R, Lippman M (1999) The influence of maternal diet on breast cancer risk among female offspring. Nutrition15:392–401

17. de Assis S, Hilakivi-Clarke L (2006) Timing of dietary estrogenic exposures and breast cancer risk. Ann N Y Acad Sci 1089:14–35

18. Troisi R, Potischman N, Hoover RN (2007) Exploring the underlying hormonal mechanisms of prenatal risk factors for breast cancer: a review and commentary. Cancer Epidemiol Biomark Prev 16:1700–1712

19. Hill J, Hodsdon W (2014) In utero exposure and breast cancer development: an epigenetic perspective. J Environ Pathol Toxicol Oncol 33:239–245

20. Bukowski R, Chlebowski RT, Thune I, Furberg AS, Hankins GD, Malone FD et al (2012) Birth weight, breast cancer and the potential mediating hormonal environment. PLoS One 7(7):e40199. https://doi.org/10.1371/journal.pone.0040199

21. Swerdlow AJ, De Stavola BL, Swanwick MA, Maconochie NE (1997) Risks of breast and testicular cancers in young adult twins in England and Wales: evidence on prenatal and genetic aetiology. Lancet 350:1723–1728

22. Xue F, Michels KB (2007) Intrauterine factors and risk of breast cancer: a systematic review and meta-analysis of current evidence. Lancet Oncol 8:1088–1100

23. Palmer JR, Wise LA, Hatch EE, Troisi R, Titus-Ernstoff L, Strohsnitter W et al (2006) Prenatal diethylstilbestrol exposure and risk of breast cancer. Cancer Epidemiol Biomark Prev 15:1509–1514

24. Cohn BA, La Merrill M, Krigbaum NY, Yeh G, Park JS, Zimmermann L et al (2015) DDT exposure in utero and breast cancer. J Clin Endocrinol Metab 100:2865–2872

25. Hilakivi-Clarke L (2014) Maternal exposure to diethylstilbestrol during pregnancy and increased breast cancer risk in daughters. Breast Cancer Res 16:208

26. MacLennan M, Ma DW (2010) Role of dietary fatty acids in mammary gland development and breast cancer. Breast Cancer Res 12:211

27. de Assis S, Khan G, Hilakivi-Clarke L (2006) High birth weight increases mammary tumorigenesis in rats. Int J Cancer 119:1537–1546

28. Fernandez-Twinn DS, Ekizoglou S, Gusterson BA, Luan J, Ozanne SE (2007) Compensatory mammary growth following protein restriction

during pregnancy and lactation increases early-onset mammary tumor incidence in rats. Carcinogenesis 28:545–552

29. Fernandez-Twinn DS, Ekizoglou S, Martin-Gronert MS, Tarry-Adkins J, Wayman AP, Warner MJ et al (2010) Poor early growth and excessive adult calorie intake independently and additively affect mitogenic signaling and increase mammary tumor susceptibility. Carcinogenesis 31:1873–1881

30. de Assis S, Wang M, Goel S, Foxworth A, Helferich W, Hilakivi-Clarke L et al (2006) Excessive weight gain during pregnancy increases carcinogen-induced mammary tumorigenesis in Sprague-Dawley and lean and obese Zucker rats. J Nutr 136:998–1004

31. de Oliveira Andrade F, de Assis S, Jin L, Fontelles CC, Barbisan LF, Purgatto E et al (2015) Lipidomic fatty acid profile and global gene expression pattern in mammary gland of rats that were exposed to lard-based high fat diet during fetal and lactation periods associated to breast cancer risk in adulthood. Chem Biol Interact 239:118–128

32. Ferguson-Smith AC, Patti ME (2011) You are what your dad ate. Cell Metab 13:115–117

33. Lane M, Robker RL, Robertson SA (2014) Parenting from before conception. Science 345:756–760

34. Soubry A, Hoyo C, Jirtle RL, Murphy SK (2014) A paternal environmental legacy: evidence for epigenetic inheritance through the male germ line. BioEssays 36:359–371

35. Hur SS, Cropley JE, Suter CM (2017) Paternal epigenetic programming: evolving metabolic disease risk. J Mol Endocrinol 58:R159–R168

36. Carrell DT, Hammoud SS (2010) The human sperm epigenome and its potential role in embryonic development. Mol Hum Reprod 16:37–47

37. de Mateo S, Sassone-Corsi P (2014) Regulation of spermatogenesis by small non-coding RNAs: role of the germ granule. Semin Cell Dev Biol 29:84–92

38. van der Heijden GW, Derijck AA, Ramos L, Giele M, van der Vlag J, de Boer P (2006) Transmission of modified nucleosomes from the mouse male germline to the zygote and subsequent remodeling of paternal chromatin. Dev Biol 298:458–469

39. Lane N, Dean W, Erhardt S, Hajkova P, Surani A, Walter J et al (2003) Resistance of IAPs to methylation reprogramming may provide a mechanism for epigenetic inheritance in the mouse. Genesis 35:88–93

40. Barlow DP, Bartolomei MS (2014) Genomic imprinting in mammals. Cold Spring Harb Perspect Biol 6(2):–pii:a018382. https://doi.org/10.1101/cshperspect.a018382

41. Gill ME, Erkek S, Peters AH (2012) Parental epigenetic control of embryogenesis: a balance between inheritance and reprogramming? Curr Opin Cell Biol 24:387–396

42. Carrell DT (2012) Epigenetics of the male gamete. Fertil Steril 97:267–274

43. Stuppia L, Franzago M, Ballerini P, Gatta V, Antonucci I (2015) Epigenetics and male reproduction: the consequences of paternal lifestyle on fertility, embryo development, and children lifetime health. Clin Epigenetics 7:120. https://doi.org/10.1186/s13148-015-0155-4

44. Ong TP, Ozanne SE (2015) Developmental programming of type 2 diabetes: early nutrition and epigenetic mechanisms. Curr Opin Clin Nutr Metab Care 18:354–360

45. Ly L, Chan D, Trasler JM (2015) Developmental windows of susceptibility for epigenetic inheritance through the male germline. Semin Cell Dev Biol 43:96–105

46. Kaati G, Bygren LO, Edvinsson S (2002) Cardiovascular and diabetes mortality determined by nutrition during parents' and grandparents' slow growth period. Eur J Hum Genet 10:682–688

47. Pembrey ME, Bygren LO, Kaati G, Edvinsson S, Northstone K, Sjöström M et al (2006) Sex-specific, male-line transgenerational responses in humans. Eur J Hum Genet 14:159–166

48. Pembrey M, Saffery R, Bygren LO (2014) Human transgenerational responses to early-life experience: potential impact on development, health and biomedical research. J Med Genet 51:563–572

49. Linn T, Loewk E, Schneider K, Federlin K (1993) Spontaneous glucose intolerance in the progeny of low dose streptozotocin-induced diabetic mice. Diabetologia 36:1245–1251

50. Anderson LM, Riffle L, Wilson R (2006) Preconceptional fasting of fathers alters serum glucose in offspring of mice. Nutrition 22:327–331

51. Ng SF, Lin RC, Laybutt DR, Barres R, Owens JA, Morris MJ (2010) Chronic high-fat diet in fathers programs β-cell dysfunction in female rat offspring. Nature 467:963–966

52. Carone BR, Fauquier L, Habib N, Shea JM, Hart CE, Li R et al (2010) Paternally-induced transgenerational environmental reprogramming of metabolic gene expression in mammals. Cell 143:1084–1096

53. Mitchell M, Bakos HW, Lane M (2011) Paternal diet-induced obesity impairs embryo development and implantation in the mouse. Fertil Steril 95:1349–1353

54. Dunn GA, Bale TL (2011) Maternal high-fat diet effects on third-generation female body size via the paternal lineage. Endocrinology 152:2228–2236

55. Harrison M, Langley-Evans SC (2009) Intergenerational programming of impaired nephrogenesis and hypertension in rats following maternal protein restriction during pregnancy. Br J Nutr 101:1020–1030

56. Ly L, Chan D, Aarabi M, Landry M, Behan NA, AJ MF et al (2017) Intergenerational impact of paternal lifetime exposures to both folic acid deficiency and supplementation on reproductive outcomes and imprinted gene methylation. Mol Hum Reprod. https://doi.org/10.1093/molehr/gax029

57. Wei Y, Yang CR, Wei YP, Zhao ZA, Hou Y, Schatten H et al (2014) Paternally induced transgenerational inheritance of susceptibility to diabetes in mammals. Proc Natl Acad Sci U S A 111:1873–1878

58. Radford EJ, Ito M, Shi H, Corish JA, Yamazawa K, Isganaitis E et al (2014) In utero undernourishment perturbs the adult sperm methylome and intergenerational metabolism. Science 345:1255903. https://doi.org/10.1126/science.1255903

59. Martínez D, Pentinat T, Ribó S, Daviaud C, Bloks VW, Cebrià J et al (2014) In utero undernutrition in male mice programs liver lipid metabolism in the second-generation offspring involving altered LXRa DNA methylation. Cell Metab 19:941–951

60. Soubry A, Schildkraut JM, Murtha A, Wang F, Huang Z, Bernal A et al (2013) Paternal obesity is associated with IGF2 hypomethylation in newborns: results from a Newborn Epigenetics Study (NEST) cohort. BMC Med 11:29. https://doi.org/10.1186/1741-7015-11-29

61. Soubry A, Murphy SK, Wang F, Huang Z, Vidal AC, Fuemmeler BF et al (2015) Newborns of obese parents have altered DNA methylation patterns at imprinted genes. Int J Obes 39:650–657

62. Sermondade N, Faure C, Fezeu L, Shayeb AG, Bonde JP, Jensen TK et al (2013) BMI in relation to sperm count: an updated systematic review and collaborative meta-analysis. Hum Reprod Update 19:221–231

63. Fariello RM, Pariz JR, Spaine DM, Cedenho AP, Bertolla RP, Fraietta R (2012) Association between obesity and alteration of sperm DNA integrity and mitochondrial activity. BJU Int 110:863–867

64. Soubry A, Guo L, Huang Z, Hoyo C, Romanus S, Price T et al (2016) Obesity-related DNA methylation at imprinted genes in human sperm: results from the TIEGER study. Clin Epigenetics 8:51. https://doi.org/10.1186/s13148-016-0217-2

65. Donkin I, Versteyhe S, Ingerslev LR, Qian K, Mechta M, Nordkap L et al (2016) Obesity and bariatric surgery drive epigenetic variation of spermatozoa in humans. Cell Metab 23:369–378

66. Aarabi M, San Gabriel MC, Chan D, Behan NA, Caron M, Pastinen T et al (2015) High-dose folic acid supplementation alters the human sperm methylome and is influenced by the MTHFR C677T polymorphism. Hum Mol Genet 24:6301–6313

67. Chan D, McGraw S, Klein K, Wallock LM, Konermann C, Plass C et al (2017) Stability of the human sperm DNA methylome to folic acid fortification and short-term supplementation. Hum Reprod 32:272–283

68. Ong TP, Moreno FS, Ross SA (2011) Targeting the epigenome with bioactive food components for cancer prevention. J Nutrigenet Nutrigenomics 4:275–292

69. Ostermeier GC, Miller D, Huntriss JD, Diamond MP, Krawetz SA (2004) Reproductive biology: delivering spermatozoan RNA to the oocyte. Nature 429:154. https://doi.org/10.1038/429154a

70. Gapp K, Jawaid A, Sarkies P, Bohacek J, Pelczar P, Prados J et al (2014) Implication of sperm RNAs in transgenerational inheritance of the effects of early trauma in mice. Nat Neurosci 17:667–669

71. Rassoulzadegan M, Cuzin F (2015) Epigenetic heredity: RNA-mediated modes of phenotypic variation. Ann N Y Acad Sci 1341:172–175

72. McPherson NO, Fullston T, Aitken RJ, Lane M (2014) Paternal obesity, interventions, and mechanistic pathways to impaired health in offspring. Ann Nutr Metab 64:231–238

73. Fullston T, Ohlsson Teague EM, Palmer NO, DeBlasio MJ, Mitchell M, Corbett M et al (2013) Paternal obesity initiates metabolic disturbances in two generations of mice with incomplete penetrance to the F2 generation and alters the transcriptional profile of testis and sperm microRNA content. FASEB J 27:4226–4243

74. Sharma U, Conine CC, Shea JM, Boskovic A, Derr AG, Bing XY et al (2016) Biogenesis and function of tRNA fragments during sperm

maturation and fertilization in mammals. Science 351:391–396

75. Chen Q, Yan M, Cao Z, Li X, Zhang Y, Shi J et al (2016) Sperm tsRNAs contribute to intergenerational inheritance of an acquired metabolic disorder. Science 351:397–400

76. McPherson NO, Owens JA, Fullston T, Lane M (2015) Preconception diet or exercise intervention in obese fathers normalizes sperm microRNA profile and metabolic syndrome in female offspring. Am J Phys 308:E805–E821

77. Choi JY, Lee KM, Park SK, Noh DY, Ahn SH, Yoo KY et al (2005) Association of paternal age at birth and the risk of breast cancer in offspring: a case control study. BMC Cancer 5:143. https://doi.org/10.1186/1471-2407-5-143

78. Titus-Ernstoff L, Egan KM, Newcomb PA, Ding J, Trentham-Dietz A, Greenberg ER et al (2002) Early life factors in relation to breast cancer risk in postmenopausal women. Cancer Epidemiol Biomark Prev 11:207–210

79. Fontelles CC, Carney E, Clarke J, Nguyen NM, Yin C, Jin L et al (2016) Paternal overweight is associated with increased breast cancer risk in daughters in a mouse model. Sci Rep 6:28602. https://doi.org/10.1038/srep28602

80. Fontelles CC, Guido LN, Rosim MP, Andrade Fde O, Jin L, Inchauspe J et al (2016) Paternal programming of breast cancer risk in daughters in a rat model: opposing effects of animal- and plant-based high-fat diets. Breast Cancer Res 18:71. https://doi.org/10.1186/s13058-016-0729-x

81. Guido LN, Fontelles CC, Rosim MP, Pires VC, Cozzolino SM, Castro IA et al (2016) Paternal

selenium deficiency but not supplementation during preconception alters mammary gland development and 7,12-dimethylbenz[a]anthracene-induced mammary carcinogenesis in female rat offspring. Int J Cancer:139, 1873–1882

82. McPherson NO, Fullston T, Kang WX, Sandeman LY, Corbett MA, Owens JA et al (2016) Paternal under-nutrition programs metabolic syndrome in offspring which can be reversed by antioxidant/vitamin food fortification in fathers. Sci Rep 6:27010. https://doi.org/10.1038/srep27010

83. Masuyama H, Mitsui T, Eguchi T, Tamada S, Hiramatsu Y (2016) The effects of paternal high-fat diet exposure on offspring metabolism with epigenetic changes in the mouse adiponectin and leptin gene promoters. Am J Physiol Endocrinol Metab 311:E236–E245

84. Troisi R, Stephansson O, Jacobsen J, Tretli S, Sørensen HT, Gissler M et al (2014) Perinatal characteristics and bone cancer risk in offspring – a Scandinavian population-based study. Acta Oncol 53:830–838

85. Lagiou P, Samoli E, Okulicz W, Xu B, Lagiou A, Lipworth L et al (2011) Maternal and cord blood hormone levels in the United States and China and the intrauterine origin of breast cancer. Ann Oncol 22:1102–1108

86. Qiu L, Low HP, Chang CI, Strohsnitter WC, Anderson M, Edmiston K et al (2012) Novel measurements of mammary stem cells in human umbilical cord blood as prospective predictors of breast cancer susceptibility in later life. Ann Oncol 23:245–250

Chapter 6

Point-of-Care Testing and Personalized Medicine for Metabolic Disorders

Francesca L. Guest and Paul C. Guest

Abstract

This chapter describes innovations in biomarker testing that can facilitate earlier and better treatment of patients who suffer from metabolic disorders. The use of new microfluidic devices along with miniaturized biosensors and transducers enables analysis of a single drop of a blood within the time frame of a typical visit to a doctor's office. Steps are underway so that these approaches will incorporate both biochemical and clinical data, resulting in unique bioprofiles for each patient. This will allow earlier, personalized, and more effective therapeutic options. In addition, smartphone apps for self-monitoring will be used increasingly for the best possible patient outcomes.

Key words Nutritional programming, Metabolic disorders, Epigenetics, Proteomics, Biomarkers, Lab-on-a-chip, Smartphone apps, Point-of-care, Personalized medicine

1 Introduction

Metabolic disorders such as cardiovascular disorders, obesity, and type 2 diabetes mellitus and metabolic syndrome affect around one third of the people in the world [1]. Cardiovascular diseases are the number 1 cause of death worldwide with an estimated 18 million each year, representing around 30% of all global deaths [2]. In 2014, more than 600 million people were classified as obese (BMI; ≥ 30 kg/m^2), and 1.3 billion were deemed to be overweight (BMI = 25–30 kg/m^2) [3]. Obesity is responsible for approximately 5% of all deaths, and the global economic impact is equivalent to around two trillion US dollars each year [4]. The global prevalence of diabetes has nearly doubled since 1980, rising from 4.7% to 8.5% in the adult population, reflecting an increase in associated risk factors such as being overweight or obese, and it was responsible for 1.5 million deaths in 2012 [5]. All of these diseases can impair quality of life, well-being, and productivity, and they can have significant secondary effects on society, the workforce, healthcare services, and the wider economy [6, 7].

Paul C. Guest (ed.), *Investigations of Early Nutrition Effects on Long-Term Health: Methods and Applications*, Methods in Molecular Biology, vol. 1735, https://doi.org/10.1007/978-1-4939-7614-0_6, © Springer Science+Business Media, LLC 2018

Over the last few years, a number of studies have focused on the effect of undernutrition or overnutrition during critical periods of offspring development as risk factors for developing metabolic diseases later in life [8, 9]. Several mechanisms have been associated with the process of abnormal programming such as epigenetic modification and perturbations in oxidative stress pathways [10, 11]. Investigations of the mechanisms involved are in high demand in the search for efficient therapeutics to counter or delay the effects of malprogramming [12–14]. In addition, such studies could lead to the identification of novel biomarkers which could be used for prediction or early disease detection, as well as for treatment monitoring [15–18].

This chapter describes the latest advances used in the development of biomarker profiling platforms for increasing our understanding about the underlying pathologies of metabolic disorders. It is also important to understand the reasons behind the heterogeneous response of individuals to treatment and susceptibilities to complications, as found in most diseases [19–21]. Together, these factors emphasize the need for biomarkers as tools for predicting disease progression and treatment success. Currently, most biomarker platforms have medium to large footprints and are technologically challenging in their operation [22–25]. There is an increasing need for more user-friendly devices that can be used in a point-of-care or clinical setting to return a result in less than the time of a typical visit to the doctor's office. There is also an urgent need to make these tests inexpensive to maximize their cost-effectiveness.

2   State of the Art

To date, most biomarker test devices are constructed using medium to large footprint instruments, considering that they are used typically in laboratory settings for screening large numbers of samples [26]. Such instruments are costly to operate and require considerable training and expertise to carry out an analysis. Furthermore, the analysis time can vary from days to weeks depending on the number of samples under investigation. However, point-of-care testing now encompasses technology ranging from medium-sized tabletop instruments to implanted, wearable, and handheld devices, which overcome most, if not all, of the above issues [26]. Handheld devices typically consist of disposable strips in cassettes which provide space for addition of the sample, carrying out the test, and generation of the result that can be either visual or displayed through an inexpensive reader. There are now devices based on this format that can be used to detect a heart attack [27], blood clotting speed [28–30], and infectious diseases [31–33] and for detecting and measuring the levels of substances such as alcohol and drugs of abuse [34]. Perhaps the most familiar example is the Clearblue Digital Pregnancy test® produced by SPD Swiss Precision Diagnostics GmbH [35, 36] which gives both a positive or negative readout and an indication of the number of weeks since conception. Another medical area in which test strips are in routine use, both in the home and in clinics, is diabetes. These include blood tests for glucose [37] and urine analyses for ketosis [38]. Furthermore, a dry chemistry test is available in either instrument or test strip formats for detection of kidney diseases, diabetes, and urinary tract infection, which can be used in prevention screening, treatment monitoring, or patient self-testing [39]. It is clear that such devices can be operated by non-experts, with a rapid return of results (often within minutes) and at low cost.

3   Clinical Needs

There is an urgent need for increased translation of molecular algorithms using point-of-care devices to improve detection of metabolic diseases with higher sensitivity and specificity [40, 41]. The use of multiple markers increases accuracy of detection in the same way that a full fingerprint is more useful than a partial one in criminology investigations. In the near future, it is likely that increased biomarker testing by clinicians will lead "bio-" signatures in individuals that reflect the molecular changes that occur in healthy or ill states. The ideal diagnostic test should have a high accuracy for identifying patients suffering from a particular illness as well as those who have another illness or who are healthy. Currently, tests have been developed on multiplex immunoassay formats for many diseases such as obesity and the metabolic syndrome [42], type 1 diabetes [43], autoimmune disorders [44], ovarian cancer [45], neurodegenerative disorders [46–48], and mental illnesses [49–53], although all of these still require validation testing and translation to clinically useful and inexpensive platforms for point-of-care use.

There is now a heightened interest for new technologies and biomarker-based approaches that are simple to use, rapid, and cost-effective, as with the test strip approaches outlined above. The use of accurate in vitro diagnostic devices that can accurately classify patients according to the type or subtype of a particular disorder will help to reduce the duration of untreated illness and improve compliance by placing patients on the treatment that is right for them as early as possible. This will help to change the prevailing strategy in medicine from reactive medical care to a more personalized medicine approach.

4   New Point-of-Care Technologies

Several platforms for improving translation of biomarker tests have been emerging over the last decade. These include portable mass spectrometry devices as found in airports and credit card-sized lab-on-a-chip implements for detection, as well as smartphone applications (apps) for analysis and monitoring. Thus far, none of these devices have been used for routine measurements of biomarkers in cases of metabolic disorders although this is likely to change in the near future considering their success in many pilot studies and the critical need for such point-of-care devices in this field of medicine.

4.1   Portable Mass Spectrometers

Traditionally, mass spectrometry readings are brought about through the use of multiple components with a liquid chromatography front end and a mass spectrometry back end. Both of these components normally have medium to large footprints and require considerable technical expertise for operation. However, smaller gas chromatography mass spectrometry instruments have now emerged which overcome these issues by allowing their use in a more portable manner or in a field research capacity [54–57]. This portability allows application of these instruments in environments such as airports, crime scenes, or disaster sites for detection of hazardous substances in a matter of seconds. The use of such instruments in airports is potentially the most familiar. This usually involves an attendant using a gauze swipe on a traveler's luggage and then inserting this into the portable instrument for a readout, which compares the sample with known hazardous substances such as those used in bomb making or with drugs of abuse including marijuana, cocaine, or steroids (Fig. 1). With these capabilities, many other types of users can benefit from this technology, including emergency response teams, all branches of the military, law enforcement agencies, and health organizations. Furthermore, these instruments can be used by forensic, research, and clinical scientists in various fields of medicine. For example, portable matrix-assisted ionization mass spectrometry instruments have

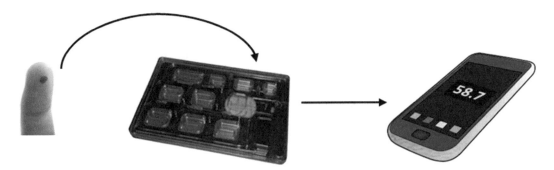

Fig. 1 Emerging technologies for point-of-care psychiatry

now been developed for rapid on-site analysis of small molecules, such as drugs, lipids, and peptides, which makes this an applicable technology for rapid assessment of biomarker patterns in bio-samples [57]. For the most direct applications in the study or monitoring of metabolic disorders, further developments are required for deployment of portable mass spectrometry devices to support testing of blood-based biomarkers.

4.2 Lab-on-a-Chip Cards

Lab-on-a-chip devices have emerged with rapid developments over the decade. These small instruments approximate the size of a credit card, and they have several advantages over existing bio-marker platforms, such as lower consumption of bio-samples, production of less waste, lower fabrication and reagent costs, faster system response times, and a higher degree of compactness due to integration and parallelization of functions [58, 59]. Several types of multiplex molecular measurements can be constructed on these devices, such as multiplex immunoassays for small molecules, peptides, and proteins [23]. In the case of patients suffering from metabolic disorders, this would translate to a shorter waiting period for results, and tests can be performed in point-of-care scenarios such as during a visit to the doctor's office. Recently a lab-on-a-chip card was developed for detection of human immunodeficiency virus through image analysis and counting of CD4-positive immune cells [60]. This card can deliver results within 20 min from application of a single drop of blood. Another lab-on-a-chip card consisting of a printed microchip and electrical sensing has been used to detect other virus types [61]. Likewise, a diagnostic immunofluorescence-based lab-on-a-chip device has been produced which can detect tuberculosis in saliva within 30 min, with an accuracy of 96% [62].

A number of biomarker tests for metabolic and immunological disorders have been developed on multiplex immunoassay platforms for analysis of serum or plasma, as described above [42–53]. Because of this, it is of high interest that multiplex immunoassays and other protein or peptide arrays have also been developed on lab-on-a-chip devices. One example of this is a card that can be used for diagnosis of prostate cancer with high accuracy using readouts based on either surface-enhanced photon scattering or voltage changes [63, 64]. The procedure involves application of a single drop of blood to a small sample well on the card, insertion of the card into a portable book-sized analyzer, and then reading of a final diagnostic "score" in less than 15 min. Potentially the main expected benefit of all of these lab-on-a-chip tests is that the more rapid analytical times will help to cut down on long waiting periods for results as seen typical of laboratory tests [25, 65]. These devices can also contain a universal serial bus, which would enable transmission of the results and other data to other devices such as computers, servers, and smartphones.

4.3 Smartphone Apps

As mentioned above, coupling the readouts of test strip or lab-on-a-chip devices to smartphone apps would be a major step forward in patient self-monitoring. This is also in-line with global companies such as Apple and Google which have both shown an increasing presence in this market through connecting diagnostic test scores with app readouts using smart devices. This would enable mobile communication and internet capabilities so that patient data can be formatted for user-friendly viewing and interpretation on handheld devices. This would also facilitate networked computing which may lead to the emergence of "bioprofiles" for each individual, consisting of molecular and physiometric data along with medical histories.

In the year 2017, the number of mobile phone users is expected to reach 4.77 billion [66], which equates to approximately 64% of the world population (as of Jan 1, 2017, this was 7.475 billion people [67]). The convergence of mobile phone technologies has resulted in a rapid evolution of the smartphone along with various apps and sensors, for wireless Internet accessibility and connectivity to other smart devices. Altogether, this has made the smartphone a versatile and user-friendly means of health and disease management [68, 69]. Clinical trials which have tested the use of smartphone apps have already shown improved health outcomes in a variety of medical areas, as well as in habit control [70]. This includes better attendance of patients at their appointments, more rapid diagnosis and treatment, and clinical elements such as improved blood sugar regulation, better medication compliance, and reduced stress levels. In terms of molecular tests, multiplex biomarker assays have now been linked to smartphone-based readouts, using an opto-mechanical interface for imaging reaction wells on a plate [71]. In this way, the readouts can be transmitted to a server or a database for analysis and interpretation and the results returned to the client in less than 1 min. This approach has already been used for detection of such diseases as mumps, measles, human immunodeficiency virus, influenza, and prostate cancer using smartphone camera optics for collection and transmission of the readouts [72–74]. It is not hard to imagine that similar approaches will begin to emerge at an increasing rate for improved healthcare of diseases such as obesity, cardiovascular disorders, and diabetes.

5 Conclusions

This chapter has described the recent movement toward improving healthcare and reducing the burden of individuals suffering from metabolic disorders through the initial study of these diseases using multiplex biomarker profiling and then translating these as blood tests on handheld microfluidic devices linked with smartphone apps for ease of use in the clinic or the doctor's office. This is currently the best approach for achieving a paradigm change in modern medicine that is more in-line with personalized medicine objectives. This will allow individuals to be treated according to their individual biomarker profiles rather than as one of many using a standard blockbuster drug that targets a specific disease. Furthermore, the increasing use of molecular blood tests on handheld devices which are capable of differentiating disease subtypes may be useful for rapid identification of those patients who are most likely to respond positively to a specific medication, therapy, or surgical procedure. This approach has already been used successfully in the treatment of certain types of cancer as well as for cancers at specific stages [75–77]. Similar approaches in the field of metabolic disorders could result in earlier, more appropriate, and, therefore, more effective treatments of patients, thereby leading to better overall outcomes.

References

1. http://www.who.int/nmh/publications/ncd_report_chapter1.pdf

2. http://www.who.int/mediacentre/factsheets/fs317/en/

3. http://www.who.int/mediacentre/factsheets/fs311/en/

4. http://www.mckinsey.com/industries/healthcare-systems-and-services/our-insights/how-the-world-could-better-fight-obesity

5. http://apps.who.int/iris/bitstream/10665/204871/1/9789241565257_eng.pdf

6. http://www3.weforum.org/docs/WEF_Harvard_HE_GlobalEconomicBurdenNonCommunicableDiseases_2011.pdf

7. Piot P, Caldwell A, Lamptey P, Nyrirenda M, Mehra S, Cahill K et al (2016) Addressing the growing burden of non-communicable disease by leveraging lessons from infectious disease management. J Glob Health 6:010304. https://doi.org/10.7189/jogh.06.010304

8. Carolan-Olah M, Duarte-Gardea M, Lechuga J (2015) A critical review: early life nutrition and prenatal programming for adult disease. J Clin Nurs 24:3716–3729

9. Lopes GA, Ribeiro VL, Barbisan LF, Marchesan Rodrigues MA (2016) Fetal developmental programing: insights from human studies and experimental models. J Matern Fetal Neonatal Med 23:1–7

10. Lee HS (2015) Impact of maternal diet on the epigenome during in utero life and the developmental programming of diseases in childhood and adulthood. Forum Nutr 7:9492–9507

11. Tarry-Adkins JL, Ozanne SE (2016) Nutrition in early life and age-associated diseases. Ageing Res Rev pii:S1568–1637(16)30179–9. https://doi.org/10.1016/j.arr.2016.08.003

12. Dixon JB (2009) Obesity and diabetes: the impact of bariatric surgery on type-2 diabetes. World J Surg 33:2014–2021

13. Khavandi K, Brownrigg J, Hankir M, Sood H, Younis N, Worth J (2014) Interrupting the natural history of diabetes mellitus: lifestyle, pharmacological and surgical strategies targeting disease progression. Curr Vasc Pharmacol 12:155–167

14. Allison BJ, Kaandorp JJ, Kane AD, Camm EJ, Lusby C, Cross CM et al (2016) Divergence of mechanistic pathways mediating cardiovascular aging and developmental programming of cardiovascular disease. FASEB J 30:1968–1975

15. Camm EJ, Martin-Gronert MS, Wright NL, Hansell JA, Ozanne SE, Giussani DA (2011) Prenatal hypoxia independent of undernutrition promotes molecular markers of insulin resistance in adult offspring. FASEB J 25:420–427

16. Martínez JA, Cordero P, Campión J, Milagro FI (2012) Interplay of early-life nutritional programming on obesity, inflammation and epigenetic outcomes. Proc Nutr Soc 71:276–283

17. Ortiz-Espejo M, Pérez-Navero JL, Olza J, Muñoz-Villanueva MC, Aguilera CM, Gil-Campos M (2013) Changes in plasma adipokines in prepubertal children with a history of extrauterine growth restriction. Nutrition 29:1321–1325

18. Tan HC, Roberts J, Catov J, Krishnamurthy R, Shypailo R, Bacha F (2015) Mother's

pre-pregnancy BMI is an important determinant of adverse cardiometabolic risk in childhood. Pediatr Diabetes 16:419–426

19. Dutton GR, Lewis CE (2015) The look AHEAD trial: implications for lifestyle intervention in type 2 diabetes mellitus. Prog Cardiovasc Dis 58:69–75

20. Paquot N (2015) From evidence-based medicine to personalized medicine: the example of type 2 diabetes. Rev Med Liege 70:299–305. [Article in French]

21. Jones PJ (2015) Inter-individual variability in response to plant sterol and stanol consumption. J AOAC Int 98:724–728

22. Tudos AJ, Besselink GJ, Schasfoort RB (2001) Trends in miniaturized total analysis systems for point-of-care testing in clinical chemistry. Lab Chip 1:83–95

23. Schumacher S, Nestler J, Otto T, Wegener M, Ehrentreich-Förster E, Michel D et al (2012) Highly-integrated lab-on-chip system for point-of-care multiparameter analysis. Lab Chip 12:464–473

24. Schumacher S, Ludecke C, Ehrentreich-Förster E, Bier FF (2013) Platform technologies for molecular diagnostics near the patient's bedside. Adv Biochem Eng Biotechnol 133:75–87

25. Peter H, Wienke J, Bier FF (2017) Lab-on-a-chip multiplex assays. Methods Mol Biol 1546:283–294

26. Guest FL, Guest PC, Martins-de-Souza D (2016) The emergence of point-of-care blood-based biomarker testing for psychiatric disorders: enabling personalized medicine. Biomark Med 10:431–443

27. Chan CP, Sum KW, Cheung KY, Glatz JF, Sanderson JE, Hempel A et al (2003) Development of a quantitative lateral-flow assay for rapid detection of fatty acid-binding protein. J Immunol Methods 279:91–100

28. Celenza A, Skinner K (2011) Comparison of emergency department point-of-care international normalised ratio (INR) testing with laboratory-based testing. Emerg Med J 28:136–140

29. van den Besselaar AM, Péquériaux NC, Ebben M, van der Feest J, de Jong K, Ganzeboom MB et al (2012) Point-of-care monitoring of vitamin K-antagonists: validation of Coagu-Chek XS test strips with international standard thromboplastin. J Clin Pathol 65:1031–1035

30. Vegt J (2017) Development of a user-friendly app for assisting anticoagulation treatment. Methods Mol Biol 1546:303–308

31. Burgess-Cassler A, Barriga Angulo G, Wade SE, Castillo Torres P, Schramm W (1996) A field test for the detection of antibodies to human immunodeficiency virus types 1 and 2 in serum or plasma. Clin Diagn Lab Immunol 3:480–482

32. Jelinek T, Grobusch MP, Schwenke S, Steidl S, von Sonnenburg F, Nothdurft HD et al (1999) Sensitivity and specificity of dipstick tests for rapid diagnosis of malaria in nonimmune travelers. J Clin Microbiol 37:721–723

33. Lee JH, Seo HS, Kwon JH, Kim HT, Kwon KC, Sim SJ et al (2015) Multiplex diagnosis of viral infectious diseases (AIDS, hepatitis C, and hepatitis A) based on point of care lateral flow assay using engineered proteinticles. Biosens Bioelectron 69:213–225

34. Wallace JA, Blum K (1982) An evaluation of the TRI Dipstick test for the detection of drugs of abuse in urine. Subst Alcohol Actions Misuse 3:129–132

35. Gnoth C, Johnson S (2014) Strips of hope: accuracy of home pregnancy tests and new developments. Geburtshilfe Frauenheilkd 74:661–669

36. Johnson S, Cushion M, Bond S, Godbert S, Pike J (2015) Comparison of analytical sensitivity and women's interpretation of home pregnancy tests. Clin Chem Lab Med 53:391–402

37. Hendey GW, Schwab T, Soliz T (1997) Urine ketone dip test as a screen for ketonemia in diabetic ketoacidosis and ketosis in the emergency department. Ann Emerg Med 29:735–738

38. Lee WC, Smith E, Chubb B, Wolden ML (2014) Frequency of blood glucose testing among insulin-treated diabetes mellitus patients in the United Kingdom. J Med Econ 17:167–175

39. http://www.cobas.com/home/product/urinalysis-testing/combur-test-strip.html

40. Schwarz E (2017) Identification and clinical translation of biomarker signatures: statistical considerations. Methods Mol Biol 1546:103–114

41. Chen J, Schwarz E (2017) Opportunities and challenges of multiplex assays: a machine learning perspective. Methods Mol Biol 1546:115–122

42. Liu MY, Xydakis AM, Hoogeveen RC, Jones PH, Smith EO, Nelson KW et al (2005) Multiplexed analysis of biomarkers related to obesity and the metabolic syndrome in human plasma, using the Luminex-100 system. Clin Chem 51:1102–1109

43. Purohit S, Sharma A, She JX (2015) Luminex and other multiplex high throughput technologies for the identification of, and host response to, environmental triggers of type

1 diabetes. Biomed Res Int 2015:326918. https://doi.org/10.1155/2015/326918

44. Op De Beéck K, Vermeersch P, Verschueren P, Westhovens R, Mariën G, Blockmans D et al (2012) Antinuclear antibody detection by automated multiplex immunoassay in untreated patients at the time of diagnosis. Autoimmun Rev 12:137–143

45. Nolen BM, Lokshin AE (2013) Biomarker testing for ovarian cancer: clinical utility of multiplex assays. Mol Diagn Ther 17:139–146

46. Kang JH, Vanderstichele H, Trojanowski JQ, Shaw LM (2012) Simultaneous analysis of cerebrospinal fluid biomarkers using microsphere-based xMAP multiplex technology for early detection of Alzheimer's disease. Methods 56:484–934

47. Schaffer C, Sarad N, DeCrumpe A, Goswami D, Herrmann S, Morales J et al (2015) Biomarkers in the diagnosis and prognosis of Alzheimer's disease. J Lab Autom 20:589–600

48. Lue LF, Schmitz CT, Snyder NL, Chen K, Walker DG, Davis KJ et al (2016) Converging mediators from immune and trophic pathways to identify Parkinson disease dementia. Neurol Neuroimmunol Neuroinflamm 3:e193. https://doi.org/10.1212/NXI.0000000000 000193

49. Schwarz E, Guest PC, Rahmoune H, Harris LW, Wang L, Leweke FM et al (2012) Identification of a biological signature for schizophrenia in serum. Mol Psychiatry 17:494–502

50. Papakostas GI, Shelton RC, Kinrys G, Henry ME, Bakow BR, Lipkin SH et al (2013) Assessment of a multi-assay, serum-based biological diagnostic test for major depressive disorder: a pilot and replication study. Mol Psychiatry 18:332–333

51. Perkins DO, Jeffries CD, Addington J, Bearden CE, Cadenhead KS, Cannon TD et al (2014) Towards a psychosis risk blood diagnostic for persons experiencing high-risk symptoms: preliminary results from the NAPLS project. Schizophr Bull 41:419–428

52. Chan MK, Krebs MO, Cox D, Guest PC, Yolken RH, Rahmoune H et al (2015) Development of a blood-based molecular biomarker test for identification of schizophrenia before disease onset. Transl Psychiatry 5:e601. https://doi.org/10.1038/tp.2015.91

53. Stelzhammer V, Haenisch F, Chan MK, Cooper JD, Steiner J, Steeb H et al (2014) Proteomic changes in serum of first onset, antidepressant drug-naïve major depression patients. Int J Neuropsychopharmacol 17 (10):1599–1608

54. Leary PE, Dobson GS, Reffner JA (2016) Development and applications of portable gas chromatography-mass spectrometry for emergency responders, the military, and law-enforcement organizations. Appl Spectrosc 70:888–896

55. Schott M, Wehrenfennig C, Gasch T, Düring RA, Vilcinskas A (2013) A portable gas chromatograph with simultaneous detection by mass spectrometry and electroantennography for the highly sensitive in situ measurement of volatiles. Anal Bioanal Chem 405:7457–7467

56. Rollman CM, Moini M (2016) Ultrafast capillary electrophoresis/mass spectrometry of controlled substances with optical isomer separation in about a minute. Rapid Commun Mass Spectrom 30:2070–2076

57. Devereaux ZJ, Reynolds CA, Foley CD, Fischer JL, DeLeeuw JL, Wager-Miller J et al (2016) Matrix-assisted ionization (MAI) on a portable mass spectrometer: analysis directly from biological and synthetic materials. Anal Chem 88(22):10831–10836

58. Pawell RS, Inglis DW, Barber TJ, Taylor RA (2013) Manufacturing and wetting low-cost microfluidic cell separation devices. Biomicrofluidics 7:056501. https://doi.org/10.1063/ 1.4821315

59. Yager P, Edwards T, Fu E, Helton K, Nelson K, Tam MR et al (2006) Microfluidic diagnostic technologies for global public health. Nature 442:412–418

60. Ermantraut E, Bickel R, Schulz T, Ullrich T Tuchscheerer J (2011) Device and method for the detection of particles. USPTO Patent US8040494. Clondiag GmbH

61. Shafiee H, Kanakasabapathy MK, Juillard F, Keser M, Sadasivam M, Yuksekkaya M (2015) Printed flexible plastic microchip for viral load measurement through quantitative detection of viruses in plasma and saliva. Sci Rep 5:9919. https://doi.org/10.1038/srep09919

62. Kim JH, Yeo WH, Shu Z, Soelberg SD, Inoue S, Kalyanasundaram D (2012) Immunosensor towards low-cost, rapid diagnosis of tuberculosis. Lab Chip 12:1437–1440

63. Gao R, Cheng Z, deMello AJ, Choo J (2016) Wash-free magnetic immunoassay of the PSA cancer marker using SERS and droplet microfluidics. Lab Chip 16:1022–1029

64. Parra-Cabrera C, Samitier J, Homs-Corbera A (2016) Multiple biomarkers biosensor with just-in-time functionalization: application to prostate cancer detection. Biosens Bioelectron 77:1192–1200

65. Guest PC (2017) Multiplex biomarker approaches to enable point-of-care testing and

personalized medicine. Methods Mol Biol 1546:311–315

66. www.statistica.com/statistics/274774/fore cast-of-mobile-phone-users-worldwide/

67. www.worldometers.info/world-population/

68. Klasnja P, Pratt W (2012) Healthcare in the pocket: mapping the space of mobile-phone health interventions. J Biomed Inform 45:184–198

69. Ventola CL (2014) Mobile devices and apps for health care professionals: uses and benefits. P T 39:356–364

70. Krishna S, Boren SA, Balas EA (2009) Health-care via cell phones: a systematic review. Telemed J E Health 15:231–240

71. Liao SC, Peng J, Mauk MG, Awasthi S, Song J, Friedman H (2016) Smart cup: a minimally-instrumented, smartphone-based point-of-care molecular diagnostic device. Sens Actuators B Chem 229:232–238

72. Yeo SJ, Choi K, Cuc BT, Hong NN, Bao DT, Ngoc NM et al (2016) Smartphone-based fluorescent diagnostic system for highly pathogenic H5N1 viruses. Theranostics 6:231–242

73. Guo T, Patnaik R, Kuhlmann K, Rai AJ, Sia SK (2015) Smartphone dongle for simultaneous measurement of hemoglobin concentration and detection of HIV antibodies. Lab Chip 15:3514–3520

74. Barbosa AI, Gehlot P, Sidapra K, Edwards AD, Reis NM (2015) Portable smartphone quantitation of prostate specific antigen (PSA) in a fluoropolymer microfluidic device. Biosens Bioelectron 70:5–14

75. Barton S, Swanton C (2011) Recent developments in treatment stratification for metastatic breast cancer. Drugs 71:2099–2113

76. Kanda M, Kodera Y (2015) Recent advances in the molecular diagnostics of gastric cancer. World J Gastroenterol 21:9838–9852

77. Sternberg IA, Vela I, Scardino PT (2016) Molecular profiles of prostate cancer: to treat or not to treat. Annu Rev Med 67:119–135

Chapter 7

Pregnancy and Lactation: A Window of Opportunity to Improve Individual Health

Guadalupe L. Rodríguez-González, Diana C. Castro-Rodríguez, and Elena Zambrano

Abstract

Human and animal studies indicate that obesity during pregnancy adversely impacts both maternal health and offspring phenotype predisposing them to chronic diseases later in life including obesity, dyslipidemia, type 2 diabetes mellitus, and hypertension. Effective interventions during human pregnancy and/or lactation are needed to improve both maternal and offspring health. This review addresses the relationship between adverse perinatal insults and its negative impact on offspring development and presents some maternal intervention studies such as diet modification, probiotic consumption, or maternal exercise, to prevent or alleviate the negative outcomes in both the mother and her child.

Key words Pregnancy, Maternal health, Offspring phenotype, Chronic disease, Intervention studies

1 Introduction

Throughout the entire life, environmental factors play an important role in the health of individuals. However, the surrounding environment has a bigger impact during embryonic and fetal life. Human epidemiological [1] and experimental animal studies [2] have suggested that perinatal insults such as placental insufficiency [3, 4], glucocorticoid exposure [5, 6], nutritional deficits/excess [7–9], stress [10, 11], as well as the maternal gut microbiome [12–14] can alter the developmental trajectory of the fetus/offspring leading to long-term unfavorable outcomes that often end in chronic noncommunicable diseases (Fig. 1). This relationship between maternal environment conditions and offspring health brings the concept of "developmental programming" or more recently called "Developmental Origins of Health and Diseases" (DOHaD) that is defined as the response to a specific challenge to the mammalian organism during a critical developmental time window that alters the trajectory of development with resulting

Paul C. Guest (ed.), *Investigations of Early Nutrition Effects on Long-Term Health: Methods and Applications*, Methods in Molecular Biology, vol. 1735, https://doi.org/10.1007/978-1-4939-7614-0_7, © Springer Science+Business Media, LLC 2018

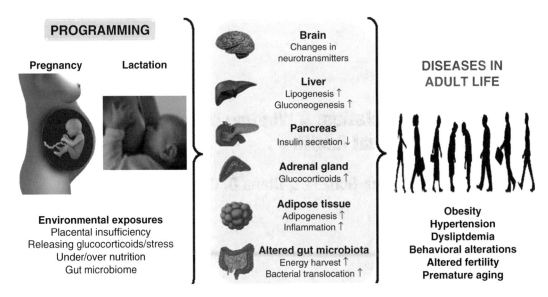

Fig. 1 Adverse perinatal insults and the programming of diseases in adult life

effects on health that persist throughout life [15]. The term "early origins of health and disease" has been used to encompass the first 1000 days of life, which span in human the period from conception to 2 years after birth [16].

A variety of adult diseases can be programmed by different environmental factors, for example, gestational under- and overnutrition [17, 18] leads to the development of obesity and metabolic alterations in the offspring. These similar phenotypic outcomes suggest that complex mechanisms are involved in the DOHaD concept including (a) changes in maternal metabolic hormones such as insulin and leptin [19, 20] and steroid hormones such as androgens, estrogens, and corticosterone that play a key role in cell differentiation, proliferation, and apoptosis [15, 21–24]; (b) impaired placental transfer of nutrients from mother to fetus [25]; (c) permanent organ or tissue structural changes like reduced nephron number in the kidney, remodeling of cardiac cells in the heart, and altered neuronal projections due to the restriction of nutrient supply during a critical period of organ development [26]; (d) increase in oxidative stress that causes damage in cell components (proteins, lipids, and DNA) [27, 28]; (e) accelerated cellular aging [26]; and (f) epigenetic alterations such as DNA methylation, histone modification, chromatin packing, and microRNA expression that can alter gene expression [29, 30] (Fig. 2).

As mentioned above, nutrition of women during pregnancy and lactation is directly connected with the growth and health of her newborn and can have a profound impact on the illness and health of both the mother and the child. In 2008, the Lancet identified the need to focus on the crucial period of pregnancy

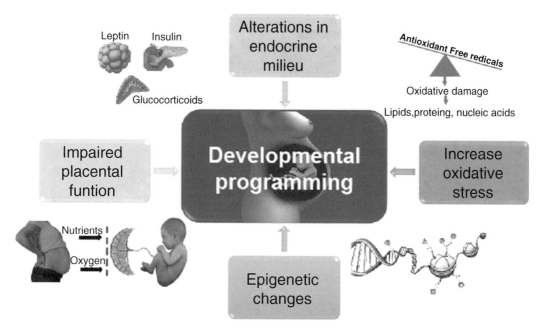

Fig. 2 Mechanism involved in developmental programming

and the first 2 years of life—the period beginning after conception to a child's second birthday—as a window of opportunity to improve individuals' health, as well as an effective intervention period to reduce the effects of under- or overnutrition. The series also called for public health strategies to be implemented by countries in order to achieve healthy nutrition and development [31–35]. This review aim at presenting some epidemiological and animal studies that proposed maternal interventions to prevent or ameliorate the negative outcomes because of adverse perinatal insults.

2   Prevention Strategies

Epidemiological studies can evaluate intrauterine exposures and have public health relevance although they cannot per se directly establish a causal relationship between a specific insult and programmed phenotypes because human studies are long length and difficult to control (mother versus child environment) and because of the high interindividual variability, limited tissue access, and ethical considerations. In addition, epidemiological studies related to the benefits of maternal interventions are only associative, because it is only implied that maternal health improvement will be beneficial for the child. Therefore, the use of animal models has several benefits such as lower costs and length of gestation, they are much more controllable than human clinical interventions, they

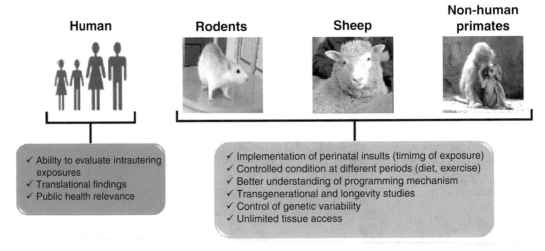

Fig. 3 Advantages of the experimental models used in developmental programming

have a greater depth of mechanistic interrogation as multiple testing, tissues can be obtained at different periods, transgenerational and longevity studies can be performed, and results can be obtained much more quickly to guide management in human pregnancy. Reproducibility and independent confirmation, the indispensable requirements of scientific certainty, are also generally easier to achieve in animal studies [1, 26, 36, 37]. The most commonly studied species of developmental programming include rats, sheep, and nonhuman primates (Fig. 3). In the following sections, epidemiological and animal studies related to different maternal intervention strategies such as nutrition, exercise, dietary supplements, or probiotic consumption will be described.

2.1 Nutritional Interventions

There are many studies which demonstrate a link between maternal nutritional insults and detrimental effects on the fetus [38, 39]. Both maternal under- and overnutrition have been extensively described with additional factors such as the timing of the insult, the gender of the offspring, and, importantly, the severity of the nutritional change leading to complex outcomes.

2.1.1 Epidemiological Studies

During pregnancy, the nutritional demand increases, and women usually respond to this requirement by incrementing their food intake. For healthy pregnant women, the acceptable macronutrient consumption of total energy intake is estimated to be 45–65% for carbohydrate, 10–35% for protein, and 20–35% for fat [40]; however, cultural beliefs, such as "eating for two," causes a higher caloric intake [41, 42]. Prepregnancy body mass index (BMI) is a basis for determining the optimal gestational weight gain (GWG) range. Based on BMI before conception, normal women (BMI, 18.5–24.9) are recommended to gain weight between 11.4 and

15.9 kg during pregnancy, overweight women (BMI, 25.0–29.9) between 6.8 and 11.4 kg, and obese women (BMI, ≥30) between 5.0 and 9.0 kg [43]. Obesity in pregnancy may lead to increased risks of high birth weight and maternal postpartum weight retention [44] and health problems like risk of cardiovascular disease [45], type 2 diabetes [46], and obesity [47]. It has been estimated that the risk of early childhood obesity increases by a factor of 1.08 per kilogram of maternal weight gained during pregnancy [48]. Pregnancy is often considered the optimal time to intervene for issues related to eating habits and physical activity to prevent excessive weight gain. GWG is a potentially modifiable risk factor for a number of adverse maternal and neonatal outcomes.

Several studies have demonstrated that dietary intervention during pregnancy results in a significant reduction in preterm births and a trend toward a reduction in the incidence of gestational diabetes [49] and a modest reduction in GWG in women of all BMIs [50]. Furthermore, lifestyle interventions can reduce or prevent weight gain for obese pregnant woman [51–53]. However, these reviews have showed some inconsistent results, which have made difficult to identify effective strategies to prevent excessive GWG among normal weight and obese women. Birth and neonatal outcomes are less certain [49, 50, 54].

One of the main environmental factors regulating fetal growth is maternal blood glucose concentrations that are influenced by maternal glucose tolerance and the amount and quality of carbohydrates consumed [55, 56]. Epidemiological studies indicate that the risks of metabolic and coronary heart disease are strongly related to foods with high-glycemic index because they are rapidly digested, absorbed, and transformed into glucose which in turn causes an accelerated and transient surge in blood glucose and insulin, earlier return of hunger sensation, and excessive caloric intake [57]. In pregnant women, the determination of glycated hemoglobin (HbA1c) levels is important in screening, diagnosing, and assessing the prognoses of the gestational abnormal glucose metabolism [58]. Ideally in pregnant women, HbA1c should be maintained below 5% during their first trimesters and below 6% during third trimester. A recent study reported that gestational diabetes mellitus causes high lactate and HbA1c levels in the mother, as well as increased fetal blood pressure and birth weight [59].

For women with type 1 diabetes, a lower gestational weight gain is generally recommended. The World Health Organization and Food and Agriculture Organization recommend low glycemic index foods to prevent common chronic diseases like obesity and type 2 diabetes, because these decrease blood glucose and insulin, promote greater fat oxidation, decrease lipogenesis, and increase satiety [57]. Therefore, a low glycemic index diet is considered safe and has shown positive effects on glycemic control and pregnancy outcomes for both healthy women, as well as those with type 2 diabetes and gestational diabetes [60]. In nonpregnant women with type 1 and type 2 diabetes, a low glycemic index diet has positive effects on postprandial blood glucose levels and HbA1c [61], while in early pregnancy, it has a positive influence on maternal glycemic index, food and nutrient intake, GWG, and infant birth weight [62]. In addition, neonates whose mothers were on a low glycemic index diet during pregnancy had lower thigh circumference in comparison with the control diet [63]. In women with gestational diabetes, metformin improves insulin sensitivity and leads to less weight gain. In pregnant women without diabetes but with a BMI > 35, metformin intake reduced maternal gestational weight gain. No adverse neonatal outcomes were reported [64].

In addition to a healthy diet, supplements during pregnancy are recommended. For example, supplementing pregnant and lactating woman with long-chain polyunsaturated fatty acids from week 18 of pregnancy until 3 months after delivery has a beneficial impact on child body composition [65]. In a study of lactating woman, it was found that docosahexaenoic acid (DHA) intake increased DHA concentrations in breast milk. This finding suggests that mothers with inadequate dietary intake of DHA should change their dietary habits to consume a diet rich in DHA or take sufficient DHA supplements to meet the average nutritional needs of infants [66]. In addition, it was reported that dietary supplementation of lactating women with n-3 long-chain polyunsaturated fatty acids increased DHA in maternal plasma, peripheral blood mononuclear cells and milk, and can modulate the concentration of cytokines in the aqueous phase of human milk and the production of cytokines by human milk cells [67]. Despite the beneficial effects of supplements, much caution and research are still needed before supplement interventions can be recommended in complicated pregnancies (Fig. 4).

2.1.2 Animal Studies

Reports of experimental interventions in animal models in the setting of obesity are scarce. In one study female rats aged 3 weeks were either fed with cafeteria diet for 34 weeks or a cafeteria diet for 26 weeks followed by a chow diet for 8 weeks. The cafeteria diet resulted in higher weight gain, increased visceral adipose tissue and liver weight, and insulin resistance. Dietary changes reduced energy intake, body weight, visceral adipose tissue and liver weight, and improved insulin sensitivity [68]. Therefore, it is necessary to develop interventions before pregnancy to prevent the adverse outcomes due to maternal obesity in both the mother and the offspring. In the pregnant rat, we have conducted studies on maternal dietary intervention before and during pregnancy and have

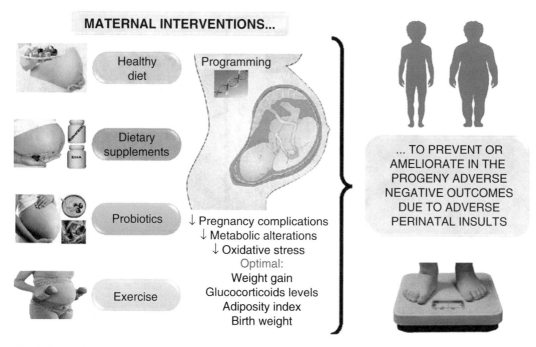

Fig. 4 Types of maternal interventions to prevent the negative effects of programming

demonstrated that maternal dietary intervention 1 month before mating and during pregnancy and lactation partially prevents some of the negative biochemical and metabolic outcomes observed in offspring from obese mothers [69]. It is of interest that it was not necessary to return maternal weight to control levels for this benefit. The study provided evidence that unwanted developmental programming effects on offspring that result from maternal obesity are at least partially prevented by dietary intervention prior to pregnancy [36, 69]. We have also evaluated the effects of maternal high-fat diet and dietary intervention at different periconceptional periods on male offspring anxiety-related behavior, exploration, learning, and motivation. Dietary intervention prevented the increased exploratory behavior displayed by offspring of obese mothers in the open field test. Additionally, dietary intervention restored normal levels of motivation [39] (Fig. 5). In pregnant ewes fed with obesogenic diet, reduction of dietary intake (from 150% to 100% of the control diet) beginning on day 28 of gestation was early enough to at least partially prevent some of the negative outcomes in the fetus such as those on fetal growth, adiposity, and glucose/insulin dynamics [70].

Resveratrol is a stilbenoid (natural phenol) naturally produced by the skin of red grapes and has gained significant interest [71]. Many studies have shown that resveratrol possesses cardiovascular-protective, antiplatelet, antioxidant, anti-inflammatory, blood glucose-lowering, and anticancer activities,

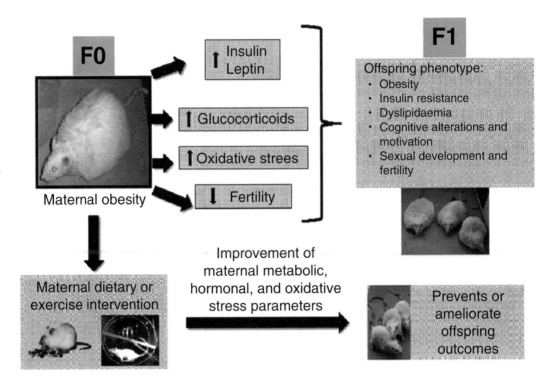

Fig. 5 Programming mechanism in a rodent model: maternal obesity and dietary or exercise interventions effects

stimulates vasodilation, and protects against some neurodegenerative diseases and obesity [72]. In our group, we showed that maternal protein restriction during rat pregnancy leads to maternal metabolic dysfunction, increased corticosterone and oxidative stress, and reduced antioxidant enzyme activity, while in the placenta and liver, it caused increased oxidative stress biomarkers and antioxidant enzyme activity. In adult life, both male and female offspring from the restricted group showed higher leptin, insulin, and corticosterone levels, whilst triglycerides were increased only in males and fat only in females. Resveratrol intervention was found to decrease maternal leptin levels and improved maternal, fetal, and placental oxidative stress markers [28]. Furthermore, male offspring of female mice fed a high-fat diet with 0.2% (w/w) resveratrol supplementation during pregnancy and lactation showed increased energy expenditure and insulin sensitivity which was correlated with enhanced brown adipose function and white adipose tissue browning, despite being fed a high-fat diet for 11 weeks [73]. A recent study assessed the role of resveratrol supplementation during pregnancy in nonhuman primates. Briefly, mothers were fed a Western-style diet consisting of 36% fat with or without resveratrol supplementation. Resveratrol improved placental inflammatory markers, maternal and fetal hepatic triglyceride accumulation, uterine blood flow, and insulin sensitivity. Resveratrol was detected in maternal plasma, demonstrating an ability to cross the placental barrier to exert effects on the fetus [74].

Conjugated linoleic acid is a mixture of isomers of linoleic acid (C18:2 n-6) and has been considered to have many potential health benefits [75]. In a study conducted in rats, the authors examined the effects of a high-fat diet consumption during pregnancy and lactation on metabolic and inflammatory profiles and whether maternal supplementation with conjugated linoleic acid could have beneficial effects on mothers and offspring. In the mothers, conjugated linoleic acid did not reduce body weight although plasma concentrations of pro-inflammatory cytokines interleukin-1β and tumor necrosis factor-α were reduced. In male and female fetuses, the displayed catch-up growth and impaired insulin sensitivity were reversed by conjugated linoleic acid maternal intervention [76]. In addition, maternal supplementation with conjugated linoleic acid reversed high-fat diet skeletal muscle atrophy and inflammation in adult male rat offspring [77].

Leptin has received a significant interest as a potential programming factor; alterations in the profile of leptin in early life are associated with altered susceptibility to obesity and metabolic disorders in adulthood. Maintenance of a critical leptin level during early development facilitates the normal maturation of tissues and signaling pathways involved in metabolic homeostasis [78]. In a murine model, pregnant rats were assigned to one of two nutritional groups: (1) undernutrition (30% of ad libitum) of a standard diet throughout gestation or (2) standard diet ad libitum throughout gestation. The authors reported that leptin treatment from postnatal day 3–13 resulted in a transient slowing of neonatal weight gain and normalized caloric intake, locomotor activity, body weight, fat mass, and fasting plasma glucose, insulin, and leptin concentrations in programmed offspring in adult life in contrast to saline-treated offspring of undernourished mothers who developed all of these features on a high-fat diet [79]. In another rat study, leptin supplementation throughout lactation partially reverted most of the developmental effects on hypothalamic structure and function caused by moderate maternal caloric restriction during gestation [80].

Experimental studies in rodents have shown that females are more susceptible to exhibiting fat expansion and metabolic disease compared with males in several models of fetal programming. A study tested the hypothesis that female rat pups exposed to maternal separation (3 h/day, postnatal days 2–14) display an exacerbated response to diet-induced obesity. A group of female rats were treated with metyrapone (inhibitor of glucocorticoid synthesis) 30 min before the daily separation and weaned onto a high-fat diet. Metyrapone treatment significantly attenuated diet-induced obesity risk factors, including elevated adiposity, hyperleptinemia, and glucose intolerance. The authors concluded that pharmacological and/or behavioral inflection of the stress levels is a potential therapeutic approach for prevention of early life stress-enhanced

Table 1
Benefits of dietary intervention

Experimental model	Study design	Intervention period	Beneficial outcomes	Reference
Human	Nondiabetic pregnant woman with a BMI > 35. Metformin intake	Pregnancy	Reduce maternal weight gain	[64]
Human	N-3 very long-chain polyunsaturated fatty acid supplementation	Pregnancy and lactation	Has beneficial impact on children's body composition	[65]
Human	Docosahexaenoic acid	Lactation	Modulate the concentration of cytokines in the aqueous phase of human milk and the production of cytokines by human milk cells	[67]
Mice	High-fat diet supplemented with resveratrol	Pregnancy and lactation	Male offspring had increased energy expenditure and insulin sensitivity and enhanced brown adipose function	[73]
Rat	High-fat diet with conjugated linoleic acid	Pregnancy and lactation	Maternal plasma concentrations of pro-inflammatory cytokines were reduced. In male and female fetuses, the displayed catch-up growth and impaired insulin sensitivity were reversed	[76]
Rat	Maternal protein restriction and oral resveratrol administration	Pregnancy	Resveratrol intervention, decreased maternal leptin, and improved maternal, fetal, and placental oxidative stress markers	[28]
Rat	Undernourishment during pregnancy and offspring leptin intervention	Neonatal life	Normalized in the offspring the caloric intake, locomotor activity, body weight, fat mass, and fasting plasma glucose, insulin, and leptin concentrations	[79]
Rat	Maternal caloric restriction in pregnancy and leptin intervention	Lactation	Ameliorates in the offspring the developmental effects on hypothalamic structure and function	[80]

(continued)

Table 1
(continued)

Experimental model	Study				Reference
	Study design	Intervention period	Beneficial outcomes		
Rat	Maternal separation. Metyrapone administration	Lactation	Metyrapone treatment attenuated diet-induced obesity risk factors, like elevated adiposity, hyperleptinemia, and glucose intolerance		[81]
Ewe	Obesogenic diet in pregnancy and reduction on dietary intake	Gestation	Prevent alterations in fetal growth, adiposity, and glucose/insulin dynamics		[70]
Nonhuman primates	Western-style diet and resveratrol supplementation	Pregnancy	Improve placental inflammatory markers, maternal and fetal hepatic triglyceride accumulation, uterine blood flow, and insulin sensitivity		[74]

obesity and metabolic disease [81]. The evidence related to maternal dietary interventions is summarized in Table 1.

2.2 Probiotic Intervention

In pregnancy and lactation, maternal transfer of microorganisms is possible; thus the mother's diet and microbiota can influence microbiota colonization in the offspring [82]. Probiotics are live microorganisms that administered in adequate amounts confer a health benefit on the host [83]. Probiotics consist of individual or multiple live bacterial species (e.g., *Lactobacillus* and *Bifidobacterium*) that during intake contribute to decrease gastrointestinal disorders [84]. Studies in humans and animals have demonstrated the beneficial effects of probiotics. During pregnancy, daily consumption of probiotics may reduce the risk of preeclampsia [85], maintain serum insulin levels [86], and reduce the frequency of gestational diabetes mellitus [87] during pregnancy. Probiotics are now widely studied for their beneficial effects in treatment of many prevailing diseases. The effects are dependent on the probiotic strain, the host, and the specific host characteristics, such as age and baseline nutritional status (Fig. 6).

2.2.1 Epidemiological Studies

Microbial colonization in the fetus has been thought to begin during birth and then develop under the influence of breastfeeding and skin-to-skin contact with the mother [88–90]. Gut microbiota composition has been suggested to be an instrumental component in the microbial, metabolic, and immunological programming of the child [91, 92]. Modification of gut microbiota by probiotics

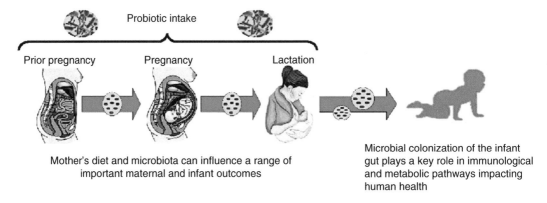

Fig. 6 Effects of maternal probiotic consumption during pregnancy

during pregnancy and lactation has the potential to reduce the risk of immune-mediated and metabolic disease in both the child and the mother [93–97]. Most studies have provided evidence of metabolic activity of gut microbiota that facilitates extraction of calories from ingested dietary substances and their storage in host adipose tissue for later use, developing allergy and obesity later in life [91, 98–100]. For example, supplementation in early pregnancy with *Lactobacillus rhamnosus* GG (ATCC53103)/*Bifidobacterium lactis* controls maternal body weight and body composition during and after pregnancy [101]. Intervention with a multispecies probiotic mixture (VSL#3 capsule) in pregnant woman with gestational diabetes reduced maternal insulin serum levels, HOMA, and there was a significant decrease in inflammatory markers [102]. In addition, early pregnancy probiotic supplementation with *Lactobacillus rhamnosus* HN001 reduced gestational diabetes mellitus prevalence, particularly among older women [103]. *Lactobacillus acidophilus/L. casei/Bifidobacterium bifidum* intervention in pregnant women with gestational diabetes mellitus had beneficial effects on glycemic control, triglycerides, and VLDL cholesterol concentrations [104]. In the Norwegian Mother and Child Cohort, the intake of milk-based products containing probiotic lactobacilli (*Lactobacillus acidophilus/Bifidobacterium lactis/Lactobacillus rhamnosus* GG) was associated with lower risk of preeclampsia in primiparous women [85]. In addition, maternal probiotic supplementation during pregnancy and breastfeeding reduced the risk of developing eczema in allergic mothers positive for skin prick test [93].

In a study, 256 women were randomized at their first trimester of pregnancy into a control and a dietary intervention group. The intervention group received dietary probiotics (*Lactobacillus rhamnosus* GG and *Bifidobacterium lactis* Bb12; diet/probiotics) or placebo (diet/placebo). Probiotic intervention reduced the frequency of gestational diabetes mellitus. The safety of this approach was attested by normal duration of pregnancies with no adverse events in mothers or children, and dietary intervention diminished the risk of larger birth size in affected cases. These results show that probiotic supplemented perinatal dietary advice could be a safe and cost-effective tool in addressing the metabolic disease epidemic [87]. Another study with humans examined whether supplementation of probiotics with dietary advice affects glucose metabolism in normoglycemic pregnant women. At the first trimester of pregnancy, 256 women were randomized to receive nutrition counseling to modify dietary intake. The dietary intervention group was further randomized to receive probiotics (*Lactobacillus rhamnosus* GG and *Bifidobacterium lactis* Bb12; diet/probiotics) or placebo (diet/placebo) in a double-blind manner, while the control group received placebo (control/placebo). Blood glucose concentrations were lowest in the diet/probiotic group during pregnancy and over the 12-month postpartum period. Better glucose tolerance in the diet/probiotic group was confirmed by a reduced risk of elevated glucose concentration compared with the control/placebo group, as well as by the lowest insulin concentration and homeostasis model assessment during the last trimester of pregnancy. The effects observed extended over the 12-month postpartum period. These findings demonstrated improved blood glucose control with probiotics in a normoglycemic population and therapeutic management of glucose disorders [97]. The composition of gut microbiota may thus be taken to represent a novel contributor to obesity above the traditional and well-known risk factors, excessive energy intake, and sedentary behavior. On this basis, the impact of perinatal probiotic intervention on childhood growth patterns and the development of overweight during a 10-year follow-up were evaluated. Altogether 159 women were randomized and double blinded to receive probiotics (*Lactobacillus rhamnosus* GG, ATCC 53103) or placebo 4 weeks before expected delivery and the intervention extending for 6 months postnatally. The perinatal probiotic intervention appeared to moderate the initial phase of excessive weight gain. The effect of intervention was also shown as a tendency to reduce the birth weight-adjusted mean BMI. The results of this study demonstrated that manipulation of early gut microbiota with probiotics may modify the growth pattern of the child by restraining excessive weight gain [105].

2.2.2 Animal Studies

Animal models are important to understand mechanisms involved in fetal programming and to propose prevention strategies before they can be used in humans. In mice, it has been shown that maternal exposure to *L. paracasei* in the last week of gestation and during lactation prevented the development of airway inflammation in offspring, as demonstrated by attenuation of eosinophil influx in the lungs [106]. Another study in animals analyzed the effects of a perinatal *Lactobacillus rhamnosus* GG (LGG) supplementation on the development of allergic disorders in the offspring. Female BALB/c mice received intragastric LGG every day before conception, during pregnancy and lactation (perinatal supplementation group), or before conception and during pregnancy only (prenatal supplementation group), and cytokine expression of placental tissues was examined. Offspring of LGG-supplemented and sham-exposed mothers were sensitized to ovalbumin (OVA), followed by aerosol allergen challenges. Development of experimental asthma was assessed by bronchoalveolar lavage analysis, lung histology, and lung function measurement, and cytokine production of splenic mononuclear cells was analyzed following in vitro stimulation. This study suggests that LGG exerted beneficial effects on the development of experimental allergic asthma, when applied in an early phase of life. Immunological effects are, at least in part, mediated via the placenta, probably by induction of pro-inflammatory cell signals [107]. In rats, an adapted diet containing long-chain polyunsaturated fatty acids, prebiotics, and probiotics was found to revert the negative imprinting of neonatal stress on both intestinal barrier function and growth [108]. Another study in rats evaluated whether maternal probiotic intervention influences the alterations in the brain-immune-gut axis induced by neonatal maternal separation and/or restraint stress in adulthood. Briefly, female rats had free access to drinking water supplemented with *Bifidobacterium animalis* subsp. *lactis* BB-12® and *Propionibacterium jensenii* 702 from 10 days before conception until weaning. Maternal probiotic intervention induced activation of neonatal stress pathways and an imbalance in gut microflora. However, it improved the immune environment of stressed animals and partly protected against stress-induced disturbances in adult gut microflora [109]. Probiotic strains of *Lactobacillus acidophilus* and *Bifidobacterium lactis* were fed to pregnant mice, rats, and sows for at least 7 days prior to vaginal delivery. The probiotic bacteria were detected in the feces and vagina of maternal mice, rats, and sows after, but not before, administration. Probiotic bacteria administered to mothers during late gestation are transferred to infants born vaginally and influence the assemblages of gastrointestinal tract bacteria. However, colonization of the neonatal gastrointestinal tract and persistence past weaning does not occur in all offspring and varies among probiotics and animal models [110]. Evidence related to maternal probiotic interventions are summarized in Table 2.

2.3 Exercise Interventions

Physical activity is defined as any voluntary movement produced by skeletal muscles that lead to energy expenditure. However, exercise is defined as a physical activity that is planned, structured, and repetitive whose objective is the improvement and/or maintenance of physical fitness [111].

Table 2
Beneficial effects of known probiotic strains

Experimental model	Strain	Health benefits	Reference
Human	*Lactobacillus rhamnosus* GG (ATCC53103)/*Bifidobacterium lactis*	Supplementations with these probiotics initiated in early pregnancy are effective in controlling the weight as well as the body composition of the mother during and after pregnancy	[101]
Human	VSL#3 (*Streptococcus thermophilus, Bifidobacterium breve, Bifidobacterium longum, Bifidobacterium infantis, Lactobacillus acidophilus, Lactobacillus plantarum, Lactobacillus paracasei/ Lactobacillus delbrueckii* subsp. *bulgaricus*)	In women with gestational diabetes mellitus, the probiotic mixture intervention modulated some inflammatory markers and had benefits on glycemic control in pregnant women	[102]
Human	*Lactobacillus rhamnosus* HN001	Reduced gestational diabetes mellitus prevalence, particularly among older women	[103]
Human	*Lactobacillus acidophilus/ L. casei/ Bifidobacterium bifidum*	Probiotic intervention in pregnant women with gestational diabetes mellitus had beneficial effects on glycemic control, triglycerides, and VLDL cholesterol concentrations	[104]
Human	*Lactobacillus acidophilus/ Bifidobacterium lactis/ Lactobacillus rhamnosus* GG	Probiotic consumption was associated with lower risk of preeclampsia in primiparous women	[85]
Human	*Lactobacillus rhamnosus* LPR/ *Bifidobacterium longum* BL999/ *L. paracasei* ST11	Probiotic administration to the pregnant and breastfeeding mother is safe and effective in reducing the risk of eczema in infants with allergic mothers positive for skin prick test	[93]
Human	*Lactobacillus rhamnosus* GG and *Bifidobacterium lactis* Bb12	Lower frequency of gestational diabetes mellitus and dietary probiotic intervention diminished the risk of larger birth size in affected cases	[105]
Human	*Lactobacillus rhamnosus* GG and *Bifidobacterium lactis* Bb12	During pregnancy and over the 12 months postpartum, blood glucose concentrations were lower in the probiotic intervened. In the last trimester of pregnancy, it reduced insulin concentration and the homeostasis model assessment	[87]

(continued)

Table 2
(continued)

Experimental model	Strain	Health benefits	Reference
Human	*Lactobacillus rhamnosus* GG, ATCC 53103	The perinatal probiotic intervention appeared to moderate the initial phase of excessive weight gain, especially among children who later became overweight, but not the second phase of excessive weight gain, the impact being most pronounced at the age of 4 years. The effect of intervention was also shown as a tendency to reduce the birth weight-adjusted mean body mass index at the age of 4 years	[97]
Mice	*L. paracasei* NCC 2461	Maternal probiotic intervention prevented the development of airway inflammation in offspring	[106]
Mice	*Lactobacillus rhamnosus* GG	Maternal supplementation with *Lactobacillus rhamnosus* GG before conception, during pregnancy and lactation exerted beneficial effects on the development of experimental allergic asthma, when applied in a very early phase of life	[107]
Rat	*Lactobacillus paracasei* NCC2461	Reverted the negative imprinting of neonatal stress on both intestinal barrier function and growth	[108]
Rat	*Bifidobacterium animalis* subsp. *lactis* BB-12® and *Propionibacterium jensenii* 702	Prebiotic intervention before and during pregnancy and lactation improved in the offspring the immune environment of stressed animals and protected, in part, against stress-induced disturbances in adult gut microflora	[109]
Mice, rats, pig	*Lactobacillus acidophilus/ Bifidobacterium lactis*	Probiotics administered to mothers during late gestation are transferred to infants born vaginally and influence the assemblages of GIT bacteria	[110]

2.3.1 Epidemiological Studies

In a randomized controlled trial (RCT) and cluster RCT, it was reported that exercise intervention during pregnancy lowered the risk of excessive GWG, caesarean delivery, macrosomia, and neonatal respiratory morbidity, particularly for high-risk women receiving combined diet and exercise interventions. Maternal hypertension was also reduced [112]. The effects of aerobic exercise have been long understood although the effects of exercise during pregnancy on the fetus and neonate are just beginning to be explored [113–116]. The hemodynamic changes of physical activity during pregnancy caused decreased oxygen delivery to the fetus and potential growth restriction. Although studies have shown a decrease in uterine circulation due to maternal exercise, many mechanisms act to maintain a relatively constant oxygen delivery to the fetus like an increased maternal hematocrit and oxygen transport in the blood, as well as a redistribution of blood flow to the placenta rather than to the uterus [117]. Current guidelines recommend that all pregnant women without contraindications engage in ≥ 30 min of moderate-intensity exercise [118]. There is growing evidence that during maternal exercise, the fetus is not at risk of hypoxia or significant bradycardia [119–121]. Also the neonates of exercising women are not at risk of being born disproportionately or underweight. Moreover, the neonates of exercising women have decreased body fat mass compared to fetuses of non-exercising mothers [122]. Previous studies have shown that the fetal heart adapts in response to maternal aerobic exercise training. For example, fetuses of exercising women had improved cardiovascular autonomic control indicated by decreased heart rates and increased HRV, relative to those of non-exercisers [115, 123]. However, the available literature evaluating physical activity during pregnancy among women who are overweight or obese is more limited and contradictory.

In a study conducted, women aged 18–40 years with a BMI ≥ 25 kg/m^2 and a singleton pregnancy <20 weeks of gestation participated in a structured home-based moderate-intensity antenatal exercise program utilizing magnetic stationary bicycles from 20 to 35 weeks of gestation. Participants received a written program prescribing frequency and duration of weekly exercises and were provided with heart rate monitors to wear only during all cycling sessions to maintain a moderate-intensity (40–59% VO$_2$ reserve). A total of 67 sessions were prescribed. The authors reported that offspring birth weight and perinatal outcomes were similar between groups. Aerobic fitness improved in the intervention group compared with controls, and there was no difference in weight gain, quality of life, pregnancy outcomes, or postnatal maternal body composition between groups [124]. A systematic study reviewed the benefits and harms of an exercise intervention for pregnant women who are overweight or obese. The studies included were randomized controlled trials comparing supervised antenatal exercise intervention with routine standard antenatal care in women who were overweight or obese during pregnancy, and six randomized controlled trials and one quasi-randomized trial were identified and included. The findings showed that a supervised antenatal exercise intervention was associated with lower gestational weight when compared with standard antenatal care. Monitored physical activity intervention appears to be successful in limiting gestational weight gain; however, the effect on maternal and infant health is less certain [125]. In the BAMBINO study, the authors investigated if an individualized exercise plan based on personal preferences and ability in obese pregnant woman could reduce gestational weight gain, improve maternal circulating lipid profile, as well as alter leptin, interleukin-8 (IL-8), and monocyte chemoattractant protein-1 (MCP-1) levels. Physical activity achieved in obese women in the exercise intervention group was not sufficient to alter gestational weight gain, leptin, MCP-1, IL-8, or circulating lipid levels [126]. The cardiorespiratory responses and work efficiency of graded treadmill exercise in healthy nonpregnant, normal weight, and obese pregnant women were compared. The results showed that healthy obese pregnant women have the aerobic capacity to undertake structured walking activities. A target heart rate of 102–124 beats per minute should be promoted for women 20–29 years of age and a rate of 101–120 beats per minute for women 30–39 years of age. Combining healthy eating with a walking plan prevents excessive weight gain during pregnancy and promotes a healthy fetal environment [127]. A systematic review and meta-analysis of randomized controlled trials showed that physical activity interventions were effective for overweight or obese pregnant as well as for postpartum women. Pregnant women in the intervention groups gained less weight and lost more body weight postpartum [128]. On the other hand, insulin resistance and hyperglycemia are the main pathophysiological factors leading to gestational diabetes mellitus type 2 diabetes in pregnant women. Physical activity can help to prevent gestational diabetes mellitus by improving insulin sensitivity and excessive gestational weight gain which, in turn, may help to prevent offspring postnatal complications, including childhood obesity and diabetes in adult life [117]. Exercise intervention in women with gestational diabetes intervention had the lowest BMI increase during late and midpregnancy and experienced a significantly lower risk of preterm birth, low birth weight, and macrosomia [129]. At 10–14-week postpartum, 68 Swedish women with a self-reported prepregnancy BMI of 25–35 kg/m^2 were randomized to a 12-week behavioral modification treatment with exercise. The goal of exercise treatment was to perform four walks of 45 min a week at 60–70% of max heart rate using a heart rate monitor. However, no differences were observed in metabolic parameters [130].

2.3.2   Animal Studies   Exercise during pregnancy has long-lasting effects on offspring health. The skeleton influences glucose handling through the actions of the bone-derived hormone osteocalcin. Offspring of exercised dams had a lower volumetric bone mineral density than controls. Serum concentrations of undercarboxylated osteocalcin were significantly greater in the male offspring of exercised animals. These results suggest that moderate exercise during pregnancy can result in lasting changes to the musculoskeletal system and adiposity in offspring [131]. Female Sprague-Dawley rats were split into sedentary and voluntary access to a running wheel in the cage prior to and during mating, pregnancy, and nursing. Adult offspring born to exercised dams had enhanced glucose disposal during glucose tolerance testing and had decreased insulin levels and hepatic glucose production during the clamp procedure. In addition, offspring from exercised dams had increased glucose uptake in skeletal muscle and decreased heart glucose. Exercise during pregnancy enhances offspring insulin sensitivity and improves offspring glucose homeostasis [132]. A study in rats determined whether maternal exercise during pregnancy leads to reduced mammary tumorigenesis in female offspring. Pregnant rats had free access to a running wheel. Female pups from exercised dams were weaned at 21 days of age and fed a high-fat diet without access to a running wheel. At 6 weeks, all pups were injected with the carcinogen *N*-methyl-*N*-nitrosourea (MNU). Pups from exercised dams had a substantially lower tumor incidence [133]. In a murine model, it was shown that maternal physical activity during pregnancy affects offspring lifelong propensity for physical activity and may have important implications for combating the worldwide epidemic of physical inactivity and obesity [134]. In addition, voluntary running significantly improved glucose homeostasis and body composition in pregnant female mice fed with high-fat diet [135].

Physical exercise is recommended not only to assist weight loss and maintenance during the treatment of obesity but also to enhance whole-body insulin sensitivity and to improve the overall metabolic profile [68]. Female mice were exposed to a high-fat diet with voluntary wheel exercise for 6 weeks before and throughout pregnancy. Methylation of the Pgc-1α promoter were assessed in skeletal muscle from neonatal and 12-month-old offspring, and glucose and insulin tolerance tests were performed in the offspring at 6, 9, and 12 months. Maternal exercise prevented the maternal high-fat diet-induced Pgc-1α hypermethylation and enhanced Pgc-1α and its target gene expression, concurrent with amelioration of age-associated metabolic dysfunction at 9 months of age. Therefore, maternal exercise intervention prevents maternal high-fat diet-induced epigenetic and metabolic dysregulation in the offspring [136]. In a study conducted in female rats (aged 3 weeks), rats were fed with a cafeteria diet for 34 weeks with or without physical exercise starting at week 26 for 8 weeks. Physical exercise improved insulin sensitivity in obese rats with a cafeteria diet [68]. In our group, we recently reported the effects of an exercise intervention in obese rats before and during pregnancy on mothers. Briefly, female rats were weaned onto a high-fat diet and, 1 month before breeding, one half of the rats were randomly selected to begin a wheel-running exercise. Mothers continued to be placed in the wheel through pregnancy. A training session lasted 15 min, which we established was the optimum running schedule that was always completed, followed by a 15 min rest period and a second 15 min run. Rats were allowed 2 days rest every 7 days. Before pregnancy, all rats completed the 30 min running, while during pregnancy, rats were placed in the wheel for only one 15 min session per day, and the amount of voluntary exercise completed varied between animals especially in late gestation. Maternal exercise intervention decreased corticosterone concentrations; however, the values were not returned to those in the control. Similar results were seen at the end of lactation and in the neonate and young adult male offspring [36]. Exercise did not change maternal weight but did completely prevent the rise in maternal triglycerides and partially prevented the increases in glucose, insulin, cholesterol, leptin, fat, and oxidative stress. Exercise also recovered in female rats the decrease fertility induced by obesity [137]. We also evaluated prepuberal offspring outcomes and found that maternal exercise intervention prevented leptin increase and ameliorate the increased triglycerides [36] (Fig. 5). Evidence related to the benefits of maternal exercise intervention are summarized in Table 3.

2.4 Offspring Interventions

Growing evidence in human [127] and animal [28, 69] studies indicates that changes in lifestyle of obese mothers, such as reducing calorie intake or increasing exercise, prevent excessive weight gain during pregnancy and the subsequent adverse outcomes in offspring health (Fig. 7). However, if these interventions cannot be applied in obese pregnant women, it is necessary to determine whether interventions in offspring are beneficial. To date, few studies have examined offspring interventions.

A randomized controlled trial reported that home-based early intervention delivered by trained community nurses was effective in reducing mean BMI for children at age 2 [138]. In animal studies, it has been reported that hypoxia in utero impairs endothelial dysfunction. Pregnant rats were exposed to hypoxic conditions from gestational day 15 to 21. Offspring from hypoxic pregnancies was randomized at 10 weeks of age to either an exercise-trained or sedentary group. Exercise-trained rats ran on a treadmill for 30 min, 5 days a week, and 6 weeks. Aerobic exercise training in offspring from hypoxic pregnancies improved endothelium-derived hyperpolarization-mediated vasodilatation in gastrocnemius muscle arteries [139]. Exercise is extensively accepted as a beneficial intervention for metabolic diseases. In one study, female rats were fed with a high-fat diet before mating and throughout pregnancy and lactation, and their offspring were weaned onto a high-fat diet. After 7 weeks, half of the rats were exercised for 5 weeks in running wheels. The authors reported that regardless of the diet consumed after weaning, a short period of exercise intervention reduced body

Table 3
Benefits of maternal exercise intervention

Animal model	Study		Intervention period	Beneficial outcomes	Reference
	Study design				
Human	Pregnant women		Pregnancy	Exercise during pregnancy lowered the risk of gestational weight gain, caesarean delivery, macrosomia, and neonatal respiratory morbidity	[112]
Human	16 weeks of moderate-intensity exercise in pregnant women (BMI ≥ 25 kg/m^2)		Pregnancy	Moderate-intensity exercise in overweight/obese pregnant women improved fitness but had no clinical effects	[124]
Human	Antenatal exercise intervention in women who are overweight or obese		Pregnancy	Lower gestational weight when compared with standard antenatal care	[125]
Human	Walking program: three to four times per week per 40 min		Pregnancy	Combining healthy eating with a walking plan prevents excessive weight gain during pregnancy and promotes a healthy fetal environment	[127]
Human	Exercise intervention in pregnant women with gestational diabetes mellitus		Pregnancy	Physical activity improved maternal insulin sensitivity and excessive gestational weight gain, which in turn may help to prevent offspring postnatal complications, including childhood obesity and diabetes in adult life	[117]
Mice	Maternal voluntary exercise in a running wheel		Before and during pregnancy	Maternal physical activity during pregnancy affects the offspring's lifelong propensity for physical activity	[134]
Mice	Female were put on a high-fat diet for 2 weeks with voluntary access to a running wheel		Prior to mating, during mating, and throughout pregnancy	Improvement of glucose homeostasis and body composition in pregnant female mice	[135]
Mice	Voluntary wheel exercise in female mice fed a high-fat diet		Before and throughout pregnancy	Prevents maternal high-fat diet-induced epigenetic and metabolic dysregulation in the offspring	[136]

(continued)

Table 3
(continued)

Animal model	Study design	Intervention period	Beneficial outcomes	Reference
	Study			
Rat	Moderate maternal exercise	Pregnancy	Offspring of exercised dams had a lower volumetric bone mineral density than controls. Serum concentrations of undercarboxylated osteocalcin were significantly greater in the male offspring of exercised	[131]
Rat	Voluntary access to a running wheel	Prior to and during mating, pregnancy, and nursing	Enhancement of offspring's insulin sensitivity and improvement of glucose homeostasis.	[132]
Rat	Maternal free access to a running wheel and offspring injection with the carcinogen N-methyl-N-nitrosourea	Pregnancy	Maternal exercise intervention in pregnancy lowered tumor incidence in offspring injected with the carcinogen	[133]
Rat	Voluntary exercise in rats fed a high-fat diet	Prior to and during mating and pregnancy	Prevent the rise in maternal corticosterone and triglycerides and partially the increases in glucose, insulin, cholesterol, leptin, fat, and oxidative stress. In the offspring, prevented leptin increase and ameliorated the increased triglycerides	[36, 137]

weight, adiposity, leptin, insulin, and triglyceride plasma levels, as well as glucose intolerance in female offspring of obese mothers [140]. In our group, we have shown that maternal obesity increases both testicular and sperm oxidative stress, reduces sperm quality, and decreases fertility rate in rat offspring [141] and that physical voluntary exercise even in old male offspring of obese mothers has beneficial effects on adiposity index, gonadal fat, oxidative stress markers, sperm quality, and fertility. Thus, regular physical exercise in male offspring from obese mothers recuperates key male reproductive functions even at advanced age. The encouraging feature of these data is the indication that it is never too late for exercise to be beneficial [142].

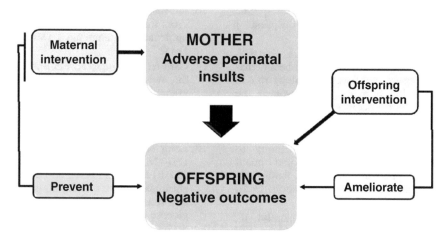

Fig. 7 Maternal interventions prevents offspring negative outcomes, while offspring interventions ameliorate or modulate the adverse effects of developmental programming

3 Conclusions

In human, the first 1000 days of life—conception through age 24 months—plays an important role in finding mechanism involved in developmental programming. Therefore, specific interventions before and during pregnancy and lactation provide a window of opportunities to promote the health not only of the mother but also of the child.

Acknowledgments

Elena Zambrano is supported by the Newton Fund RCUK-CONACyT (Research Councils UK—Consejo Nacional de Ciencia y Tecnología) with the project entitle "Interventions to improve maternal metabolic profile and prevent cardio-metabolic and behavioural deficits in future generations due to programming by maternal obesity."

References

1. Tain YL, Hsu CN (2017) Developmental origins of chronic kidney disease: should we focus on early life? Int J Mol Sci 18(2):E381

2. Dickinson H, Moss TJ, Gatford KL, Moritz KM, Akison L, Fullston T et al (2016) A review of fundamental principles for animal models of DOHaD research: an Australian perspective. J Dev Orig Health Dis 7:449–472

3. Briffa JF, O'Dowd R, Moritz KM, Romano T, Jedwab LR, McAinch AJ et al (2017) Uteroplacental insufficiency reduces rat plasma leptin concentrations and alters placental leptin transporters: ameliorated with enhanced milk intake and nutrition. J Physiol 595:3389–3407

4. Burton GJ, Fowden AL, Thornburg KL (2016) Placental origins of chronic disease. Physiol Rev 96:1509–1565

5. Cuffe JS, Turton EL, Akison LK, Bielefeldt-Ohmann H, Moritz KM (2017) Prenatal

corticosterone exposure programs sex-specific adrenal adaptations in mouse offspring. J Endocrinol 232:37–48

6. Savoy C, Ferro MA, Schmidt LA, Saigal S, Van Lieshout RJ (2016) Prenatal betamethasone exposure and psychopathology risk in extremely low birth weight survivors in the third and fourth decades of life. Psychoneuroendocrinology 74:278–285

7. Arabin B, Baschat AA (2017) Pregnancy: an underutilized window of opportunity to improve long-term maternal and infant health-an appeal for continuous family care and interdisciplinary communication. Front Pediatr 5:69

8. Pankey CL, Walton MW, Odhiambo JF, Smith AM, Ghnenis AB, Nathanielsz PW et al (2017) Intergenerational impact of maternal overnutrition and obesity throughout pregnancy in sheep on metabolic syndrome in grandsons and granddaughters. Domest Anim Endocrinol 60:67–74

9. Rodriguez-Rodriguez P, Lopez de Pablo AL, Garcia-Prieto CF, Somoza B, Quintana-Villamandos B, Gomez de Diego JJ et al (2017) Long term effects of fetal undernutrition on rat heart. Role of hypertension and oxidative stress. PLoS One 12(2):e0171544

10. Entringer S, Buss C, Wadhwa PD (2010) Prenatal stress and developmental programming of human health and disease risk: concepts and integration of empirical findings. Curr Opin Endocrinol Diabetes Obes 17:507–516

11. Lesage J, Del-Favero F, Leonhardt M, Louvart H, Maccari S, Vieau D (2004) Prenatal stress induces intrauterine growth restriction and programmes glucose intolerance and feeding behaviour disturbances in the aged rat. J Endocrinol 181:291–296

12. Koren O, Goodrich JK, Cullender TC, Spor A, Laitinen K, Backhed HK et al (2012) Host remodeling of the gut microbiome and metabolic changes during pregnancy. Cell 150:470–480

13. Ma J, Prince AL, Bader D, Hu M, Ganu R, Baquero K et al (2014) High-fat maternal diet during pregnancy persistently alters the offspring microbiome in a primate model. Nat Commun 5:3889

14. Wesolowski SR, Kasmi KC, Jonscher KR, Friedman JE (2017) Developmental origins of NAFLD: a womb with a clue. Nat Rev Gastroenterol Hepatol 14:81–96

15. Zambrano E, Nathanielsz PW (2013) Mechanisms by which maternal obesity programs offspring for obesity: evidence from animal studies. Nutr Rev 71(Suppl 1):S42–S54

16. Rabadan-Diehl C, Nathanielsz P (2013) From mice to men: research models of developmental programming. J Dev Orig Health Dis 4:3–9

17. Gardner DS, Tingey K, Van Bon BW, Ozanne SE, Wilson V, Dandrea J et al (2005) Programming of glucose-insulin metabolism in adult sheep after maternal undernutrition. Am J Physiol Regul Integr Comp Physiol 289:R947–R954

18. Long NM, George LA, Uthlaut AB, Smith DT, Nijland MJ, Nathanielsz PW et al (2010) Maternal obesity and increased nutrient intake before and during gestation in the ewe results in altered growth, adiposity, and glucose tolerance in adult offspring. J Anim Sci 88:3546–3553

19. Forhead AJ, Fowden AL (2009) The hungry fetus? Role of leptin as a nutritional signal before birth. J Physiol 587:1145–1152

20. Nakae J, Kido Y, Accili D (2001) Distinct and overlapping functions of insulin and IGF-I receptors. Endocr Rev 22:818–835

21. Bispham J, Gopalakrishnan GS, Dandrea J, Wilson V, Budge H, Keisler DH et al (2003) Maternal endocrine adaptation throughout pregnancy to nutritional manipulation: consequences for maternal plasma leptin and cortisol and the programming of fetal adipose tissue development. Endocrinology 144:3575–3585

22. Bondesson M, Hao R, Lin CY, Williams C, Gustafsson JA (2015) Estrogen receptor signaling during vertebrate development. Biochim Biophys Acta 1849:142–151

23. Hiort O (2013) The differential role of androgens in early human sex development. BMC Med 11:152

24. Zambrano E, Reyes-Castro LA, Nathanielsz PW (2015) Aging, glucocorticoids and developmental programming. Age 37:9774

25. Larque E, Pagan A, Prieto MT, Blanco JE, Gil-Sanchez A, Zornoza-Moreno M et al (2014) Placental fatty acid transfer: a key factor in fetal growth. Ann Nutr Metab 64:247–253

26. Sutton EF, Gilmore LA, Dunger DB, Heijmans BT, Hivert MF, Ling C et al (2016) Developmental programming: state-of-the-science and future directions-summary from a Pennington biomedical symposium. Obesity 24:1018–1026

27. Luo ZC, Fraser WD, Julien P, Deal CL, Audibert F, Smith GN et al (2006) Tracing the origins of "fetal origins" of adult diseases: programming by oxidative stress? Med Hypotheses 66:38–44

28. Vega CC, Reyes-Castro LA, Rodriguez-Gonzalez GL, Bautista CJ, Vazquez-Martinez M, Larrea F et al (2016) Resveratrol partially prevents oxidative stress and metabolic dysfunction in pregnant rats fed a low protein diet and their offspring. J Physiol 594:1483–1499

29. Lane RH (2014) Fetal programming, epigenetics, and adult onset disease. Clin Perinatol 41:815–831

30. Perera F, Herbstman J (2011) Prenatal environmental exposures, epigenetics, and disease. Reprod Toxicol 31:363–373

31. Bhutta ZA, Ahmed T, Black RE, Cousens S, Dewey K, Giugliani E et al (2008) What works? Interventions for maternal and child undernutrition and survival. Lancet 371 (9610):417–440

32. Black RE, Allen LH, Bhutta ZA, Caulfield LE, de Onis M, Ezzati M et al (2008) Maternal and child undernutrition: global and regional exposures and health consequences. Lancet 371(9608):243–260

33. Bryce J, Coitinho D, Darnton-Hill I, Pelletier D, Pinstrup-Andersen P, Maternal and Child Undernutrition Study Group (2008) Maternal and child undernutrition: effective action at national level. Lancet 371 (9611):510–526

34. Morris SS, Cogill B, Uauy R, Maternal and Child Undernutrition Study Group (2008) Effective international action against undernutrition: why has it proven so difficult and what can be done to accelerate progress? Lancet 371(9612):608–621

35. Victora CG, Adair L, Fall C, Hallal PC, Martorell R, Richter L et al (2008) Maternal and child undernutrition: consequences for adult health and human capital. Lancet 371 (9609):340–357

36. Nathanielsz PW, Ford SP, Long NM, Vega CC, Reyes-Castro LA, Zambrano E (2013) Interventions to prevent adverse fetal programming due to maternal obesity during pregnancy. Nutr Rev 71(Suppl 1):S78–S87

37. Zambrano E, Ibanez C, Martinez-Samayoa PM, Lomas-Soria C, Durand-Carbajal M, Rodriguez-Gonzalez GL (2016) Maternal obesity: lifelong metabolic outcomes for offspring from poor developmental trajectories during the perinatal period. Arch Med Res 47:1–12

38. Poston L (2012) Maternal obesity, gestational weight gain and diet as determinants of offspring long term health. Best Pract Res Clin Endocrinol Metab 26:627–639

39. Rodriguez JS, Rodriguez-Gonzalez GL, Reyes-Castro LA, Ibanez C, Ramirez A, Chavira R et al (2012) Maternal obesity in the rat programs male offspring exploratory, learning and motivation behavior: prevention by dietary intervention pre-gestation or in gestation. Int J Dev Neurosci 30:75–81

40. Mathiesen ER, Vaz JA (2008) Insulin treatment in diabetic pregnancy. Diabetes Metab Res Rev 24(Suppl 2):S3–20

41. Carruth BR, Skinner JD (1991) Practitioners beware: regional differences in beliefs about nutrition during pregnancy. J Am Diet Assoc 91(4):435–440

42. Clark M, Ogden J (1999) The impact of pregnancy on eating behaviour and aspects of weight concern. Int J Obes Relat Metab Disord 23:18–24

43. National Research Council, Institute of Medicine, Youth, and Families Board on Children, Food and Nutrition Board, Committee to Reexamine IOM Pregnancy Weight Guidelines (2009) In: Yaktine AL, Rasmussen KM (eds) Weight gain during pregnancy: reexamining the guidelines. National Academies Press, Washington DC. ISBN-10: 0309131138

44. Siega-Riz AM, Viswanathan M, Moos MK, Deierlein A, Mumford S, Knaack J et al (2009) A systematic review of outcomes of maternal weight gain according to the Institute of Medicine recommendations: birthweight, fetal growth, and postpartum weight retention. Am J Obstet Gynecol 201(339): e331–e314

45. Shah BR, Retnakaran R, Booth GL (2008) Increased risk of cardiovascular disease in young women following gestational diabetes mellitus. Diabetes Care 31:1668–1669

46. Hedderson MM, Gunderson EP, Ferrara A (2010) Gestational weight gain and risk of gestational diabetes mellitus. Obstet Gynecol 115:597–604

47. Rooney BL, Schauberger CW, Mathiason MA (2005) Impact of perinatal weight change on long-term obesity and obesity-related illnesses. Obstet Gynecol 106:1349–1356

48. Schack-Nielsen L, Michaelsen KF, Gamborg M, Mortensen EL, Sorensen TI (2010) Gestational weight gain in relation to offspring body mass index and obesity from infancy through adulthood. Int J Obes 34:67–74

49. Thangaratinam S, Rogozinska E, Jolly K, Glinkowski S, Duda W, Borowiack E et al (2012) Interventions to reduce or prevent obesity in pregnant women: a systematic

review. Health Technol Assess 16(31):iii–iv, 1–191

50. Thangaratinam S, Rogozinska E, Jolly K, Glinkowski S, Roseboom T, Tomlinson JW et al (2012) Effects of interventions in pregnancy on maternal weight and obstetric outcomes: meta-analysis of randomised evidence. BMJ 344:e2088

51. Skouteris H, Hartley-Clark L, McCabe M, Milgrom J, Kent B, Herring SJ et al (2010) Preventing excessive gestational weight gain: a systematic review of interventions. Obes Rev 11:757–768

52. Dodd JM, Grivell RM, Crowther CA, Robinson JS (2010) Antenatal interventions for overweight or obese pregnant women: a systematic review of randomised trials. BJOG 117:1316–1326

53. Dodd JM, Turnbull D, McPhee AJ, Deussen AR, Grivell RM, Yelland LN et al (2014) Antenatal lifestyle advice for women who are overweight or obese: LIMIT randomised trial. BMJ 348:g1285

54. Dodd JM, McPhee AJ, Turnbull D, Yelland LN, Deussen AR, Grivell RM et al (2014) The effects of antenatal dietary and lifestyle advice for women who are overweight or obese on neonatal health outcomes: the LIMIT randomised trial. BMC Med 12:163

55. McGowan CA, McAuliffe FM (2010) The influence of maternal glycaemia and dietary glycaemic index on pregnancy outcome in healthy mothers. Br J Nutr 104:153–159

56. Walsh JM, Mahony R, Byrne J, Foley M, McAuliffe FM (2011) The association of maternal and fetal glucose homeostasis with fetal adiposity and birthweight. Eur J Obstet Gynecol Reprod Biol 159:338–341

57. Kong AP, Chan RS, Nelson EA, Chan JC (2011) Role of low-glycemic index diet in management of childhood obesity. Obes Rev 12:492–498

58. Zhang XM, Ding YL (2008) Glycosylated hemoglobin test in gestational abnormal glucose metabolism. Zhong Nan Da Xue Xue Bao Yi Xue Ban 33:85–88

59. Nagalakshmi CS, Santhosh NU, Krishnamurthy N, Chethan C, Shilpashree MK (2016) Role of altered venous blood lactate and HbA1c in women with gestational diabetes mellitus. J Clin Diagn Res 10(12): BC18–BC20

60. Roskjaer AB, Andersen JR, Ronneby H, Damm P, Mathiesen ER (2015) Dietary advices on carbohydrate intake for pregnant women with type 1 diabetes. J Matern Fetal Neonatal Med 28:229–233

61. Burani J, Longo PJ (2006) Low-glycemic index carbohydrates: an effective behavioral change for glycemic control and weight management in patients with type 1 and 2 diabetes. Diabetes Educ 32:78–88

62. McGowan CA, Walsh JM, Byrne J, Curran S, McAuliffe FM (2013) The influence of a low glycemic index dietary intervention on maternal dietary intake, glycemic index and gestational weight gain during pregnancy: a randomized controlled trial. Nutr J 12(1):140

63. Donnelly JM, Walsh JM, Byrne J, Molloy EJ, McAuliffe FM (2015) Impact of maternal diet on neonatal anthropometry: a randomized controlled trial. Pediatr Obes 10:52–56

64. Syngelaki A, Nicolaides KH, Balani J, Hyer S, Akolekar R, Kotecha R et al (2016) Metformin versus placebo in obese pregnant women without diabetes mellitus. N Engl J Med 374:434–443

65. Helland IB, Smith L, Blomen B, Saarem K, Saugstad OD, Drevon CA (2008) Effect of supplementing pregnant and lactating mothers with n-3 very-long-chain fatty acids on children's IQ and body mass index at 7 years of age. Pediatrics 122(2):e472–e479

66. Deng J, Li X, Ding Z, Wu Y, Chen X, Xie L (2017) Effect of DHA supplements during pregnancy on the concentration of PUFA in breast milk of Chinese lactating mothers. J Perinat Med 45:437–441

67. Hawkes JS, Bryan DL, Makrides M, Neumann MA, Gibson RA (2002) A randomized trial of supplementation with docosahexaenoic acid-rich tuna oil and its effects on the human milk cytokines interleukin 1 beta, interleukin 6, and tumor necrosis factor alpha. Am J Clin Nutr 75:754–760

68. Goularte JF, Ferreira MB, Sanvitto GL (2012) Effects of food pattern change and physical exercise on cafeteria diet-induced obesity in female rats. Br J Nutr 108:1511–1518

69. Zambrano E, Martinez-Samayoa PM, Rodriguez-Gonzalez GL, Nathanielsz PW (2010) Dietary intervention prior to pregnancy reverses metabolic programming in male offspring of obese rats. J Physiol 588:1791–1799

70. Tuersunjiang N, Odhiambo JF, Long NM, Shasa DR, Nathanielsz PW, Ford SP (2013) Diet reduction to requirements in obese/overfed ewes from early gestation prevents glucose/insulin dysregulation and returns fetal adiposity and organ development to control levels. Am J Physiol Endocrinol Metab 305:E868–E878

71. Segovia SA, Vickers MH, Gray C, Reynolds CM (2014) Maternal obesity, inflammation, and developmental programming. Biomed Res Int 2014:418975

72. Kursvietiene L, Staneviciene I, Mongirdiene A, Bernatoniene J (2016) Multiplicity of effects and health benefits of resveratrol. Medicina 52:148–155

73. Zou T, Chen D, Yang Q, Wang B, Zhu MJ, Nathanielsz PW et al (2017) Resveratrol supplementation of high-fat diet-fed pregnant mice promotes brown and beige adipocyte development and prevents obesity in male offspring. J Physiol 595:1547–1562

74. Roberts VH, Pound LD, Thorn SR, Gillingham MB, Thornburg KL, Friedman JE et al (2014) Beneficial and cautionary outcomes of resveratrol supplementation in pregnant non-human primates. FASEB J 28:2466–2477

75. Shokryzadan P, Rajion MA, Meng GY, Boo LJ, Ebrahimi M, Royan M et al (2017) Conjugated linoleic acid: a potent fatty acid linked to animal and human health. Crit Rev Food Sci Nutr 57:2737–2748

76. Segovia SA, Vickers MH, Zhang XD, Gray C, Reynolds CM (2015) Maternal supplementation with conjugated linoleic acid in the setting of diet-induced obesity normalises the inflammatory phenotype in mothers and reverses metabolic dysfunction and impaired insulin sensitivity in offspring. J Nutr Biochem 26:1448–1457

77. Pileggi CA, Segovia SA, Markworth JF, Gray C, Zhang XD, Milan AM et al (2016) Maternal conjugated linoleic acid supplementation reverses high-fat diet-induced skeletal muscle atrophy and inflammation in adult male rat offspring. Am J Physiol Regul Integr Comp Physiol 310(5):R432–R439

78. Vickers MH, Sloboda DM (2012) Leptin as mediator of the effects of developmental programming. Best Pract Res Clin Endocrinol Metab 26:677–687

79. Vickers MH, Gluckman PD, Coveny AH, Hofman PL, Cutfield WS, Gertler A et al (2005) Neonatal leptin treatment reverses developmental programming. Endocrinology 146:4211–4216

80. Konieczna J, Garcia AP, Sanchez J, Palou M, Palou A, Pico C (2013) Oral leptin treatment in suckling rats ameliorates detrimental effects in hypothalamic structure and function caused by maternal caloric restriction during gestation. PLoS One 8(11):e81906

81. Murphy MO, Herald JB, Wills CT, Unfried SG, Cohn DM, Loria AS (2017) Postnatal treatment with metyrapone attenuates the effects of diet-induced obesity in female rats exposed to early-life stress. Am J Physiol Endocrinol Metab 312(2):E98–E108

82. Hashemi A, Villa CR, Comelli EM (2016) Probiotics in early life: a preventative and treatment approach. Food Funct 7:1752–1768

83. Hill C, Guarner F, Reid G, Gibson GR, Merenstein DJ, Pot B et al (2014) Expert consensus document. The International Scientific Association for Probiotics and Prebiotics consensus statement on the scope and appropriate use of the term probiotic. Nat Rev Gastroenterol Hepatol 11:506–514

84. Correia MI, Liboredo JC, Consoli ML (2012) The role of probiotics in gastrointestinal surgery. Nutrition 28:230–234

85. Brantsaeter AL, Myhre R, Haugen M, Myking S, Sengpiel V, Magnus P et al (2011) Intake of probiotic food and risk of preeclampsia in primiparous women: the Norwegian mother and child cohort study. Am J Epidemiol 174:807–815

86. Asemi Z, Samimi M, Tabassi Z, Naghibi Rad M, Rahimi Foroushani A, Khorammian H et al (2013) Effect of daily consumption of probiotic yoghurt on insulin resistance in pregnant women: a randomized controlled trial. Eur J Clin Nutr 67:71–74

87. Luoto R, Laitinen K, Nermes M, Isolauri E (2010) Impact of maternal probiotic-supplemented dietary counselling on pregnancy outcome and prenatal and postnatal growth: a double-blind, placebo-controlled study. Br J Nutr 103:1792–1799

88. Matsumiya Y, Kato N, Watanabe K, Kato H (2002) Molecular epidemiological study of vertical transmission of vaginal *Lactobacillus* species from mothers to newborn infants in Japanese, by arbitrarily primed polymerase chain reaction. J Infect Chemother 8:43–49

89. Dominguez-Bello MG, Costello EK, Contreras M, Magris M, Hidalgo G, Fierer N et al (2010) Delivery mode shapes the acquisition and structure of the initial microbiota across multiple body habitats in newborns. Proc Natl Acad Sci U S A 107:11971–11975

90. Gajer P, Brotman RM, Bai G, Sakamoto J, Schutte UM, Zhong X et al (2012) Temporal dynamics of the human vaginal microbiota. Sci Transl Med 4:132ra152

91. Backhed F, Ding H, Wang T, Hooper LV, Koh GY, Nagy A et al (2004) The gut microbiota as an environmental factor that regulates fat storage. Proc Natl Acad Sci U S A 101:15718–15723

92. Cani PD, Delzenne NM (2007) Gut microflora as a target for energy and metabolic homeostasis. Curr Opin Clin Nutr Metab Care 10:729–734

93. Rautava S, Kainonen E, Salminen S, Isolauri E (2012) Maternal probiotic supplementation during pregnancy and breast-feeding reduces the risk of eczema in the infant. J Allergy Clin Immunol 130:1355–1360

94. Agarwal R, Sharma N, Chaudhry R, Deorari A, Paul VK, Gewolb IH et al (2003) Effects of oral *Lactobacillus* GG on enteric microflora in low-birth-weight neonates. J Pediatr Gastroenterol Nutr 36:397–402

95. Mohan R, Koebnick C, Schildt J, Schmidt S, Mueller M, Possner M et al (2006) Effects of *Bifidobacterium lactis* Bb12 supplementation on intestinal microbiota of preterm infants: a double-blind, placebo-controlled, randomized study. J Clin Microbiol 44:4025–4031

96. Gueimonde M, Sakata S, Kalliomaki M, Isolauri E, Benno Y, Salminen S (2006) Effect of maternal consumption of *lactobacillus* GG on transfer and establishment of fecal bifidobacterial microbiota in neonates. J Pediatr Gastroenterol Nutr 42:166–170

97. Laitinen K, Poussa T, Isolauri E, Nutrition, Allergy, Mucosal Immunology and Intestinal Microbiota Group (2009) Probiotics and dietary counselling contribute to glucose regulation during and after pregnancy: a randomised controlled trial. Br J Nutr 101:1679–1687

98. Backhed F, Manchester JK, Semenkovich CF, Gordon JI (2007) Mechanisms underlying the resistance to diet-induced obesity in germ-free mice. Proc Natl Acad Sci U S A 104:979–984

99. Kalliomaki M, Kirjavainen P, Eerola E, Kero P, Salminen S, Isolauri E (2001) Distinct patterns of neonatal gut microflora in infants in whom atopy was and was not developing. J Allergy Clin Immunol 107:129–134

100. Kalliomaki M, Collado MC, Salminen S, Isolauri E (2008) Early differences in fecal microbiota composition in children may predict overweight. Am J Clin Nutr 87:534–538

101. Ilmonen J, Isolauri E, Poussa T, Laitinen K (2011) Impact of dietary counselling and probiotic intervention on maternal anthropometric measurements during and after pregnancy: a randomized placebo-controlled trial. Clin Nutr 30:156–164

102. Jafarnejad S, Saremi S, Jafarnejad F, Arab A (2016) Effects of a multispecies probiotic mixture on glycemic control and inflammatory status in women with gestational diabetes: a randomized controlled clinical trial. J Nutr Metab 2016:5190846

103. Wickens KL, Barthow CA, Murphy R, Abels PR, Maude RM, Stone PR et al (2017) Early pregnancy probiotic supplementation with *Lactobacillus rhamnosus* HN001 may reduce the prevalence of gestational diabetes mellitus: a randomised controlled trial. Br J Nutr 117:804–813

104. Karamali M, Dadkhah F, Sadrkhanlou M, Jamilian M, Ahmadi S, Tajabadi-Ebrahimi M et al (2016) Effects of probiotic supplementation on glycaemic control and lipid profiles in gestational diabetes: a randomized, double-blind, placebo-controlled trial. Diabetes Metab 42:234–241

105. Luoto R, Kalliomaki M, Laitinen K, Isolauri E (2010) The impact of perinatal probiotic intervention on the development of overweight and obesity: follow-up study from birth to 10 years. Int J Obes 34:1531–1537

106. Schabussova I, Hufnagl K, Tang ML, Hoflehner E, Wagner A, Loupal G et al (2012) Perinatal maternal administration of *Lactobacillus paracasei* NCC 2461 prevents allergic inflammation in a mouse model of birch pollen allergy. PLoS One 7(7):e40271

107. Blumer N, Sel S, Virna S, Patrascan CC, Zimmermann S, Herz U et al (2007) Perinatal maternal application of *Lactobacillus rhamnosus* GG suppresses allergic airway inflammation in mouse offspring. Clin Exp Allergy 37:348–357

108. Garcia-Rodenas CL, Bergonzelli GE, Nutten S, Schumann A, Cherbut C, Turini M et al (2006) Nutritional approach to restore impaired intestinal barrier function and growth after neonatal stress in rats. J Pediatr Gastroenterol Nutr 43:16–24

109. Barouei J, Moussavi M, Hodgson DM (2012) Effect of maternal probiotic intervention on HPA axis, immunity and gut microbiota in a rat model of irritable bowel syndrome. PLoS One 7(10):e46051

110. Buddington RK, Williams CH, Kostek BM, Buddington KK, Kullen MJ (2010) Maternal-to-infant transmission of probiotics: concept validation in mice, rats, and pigs. Neonatology 97:250–256

111. Caspersen CJ, Powell KE, Christenson GM (1985) Physical activity, exercise, and physical fitness: definitions and distinctions for health-related research. Public Health Rep 100:126–131

112. Muktabhant B, Lawrie TA, Lumbiganon P, Laopaiboon M (2015) Diet or exercise, or both, for preventing excessive weight gain in

pregnancy. Cochrane Database Syst Rev 6: CD007145

113. Clapp JF III, Simonian S, Lopez B, Appleby-Wineberg S, Harcar-Sevcik R (1998) The one-year morphometric and neurodevelopmental outcome of the offspring of women who continued to exercise regularly throughout pregnancy. Am J Obstet Gynecol 178:594–599

114. Clapp JF III (2003) The effects of maternal exercise on fetal oxygenation and fetoplacental growth. Eur J Obstet Gynecol Reprod Biol 110(Suppl 1):S80–S85

115. May LE, Glaros A, Yeh HW, Clapp JF III, Gustafson KM (2010) Aerobic exercise during pregnancy influences fetal cardiac autonomic control of heart rate and heart rate variability. Early Hum Dev 86:213–217

116. May LE, Scholtz SA, Suminski R, Gustafson KM (2014) Aerobic exercise during pregnancy influences infant heart rate variability at one month of age. Early Hum Dev 90:33–38

117. Cid M, Gonzalez M (2016) Potential benefits of physical activity during pregnancy for the reduction of gestational diabetes prevalence and oxidative stress. Early Hum Dev 94:57–62

118. Seneviratne SN, McCowan LM, Cutfield WS, Derraik JG, Hofman PL (2015) Exercise in pregnancies complicated by obesity: achieving benefits and overcoming barriers. Am J Obstet Gynecol 212:442–449

119. Clapp JF III, Little KD, Appleby-Wineberg SK, Widness JA (1995) The effect of regular maternal exercise on erythropoietin in cord blood and amniotic fluid. Am J Obstet Gynecol 172:1445–1451

120. Clapp JF III, Stepanchak W, Tomaselli J, Kortan M, Faneslow S (2000) Portal vein blood flow-effects of pregnancy, gravity, and exercise. Am J Obstet Gynecol 183:167–172

121. Kennelly MM, McCaffrey N, McLoughlin P, Lyons S, McKenna P (2002) Fetal heart rate response to strenuous maternal exercise: not a predictor of fetal distress. Am J Obstet Gynecol 187:811–816

122. Clapp JF III, Capeless EL (1990) Neonatal morphometrics after endurance exercise during pregnancy. Am J Obstet Gynecol 163:1805–1811

123. May LE, Suminski RR, Langaker MD, Yeh HW, Gustafson KM (2012) Regular maternal exercise dose and fetal heart outcome. Med Sci Sports Exerc 44:1252–1258

124. Seneviratne SN, Jiang Y, Derraik J, McCowan L, Parry GK, Biggs JB et al

125. Sui Z, Grivell RM, Dodd JM (2012) Antenatal exercise to improve outcomes in overweight or obese women: a systematic review. Acta Obstet Gynecol Scand 91(5):538–545

126. Dekker Nitert M, Barrett HL, Denny KJ, McIntyre HD, Callaway LK, BAMBINO group (2015) Exercise in pregnancy does not alter gestational weight gain, MCP-1 or leptin in obese women. Aust N Z J Obstet Gynaecol 55:27–33

127. Mottola MF (2013) Physical activity and maternal obesity: cardiovascular adaptations, exercise recommendations, and pregnancy outcomes. Nutr Rev 71(Suppl 1):S31–S36

128. Choi J, Fukuoka Y, Lee JH (2013) The effects of physical activity and physical activity plus diet interventions on body weight in overweight or obese women who are pregnant or in postpartum: a systematic review and meta-analysis of randomized controlled trials. Prev Med 56:351–364

129. Wang C, Zhu W, Wei Y, Feng H, Su R, Yang H (2015) Exercise intervention during pregnancy can be used to manage weight gain and improve pregnancy outcomes in women with gestational diabetes mellitus. BMC Pregnancy Childbirth 15:255

130. Brekke HK, Bertz F, Rasmussen KM, Bosaeus I, Ellegard L, Winkvist A (2014) Diet and exercise interventions among overweight and obese lactating women: randomized trial of effects on cardiovascular risk factors. PLoS One 9(2):e88250

131. Rosa BV, Blair HT, Vickers MH, Dittmer KE, Morel PC, Knight CG et al (2013) Moderate exercise during pregnancy in Wistar rats alters bone and body composition of the adult offspring in a sex-dependent manner. PLoS One 8(12):e82378

132. Carter LG, Qi NR, De Cabo R, Pearson KJ (2013) Maternal exercise improves insulin sensitivity in mature rat offspring. Med Sci Sports Exerc 45:832–840

133. Camarillo IG, Clah L, Zheng W, Zhou X, Larrick B, Blaize N et al (2014) Maternal exercise during pregnancy reduces risk of mammary tumorigenesis in rat offspring. Eur J Cancer Prev 23:502–505

134. Eclarinal JD, Zhu S, Baker MS, Piyarathna DB, Coarfa C, Fiorotto ML et al (2016) Maternal exercise during pregnancy promotes physical activity in adult offspring. FASEB J 30:2541–2548

135. Carter LG, Ngo Tenlep SY, Woollett LA, Pearson KJ (2015) Exercise improves glucose disposal and insulin signaling in pregnant mice fed a high fat diet. J Diabetes Metab 6 (12)

136. Laker RC, Lillard TS, Okutsu M, Zhang M, Hoehn KL, Connelly JJ et al (2014) Exercise prevents maternal high-fat diet-induced hypermethylation of the Pgc-1alpha gene and age-dependent metabolic dysfunction in the offspring. Diabetes 63:1605–1611

137. Vega CC, Reyes-Castro LA, Bautista CJ, Larrea F, Nathanielsz PW, Zambrano E (2015) Exercise in obese female rats has beneficial effects on maternal and male and female offspring metabolism. Int J Obes 39:712–719

138. Wen LM, Baur LA, Simpson JM, Rissel C, Wardle K, Flood VM (2012) Effectiveness of home based early intervention on children's BMI at age 2: randomised controlled trial. BMJ 344:e3732

139. Williams SJ, Hemmings DG, Mitchell JM, McMillen IC, Davidge ST (2005) Effects of maternal hypoxia or nutrient restriction during pregnancy on endothelial function in adult male rat offspring. J Physiol 565:125–135

140. Bahari H, Caruso V, Morris MJ (2013) Late-onset exercise in female rat offspring ameliorates the detrimental metabolic impact of maternal obesity. Endocrinology 154:3610–3621

141. Rodriguez-Gonzalez GL, Vega CC, Boeck L, Vazquez M, Bautista CJ, Reyes-Castro LA et al (2015) Maternal obesity and overnutrition increase oxidative stress in male rat offspring reproductive system and decrease fertility. Int J Obes 39:549–556

142. Santos M, Rodriguez-Gonzalez GL, Ibanez C, Vega CC, Nathanielsz PW, Zambrano E (2015) Adult exercise effects on oxidative stress and reproductive programming in male offspring of obese rats. Am J Physiol Regul Integr Comp Physiol 308:R219–R225

Chapter 8

Utility of Small Animal Models of Developmental Programming

Clare M. Reynolds and Mark H. Vickers

Abstract

Any effective strategy to tackle the global obesity and rising noncommunicable disease epidemic requires an in-depth understanding of the mechanisms that underlie these conditions that manifest as a consequence of complex gene-environment interactions. In this context, it is now well established that alterations in the early life environment, including suboptimal nutrition, can result in an increased risk for a range of metabolic, cardiovascular, and behavioral disorders in later life, a process preferentially termed developmental programming. To date, most of the mechanistic knowledge around the processes underpinning development programming has been derived from preclinical research performed mostly, but not exclusively, in laboratory mouse and rat strains. This review will cover the utility of small animal models in developmental programming, the limitations of such models, and potential future directions that are required to fully maximize information derived from preclinical models in order to effectively translate to clinical use.

Key words Animal models, Developmental programming, Undernutrition, Overnutrition, Noncommunicable disease

1 Background

It is now clear from a range of epidemiological, clinical, and experimental observations that metabolic, cardiovascular, and behavioral disorders which commonly manifest in adult life may have their roots before birth. Moreover, this process, preferentially termed "developmental programming," needs to be viewed as a transgenerational phenomenon with evidence of transmission of disease traits across multiple generations [1]. The developmental origins of health and disease (DOHaD) hypothesis has been clearly demonstrated by a range of epidemiological studies dating from the late 1980s. Initial work by David Barker linked low birth weight to increased mortality from ischemic heart disease in the Hertfordshire Cohort [2, 3]. It was this landmark study which gave rise to the theory that early life adversity can predispose to increased risk of

Paul C. Guest (ed.), *Investigations of Early Nutrition Effects on Long-Term Health: Methods and Applications*, Methods in Molecular Biology, vol. 1735, https://doi.org/10.1007/978-1-4939-7614-0_8, © Springer Science+Business Media, LLC 2018

noncommunicable disease during adulthood. The DOHaD paradigm has evolved significantly since these early epidemiological studies. Cohorts detailing malnutrition during distinct periods of gestation, such as the Dutch Hunger Winter [4], the Siege of Leningrad [5], and Chinese Famine [6], have been useful for validation and further developing the initial findings of Barker.

However, human studies are limited by a range of factors, including quality of data records for retrospective studies and avoidance of interacting experimental confounders, long generation times, availability of accurate methods to measure nutritional intakes, and investigation of effects at the level of individual tissues. Most evidence for the mechanistic underpinnings of developmental programming has therefore been derived via observations from experimental animal models, particularly the rodent. By mimicking the common problems associated with human pregnancy and early life in animal models, the molecular complexities which contribute to the developmental programming of obesity and cardiometabolic disease have been uncovered. Small animal models allow the opportunity to analyze discrete developmental windows of plasticity and interactions therein and also the impact of inappropriate predictive adaptive responses [7] in response to early life cues that are maladaptive for the actual postnatal environment (i.e., development of a thrifty phenotype) [8]. Work in small animal models of programming, particularly around altered early life nutrition, has now clearly shown that a range of challenges during the peri-conceptual period, pregnancy, and/or neonatal life results in development of an aberrant cardiometabolic and behavioral phenotype, effects of which can be passed across generations.

2 Study Design

It has been suggested that small animal experiments be designed to be more akin to human randomized controlled trials [7]. Part of the rationale for this was based on the vast number of preclinical studies undertaken in some research domains, such as neuroprotection in stroke, that have not translated to any proven clinical efficacy. The authors cited high false-positive rates from poorly performed preclinical trials and the lack of standards required for reporting animal models as compared to that for human clinical studies [7]. In the DOHaD field, the variety of small animal models utilized have been relatively consistent in producing a phenotype that displays some form of aberrant metabolic phenotype reflecting alterations in the early life environment. Although this confirms and reinforces the programming paradigm, there are potential refinements to the way these studies are undertaken and reported that will further aid in the evaluation of the key processes involved. In 2010, a set of guidelines around reporting of in vivo experiments was produced (ARRIVE, Animal Research: Reporting In Vivo Experiments) [8, 9]. The ARRIVE guidelines comprise an itemized checklist that details the information that research publications that utilize animal models should incorporate.

In addition to animal numbers, these include specific characteristics of the animals used (including species, strain, sex, and genetic background); housing conditions (e.g., singleton versus group housing) and husbandry (including number of true biological replicates used); and the experimental, statistical, and analytical methods (including information on the methods used to reduce experimental bias such as blinding and randomization). As per the ARRIVE guidelines, "all the items in the checklist have been included to promote high-quality, comprehensive reporting to allow an accurate critical review of what was done and what was found."

3   Rodent Models

Rodents, specifically rats (*Rattus norvegicus*) and mice (*Mus musculus*), are the most common animal models used for modern biomedical research. While logistical reasons such as cost and space effectiveness play a role, there are also key physiological reasons which make rodents an attractive model for examining the developmental programming paradigm. Rodents have relatively short gestational times (19–22 days), wean within 3–4 weeks, undergo puberty by 5–6 weeks, and live for relatively short time periods (2–4 years) making it feasible to study the complete lifespan and indeed transgenerational effects (which require three generations post insult in females and two generations in males) [9]. While the production of large litters in rodents (numbers can vary based on species and strain) can be seen as a disadvantage given the deviation from human physiology, this factor can be beneficial in some instances. Large litter sizes provide the capacity to examine both male and female offspring exposed to the same in utero stressor and the ability to obtain tissues from several distinct timepoints within one litter. Another major benefit of working with rodents is the molecular tools available based on sequencing of the mouse [10] and rat [11] genomes.

Hemochorial placentation is another feature which makes rodents a useful model for the study of DOHaD. As the organ which bridges maternal and fetal environments, the placenta is critical to understanding the mechanisms through which developmental programming occurs. Use of rodent placentae has been instrumental in our understanding of the regulatory mechanisms which govern placental development and its impact on health and disease in the mother and offspring. However, it is important to note that placental development in rodents, while sharing similar functional and structural characteristics with humans, have distinct features. Rodents have an inverted yolk sac placenta which persists throughout the duration of pregnancy and is involved in feto-maternal exchange [12]. In humans, the yolk sac precedes placental development and acts as a rudimentary circulatory system, but it is obsolete by the end of the first trimester. Another major difference between rodent and human placenta is the endocrine profile. In humans, the endocrine capacity is mediated by the corpus luteum (develops from the ovarian follicle) during early pregnancy; this is stimulated by human chorionic gonadotropin (hCG) to produce progesterone [13]. As pregnancy progresses, the placental syncytiotrophoblast cells produce hormones such as hCG, progesterone, estrogen, placental lactogen, and placental growth hormone to maintain pregnancy [14]. However, in rodents the corpus luteum produces progesterone throughout pregnancy, while other hormones such as lactogen and growth hormone are produced from the pituitary [15].

While sheep models are still commonly used in DOHaD research, a significant proportion of studies are carried out in rodents, particularly the rat. The experimental approaches which are commonly used are detailed below.

4   Undernutrition

Initial animal studies predominantly focused on early life nutritional restriction to mimic the epidemiological data from which the DOHaD hypothesis was generated. There are a wide range of factors which can influence nutrient availability in utero including maternal dietary intake, maternal metabolic health, and placental function. Deficiencies in any of these factors can have a detrimental effect on fetal growth, organ development, and long-term health outcomes. Several different approaches have been established in rodent models to examine the impact of suboptimal early life nutrient availability.

4.1 Global Undernutrition

Global nutrition restriction is a well-established model of developmental programming which displays similarities to the famine cohorts from which early DOHaD theories were developed [4–6]. The degree of restriction in rat models varies widely and ranges from moderate (30% kcal reduction) to severe (70% kcal reduction) restriction of ad libitum food intakes. In general, the offspring from these malnourished mothers display intrauterine growth restriction and reduced birth weight followed by catch-up growth upon weaning culminating in obesity and cardiometabolic dysfunction during adulthood. This model is most likely the most physiologically relevant for the study of starvation during pregnancy and provides a unique insight into the effects of extreme nutrition deprivation into the next generation. Its most useful application is demonstrating the effect of the predictive adaptive response (PAR) hypothesis in vivo [7]. Application of a high-fat diet (HFD) in the postnatal period has shown that a "second hit" is often required to fully appreciate the impact of maternal malnutrition on offspring outcomes [16, 17].

Reynolds et al. demonstrated that 50% maternal undernutrition in Sprague-Dawley rats resulted in increased weight accompanied by PPARγ-mediated adipose tissue hypertrophy and insulin resistance. These effects were significantly exacerbated by a postnatal HFD. Furthermore these animals displayed significantly increased adipose tissue inflammation, both in the adipocytes themselves and the stromal vascular fraction [17]. There is ample evidence that both male and female offspring exposed to global maternal undernutrition in utero develop obesity and insulin resistance. Adipogenic processes appear to be a common consequence of maternal undernutrition [18–20]. Thompson et al. recently demonstrated that adipocyte hypertrophy in male adult offspring of 30% nutrient-restricted Wistar mothers occurred in multiple adipose tissue depots. These morphological alterations were linked to increased capacity for de novo fatty acid synthesis [21]. Further evidence in a Sprague-Dawley moderate (50%) undernutrition model demonstrates that histone modifications in the glucose transporter GLUT4 may be a causative factor in the development of insulin resistance in skeletal muscle [22] potentially implicating epigenetic processes in programmed metabolic perturbations.

4.2 Low Protein

In addition to global nutrient restriction during pregnancy, restriction of specific macronutrients represents another common method used for studying the molecular mechanisms associated with developmental programming. Protein consumption during pregnancy is essential for growth and development of the fetus and deficiencies can result in permanent alterations in the structure and function of key organs such as the pancreas and kidney [23]. When considered in relation to human situations, this model can provide a useful assessment of health outcomes in offspring from vegetarian mothers. For example, a study in India reported that offspring from mothers consuming a low-protein diet during pregnancy displayed increased visceral fat deposition and a predisposition to insulin resistance [24]. Rodent chow diets are composed of approximately 15–20% protein. Low-protein models typically reduce protein composition by 50% with the Hope Farm Diet [25], one of the first low-protein diets used in programming studies (isocaloric diet with 8% protein).

This type of model has been extensively utilized in the study of programmed hypertension, insulin resistance, and type 2 diabetes in the rat. Reduction in the number of fetal pancreatic beta cells is observed in low-protein models [26] and likely mediates impaired glucose tolerance in adult offspring [27–29], notably this is driven by different mechanisms in males and females [29]. Altered mitochondrial function is another mechanism likely to contribute to altered metabolic health in low-protein models. Several studies have observed reduced mitochondrial copy number and expression of genes relating to mitochondrial respiration [30–33]. These effects are commonly observed in humans prior to the onset of T2DM and interestingly are correlated with low-birthweight [34]. However, based on the composition of the protein restriction, both birthweights and adult offspring health outcomes can vary widely [35]. Differences in the proportion of carbohydrate and lipid to adjust the caloric content of the diet following protein restriction can represent another major confounder in this type of study. Therefore, care must be exercised when formulating diets in this type of study.

4.3   Uterine Ligation

Surgical methods such as uterine artery ligation are also utilized to examine the DOHaD paradigm. This procedure was developed in the early 1960s by Wigglesworth et al. [36] as a model of fetal growth restriction and has since been employed to mimic uteroplacental insufficiency induced-uterine growth restriction (IUGR). While maternal undernutrition is the most likely cause of IUGR in developing nations, maternal physiological issues resulting in reduced placental blood flow generally results in IUGR. Therefore, uterine artery ligation represents the one of most appropriate models for the study of IUGR relevant to developed nations. There are two main study designs to consider in models of uterine ligation. In the case of bilateral uterine ligation during the latter stages of pregnancy, a sham procedure is carried out on control animals to ensure the effects observed are not as a result of maternal stress. This is the most commonly used protocol. Unilateral uterine artery ligation utilizes pups from the unligated uterine horn as control animals. While it is useful to have controls exposed to the same maternal conditions, factors such as hyperperfusion of the unligated artery can result in overgrowth of control pups [37]. A further consideration is the late nature of this insult. In order to produce viable pups, the ligation surgery can only be carried out in the last few days of pregnancy [37]. The window of maternal insult is an important consideration in developmental programming studies; therefore this experimental paradigm can only address programming effects related to late pregnancy.

This model has been a useful tool in deciphering the mechanisms of metabolic programming following IUGR. However, there are inconsistencies in the occurrence of IUGR in pups from ligated mothers. A recent meta-analysis indicated that there was no overall effect of uterine artery ligation on birthweight or catch-up growth [38]. Furthermore there is evidence that the sham control operation can induce alterations in lipid homeostasis and therefore may not be the most appropriate model for study of lipid metabolism [39]. This indicates a need to include a nonoperated negative control. However, this study did indicate a negative effect of uterine ligation on offspring glucose tolerance. Several studies have reported adverse effects on insulin sensitivity in offspring. Simmons et al. indicated that ligated offspring demonstrate hyperglycemia and hyperinsulinemia by 7 weeks of age which progressed to insulin resistance by 26 weeks. It is likely that a reduction in pancreatic beta cell mass influences these effects [40]. Furthermore, epigenetics has been implicated in the conference of alterations in growth and insulin sensitivity. Changes in hepatic histone activity in the IGF-1 gene influence the growth trajectory of ligated offspring [41]. Histone deacetylation and DNA methylation of PDX-1, a key gene in the development of the pancreas, has also been implicated in reduced beta-cell development and functionality [42, 43]. In addition to insulin resistance, there is also evidence of hypertension and cardiorenal dysfunction [44]. Reduced nephron endowment [45], alterations in the renin-angiotensin system [46], and reduced cardiomyocyte numbers [47] contribute to these effects.

5 Overnutrition

5.1 Diets

While most of the initial work conducted in the developmental programming field has focused on models of undernutrition which support the original findings of David Barker, research has now turned to address the worldwide problem of maternal obesity. Similarly to models of maternal nutrient restriction during pregnancy, there are several distinct dietary overnutrition models during a range of developmental windows. The most extensively studied models are the cafeteria diet and the single nutrient-enriched diets. However, several models of genetically induced obesity have been examined in recent years.

There has been some debate over the fat source and composition used in maternal obesogenic models and their utility as compared to cafeteria style diets. Purified high-fat diets are typically successful in inducing weight gain in the rodent, and some have argued that a "more robust" diet-induced obesity model is that based on feeding rodents a diet that consists of highly palatable, energy-dense human junk food—the so-called "cafeteria" diet [48, 49]. The single source fat diet has the benefit that these diets are typically "open source" and will not change in composition over time thus are highly reproducible. The cafeteria diet approach can present logistical issues in that they suffer from a lack of standardization and also the difficulty in accurately assessing nutritional intake given the range of foods offered. The recent work by Barrett et al. covers this area well [50]. However, both approaches result in a phenotype commonly characterized by increased adiposity, cardiometabolic and reproductive dysfunction, and disorders of energy balance; thus both are useful tools to further understanding of mechanisms involved in aberrant programming. It is noteworthy that models of both maternal undernutrition and maternal obesity can result in common phenotypic outcomes in offspring. Although the mechanisms may not be the same, it must be remembered that in some cases, "overnutrition" may in fact represent malnutrition due to micronutrient deficiencies in high-calorie diets and thus may account for some of the similarities observed across models.

5.2   Cafeteria Diet

This model was designed to mimic the highly palatable, calorie-dense foods which are prevalent in Western society and are known to be associated with fat accretion and metabolic dysfunction. This model exposes animals ad libitum to foods such as crisps, chocolate, condensed milk, biscuits, cheese, and deli meats inducing increased weight gain and food intake resulting in an inflammatory response and insulin resistance in the adipose tissue and liver. While fat and sugar intakes are significantly increased, protein consumption remains relatively consistent. Despite similarities to Western diets, cafeteria diets are not as widely used in developmental programming research as single nutrient diets. This may be due to the complex nature of cafeteria diets and the difficulty disentangling the specific contributions of any one nutrient. Nonetheless, this model is a useful tool for inducing maternal obesity during gestation, although results are not as consistent as single nutrient models.

This model has significant effects on maternal metabolism with increased adiposity, hyperglycemia, and insulin resistance in both rats and mice [51, 52]. This has been shown to program adiposity and cardiometabolic dysfunction in offspring via a range of molecular mechanisms such as increased expression of adipogenic genes [53, 54], global DNA methylation in the liver [55], and increased expression of orexigenic neuropeptides in the hypothalamus [56]. However, several studies have also demonstrated that maternal exposure to a cafeteria diet did not elicit metabolic derangements in either male or female offspring despite significant weight gain during pregnancy. However, the effects of exposure were observed in both the second generation with significantly increased indices of insulin resistance and increased fat mass [51] or following a postnatal HFD [57].

5.3   Single Nutrient

Despite the relevance of the cafeteria diet to modern dietary intakes, it is difficult to determine the specific contribution of individual nutrients. Therefore single nutrient-enriched diets are the most common models of maternal overnutrition in the DOHaD field. High-fat diets remain the most extensively studied single nutrient diet in the developmental field. The fat content in rodent diets is typically obtained from plant-based sources and accounts for 6–10% of total caloric content with high-fat diets ranging from 20 to 60% kcal from a variety of sources. There are several factors which must be taken into consideration when designing HFD-induced developmental programming studies. The source of the fat is important [58] as plant-based oils such as soybean oil can contain phytoestrogens which can impact the lipogenic pathways and glucose homeostasis [59], whereas animal-based fats such as lard can activate toll-like receptor (TLR) pathways contributing to a pro-inflammatory phenotype [60]. The content of other macronutrients in high-fat diets is also important, while protein content is rarely changed; the ratio between fat and carbohydrate is another consideration in this type of study. For example, both sucrose and fructose have been associated with metabolic derangements in rodents particularly during pregnancy and lactation [61, 62] with detrimental effects on offspring health. As chow diets are vastly different in macronutrient composition to high-fat diets, it is essential that control diets are matched in micronutrient content as well as fat and carbohydrate source.

Another factor which should be taken into consideration is the species and strain of experimental model as some common strains such as the Wistar and Sprague-Dawley rat [63] as well as C57BL/6 [64] mice can be prone to obesity. The phenotype of offspring from HFD fed mothers is closely related to the metabolic syndrome in humans with evidence of obesity, insulin resistance, hyperglycemia, hyperlipidemia, endothelial dysfunction, and hypertension [16, 65–68]. However a recent meta-analysis demonstrated only marginal differences in birthweight when compared to models of undernutrition, notably there were differential effects between mice and rat strains [69]. In addition to high-fat models of single nutrient excess, high-sugar models are being incorporated in developmental programming research. While increased fat consumption has been traditionally associated with obesity onset, there is increasing evidence that sugar-sweetened products such as soft drinks, particularly those high in high-fructose corn syrup, contribute significantly to the current obesity epidemic [70, 71]. Therefore, models of maternal sugar consumption during pregnancy have become ever more relevant as a tool to delineate the pathways which lead to developmental programming of adiposity and metabolic dysfunction in offspring. Indeed the detrimental effects of fructose consumption during pregnancy on the health of the mother and offspring have been clearly outlined [61, 72, 73].

6 Genetic Manipulation

Overall in biomedical research, mice are the species of choice for many experimental models given the ease of genetic manipulation. However, these models have been relatively underutilized in the developmental programming field. The models which have been used are designed to mimic specific pregnancy complications rather than nutrition- or stress-related DOHaD paradigms. While the specific gene defects examined may not be reflective of normal human physiology, these models are important for determining the underlying mechanisms and contribution of specific targets related to maternal pathology on offspring health in later life. Examples include the *db/db* model of gestational diabetes which contains a mutation in the leptin receptor. While homozygotes are infertile, heterozygotes are healthy and only develop diabetes during pregnancy [74]. The offspring of *db/db* heterozygotes demonstrate similarities with children from mothers who had gestational diabetes and are predisposed to obesity and insulin resistance in adulthood [75]. However, the utility of this model has come into question with several groups reporting a loss of the gestational diabetes phenotype in these mice [76, 77]. Furthermore, IGF1 knockouts and placental IGF-2 knockout has demonstrated the importance of this class of hormones on fetal growth [78, 79], a finding with important implications for the understanding of IUGR. Interestingly, recent advances have provided the technology to edit the rat genome (CRISPR/Cas-mediated gene editing). However, this technology has not been utilized in DOHaD research to date.

7   Interventions

A number of intervention strategies across a range of animal models have already shown efficacy in reversing or ameliorating programmed disorders, but little has translated to clinical utility. Interventions have covered pharmacologic (e.g., leptin [80–82], growth hormone [17, 83, 84]), nutritional (e.g., methyl donors [85, 86], taurine [87, 88], lipid supplements [67, 89, 90]), and exercise paradigms [91, 92]. However, it is clear from data to date that it is not a "one-size-fits-all," and those at risk need to be identified; treatment to offspring from control pregnancies, for example, has the potential to induce metabolic derangement due to interfering with already-replete systems.

This is difficult as programming represents a continuum and risk potential is not just in those at either end of the birth weight spectrum as programming can occur across normal birth weights. Compounding this is the potential for sexually dimorphic responses to treatment, dosing/duration effects, and additional unforeseen effects in mother and/or offspring. An example of this is maternal folic acid supplementation which has shown efficacy in small animals models in reversing epigenetic effects associated with a maternal low-protein diet (e.g., hypomethylation of key genes such as the glucocorticoid receptor) [85] but, given its broad role in one-carbon metabolism, also has the potential to adversely impact on other outcomes including insulin sensitivity and mammary tumorigenesis [93]. A further example is that of neonatal leptin treatment whereby leptin treatment to male rat pups born to mothers with normal pregnancies resulted in an adverse metabolic response in these offspring as adults [94], which may be a consequence of inducing an amplified and prolonged leptin surge, a feature that has been previously shown to be a characteristic of offspring born to obese mothers [95]. Given that a one-size-fits-all approach may indeed induce harm, this has brought into focus the role of biomarkers in later risk prediction—an example being the utility of gene promoter methylation in cord tissue as a predictor of later childhood obesity [96]. Such work, although largely associative in nature, suggests that perinatal epigenetic analysis has utility in identifying vulnerability at the level of an individual for later risk of obesity and metabolic disease. There is also a need to clearly identify the source and quality of the supplement used. As an example of this, outcomes related to efficacy of omega-3/fish oil supplements are widely discrepant with some studies reporting beneficial effects, while others report no change compared to unsupplemented controls. Recent evidence would suggest that the quality of the supplement used may be the source of such experimental variability with many supplements shown to be oxidized and thus will be of little efficacy and even potentially harmful [97, 98]. Ideally, analysis of any supplements used should be incorporated into any publication to verify the integrity of the product used.

As with nutritional supplements, research outcomes can also be impacted by the source of the pharmacologic agent used and whether the treatment paradigm utilizes a homologous approach. An example of this is the adipokine leptin as utilized in rat models whereby use of recombinant rat leptin avoids potential confounds associated with use of human leptin in the rat including altered natriuresis and renal function (human leptin has natriuretic activity in the rat) and altered responsiveness to diet-induced obesity [99–101]. Differences observed across laboratories in experimental outcomes derived from similar animal studies are often ascribed to differences in experimental detail, e.g., dietary composition or feeding regime [50]. Careful consideration should be given in studies of dietary manipulation to try and standardize as much as is possible and thus reduced inherent variability—all papers in this field should ideally detail the composition of the experimental diets utilized. As such, "open-source" high-fat diets are standardized and are thus commonly used, whereas cafeteria-style diets will inevitability incur more variability across studies given the wide range of foods offered. As an example, there is a move away from traditional "chow"-style control diets to semi-purified control diets that represent greater consistency in composition over time. Greater investigation of food intake behaviors should also be undertaken (i.e., meal patterns).

However, layered on this are baseline differences within even the most commonly used rodent models used including the Wistar and Sprague-Dawley rat across laboratories and even the effects of experimental handling differing between research groups and potential impact upon experimental outcomes.

8   Transgenerational Studies

Where small animal models represent great utility is in the area of transgenerational studies and propagation of the programmed phenotype [1]. Given the short generation times required, rodent studies are invaluable in assessment of true transgenerational outcomes from both the paternal and maternal lineages to the F3 generation and beyond. Although many studies have reported programming effects through to F2, the F2 offspring still represent a memory of the original environmental insult and thus only F3 and beyond can be considered truly transgenerational and of those studies to date almost half report an amelioration or indeed reversal of the programmed phenotype.

9   Sex-Specific Differences

While developmental programming occurs in both males and females, it is clear from experimental models to date that similar environmental exposures at the same stage of development can have very different effects in both male and female offspring. Of note, the National Institutes of Health (NIH) recently issued a statement recognizing the need to interrogate such sexual dimorphic responses in research [102]: "NIH expects that sex as a biological variable will be factored into research designs, analyses, and reporting in vertebrate animal and human studies. Strong justification from the scientific literature, preliminary data, or other relevant considerations must be provided for applications proposing to study only one sex."

As an example, early life insults, including maternal stressors, are well established to result in an increased risk for neurodevelopmental disorders in offspring but are known to impact upon males significantly more than females [103]. There are also sex differences in the placenta, which will lead to differences in response to environmental insults and to sex-specific signals transmitted to the developing fetus as recently shown in rat placentae following maternal exposure to a HF/high salt diet [104]. Many studies are undertaken in male offspring to avoid the potential confounds of estrus given the short cycling period in the rodent [105]. However, these limitations can be potentially met by staging the females at the time of sampling and using this as a factor in any subsequent analysis. Accounting for sexual dimorphism in programming is important for translation as data to date imply that the optimal conditions for fetal/infant development, and therefore advice offered to pregnant women, may differ according to the sex of their offspring.

In order to efficiently determine true sex-specific effects, there are several considerations which should be noted. The lactational period represents a critical programming period, and many studies have identified that nutritional insults during this time can influence adult-onset disease [106]. Therefore, offspring studied should be standardized to equal numbers of males and females after birth to normalize early life nutrition. Furthermore, animals studied should be littermates to ensure that genetic (or indeed epigenetic) factors do not influence the outcomes. There are several possible mechanisms which may mediate the sexually dimorphic effects observed in developmental programming models. The first of which is differences in sex hormones where estrogen is thought to contribute to the protective effects in females. Ojeda et al. demonstrated that placental insufficiency which mediated developmental programming of hypertension in adult male rats could be reversed by castration [107]. Similarly, females were protected from hypertension, but ovariectomy increased blood pressure significantly [46]. Furthermore, restoration of estrogen could normalize increased blood pressure [108]. These studies and others demonstrate that cardiometabolic tissues are sensitive to sex hormones and may influence the sexually dimorphic phenotypes observed across programming studies.

However, there is evidence that sex-specific changes can occur prior to birth during a timeframe when sex hormones are not developed and therefore cannot account for differences between males and females. Another factor which may play a role is the independent effect of XX/XY sex chromosomes [109]. While females have two X chromosomes, one of these is randomly inactivated thereby preventing gene expression. However some of these genes escape this process and are more highly expressed in females. While manipulation of the gonad inducing Sry gene in the four core genotypes (FCG) model [110] and the pseudoautosomal region of the Y chromosome in the XY*model [111, 112] have been informative in developing the concept of sex chromosome-based sexually dimorphic phenotypes, there is little evidence of how this paradigm relates to sexually dimorphic effects stemming from adverse early life conditions.

10   Summary

Small animal models have been invaluable in understanding some of the mechanisms underpinning developmental programming. However, careful consideration needs to be taken when planning such studies to optimize effective translation to the human setting. These include incorporation of sex-specific effects, accounting for potential differences in timing of developmental processes (rodents are altricial so some developmental processes are finalized in the neonatal period as opposed to in utero in the human), fully balanced experimental design, and use of appropriate control diets. Small animal models can also confer considerable advantages over other model species including the ease of genetic manipulation, more so now with greater access to CRISPR-based technologies, and are used in transgenerational studies to examine true trait transmission across generation (i.e., F3 and beyond). Although some intervention strategies have shown to be highly efficacious in the rodent at reversing programming effects, it is clear that a "one-size-fits-all" approach will not work, and thus further understanding of the mechanisms involved, including the identification and use of biomarkers, is needed before safe and effective translation to clinical use can be undertaken.

References

1. Aiken CE, Ozanne SE (2014) Transgenerational developmental programming. Hum Reprod Update 20:63–75

2. Barker DJ, Winter PD, Osmond C, Margetts B, Simmonds SJ (1989) Weight in infancy and death from ischaemic heart disease. Lancet 2:577–580

3. Barker DJ (1990) The fetal and infant origins of adult disease. BMJ 301:1111

4. Lumey LH, Van Poppel FW (1994) The Dutch famine of 1944–45: mortality and morbidity in past and present generations. Soc Hist Med 7:229–246

5. Stanner SA, Yudkin JS (2001) Fetal programming and the Leningrad Siege study. Twin Res 4:287–292

6. Wang N, Wang X, Li Q, Han B, Chen Y, Zhu C et al (2017) The famine exposure in early life and metabolic syndrome in adulthood. Clin Nutr 36:253–259

7. Gluckman PD, Hanson MA, Beedle AS, Spencer HG (2008) Predictive adaptive responses in perspective. Trends Endocrinol Metab 19:109–110

8. Hales CN, Barker DJ (2001) The thrifty phenotype hypothesis. Br Med Bull 60:5–20

9. Dickinson H, Moss TJ, Gatford KL, Moritz KM, Akison L, Fullston T et al (2016) A review of fundamental principles for animal models of DOHaD research: an Australian perspective. J Dev Orig Health Dis 7:449–472

10. Mouse Genome Sequencing Consortium et al (2002) Initial sequencing and comparative analysis of the mouse genome. Nature 420:520–562

11. Gibbs RA, Weinstock GM, Metzker ML, Muzny DM, Sodergren EJ, Scherer S et al (2004) Genome sequence of the Brown Norway rat yields insights into mammalian evolution. Nature 428:493–521

12. Carter AM, Enders AC (2016) Placentation in mammals: definitive placenta, yolk sac, and paraplacenta. Theriogenology 86:278–287

13. Cole LA (2012) hCG, the wonder of today's science. Reprod Biol Endocrinol 10:24. https://doi.org/10.1186/1477-7827-10-24

14. Schmidt A, Morales-Prieto DM, Pastuschek J, Fröhlich K, Markert UR (2015) Only humans

have human placentas: molecular differences between mice and humans. J Reprod Immunol 108:65–71

15. Forsyth IA (1994) Comparative aspects of placental lactogens: structure and function. Exp Clin Endocrinol 102:244–251

16. Howie GJ, Sloboda DM, Reynolds CM, Vickers MH (2013) Timing of maternal exposure to a high fat diet and development of obesity and hyperinsulinemia in male rat offspring: same metabolic phenotype, different developmental pathways? J Nutr Metab 2013:517384. https://doi.org/10.1155/2013/517384

17. Reynolds CM, Li M, Gray C, Vickers MH (2013) Preweaning growth hormone treatment ameliorates adipose tissue insulin resistance and inflammation in adult male offspring following maternal undernutrition. Endocrinology 154:2676–2686

18. Lecoutre S, Marousez L, Drougard A, Knauf C, Guinez C, Eberlé D et al (2017) Maternal undernutrition programs the apelinergic system of adipose in adult male rat offspring. J Dev Orig Health Dis 8:3–7

19. Lecoutre S, Breton C (2014) The cellularity of offspring's adipose tissue is programmed by maternal nutritional manipulations. Adipocytes 3:256–262

20. Howie GJ, Sloboda DM, Vickers MH (2012) Maternal undernutrition during critical windows of development results in differential and sex-specific effects on postnatal adiposity and related metabolic profiles in adult rat offspring. Br J Nutr 108:298–307

21. Thompson N et al (2014) Metabolic programming of adipose tissue structure and function in male rat offspring by prenatal undernutrition. Nutr Metab (Lond) 11:50. https://doi.org/10.1186/1743-7075-11-50

22. Raychaudhuri N, Raychaudhuri S, Thamotharan M, Devaskar SU (2008) Histone code modifications repress glucose transporter 4 expression in the intrauterine growth-restricted offspring. J Biol Chem 283:3611–13626

23. Jahan-Mihan A, Rodriguez J, Christie C, Sadeghi M, Zerbe T (2015) The role of maternal dietary proteins in development of metabolic syndrome in offspring. Nutrients 7:9185–9217

24. Yajnik CS, Fall CH, Coyaji KJ, Hirve SS, Rao S, Barker DJ et al (2003) Neonatal anthropometry: the thin-fat Indian baby. The Pune Maternal Nutrition Study. Int J Obes Relat Metab Disord 27:173–180

25. Snoeck A, Remacle C, Reusens B, Hoet JJ (1990) Effect of a low protein diet during pregnancy on the fetal rat endocrine pancreas. Biol Neonate 57:107–118

26. Sparre T, Reusens B, Cherif H, Larsen MR, Roepstorff P, Fey SJ et al (2003) Intrauterine programming of fetal islet gene expression in rats—effects of maternal protein restriction during gestation revealed by proteome analysis. Diabetologia 46:1497–1511

27. Langley SC, Browne RF, Jackson AA (1994) Altered glucose tolerance in rats exposed to maternal low protein diets in utero. Comp Biochem Physiol Physiol 109:223–229

28. Zambrano E, Bautista CJ, Deás M, Martínez-Samayoa PM, González-Zamorano M, Ledesma H et al (2006) A low maternal protein diet during pregnancy and lactation has sex and window of exposure-specific effects on offspring growth and food intake, glucose metabolism and serum leptin in the rat. J Physiol 571:221–230

29. Chamson-Reig A, Thyssen SM, Hill DJ, Arany E (2009) Exposure of the pregnant rat to low protein diet causes impaired glucose homeostasis in the young adult offspring by different mechanisms in males and females. Exp Biol Med (Maywood) 234:1425–1436

30. Claycombe KJ, Vomhof-DeKrey EE, Garcia R, Johnson WT, Uthus E, Roemmich JN et al (2016) Decreased beige adipocyte number and mitochondrial respiration coincide with increased histone methyl transferase (G9a) and reduced FGF21 gene expression in Sprague-Dawley rats fed prenatal low protein and postnatal high-fat diets. J Nutr Biochem 31:113–121

31. Ferreira DJS, da Silva Pedroza AA, Braz GR, da Silva-Filho RC, Lima TA, Fernandes MP et al (2016) Mitochondrial bioenergetics and oxidative status disruption in brainstem of weaned rats: immediate response to maternal protein restriction. Brain Res 1642:553–561

32. Claycombe KJ, Roemmich JN, Johnson L, Vomhof-DeKrey EE, Johnson WT (2015) Skeletal muscle Sirt3 expression and mitochondrial respiration are regulated by a prenatal low-protein diet. J Nutr Biochem 26:184–189

33. Moraes C, Rebelato HJ, Amaral ME, Resende TM, Silva EV, Esquisatto MA et al (2014) Effect of maternal protein restriction on liver metabolism in rat offspring. J Physiol Sci 64:347–355

34. Brøns C, Jensen CB, Storgaard H, Alibegovic A, Jacobsen S, Nilsson E et al (2008) Mitochondrial function in skeletal

muscle is normal and unrelated to insulin action in young men born with low birth weight. J Clin Endocrinol Metab 93:3885–3892

35. Armitage JA, Poston L, Taylor PD (2008) Developmental origins of obesity and the metabolic syndrome: the role of maternal obesity. Front Horm Res 36:73–84

36. Wigglesworth JS (1964) Experimental growth retardation in the foetal rat. J Pathol Bacteriol 88:1–13

37. Kollée LA, Monnens LA, Trijbels JM, Veerkamp JH, Janssen AJ (1979) Experimental intrauterine growth retardation in the rat. Evaluation of the Wigglesworth model. Early Hum Dev 3:295–300

38. Neitzke U, Harder T, Schellong K, Melchior K, Ziska T, Rodekamp E et al (2008) Intrauterine growth restriction in a rodent model and developmental programming of the metabolic syndrome: a critical appraisal of the experimental evidence. Placenta 29:246–254

39. Nüsken K-D, Dötsch J, Rauh M, Rascher W, Schneider H (2008) Uteroplacental insufficiency after bilateral uterine artery ligation in the rat: impact on postnatal glucose and lipid metabolism and evidence for metabolic programming of the offspring by sham operation. Endocrinology 149:1056–1063

40. Simmons RA, Templeton LJ, Gertz SJ (2001) Intrauterine growth retardation leads to the development of type 2 diabetes in the rat. Diabetes 50:2279–2286

41. Fu Q, Yu X, Callaway CW, Lane RH, McKnight RA (2009) Epigenetics: intrauterine growth retardation (IUGR) modifies the histone code along the rat hepatic IGF-1 gene. FASEB J 23:2438–2449

42. Park JH, Stoffers DA, Nicholls RD, Simmons RA (2008) Development of type 2 diabetes following intrauterine growth retardation in rats is associated with progressive epigenetic silencing of Pdx1. J Clin Invest 118:2316–2324

43. Thompson RF, Fazzari MJ, Niu H, Barzilai N, Simmons RA, Greally JM (2010) Experimental intrauterine growth restriction induces alterations in DNA methylation and gene expression in pancreatic islets of rats. J Biol Chem 285:15111–15118

44. Cheong JN, Cuffe JSM, Jefferies AJ, Moritz KM, Wlodek ME (2016) Adrenal, metabolic and cardio-renal dysfunction develops after pregnancy in rats born small or stressed by physiological measurements during pregnancy. J Physiol 594:6055–6068

45. Moritz KM, Mazzuca MQ, Siebel AL, Mibus A, Arena D, Tare M et al (2009) Uteroplacental insufficiency causes a nephron deficit, modest renal insufficiency but no hypertension with ageing in female rats. J Physiol 587:2635–2646

46. Ojeda NB et al (2011) Hypersensitivity to acute ANG II in female growth-restricted offspring is exacerbated by ovariectomy. Am J Physiol Regul Integr Comp Physiol 301:R199–1205

47. Black MJ, Siebel AL, Gezmish O, Moritz KM, Wlodek ME (2012) Normal lactational environment restores cardiomyocyte number after uteroplacental insufficiency: implications for the preterm neonate. Am J Physiol Regul Integr Comp Physiol 302:R1101–R1110

48. Johnson AR, Wilkerson MD, Sampey BP, Troester MA, Hayes DN, Makowski L et al (2016) Cafeteria diet-induced obesity causes oxidative damage in white adipose. Biochem Biophys Res Commun 473:545–550

49. Sampey BP, Vanhoose AM, Winfield HM, Freemerman AJ, Muehlbauer MJ, Fueger PT et al (2011) Cafeteria diet is a robust model of human metabolic syndrome with liver and adipose inflammation: comparison to high-fat diet. Obesity (Silver Spring) 19:1109–1117

50. Barrett P, Mercer JG, Morgan PJ (2016) Preclinical models for obesity research. Dis Model Mech 9:1245–1255

51. King V, Dakin RS, Liu L, Hadoke PW, Walker BR, Seckl JR et al (2013) Maternal obesity has little effect on the immediate offspring but impacts on the next generation. Endocrinology 154:2514–2524

52. Crew RC, Waddell BJ, Mark PJ (2016) Maternal obesity induced by a 'cafeteria' diet in the rat does not increase inflammation in maternal, placental or fetal tissues in late gestation. Placenta 39:33–40

53. Bayol SA, Simbi BH, Bertrand JA, Stickland NC (2008) Offspring from mothers fed a 'junk food' diet in pregnancy and lactation exhibit exacerbated adiposity that is more pronounced in females. J Physiol 586:3219–3230

54. Samuelsson A-M, Matthews PA, Argenton M, Christie MR, McConnell JM, Jansen EH et al (2008) Diet-induced obesity in female mice leads to offspring hyperphagia, adiposity, hypertension, and insulin resistance. Hypertension 51:383–392

55. Daniel ZC, Akyol A, McMullen S, Langley-Evans SC (2014) Exposure of neonatal rats to maternal cafeteria feeding during suckling

alters hepatic gene expression and DNA methylation in the insulin signalling pathway. Genes Nutr 9:365. https://doi.org/10.1007/s12263-013-0365-3

56. Chen H, Morris MJ (2009) Differential responses of orexigenic neuropeptides to fasting in offspring of obese mothers. Obesity (Silver Spring) 17:1356–1362

57. Mucellini AB et al (2014) Effects of exposure to a cafeteria diet during gestation and after weaning on the metabolism and body weight of adult male offspring in rats. Br J Nutr 111:1499–1506

58. Buettner R, Schölmerich J, Bollheimer LC (2007) High-fat diets: modeling the metabolic disorders of human obesity in rodents. Obesity 15:798–808

59. Lephart ED, Setchell KDR, Handa RJ, Lund TD (2004) Behavioral effects of endocrine-disrupting substances: phytoestrogens. ILAR J 45:443–454

60. Velloso LA, Folli F, Saad MJ (2015) TLR4 at the crossroads of nutrients, gut microbiota, and metabolic inflammation. Endocr Rev 36:245–271

61. Vickers MH, Clayton ZE, Yap C, Sloboda DM (2011) Maternal fructose intake during pregnancy and lactation alters placental growth and leads to sex-specific changes in fetal and neonatal endocrine function. Endocrinology 152:1378–1387

62. Tamura K, Ohki K, Kobayashi R, Uneda K, Azushima K, Ohsawa M et al (2014) Fetal programming by high-sucrose diet during pregnancy affects the vascular angiotensin II receptor–PKC–L-type Ca2+ channels (Cav1.2) axis to enhance pressor responses. Hypertens Res 37:796–798

63. Marques C, Meireles M, Norberto S, Leite J, Freitas J, Pestana D et al (2015) High-fat diet-induced obesity Rat model: a comparison between Wistar and Sprague-Dawley Rat. Adipocytes 5:11–21

64. West DB, Boozer CN, Moody DL, Atkinson RL (1992) Dietary obesity in nine inbred mouse strains. Am J Phys 262:R1025–R1032

65. Gray C, Harrison CJ, Segovia SA, Reynolds CM, Vickers MH (2015) Maternal salt and fat intake causes hypertension and sustained endothelial dysfunction in fetal, weanling and adult male resistance vessels. Sci Rep 5:9753. https://doi.org/10.1038/srep09753

66. Reynolds CM, Gray C, Li M, Segovia SA, Vickers MH (2015) Early life nutrition and energy balance disorders in offspring in later life. Nutrients 7:8090–8111

67. Reynolds CM, Segovia SA, Zhang XD, Gray C, Vickers MH (2015) Conjugated linoleic acid supplementation during pregnancy and lactation reduces maternal high-fat-diet-induced programming of early-onset puberty and hyperlipidemia in female rat offspring. Biol Reprod 92:40. https://doi.org/10.1095/biolreprod.114.125047

68. Alfaradhi MZ, Ozanne SE (2011) Developmental programming in response to maternal overnutrition. Front Genet 2:27. https://doi.org/10.3389/fgene.2011.00027

69. Ribaroff GA, Wastnedge E, Drake AJ, Sharpe RM, Chambers TJG (2017) Animal models of maternal high fat diet exposure and effects on metabolism in offspring: a meta-regression analysis. Obes Rev 18:673–686

70. Shapiro A, Mu W, Roncal C, Cheng KY, Johnson RJ, Scarpace PJ (2008) Fructose-induced leptin resistance exacerbates weight gain in response to subsequent high-fat feeding. Am J Physiol Regul Integr Comp Physiol 295:R1370–R1375

71. Johnson RJ et al (2013) Sugar, uric acid, and the etiology of diabetes and obesity. Diabetes 62:3307–3315

72. Mukai Y, Kumazawa M, Sato S (2013) Fructose intake during pregnancy up-regulates the expression of maternal and fetal hepatic sterol regulatory element-binding protein-1c in rats. Endocrine 44:79–86

73. Zou M et al (2012) Fructose consumption during pregnancy and lactation induces fatty liver and glucose intolerance in rats. Nutr Res 32(8):588–598

74. Kaufmann RC, Amankwah KS, Dunaway G, Maroun L, Arbuthnot J, Roddick JW Jr (1981) An animal model of gestational diabetes. Am J Obstet Gynecol 141:479–482

75. Yamashita H, Shao J, Qiao L, Pagliassotti M, Friedman JE (2003) Effect of spontaneous gestational diabetes on fetal and postnatal hepatic insulin resistance in Lepr(db/+) mice. Pediatr Res 53:411–418

76. Plows JF, Yu X, Broadhurst R, Vickers MH, Tong C, Zhang H et al (2017) Absence of a gestational diabetes phenotype in the LepRdb/+ mouse is independent of control strain, diet, misty allele, or parity. Sci Rep 7:45130. https://doi.org/10.1038/srep45130

77. Pollock KE, Stevens D, Pennington KA, Thaisrivongs R, Kaiser J, Ellersieck MR et al (2015) Hyperleptinemia during pregnancy decreases adult weight of offspring and is associated with increased offspring locomotor activity in mice. Endocrinology 156:3777–3790

78. Constância M, Hemberger M, Hughes J, Dean W, Ferguson-Smith A, Fundele R et al (2002) Placental-specific IGF-II is a major modulator of placental and fetal growth. Nature 417:945–948

79. Liu JP, Baker J, Perkins AS, Robertson EJ, Efstratiadis A (1993) Mice carrying null mutations of the genes encoding insulin-like growth factor I (Igf-1) and type 1 IGF receptor (Igf1r). Cell 7:59–72

80. Vickers MH, Gluckman PD, Coveny AH, Hofman PL, Cutfield WS, Gertler A et al (2005) Neonatal leptin treatment reverses developmental programming. Endocrinology 146:4211–4216

81. Itoh H, Yura S, Sagawa N, Kanayama N, Konihi I, Hamamatsu Birth Cohort for Mothers and Children (HBC) Study Team (2011) Neonatal exposure to leptin reduces glucose tolerance in adult mice. Acta Physiol 202:159–164

82. Vickers MH, Sloboda DM (2012) Leptin as mediator of the effects of developmental programming. Best Pract Res Clin Endocrinol Metab 26:677–687

83. Li M, Reynolds CM, Gray C, Vickers MH (2015) Preweaning GH treatment normalizes body growth trajectory and reverses metabolic dysregulation in adult offspring after maternal undernutrition. Endocrinology 156:3228–3238

84. Reynolds CM, Li M, Gray C, Vickers MH (2013) Pre-weaning growth hormone treatment ameliorates bone marrow macrophage inflammation in adult male rat offspring following maternal undernutrition. PLoS One 8: e68262. https://doi.org/10.1371/journal. pone.0068262

85. Lillycrop KA, Phillips ES, Jackson AA, Hanson MA, Burdge GC (2005) Dietary protein restriction of pregnant rats induces and folic acid supplementation prevents epigenetic modification of hepatic gene expression in the offspring. J Nutr 135:1382–1386

86. Bai SY, Briggs DI, Vickers MH (2012) Increased systolic blood pressure in rat offspring following a maternal low-protein diet is normalized by maternal dietary choline supplementation. J Dev Orig Health Dis 3:342–349

87. Li M, Reynolds CM, Sloboda DM, Gray C, Vickers MH (2015) Maternal taurine supplementation attenuates maternal fructose-induced metabolic and inflammatory dysregulation and partially reverses adverse metabolic programming in offspring. J Nutr Biochem 26:267–276

88. Boujendar S, Arany E, Hill D, Remacle C, Reusens B (2003) Taurine supplementation of a low protein diet fed to rat dams normalizes the vascularization of the fetal endocrine pancreas. J Nutr 133:2820–2825

89. Gray C, Vickers MH, Segovia SA, Zhang XD, Reynolds CM (2015) A maternal high fat diet programmes endothelial function and cardiovascular status in adult male offspring independent of body weight, which is reversed by maternal conjugated linoleic acid (CLA) supplementation. PLoS One 10:e0115994. https://doi.org/10.1371/journal.pone. 0115994

90. Wyrwoll CS, Mark PJ, Mori TA, Puddey IB, Waddell BJ (2006) Prevention of programmed hyperleptinemia and hypertension by postnatal dietary omega-3 fatty acids. Endocrinology 147:599–606

91. Vega CC, Reyes-Castro LA, Bautista CJ, Larrea F, Nathanielsz PW, Zambrano E (2015) Exercise in obese female rats has beneficial effects on maternal and male and female offspring metabolism. Int J Obes 39:712–719

92. Raipuria M, Bahari H, Morris MJ (2015) Effects of maternal diet and exercise during pregnancy on glucose metabolism in skeletal muscle and fat of weanling rats. PLoS One 10: e0120980. https://doi.org/10.1371/journal.pone.0120980

93. Burdge GC, Lillycrop KA (2012) Folic acid supplementation in pregnancy: are there devils in the detail? Br J Nutr 108:1924–1930

94. Vickers MH, Gluckman PD, Coveny AH, Hofman PL, Cutfield WS, Gertler A et al (2008) The effect of neonatal leptin treatment on postnatal weight gain in male rats is dependent on maternal nutritional status during pregnancy. Endocrinology 149:1906–1913

95. Kirk SL et al (2009) Maternal obesity induced by diet in rats permanently influences central processes regulating food intake in offspring. PLoS One 4:e5870. https://doi.org/10.1371/journal.pone.0005870

96. Godfrey KM, Sheppard A, Gluckman PD, Lillycrop KA, Burdge GC, McLean C et al (2011) Epigenetic gene promoter methylation at birth is associated with child's later adiposity. Diabetes 60:1528–1534

97. Albert BB, Derraik JG, Cameron-Smith D, Hofman PL, Tumanov S, Villas-Boas SG et al (2015) Fish oil supplements in New Zealand are highly oxidised and do not meet label content of n-3 PUFA. Sci Rep 5:7928. https://doi.org/10.1038/srep07928

98. Albert BB, Vickers MH, Gray C, Reynolds CM, Segovia SA, Derraik JG et al (2016) Oxidized fish oil in rat pregnancy causes high newborn mortality and increases maternal insulin resistance. Am J Physiol Regul Integr Comp Physiol 311:R497–R504

99. Bełtowski J, Wójcicka G, Borkowska E (2002) Human leptin stimulates systemic nitric oxide production in the rat. Obes Res 10:939–946

100. Bełtowski J, Wjcicka G, Górny D, Marciniak A (2002) Human leptin administered intra-peritoneally stimulates natriuresis and decreases renal medullary Na+, K+-ATPase activity in the rat—impaired effect in dietary-induced obesity. Med Sci Monit 8: BR221–BR229

101. Jackson EK, Herzer WA (1999) A comparison of the natriuretic/diuretic effects of rat vs. human leptin in the rat. Am J Phys 277: F761–F765

102. Tannenbaum C, Schwarz JM, Clayton JA, de Vries GJ, Sullivan C (2016) Evaluating sex as a biological variable in preclinical research: the devil in the details. Biol Sex Differ 7:13. https://doi.org/10.1186/s13293-016-0066-x

103. Bale TL (2016) The placenta and neurodevelopment: sex differences in prenatal vulnerability. Dialogues Clin Neurosci 18:459–464

104. Reynolds CM, Vickers MH, Harrison CJ, Segovia SA, Gray C (2015) Maternal high fat and/or salt consumption induces sex-specific inflammatory and nutrient transport in the rat placenta. Physiol Rep 3. pii: e12399. doi: 10.14814/phy2.12399

105. Mauvais-Jarvis F, Arnold AP, Reue K (2017) A guide for the design of pre-clinical studies on sex differences in metabolism. Cell Metab 25:1216–1230

106. Li M, Reynolds CM, Sloboda DM, Gray C, Vickers MH (2013) Effects of taurine supplementation on hepatic markers of inflammation and lipid metabolism in mothers and offspring in the setting of maternal obesity. PLoS One 8:e76961. https://doi.org/10.1371/journal.pone.0076961

107. Ojeda NB, Grigore D, Yanes LL, Iliescu R, Robertson EB, Zhang H et al (2007) Testosterone contributes to marked elevations in mean arterial pressure in adult male intrauterine growth restricted offspring. Am J Physiol Regul Integr Comp Physiol 292: R758–R763. https://doi.org/10.1152/ajpregu.00311.2006

108. Sampson AK et al (2012) The arterial depressor response to chronic low-dose angiotensin II infusion in female rats is estrogen dependent. Am J Physiol Regul Integr Comp Physiol 302:R159–R165. https://doi.org/10.1152/ajpregu.00256.2011

109. Link JC, Chen X, Arnold AP, Reue K (2013) Metabolic impact of sex chromosomes. Adipocytes 2:74–79

110. Arnold AP, Chen X (2009) What does the 'four core genotypes' mouse model tell us about sex differences in the brain and other tissues? Front Neuroendocrinol 30:1–9

111. Arnold AP (2014) Conceptual frameworks and mouse models for studying sex differences in physiology and disease: why compensation changes the game. Exp Neurol 259:2–9

112. Burgoyne PS, Mahadevaiah SK, Perry J, Palmer SJ, Ashworth A (1998) The Y* rearrangement in mice: new insights into a perplexing PAR. Cytogenet Cell Genet 80:37–40

Part II

Protocols

Chapter 9

Generation of Maternal Obesity Models in Studies of Developmental Programming in Rodents

Paul D. Taylor, Phillippa A. Matthews, Imran Y. Khan, Douglas Rees, Nozomi Itani, and Lucilla Poston

Abstract

Mother-child cohort studies have established that both pre-pregnancy body mass index (BMI) and gestational weight gain (GWG) are independently associated with cardio-metabolic risk factors in juvenile and adult offspring, including systolic and diastolic blood pressure. In rodent studies maternal obesity confers many facets of the metabolic syndrome including a persistent sympathy-excitatory hyperresponsiveness and hypertension acquired in the early stages of development. Insight from these animal models raises the possibility that early life exposure to the nutritional and hormonal environment of obesity in pregnancy in humans may lead to early onset of metabolic syndrome and/or essential hypertension. This chapter will address the development of rodent models of maternal overnutrition and obesity, which have proved invaluable in generating testable hypotheses for clinical translation and the development of intervention strategies to stem the swelling tide of obesity and its comorbidities predicted for future generations.

Key words Maternal obesity, Developmental programming, Rodents, Diet, Metabolic syndrome

---

1 Introduction

In the context of a global obesity epidemic, there is now a clear consensus emerging from mother-child cohort studies that pre-pregnancy body mass index (BMI) as well as gestational weight gain (GWG) can precipitate, particularly in genetically susceptible individuals, a cardiovascular and metabolic phenotype with a trajectory toward premature death, through all-cause mortality, but predominantly cardiovascular disease [1–4].

The Health Survey for England (HSE, 2013) currently estimates rates for obesity of 26% and 24% for men and women, respectively. If current trends continue, rates are estimated to increase to as much as 48% in men and 43% in women by 2030 (*Statistics on Obesity, Physical Activity and Diet: England 2015. hscic*). In line with these trends, the prevalence of obesity in pregnancy has more than doubled in the past 20 years, with an

Paul C. Guest (ed.), *Investigations of Early Nutrition Effects on Long-Term Health: Methods and Applications*, Methods in Molecular Biology, vol. 1735, https://doi.org/10.1007/978-1-4939-7614-0_9, © Springer Science+Business Media, LLC 2018

estimated 20% of pregnant women in the United Kingdom classified as obese [5, 6]. Given that maternal obesity now constitutes the single biggest obstetric risk factor for both maternal and fetal health and the mounting concerns over long-term morbidity and mortality in offspring born to obese pregnant women, the need to understand the underlying mechanisms and risk factors associated with obesity in pregnancy and developmental programming effects on offspring has never been more pressing. This chapter will address the development of rodent models of maternal obesity, which have proved invaluable in generating testable hypotheses for clinical translation and the development of intervention strategies to stem the swelling tide of obesity and its comorbidities predicted for future generations.

Although the underlying mechanisms are not yet understood, numerous studies have demonstrated the association between maternal BMI and offspring adiposity or BMI, supporting the concept of a transgenerational acceleration of obesity [7–9]. Estimates from meta-analyses suggest a threefold increase in risk of childhood obesity associated with maternal pre-pregnancy obesity [10] and a 33% increased risk associated with gestational weight gain above Institute of Medicine (IOM) guidelines (USA). In addition, maternal pre-pregnancy BMI and GWG are also associated with dyslipidemia, insulin resistance, and inflammatory markers which contribute further to cardio-metabolic risk profiles in childhood [11–14]. These obesity-related risk factors will ultimately increase blood pressure in childhood, and for the most part, the reported associations between maternal BMI and offspring hypertension appear to be largely mediated by the offspring's BMI at the time of study [15]. However, emerging evidence in mother-child cohorts, particularly of younger children, suggest an independent relationship between maternal BMI and offspring blood pressure [15–19].

The Amsterdam Born Children and their Development (ABCD) study reported that pre-pregnancy BMI in over 3000 women was positively associated with offspring diastolic blood pressure (DBP) and systolic blood pressure (SBP) at age 5–6 years [16]. After adding birth weight and the child's current BMI to the model, the independent effect size of pre-pregnancy BMI on blood pressure decreased by approximately 50%. However, the relationship still held, indicating an independent relationship between maternal BMI and childhood blood pressure.

Candidate "vectors" in the transmission of obesogenic and cardiovascular risk to the developing fetus include macronutrients such as glucose and lipids (fetal over nutrition hypothesis) which can initiate reactive insulin, leptin, and glucocorticoid signaling pathways in the fetus or neonate. Immunological and inflammatory mediators can also interfere with developmental processes during periods of developmental plasticity. Increased GWG, particularly in the first trimester, which is a critical period for placental development, is associated with increased risks of childhood overweight and a clustering of cardio-metabolic risk factors [12, 20]. Indeed, leptin may play an important role in early placentation by stimulating several genes involved in angiogenic signaling pathways and fatty acid metabolism [21, 22].

Perhaps some of the best available evidence in support of the association between maternal obesity and offspring cardio-metabolic risk is derived from "sibling-pair" studies performed in children born to mothers before and after bariatric weight loss surgery [23, 24]. The prevalence of overweight and obesity was higher in the children born before, compared to siblings born after maternal biliopancreatic diversion bariatric surgery. Children born after maternal surgery (AMS) exhibited threefold lower prevalence of severe obesity, greater insulin sensitivity, improved lipid profile, and lower C-reactive protein and leptin compared to children born before maternal surgery. These studies demonstrate the powerful benefits of weight reduction in obese pregnancy for offspring cardio-metabolic risk which is sustained into adolescence and are most likely attributable to an improved intrauterine environment. More recent studies employing this paradigm suggest that improved maternal gestational lipid profile and carbohydrate metabolism, as a consequence of maternal surgery, interacts with offspring gene variations to modulate gene expression levels and improve cardio-metabolic risk profiles in older siblings, specifically through differential methylation of genes involved in immune and inflammatory pathways [25–27].

A limitation in these sib-pair studies is the potential influence of an altered "shared" postnatal maternal nutritional environment post-maternal surgery impacting childhood. While impressive, these intervention studies are still essentially observational and do not carry the same weight of evidence as randomized controlled trials (RCT) in establishing causality. In the absence of long-term follow-up studies from ongoing RCTs, which are eagerly awaited [28], animal models can, to a large degree, avoid confounding variables associated with human epidemiological studies, and rodent studies, in particular, provide mechanistic insight into the effects of obesity in pregnancy and have generated testable hypotheses that can be translated back in to clinical studies.

Numerous animal models have been developed to recreate the conditions described in the early epidemiological association studies that first generated the DOHaD hypothesis. Such studies allow mechanistic investigation of the nutritional and hormonal factors that can shape offspring phenotype. Animal studies have certain advantages over the human cohort studies in establishing cause and effect:

1. Rodents are mammals and share all but 1% of our genes, together with highly conserved physiological systems and similar placentation, which despite altricial versus precocial species differences in developmental stage at birth, make them an excellent model for human pregnancy.

2. Rodent models are particularly amenable to developmental programming and life course studies due to the relatively short life cycles. Rats and mice reach sexual maturity in a little over 1 month of age, which means that the consequences of environmental influences in development on the adult phenotype can be studied within a reasonable timeframe.

3. Rodent models can avoid many of the residual confounding factors observed in human population studies by reducing genetic variability in subjects (through the use of inbred strains and genetically identical animals).

4. Environmental conditions can be tightly controlled (e.g., standardizing animal husbandry).

5. Experimental diets can be tested that could not ethically or practically be tested in human cohorts.

6. Rodent models facilitate the investigation of underlying physiological, cellular, and molecular mechanisms during critical periods of development not easily available to clinical researchers.

Animal studies support an influence of a maternal obesogenic environment in pregnancy on determinants of cardiovascular control, independent of the programming of obesity in offspring. Rodent models in particular suggest that early life exposure to hyperleptinemia may directly predispose to early onset hypertension, hyperphagia, and cardiac dysfunction (Fig. 1).

Early models of fat feeding in pregnancy attempted to mimic the dietary intake of Western societies. Several groups have, therefore, investigated the effects of a diet high in fat or cholesterol, which may be more relevant to the Western diet and may give insight particularly into the developmental programming and etiology of cardiovascular disease [29–33]. Table 1 summarizes some of the maternal diets employed which generally do not cause marked weight gain and tend to address the influence of diet rather than obesity per se. In our own model of maternal fat feeding, in which dietary carbohydrate is substituted 20% w/w with lard (supplemented with vitamins, essential micronutrients, and protein to control values (Fig. 2), maternal weight gain is modest (Fig. 3), and plasma leptin, lipids, and 8-epi-PGF2α are normal. However, the maternal metabolic profile at gestation day 20 is characterized by markedly increased insulin and corticosterone concentrations [48] associated with offspring cardiovascular dysfunction [31, 32, 49–51].

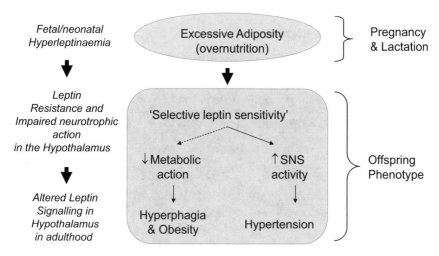

Fig. 1 Developmental programming of obesity and hypertension secondary to maternal obesity and/or neonatal hyperleptinemia. The schematic shows the proposed developmental origins of "selective leptin sensitivity" in which the anorexic actions of leptin are lost while the pressor effect is enhanced

In addition to commercial high-fat diets and standard chow diet supplemented with various forms of fat [48, 52, 53], a "cafeteria" diet has been employed in rodent studies worldwide for modeling maternal obesity. Cafeteria diets commonly involve providing a selection of highly palatable human food stuffs that are rich in fat, sugar, and salt [54] in addition to standard rodent chow, and these experimental diets more accurately reflect the obesogenic Western-style diet consumed by human [55]. Independent studies consistently report noticeable preference of animals to cafeteria foods when offered concurrently with standard chow, which subsequently leads to hyperphagia and a progressive increase in caloric intake [54, 56, 57].

The range of food products in a cafeteria diet employed in animal studies is variable and culture-dependent [56, 58–60] but often includes sweet snacks, cakes, pastries, processed meat, dairy products, potato crisps, and chocolates (Table 2). A typical feeding regimen involves a random selection of four of the cafeteria foods in excess quantities along with standard chow, and the cafeteria foods are replaced with new items regularly such that the animals do not receive the same foods consecutively [36, 56, 59]. Alternatively, cafeteria foods may be premixed at a known proportion [65]. Although the nature of cafeteria diets allows less control over the intake of each nutritional component compared to commercial high-fat diets, nutrition can be calculated retrospectively through accurate weighing and detailed analysis of cafeteria foods consumed by each individual animal. Studies report that a cafeteria diet in pre-mating does promotes an increase in daily energy intake by 100–300 kJ [36, 56, 64, 66] compared to standard chow-fed animals, with a marked increase in fat intake [56, 64]. Furthermore,

Table 1
Rodent models of fat feeding in pregnancy

Species	Breed	HFD composition	Control diet composition	HFD versus control weight preconception	Reference
Rat	Sprague-Dawley	45% fat, 35% CHO, 20% protein	10% fat, 70% CHO, 20% protein	276.8 ± 2.9 g versus 267.6 ± 3.2 g ($p < 0.05$)	[34]
Rat	Sprague-Dawley	60% fat, 20% CHO, 20% protein	17% fat	N/A	[35]
Rat	Sprague-Dawley	34% fat, 47% CHO 13% protein	14% fat, 65% CHO, 21% protein	231.2 ± 6.7 versus 285.0 ± 11.1 ($p < 0.01$)	[36]
Rat	Sprague-Dawley	60% fat, 24% CHO, 16% protein	10% fat, 70% CHO, 19% protein	389 ± 12 versus 328 ± 10 ($p < 0.05$)	[37]
Rat	Sprague-Dawley	HFD: 30% lard, 18% protein, 40% carbohydrate, 3% fiber	4% fat, 22% protein, 51% carbohydrate, 5% fiber	N/A	[38]
Rat	Wistar	45% fat, 8% protein	17% fat	295 ± 1.2 g versus 265 ± 3.5 g ($p < 0.05$)	[39]
Rat	Wistar	65% fat, 23% CHO, 13% protein	12% fat, 66% CHO, 20% protein	285 ± 4 g versus 272 ± 3 g ($p < 0.05$)	[40]
Rat	Wistar	40% fat, 46% CHO, 14% protein	10% fat, 75% CHO, 15% protein	N/A	[41]
Rat	Wistar	59% fat, 20% CHO, 21% protein	22% fat, 66% CHO, 23% protein	N/A	[42]
Rat	Wistar	40% fat, 46% CHO, 14% protein	10% fat, 75% CHO, 15% protein	N/A	[43]
Rat	Wistar	59% fat, 28% CHO, 15% protein	8% fat, 80% CHO, 14% protein	N/A	[44]
Rat	Wistar	40% fat	5% fat	N/A	[45]
Mouse	Balb/cByJ	60% fat	10% fat	N/A	[46]
Mouse	C57BL/6J	31% fat, 45% carbohydrate, 24% protein	12% fat, 59% carbohydrate, 29% protein	N/A	[47]

cafeteria diet-induced voluntary hyperphagia and the increased energy intake lead to an accelerated increase in pre-mating body weight, with 20–120% more weight gain compared to control

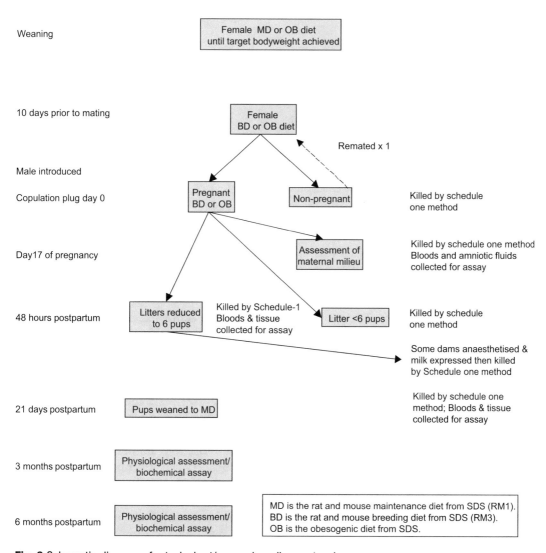

Fig. 2 Schematic diagram of a typical rat/mouse breeding protocol

animals, depending on the timing and duration of exposure to the obesogenic diet [36, 56, 61, 62, 66]. This increase in caloric intake and weight gain is sustained during pregnancy and/or lactation in some studies [54, 56, 65, 66] and not in others [58, 64]. The difference may be attributable to different experimental designs and whether or not animals were exposed to cafeteria foods before pregnancy (Table 2). Importantly, cafeteria diet-induced obesity is not only associated with increased adiposity but also with higher circulating levels of glucose, insulin, leptin, cholesterol, and triglycerides [36, 63, 65]. Therefore, cafeteria diets produce a phenotype of obesity with alterations in metabolic factors and impaired nutritional homeostasis that closely resemble the situation seen in overweight or obese women of reproductive age.

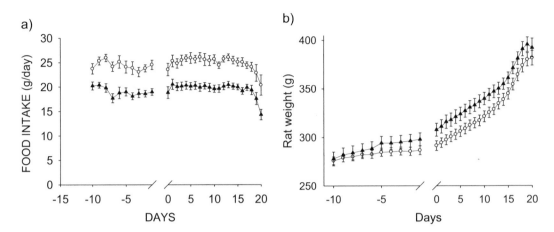

Fig. 3 (**a**) Maternal food intake and (**b**) maternal body weight for all dietary groups from 10 days prior to mating - throughout pregnancy (0–20 days). Each point represents the mean of 20 rat dams. Error bars represent SEM for control

The gut microbiome, with an estimated biomass of 2–3 kg in humans, has evolved symbiotically to perform essential biological functions such as the digestion of complex carbohydrates, but these microbial genes can also influence innate and adaptive immunity in the host and may have key regulatory functions in metabolic pathways in health and disease [67–69]. The microbiome, which is passed on from mother to offspring during normal vaginal delivery and during suckling, is therefore emerging as another "vector" for inheritance of epigenetic traits that may influence obesity risk in offspring [70]. The mode of delivery can affect phenotypic traits in the offspring, and babies born particularly by pre-labor cesarean section (CS) experience suboptimal microbial colonization of the intestinal tract [71, 72] which can influence immunity, feeding behavior, cardio-metabolic risk metabolism, neurological and stress-related conditions [73], and up to 46% increased risk of childhood obesity [74].

Inoculation studies in mice demonstrate how transfer of fecal material from affected mice to germ-free wild-type mice can confer phenotypic traits such as hyperphagia, obesity, and hypertension associated with adysbiotic gut microbiota [75]. Diet can also influence composition of the gut microbiome, with Western high-fat diets favoring the *Bacteroides*, a genus of gram-negative, obligate anaerobic bacteria associated with obesity [76–78]. These studies demonstrate how diet and/or obesity in pregnancy could establish dysbiosis in the offspring through an inherited microbiome but also how easily, especially in coprophilic species, that a particular microbiome could be spread throughout an animal unit.

The following sections describe the animal models and methods employed in our laboratory and others to investigate the influence of maternal overnutrition and obesity on maternal and

Table 2
Rodents models of a cafeteria diet-induced obesity in pregnancy

Species/ breed	Study design	Cafeteria diet	Effects on maternal BW	Other effects	Reference
Rat Wistar	CAF from postpartum day 1 until day 21	Standard chow + cookies with liver pate and sobrasada (typical Majorcan sausage), candies, fresh bacon, biscuits, chocolate, salted peanuts, cheese, milk containing 20% (w/v) sucrose and ensaimada (typical Majorcan pastry)	CON dams' BW increased during lactating period (absent in the CAF dams). CAF dams were significantly lighter than CONs for day 11–21 postpartum. Despite lower BW CAF dams has increased % body fat and decreased % lean mass		[58]
Rat Wistar	CAF from 10- to 100-day-old then on standard chow for 30 days and mated	(35.2% carbohydrate, 23.4% lipid, 11.7% protein, 28.4% water, and 1.31% fiber)	CAF increased BW from 42 days due to increased fat. BW decreased post-CAF but remained significantly higher at mating on 130 days. BW no longer different during lactation, but % fat remained higher and % lean mass remained lower	Milk and plasma leptin and adiponectin levels were increased in post-CAF dams during lactation	[61]
Rat Wistar	CAF from 3- to 11-week-old, mated then reallocated to CON or CAF during gestation, then again reallocated to CON or CAF during lactation	Standard chow + random (4) of biscuits, potato crisps, fruit, nut chocolate, Mars bars, cheddar cheese, golden syrup cake, pork pie, cocktail sausages, liver and bacon	CAF increased gestational weight gain irrespective of the pre-gestational CON or CAF diet, mainly due to increased fat pad mass	No difference in plasma glucose, triacylglycerol, or cholesterol on day 5 or day 20 gestation	[56]
Rat Wistar	CAF from 3- to 10-week-old, mated then reallocated to CON or CAF during gestation, then again reallocated to CON or CAF during lactation	pâté, strawberry jam, and peanuts	CAF pre-mating increased BW, then CAF during pregnancy increased weight gain irrespective of its pre-mating diet		[62]

(continued)

Table 2
(continued)

Species/breed	Study design	Cafeteria diet	Effects on maternal BW	Other effects	Reference
Rat Wistar	CAF from 21- to 90-day-old, mated then reallocated to CON or CAF during gestation and lactation	Standard chow + random (4) of salami, bread, snack Yokitos and Fratelli (Brazilian snacks), Jelly bean, Coca-Cola, smoked sausage, chocolate cake, biscuit, marshmallow, ham, chocolate wafer, gumdrop	CAF pre-mating increased BW, then CAF during pregnancy increased weight gain irrespective of its pre-mating diet	At weaning of pups plasma insulin, leptin, and triglyceride levels were increased in CON/CAF and CAF/CAF animals	[59]
Rat Wistar	CAF from 21-day-old and maintained on CAF throughout mating, pregnancy and lactation between 170- and 212-day-old		CAF increased BW by the end of the lactation period	Relative visceral and retroperitoneal fat mass increased. Total cholesterol, insulin, and leptin levels increased	[63]
Rat Wistar	200 g (age unknown) does on CAF for 4 weeks, mated and maintained on CAF throughout pregnancy and lactation	Standard chow + peanut butter, hazelnut spread, chocolate-flavored biscuits (cookies), extruded savory snacks, sweetened multigrain breakfast cereal, ham- and chicken-flavored processed meat, lard	CAF increased BW after 1 week and remained heavier during pregnancy and lactation	Pup cannibalism in CAF dams	[64]
Rat Wistar	180–200 g (age unknown) pregnant dams on CAF until the end of lactation	Standard diet + pâté, cheese, bacon, potato chips, cookies and chocolate (in a proportion of 2:2:2:1:1:1, by weight. Contains 420 kJ with 23% protein, 42% lipids and 35% carbohydrate)	CAF increased weight gain (100 versus 180 g) during pregnancy	Increased relative fat mass at parturition and at weaning. Increased plasma glucose, leptin, insulin, cholesterol, and triglycerides at parturition and at weaning	[65]

Rat Sprague-Dawley	CAF from 8- to 13-week-old, mated and maintained on CAF throughout pregnancy and lactation	Standard chow + sweetened condensed milk, skim milk powder, saturated animal fat, + random (4) of noodle, meat pie, cakes, and biscuits (1533 kJ, with 34% fat, 19% protein and 47% carbohydrate)	5 weeks of CAF pre-mating increased maternal BW at mating and increased weight gain during pregnancy	Increased plasma triglyceride, fasting insulin and leptin, and an index of insulin resistance (HOMA)	[36]
Rat Wistar	CAF from mating then given either CON or CAF during lactation	Standard chow + chocolate chip muffins, jam doughnuts, butter flapjacks, biscuits, cheese, marshmallow, potato crisps, and caramel chocolate bars (1743.8 kJ, 9.17% protein, 47.04% carbohydrates including 24.95% sugars, 19.15% fat including 8.54% saturated fat, 3.02% fibers, and 0.38% sodium)	CAF dams were significantly heavier at term but no longer different at weaning	Hyperphagia with a marked preference for CAF and reduced physical activity	[54]
Rat Sprague-Dawley	40–45-day-old does on CAF for 22 days, mated and maintained on CAF throughout pregnancy and lactation	Standard chow + condensed milk, sucrose, muffins, croissants, powdered milk, lard (1863 kJ) with 11.4% water, 17.1% fat, 3.38% fiber, 10.6% protein, 2.47% ashes)	CAF increased BW at mating but no longer different during and after pregnancy	Increased lumbar adipose tissue weight at weaning increased plasma triacylglycerol, nonesterified fatty acids, glycerol levels at weaning	[66]

offspring phenotype [all experiments were carried out in accordance with the UK Animals (Scientific Procedures) Act, 1986 and approved by the local AWERB ethic committee at King's College London]. Much like the early epidemiological studies in this area, animal models have largely focused on the effects on maternal undernutrition and low birth weight on the adult offspring phenotype. As interest developed in the other end of the birth weight spectrum and in the influence of maternal obesity and GDM on long-term offspring health, animal models of maternal overnutrition began to emerge [79]. We initially investigated the effect of a lard-rich diet on maternal and offspring cardiovascular phenotype [31] and hypothesized that feeding a diet rich in animal fat to rat dams before and during pregnancy would lead to abnormal uterine artery function and so contribute to the in utero programming of adulthood cardiovascular dysfunction in the offspring.

2   Materials

2.1   High Fat "Lard-Supplemented" Experimental Diet

1. Female Sprague-Dawley rats (100–120 days old) (see **Note 1**).
2. Control rat chow: 4% fat (corn oil), 21% protein, and 51% carbohydrate.
3. Experimental high-fat (HF) diet: standard chow supplemented with 20% (w/w) animal lard and 20% additional vitamins, minerals, protein, inositol, and choline to correct for the dilution (see **Note 1**).

2.2   Obesogenic Diet

1. C57BL/6J mice (6 weeks of age) (Charles River Laboratories, UK).
2. Control chow: Rat & Mouse Maintenance Diet [RM No. 1; Special Diets Services (SDS)].
3. Obesogenic diet: high-fat pellets supplemented with sweetened condensed milk (see **Notes 2** and **3**).

2.3   Established Murine Model of Maternal Obesity

1. Proven male C57BL/6J mice (8–10 weeks of age) (Charles River Laboratories; Saffron Walden, UK) (see **Note 3**).
2. Oxytocin.
3. Anesthesia: isoflurane.

2.4   Obesogenic Diet in Rats

1. Female Sprague-Dawley rats (Banting & Kingman; Hull, UK).
2. Obesogenic diet.
3. Standard breeding diet (RM3).

2.5   Maternal High-Sugar Diet

1. Female C57BL/6J mice (Charles River Laboratories, UK, $n = 24$), proven breeders (one previous litter).

2. Standard chow (RM1).

3. High-sugar diet: standard chow with access to sweetened condensed milk [55% simple sugars, 10% fat, 9% protein (w/w) 3.5 kcal/g, Nestle®, SZ] fortified with added micronutrient mineral and vitamin mix (AIN93G, Special Dietary Services) to achieve similar levels as standard chow.

3 Methods

3.1 High-Fat Diet

1. Feed rats for 10 days prior to mating and throughout pregnancy either with the control or HF diet ad libitum.

2. Random assign female rats to either the control or lard diet 10 days prior to mating.

3. Following the 10-day run-in period, allow animals a 7-day period to breed.

4. Mark day 0 of pregnancy by the appearance of a copulation plug.

5. Female rats that fail to become pregnant after the 7-day period are killed by a schedule one method.

6. House pregnant animals according to 12 h light-dark cycle with food and water ad libitum.

7. Monitor intake and body weight daily.

8. Maintain rats on their respective diets throughout pregnancy (21–22 days) and during the weaning period (21–22 days).

9. Following weaning, maintain all offspring from all groups on standard rat chow (Table 3).

10. Reduce litters 48 h postpartum to eight pups (four males and four females if possible) in order to standardize pup milk demand and intake and to avoid cannibalization of the pups.

11. A separate cohort of pregnant rats are killed at day 20–21 of gestation by CO_2 inhalation and cervical dislocation.

12. Obtain plasma samples from these animals by cardiac puncture.

13. Collect maternal organs, placentae, amniotic fluid and fetuses.

14. Alternatively dams are killed, and tissue is collected after weaning.

15. Dams with too few offspring (below six pups/litter) are killed on day 1 postpartum.

16. Before sexual maturity is reached, separate male and female offspring and study longitudinally for phenotypic characterization.

17. Measure body weight and food intake daily or weekly (see **Note 4**).

18. At necropsy, weigh all animals prior to dissection and take blood samples by cardiac puncture.

Table 3
Independent analysis of control and lard supplemented diets (Eclipse Scientific Group, Cambridge)

	Control	Lard supplemented diet
Moisture	10.1	8.2
Protein	21.2	19.5
Carbohydrate	57.4	41.3
Fat	5.3	25.7
Crude fiber	4.6	3.5
Ash	6.0	5.3
Calcium	1.06	0.70
Potassium	0.87	1.01
Sodium	0.24	0.20
Chloride	0.46	0.35
Manganese	68.9	83.4
Copper	13.3	20.2
Iron	152	217
Magnesium	0.19	0.19
Zinc	52.9	48.0
Phosphorus	0.80	0.69
Vitamin A	5.61	9.52
Vitamin E	63.2	110
Vitamin B1	10.5	13.7
Vitamin B2	5.49	5.83
Vitamin B6	5.80	6.2
Vitamin B12	0.031	0.036
Nicotinic acid	70	91
Pantothenic acid	27	32
Folic acid	0.50	1.3
Biotin	0.14	0.20
Vitamin C	<10	<10
Vitamin D3	1.64	3.08

19. Weigh or measure plasma and dissected tissue samples and organs.

20. Catalogue these and store at $-70\,°C$.

21. Analyze food intake and maternal body weights for 20 pregnant rats subjected to each of the two dietary regimes (Fig. 2) (*see* **Note 5**).

22. Analyze biomarkers (*see* **Note 6**).

3.2 Semi-synthetic "Obesogenic" Diet (See Note 7)

1. House C57BL/6J mice individually for the purpose of monitoring dietary intake.

2. Allow mice to acclimatize for 1 week before being randomized to either standard chow or the "obesogenic diet."

3. Provide mice access to food and water ad libitum, and maintain on a 12 h light-dark cycle, in a thermostatically controlled environment (25 °C) throughout.

4. Record body weight and food intake weekly.

5. Analyze body weight data by Bonferroni post hoc tests to compare between the two groups at each time point (*see* **Note 8**).

3.3 Established Murine Model of Maternal Obesity

1. Allow proven male C57BL/6J mice to acclimatize for 1 week before mating with females (*see* **Note 1**).

2. At the end of the 6-week feeding protocol, transfer female mice from each group into new cages with two females (from the same dietary group) and one proven male per cage (Fig. 4) (*see* **Notes 9** and **10**).

3. Examine females daily for evidence of a vaginal copulation plug and mark as day 0.

4. Separate the female from the male.

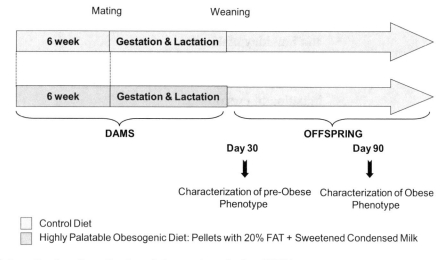

Fig. 4 Schematic: Breeding offspring of obese rats and mice (OffOb)

5. Remate females that fail to become pregnant after developing a copulation plug (determined by a failure to show a significant (10%) increase in body weight by day 10).

6. Females in which a copulation plug was not observed remained with the male and are examined frequently for evidence of pregnancy.

7. For those with repeated failure to become pregnant, kill by a schedule one method.

8. Separate females from males for successful pregnancies.

9. Maintain dams on their pre-pregnancy diet throughout gestation and lactation.

10. Forty-eight hours postpartum, weigh litters and reduce to six pups (three male, three female) where appropriate (*see* **Note 11**).

11. Exclude litters of less than four pups from the study.

12. At 21 days of age, wean pups onto standard chow RM1 and maintain on this diet.

13. At weaning, in a subgroup of dams, administer oxytocin by injection (4 IU, i.p, $n = 3$ per group) to stimulate milk production.

14. Obtain samples of milk under anesthesia without recovery, using an adapted mouth pipette.

15. Store milk samples at $-20\ °C$ prior to assay (*see* **Note 12**).

16. Assess offspring body weights and food intake weekly following weaning (*see* **Note 13**).

17. Assess maternal and fetal characteristics at gestational day 18 (*see* **Note 14**).

3.4 The Obesogenic Diet in Rats

1. House female Sprague-Dawley rats individually under standard laboratory conditions on a 12 h light-dark cycle (lights on at 07:00) in a temperature-controlled environment at $21 \pm 2\ °C$ and humidity of 40–50%.

2. Allow animals to have access to food and water ad libitum.

3. Allow animals to habituate to the animal unit for 1 week before initiation of experiments.

4. Feed rats either the standard maintenance diet (Rat & Mouse No. 1) or the obesogenic diet for 6 weeks before mating and throughout pregnancy and lactation (Fig. 5c, d).

5. Present the condensed milk separately from the pellets in a stainless steel coop cup attached to the side of the cage with a wire dish holder to prevent spillage.

6. Ensure that the control rats receive the standard maintenance diet (RM1) until 10 days before mating, when they are given the standard breeding diet (RM3) until weaning (*see* **Note 15**).

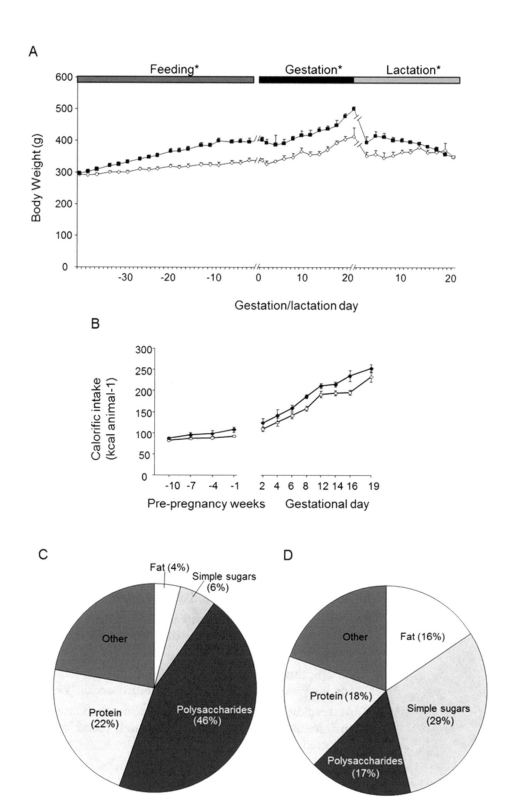

Fig. 5 Maternal serum analytes at (**a**) gestational day 18, (**b**) amniotic fluid analytes at gestational day 18, and (**c**) maternal serum analystes at weaning, in control and obese C57BL/6J mouse dams and (**d**) DEXA scans of representative dams at weaning. Values expressed as mean ± SEM. *$p < 0.05$, ***$p < 0.001$ versus control unpaired t-test

7. Standardize litter size to eight pups (four male, four female) 48 h after birth.

8. Wean all offspring at day 21 and feed RM1 diet ad libitum.

9. One male and one female from each litter are typically used for phenotypic characterization at each time point in a given protocol.

10. Record organ weights and store serum at −80 °C for future analysis.

11. After weaning of their pups, cull dams following an overnight fast.

12. Collect blood and prepare serum and store at −80 °C.

13. In a separate cohort of animals, kill litters of control and obese dams at several postnatal stages from day 2 to day 18 (*see* **Note 16**).

14. Collect blood samples (trunk blood), abdominal fat pads and stomach content, and store at −80 °C until analyzed.

3.5 Maternal "High-Sugar" Diet Alone (See Note 17)

1. Maintain female C57BL/6J mice under controlled conditions (20 °C and 60% humidity; light-dark cycle 12 h) with ad libitum access to food and water.

2. After 1 week of acclimatization, feed female mice either standard chow or standard chow supplemented with ad libitum access to fortified sweetened condensed milk.

3. Calculate macronutrient and caloric intake from measured daily intake of the diet (approx. 5% fat, 26% simple sugars, 12% protein, energy 3.4 kcal/g).

4. Maintain animals on the experimental or control diet for 6 weeks before conception and throughout pregnancy and suckling.

5. Record maternal weight and dietary intake daily.

6. At 48 h postpartum, litters are reduced to three male and three female pups to standardize the dam's milk supply during suckling.

7. All offspring are weaned at 21 days of age onto standard chow and body weight and food intake recorded weekly (*see* **Note 18**).

8. At weaning, one male and one female from each litter are fasted and humanely killed, and tissue and blood samples are collected.

9. Store samples at −80 °C for future analysis and record the weight of organs (*see* **Note 19**).

4 Notes

1. Upon arrival at the institution's animal facility, animals are switched to the standard rat chow for at least 1 week to acclimatize. The diet may differ from the supplier's maintenance diet, and both the change in diet and the transportation of animals are potentially a stressful environmental insult. Housing animals individually is also stressful and should be avoided unless individual dietary intakes are being measured. The composition of the diet was confirmed by independent analysis and is given in Table 3 (Eclipse Scientific Group, Cambridge). The high-fat diet consists of 20% lard (fatty acid constituents: 2.7% palmitoleic, 32.8% oleic, 8.1% linoleic, 0.4% linolenic, 1.55% myristic, 21.2% palmitic, and 9.6% stearic) and 4% corn oil (corn oil fatty acid constituents: 23.4% oleic, 42.9% linoleic, 0.8% linolenic, 9.8% palmitic, and 2% stearic). The lard-rich diet is supplemented with essential micronutrients and vitamins to ensure similar final contents (w/w) as in normal chow to AIN93G standard. The composition of the obesogenic diet can be modified by the preparation of "diet balls" in which the ratio of powdered "pellet" to condensed milk can be varied.

2. Semi-synthetic energy-rich and highly palatable pelleted diet [20% animal lard, 10% simple sugars, 28% polysaccharide, 23% protein (w/w), energy 4.5 kcal/g, Special Dietary Services, Witham, UK], supplemented with sweetened condensed milk [55% simple sugars, 10% fat, 9% protein (w/w) 3.5 kcal/g, Nestle®, SZ] fortified with 3.5% mineral mix and 1% vitamin mix (w/w) (AIN 93G, Special Diets Services)]. Sweetened condensed milk was supplied in a separate container, with the OB2 solid diet provided in the cage hopper as usual. During the first 2 weeks of feeding, body weights of both control and obese mice can sometimes fall. This may be a consequence of reduced food intake in both groups of mice due to neophobia arising from the presence of a new diet (control mice were also switched from a different diet), which can lead to a disinclination to eat. Alternatively, or in addition, the drop in body weights may be a consequence of the lower protein content of the control and obesogenic diets compared to the commercial breeder's diet on which the mice were maintained prior to the beginning of the protocol.

3. Proven by production of at least one previous litter.

4. This can be achieved either by manually monitored weight difference over time or by employing Labmaster Calocages which can measure both body weight and weight of the food hopper through sensitive force transducers (TSE Germany).

5. In both control and lard-fed dams, food intake did not alter with gestation over the first 15 days of pregnancy but fell sharply toward term (21–22 days). Dams on the lard-enriched diet consumed less food daily than those on the standard rat chow [average daily intake day 0–20, 25.3 ± 0.66 g/day for control ($n = 20$) versus 19.82 ± 0.60 g/day for those on the lard diet ($n = 20$, $p < 0.001$)]. The average daily gross energy intake in the controls [379.0 ± 9.9 kJ ($n = 20$)] was not different from dams on the lard diet [388.3 ± 11.8 kJ ($n = 20$) $p = 0.55$]. Despite the isocaloric intake, there was a fourfold increase in fat intake in the lard group. The lard group was heavier than controls from 0 to 15 days (lard versus controls, $p < 0.02$). Final body weights at termination of pregnancy were not different between the two groups. Dams on the high-fat diet had significantly larger spleens than control dams both by absolute weight ($p = 0.02$) and as a percentage of body weight ($p = 0.01$). Litter sizes were not significantly different between groups. Dams in the lard-fed group had smaller fetuses both in terms of weight ($p = 0.03$) and length ($p = 0.04$) and also tended to have smaller placentas ($p = 0.05$). Finally, 3% of lard-fed dams cannibalized their pups compared to none in the control group. In the first set of dams to go through this protocol, a high rate of cannibalism (50%) of the pups was observed in obese dams, whereas no evidence of cannibalism was seen in the control mice. This is similar to observations made by Guo and Jen using a high-fat diet, who found that this diet in pregnant rats was associated with a higher rate of cannibalism of their pups compared to control animals [45]. Steps to improve the environment were taken including noise reduction, improved bedding materials, and reduced handling of the mice. Initially, maternal body weights and food intake had been recorded daily (excluding weekends) throughout pregnancy and lactation. Subsequently, dams were weighed less frequently. These measures appeared to reduce the rate of cannibalism in these mice to approximately 30%.

6. There was no significant difference in the maternal plasma concentrations of 8-epi PGF2α, lipids or leptin on gestation day 20 between the two groups (Table 4). Maternal plasma insulin and corticosterone concentrations on day 20 gestation were significantly raised in the dams on the lard diet compared to controls.

7. It is known that food rich in fat and sugar is highly palatable and stimulates increased food intake, which can subsequently lead to a positive energy balance and obesity. This is believed to occur through alterations in the signals controlling appetite, with the result being an imbalance in these signals, favoring prolonged eating [80]. Activation of the opioid reward system by palatable foods is also likely to contribute to the increased

Table 4
Parameters evaluated in maternal plasma from lard-fed and control dams at day 20 gestation

	Control Mean ± SEM ($n = 13$)	Lard-fed Mean ± SEM ($n = 13$)	p value
Total cholesterol (mmol/L)	2.91 ± 0.10	3.23 ± 0.19	0.16
Triglycerides (mmol/L)	2.45 ± 0.35	2.06 ± 0.23	0.35
HDL (mmol/L)	1.65 ± 0.12	1.86 ± 0.10	0.19
Insulin (ng/mL)	1.35 ± 0.37	8.04 ± 0.47***	<0.0001
Leptin (ng/mL)	7.57 ± 0.83	6.29 ± 0.83	0.29
8-isoPGF2α (pg/mL)	283.0 ± 47.3	236.8 ± 30.7	0.41
Corticosterone (ng/mL)	541.9 ± 96.3	1164 ± 170.9**	0.005

Data are expressed as mean ± SEM with number of observations in parenthesis where **$p < 0.005$ and ***$p < 0.0001$ for comparisons of control versus lard-fed group

Table 5
Nutritional composition of sweetened condensed milk. Percentages are by weight (Nestlé)

Component	Quantity
Protein (%)	7.80
Total carbohydrate (%)	55.30
Polysaccharides (%)	0.00
Simple sugars (%)	55.30
Fat (%)	8.10
Fiber (%)	Trace
Total other (%)	28.00
Energy (kcal/g)	3.25

food consumption [81]. Various commercial diets were piloted in combination with different sugar sources in order to establish palatability caloric intake and rapid weight gain. Supplementation of a commercially available high-fat diet (SDS, England) with sweetened condensed milk [55% simple sugars, 10% fat, 9% protein (w/w) 3.5 kcal/g, Nestle®, SZ; Table 5], provided a "highly palatable" diet, high in both fat and sugar.

8. Body weights begin to diverge after 2 weeks. Over the duration of the protocol, consuming the obesogenic diet led to a significantly greater body weight compared to the controls ($p < 0.05$; RM ANOVA; Fig. 6a). Bonferroni post hoc tests

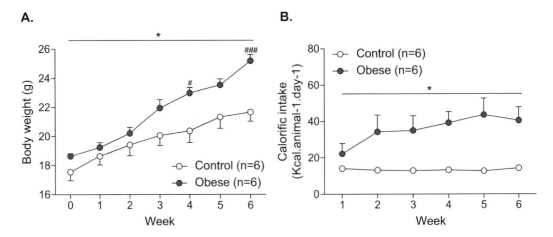

Fig. 6 Body weights (**a**) and caloric intakes (**b**) of female mice fed the obesogenic or control diet over 6 weeks. Values expressed as mean ± SEM. ***$p < 0.001$ RM ANOVA. #$p < 0.05$, ##$p < 0.01$, ###$p < 0.001$, Bonferroni post hoc tests following RM ANOVA

revealed significant differences in body weight at the end of weeks 4 and 6 (Fig. 6a). Average caloric intake was significantly increased in mice fed with the obesogenic diet compared to controls over the 8-week period ($p < 0.05$; RM ANOVA; Fig. 6b).

9. It is a good practice to only study offspring from second litters. This is because birth weight is often reduced in first pregnancies, and first litters of C57BL/6J mice are frequently lost due to cannibalism. Therefore proven females, which have raised one previous litter, are used for breeding.

10. Proven male breeders are maintained prior to mating on standard chow. A limitation of this protocol is that males have access to the experimental breeding diets during the mating period, which although a brief dietary exposure could conceivably cause epigenetic modification of the male germ line.

11. Standardizing the litter size by the culling of pups is important to standardize milk supply and demand. Pups reared in small litters have greater milk availability and show rapid weight gain, an established model of neonatal overfeeding.

12. An alternative method of milk collection is to reconstitute the stomach contents of suckling pups at necropsy. There are limitations to this method due to the partial digestion of milk and the presence of gut-derived hormones such as leptin which may contaminate the milk composition.

13. Female mice fed with the highly palatable obesogenic diet had significantly greater caloric intake compared to the control mice throughout the 6-week feeding period, gestation, and lactation (for each period, $p < 0.001$, RM ANOVA; Fig. 7b).

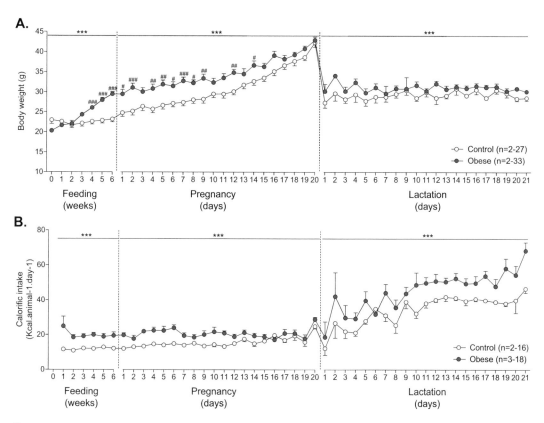

Fig. 7 Maternal body weights and caloric intakes throughout the protocol in the established model. Body weights (**a**) and caloric intakes (**b**) of female C57BL/6J mice on control and obesogenic ("obese") diets, throughout the whole protocol from the initial feeding period, through pregnancy and lactation. Values expressed as mean ± SEM. ***$p < 0.001$ RM ANOVA, #$p < 0.05$, ##$p < 0.01$, ###$p < 0.001$, Bonferroni post hoc tests following RM ANOVA

Obesity is associated with a higher intake of fat and simple sugars, while protein intake remains similar to controls (Fig. 8; macronutrient intakes). Maternal body weights of mice on the "obesogenic" diet ("obese" dams) are significantly greater than those of control animals throughout the feeding, pregnancy, and lactation periods (for each period, $p < 0.001$, RM ANOVA; Fig. 7a). During the feeding period, female mice fed with the obesogenic diet are significantly heavier from the end of the fourth week. At the time of mating, obese dams are, on average, approximately 6 g heavier than controls (30%). Obese dams remain significantly heavier throughout most of pregnancy, although body weights converge toward the end of gestation (Fig. 7a). Abdominal white adipose tissue (WAT) mass at gestation day 18 was significantly greater in dams fed with the obesogenic diet compared to control dams (Table 6). Postpartum and throughout the lactation period, body weights

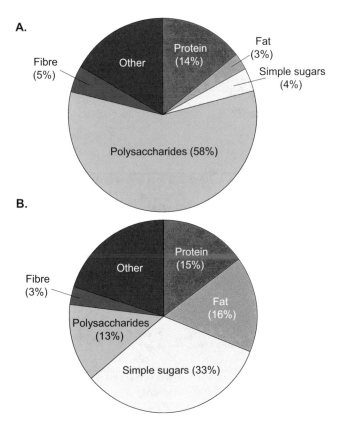

Fig. 8 Macronutrient content of food ingested by female C57BL/6J mice. Average macronutrient content of food ingested by female C57BL/6J mice on control (**a**) and obesogenic (**b**) diet during the pre-pregnancy feeding period expressed as percentage by weight

of dams on the obesogenic diet returned to being significantly greater than those of control dams (Fig. 7a).

14. Conception rates are decreased in obese dams compared to controls (75% versus 90%, respectively). Reduced conception rates in obese dams mirror the human situation, in which obesity is associated with reduced fertility [82]. White adipose tissue (WAT), brown adipose tissue (BAT) and liver mass are all significantly increased in obese dams compared to control dams (Table 6). At this stage of pregnancy, obese dams are both hyperinsulinemic and hyperleptinemic compared to controls, while there are no observed differences in serum triglycerides, cholesterol, or glucose (Fig. 9a). No significant differences are observed between the two groups in terms of litter size or weight, fetal weight or length, or placental weight (Table 7). The fetal to placental weight ratio is significantly reduced in obese dams compared to controls ($p < 0.01$; unpaired t-test). The leptin concentration in amniotic fluid is

Table 6
Maternal organ weights at gestational day 18 in control and obese C57BL/6J mouse dams

	Control (n = 9)	Obese (n = 9)
Abdominal WAT (mg)	1016.03 ± 82.95	4001.01 ± 423.01[***]
BAT (mg)	101.27 ± 4.92	149.93 ± 17.29[*]
Heart (mg)	148.00 ± 5.83	160.31 ± 3.84
Kidney (mg)	288.54 ± 14.66	317.23 ± 7.02
Liver (mg)	1755.59 ± 24.10	2151.60 ± 43.02[***]

Values expressed as mean ± SEM. $*p < 0.05$; $***p < 0.001$ versus control (unpaired t-test)

significantly increased in offspring of obese dams compared to controls ($p < 0.05$, unpaired t-test; Fig. 9b). No differences were found in amniotic fluid triglyceride, cholesterol, or glucose concentration (Fig. 9b). In the early postnatal period, no significant difference in the length of gestation is observed between control and obese dams [gestation length (days) (mean ± SEM) control: 19.6 ± 0.3, $n = 9$ versus obese 19.7 ± 0.4, $n = 7$, unpaired t-test]. At 48 h OffOb are significantly heavier than offspring of control dams (OffCon) (pup weight (males and females combined) (g) (mean ± SEM) OffCon 1.31 ± 0.04, $n = 16$ versus OffOb 1.66 ± 0.05, $n = 16$, $p < 0.01$; unpaired t-test]. At weaning, obese dams had significantly greater WAT (abdominal), BAT, heart, kidney, and liver masses compared to control dams (Table 8). DEXA scans of a subgroup of dams at weaning confirmed a two- to threefold increase in adiposity of obese dams compared to controls (Fig. 9d). At this stage, serum leptin, insulin, glucose, and cholesterol are significantly raised in obese dams compared to controls (Fig. 9c). Milk leptin concentration is significantly greater in obese dams compared to controls at weaning [leptin (ng/mL; mean ± SEM) control, 1.9 ± 0.2, $n = 3$ versus obese, 4.7 ± 1.5, $n = 3$, $p < 0.05$; unpaired t-test].

15. Pregnancy was established, within a week of cohabitation with a male, in 100% of the females on the control diet and in 83% of the females on the obesogenic diet. Average litter size was greater for the obese dams (mean litter size ± SEM: 12.3 ± 0.60, OffOb, versus 10.4 ± 0.64, OffCon, $p < 0.05$).

16. At each time point (days 2, 7, 8, 9, 11, 13, 14, 15, and 18), litters are killed by decapitation between 0800 and 1100 h.

17. While obesity contributes to gestational diabetes, diet itself is likely to play an important role in the perturbation of maternal glucose homeostasis. Several studies have suggested that diets high in simple sugars which exert a high glycemic load (GI) are

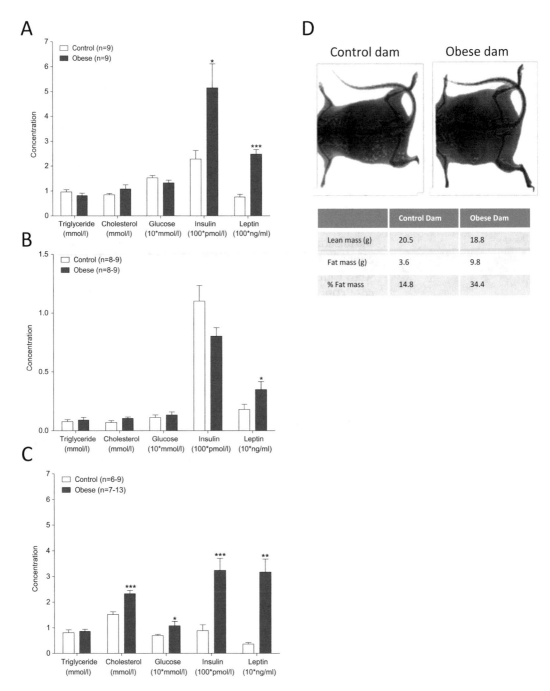

Fig. 9 Maternal body weight and food intake in rats fed control or obesogenic diet. Body weight (**A**) was recorded for 6 weeks prior to pregnancy and throughout pregnancy and lactation for the animals on the control (open symbols) or obesogenic (closed symbols) diet. Caloric intake was recorded throughout pregnancy and lactation (**B**) Macronutrient content of ingested food (expressed as percentage by weight) for control (**C**) or obese (**D**) dams during lactation. "Others" include cellulose, ash, water, etc.; *$p < 0.05$ and **$p < 0.01$ versus control dams ($n = 11$–12)

Table 7
Litter characteristics at gestational day 18 in control and obese C57BL/6J mouse dams

	Control ($n = 9$)	Obese ($n = 9$)
Litter size (number of fetuses)	7.67 ± 0.41	8.11 ± 0.61
Litter weight (g)	7.83 ± 0.57	7.21 ± 0.80
Average fetal weight (g)	1.07 ± 0.07	0.92 ± 0.05
Average fetal length (cm)	2.08 ± 0.07	1.94 ± 0.06
Average placental weight (g)	0.09 ± 0.005	0.11 ± 0.006
Fetal weight/placental weight	11.34 ± 0.55	**8.70 ± 0.66**[**]

Values expressed as mean ± SEM. **$p < 0.01$ versus control, unpaired t-test

Table 8
Maternal organ weights at weaning in control and obese C57BL/6J mouse dams

	Control ($n = 8$)	Obese ($n = 11$)
Abdominal WAT (mg)	283.6 ± 39.9	1434.0 ± 99.8***
BAT (mg)	63.6 ± 3.6	115.6 ± 12.7**
Heart (mg)	156.8 ± 11.1	216.2 ± 18.7*
Kidney (mg)	335.5 ± 6.1	460.1 ± 35.6**
Liver(mg)	922.1 ± 52.4	1319 ± 73.8**

Values expressed as mean ± SEM. *$p < 0.05$; **$p < 0.01$; ***$p < 0.001$ versus control unpaired t-test

a major cause of obesity and metabolic syndrome [83, 84]. The consumption of simple sugars (e.g., sucrose and fructose) found in soft drinks has gradually increased in the United Kingdom [85, 86], and recent studies reveal links between soft drink consumption and the incidence of metabolic syndrome. In pregnant women, it is also known that an increased dietary glycemic load is associated with increased risk of maternal obesity and gestational diabetes mellitus (GDM) [87]. Low-glycemic diet during pregnancy reduces maternal glucose levels and normalizes infant birth weight [88]. Surprisingly, considering this gradual rise in sucrose intake throughout the developed world and the growing interest in the role of a high-GI diet in obesity, the effect of exposure to a high-sucrose diet in utero on metabolic and cardiovascular complications in the offspring has scarcely been investigated. The following protocol was designed to investigate the developmental programming effects of the high-sugar component of the "obesogenic diet" detailed above by employing a maternal diet rich in sugar but low in fat.

18. Caloric intake and body weight are significantly greater in sucrose-fed dams (SF) compared with control chow-fed dams, during pre-pregnancy and throughout gestation (Fig. 10). At the end of pregnancy, food intake and body weight converge, but at weaning SF dams are again heavier [body weight (g) SF, 27.5 ± 0.8 versus C, 25.7 ± 0.2, $n = 8$, $p < 0.05$] and show a significant increase in abdominal fat mass relative to body weight, compared with controls [maternal inguinal WAT weight (mg/g BW) SF, 15.5 ± 1.0 versus C, 11.1 ± 1.6, $n = 8$, $p < 0.05$].

19. At weaning, SF dams demonstrate a fivefold increase in fasting serum insulin, which is associated with a lower blood glucose concentration than control-fed dams. No apparent difference in maternal serum leptin concentration was observed between groups at weaning (Table 9). There were no apparent

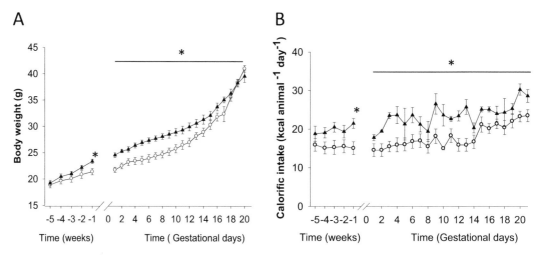

Fig. 10 Maternal body weight (**a**) (g) and (**b**) caloric intake (kcal/animal/day) in C57BL/6J mouse fed either control (C, open symbols) or a sucrose rich diet (SF, closed symbols). *$p < 0.05$ versus control using repeated measures ANOVA, $n = 6$–8 per group

Table 9
Maternal fasting serum cholesterol, triglyceride, glucose, insulin and leptin concentrations at weaning in dams fed either control ($n = 8$) or a sucrose-rich diet ($n = 6$)

Group	Cholesterol (mmol/L)	Triglyceride (mmol/L)	Glucose (μmol/L)	Insulin (pmol/L)	Leptin (ng/ml)
Control	1.52 ± 0.11	0.81 ± 0.11	7.0 ± 0.4	88.1 ± 8.3	3.7 ± 0.6
Sucrose-fed	1.89 ± 0.17	0.97 ± 0.15	$5.3 \pm 0.4*$	$475.4 \pm 14.5†$	3.0 ± 0.4

C control dams, SF sucrose fed dams
*$p < 0.05$, †$p < 0.001$ versus offspring of control dams (one-way ANOVA). Values are mean $\pm$ SEM, $n = 6$–8 per group

differences in cholesterol or triglyceride concentrations between SF and control dams. Milk leptin concentration is significantly reduced in SF dams compared with controls [Leptin (ng/mL) SF, 1.2 ± 0.9 versus C, 1.9 ± 0.2, $n = 3$, $p < 0.05$]. The maternal high-sugar diet has no effect on litter size or pup survival rates. The newborn offspring of SF dams (OSF) are heavier than the offspring of chow-fed dams [weight (g) OSF, 1.52 ± 0.14 versus OC, 1.31 ± 0.04, $n = 6$, $p < 0.01$] but are of similar weight at weaning. At weaning, female offspring of SF dams are hyperinsulinemic [insulin (pmol/L) OSF, 48.8 ± 4.7 versus OC, 28.3 ± 5.3, $n = 6$, $p < 0.05$) with increased serum IL-6 (pg/mL) (OSF, 8.7 ± 1.9 versus OC, 4.2 ± 0.5, $n = 6$, $p < 0.01$). Male OSF had increased plasminogen activator inhibitor concentrations [t-PAI (ng/mL) OSF, 2.3 ± 0.4 versus OC, 1.5 ± 0.1, $p < 0.05$, $n = 6$]. Serum resistin (ng/mL) was decreased in female offspring of SF dams (OSF, 3.0 ± 0.5 versus OC, 4.6 ± 0.4, $n = 6$, $p < 0.05$). There is no apparent difference in the serum leptin concentrations between groups [leptin (ng/mL) OSF males 1.0 ± 0.2 versus OC, 0.7 ± 0.1, $n = 6$, females, OSF 0.9 ± 0.1 versus OC, 1.0 ± 0.2, $n = 6$].

Acknowledgments

This work was funded by the British Heart Foundation and the EU Framework 6 Project, EARNEST.

References

1. Reynolds RM, Allan KM, Raja EA, Bhattacharya S, McNeill G, Hannaford PC et al (2013) Maternal obesity during pregnancy and premature mortality from cardiovascular event in adult offspring: follow-up of 1 323 275 person years. BMJ 347:f4539. https://doi.org/10.1136/bmj.f4539

2. Poston L (2012) Maternal obesity, gestational weight gain and diet as determinants of offspring long term health. Best Pract Res Clin Endocrinol Metab 26(5):627–639

3. O'Reilly JR, Reynolds RM (2013) The risk of maternal obesity to the long-term health of the offspring. Clin Endocrinol 78(1):9–16

4. Drake AJ, Reynolds RM (2010) Impact of maternal obesity on offspring obesity and cardiometabolic disease risk. Reproduction 140 (3):387–598

5. Heslehurst N, Rankin J, Wilkinson JR, Summerbell CD (2010) A nationally representative study of maternal obesity in England, UK:

trends in incidence and demographic inequalities in 619 323 births, 1989–2007. Int J Obes 34(3):420–428

6. Heslehurst N, Simpson H, Ells LJ, Rankin J, Wilkinson J, Lang R et al (2008) The impact of maternal BMI status on pregnancy outcomes with immediate short-term obstetric resource implications: a meta-analysis. Obes Rev 9 (6):635–683

7. Castillo-Laura H, Santos IS, Quadros LC, Matijasevich A (2015) Maternal obesity and offspring body composition by indirect methods: a systematic review and meta-analysis. Cad Saude Publica 31(10):2073–2092. https://doi.org/10.1590/0102-311X00159914

8. Catalano P, deMouzon SH (2015) Maternal obesity and metabolic risk to the offspring: why lifestyle interventions may have not achieved the desired outcomes. Int J Obes 39 (4):642–649

9. Zhao YL, Ma RM, Lao TT, Chen Z, Du MY, Liang K et al (2015) Maternal gestational diabetes mellitus and overweight and obesity in offspring: a study in Chinese children. J Dev Orig Health Dis 6(6):479–484

10. Yu Z, Han S, Zhu J, Sun X, Ji C, Guo X (2013) Pre-pregnancy body mass index in relation to infant birth weight and offspring overweight/obesity: a systematic review and meta-analysis. PLoS One 8(4):e61627. https://doi.org/10.1371/journal.pone.0061627

11. Gaillard R, Steegers EA, Duijts L, Felix JF, Hofman A, Franco OH et al (2014) Childhood cardiometabolic outcomes of maternal obesity during pregnancy: the generation R study. Hypertension 63(4):683–691

12. Gaillard R, Steegers EA, Franco OH, Hofman A, Jaddoe VW (2015) Maternal weight gain in different periods of pregnancy and childhood cardio-metabolic outcomes. The generation R study. Int J Obes 39(4):677–685

13. Fraser A, Tilling K, Macdonald-Wallis C, Sattar N, Brion MJ, Benfield L et al (2010) Association of maternal weight gain in pregnancy with offspring obesity and metabolic and vascular traits in childhood. Circulation 121(23):2557–2564

14. Oostvogels AJ, Stronks K, Roseboom TJ, van der Post JA, van Eijsden M, Vrijkotte TG (2014) Maternal prepregnancy BMI, offspring's early postnatal growth, and metabolic profile at age 5-6 years: the ABCD study. J Clin Endocrinol Metab 99(10):3845–3854

15. Hochner H, Friedlander Y, Calderon-Margalit-R, Meiner V, Sagy Y, Avgil-Tsadok M et al (2012) Associations of maternal prepregnancy body mass index and gestational weight gain with adult offspring cardiometabolic risk factors: the Jerusalem perinatal family follow-up study. Circulation 125(11):1381–1389

16. Gademan MG, van Eijsden M, Roseboom TJ, van der Post JA, Stronks K, Vrijkotte TG (2013) Maternal prepregnancy body mass index and their children's blood pressure and resting cardiac autonomic balance at age 5 to 6 years. Hypertension 62(3):641–647

17. Filler G, Yasin A, Kesarwani P, Garg AX, Lindsay R, Sharma AP (2011) Big mother or small baby: which predicts hypertension? J Clin Hypertens (Greenwich) 13(1):35–41

18. Wen X, Triche EW, Hogan JW, Shenassa ED, Buka SL (2011) Prenatal factors for childhood blood pressure mediated by intrauterine and/or childhood growth? Pediatrics 127(3):e713–e721

19. Taylor PD, Samuelsson AM, Poston L (2014) Maternal obesity and the developmental programming of hypertension: a role for leptin. Acta Physiol (Oxf) 210(3):508–523

20. Karachaliou M, Georgiou V, Roumeliotaki T, Chalkiadaki G, Daraki V, Koinaki S et al (2015) Association of trimester-specific gestational weight gain with fetal growth, offspring obesity, and cardiometabolic traits in early childhood. Am J Obstet Gynecol 212(4):502 e1–502 14

21. Basak S, Duttaroy AK (2012) Leptin induces tube formation in first-trimester extravillous trophoblast cells. Eur J Obstet Gynecol Reprod Biol 164(1):24–29

22. Samolis S, Papastefanou I, Panagopoulos P, Galazios G, Kouskoukis A, Maroulis G (2010) Relation between first trimester maternal serum leptin levels and body mass index in normotensive and pre-eclamptic pregnancies—role of leptin as a marker of pre-eclampsia: a prospective case-control study. Gynecol Endocrinol 26(5):338–343

23. Kral JG, Biron S, Simard S, Hould FS, Lebel S, Marceau S et al (2006) Large maternal weight loss from obesity surgery prevents transmission of obesity to children who were followed for 2 to 18 years. Pediatrics 118(6):e1644–e1649

24. Smith J, Cianflone K, Biron S, Hould FS, Lebel S, Marceau S et al (2009) Effects of maternal surgical weight loss in mothers on intergenerational transmission of obesity. J Clin Endocrinol Metab 94(11):4275–4283

25. Guenard F, Deshaies Y, Cianflone K, Kral JG, Marceau P, Vohl MC (2013) Differential methylation in glucoregulatory genes of offspring born before vs. after maternal gastrointestinal bypass surgery. Proc Natl Acad Sci U S A 110(28):11439–11444

26. Guenard F, Lamontagne M, Bosse Y, Deshaies Y, Cianflone K, Kral JG et al (2015) Influences of gestational obesity on associations between genotypes and gene expression levels in offspring following maternal gastrointestinal bypass surgery for obesity. PLoS One 10(1):e0117011. https://doi.org/10.1371/journal.pone.0117011

27. Guenard F, Tchernof A, Deshaies Y, Cianflone K, Kral JG, Marceau P et al (2013) Methylation and expression of immune and inflammatory genes in the offspring of bariatric bypass surgery patients. J Obes 2013:492170. https://doi.org/10.1155/2013/492170

28. Patel N, Godfrey KM, Pasupathy D, Levin J, Flynn AC, Hayes L et al (2017) Infant adiposity following a randomised controlled trial of a behavioural intervention in obese pregnancy. Int J Obes 41(7):1018–1026

29. Palinski W, D'Armiento FP, Witztum JL, de Nigris F, Casanada F, Condorelli M et al (2001) Maternal hypercholesterolemia and treatment during pregnancy influence the long-term progression of atherosclerosis in offspring of rabbits. Circ Res 89(11):991–996

30. Napoli C, Witztum JL, Calara F, de Nigris F, Palinski W (2000) Maternal hypercholesterolemia enhances atherogenesis in normocholesterolemic rabbits, which is inhibited by antioxidant or lipid-lowering intervention during pregnancy: an experimental model of atherogenic mechanisms in human fetuses. Circ Res 87(10):946–952

31. Khan IY, Taylor PD, Dekou V, Seed PT, Lakasing L, Graham D et al (2003) Gender-linked hypertension in offspring of lard-fed pregnant rats. Hypertension 41(1):168–175

32. Khan IY, Dekou V, Hanson M, Poston L, Taylor PD (2004) Predictive adaptive responses to maternal high fat diet prevent endothelial dysfunction but not hypertension in adult rat offspring. Circulation 110(9):1097–1102

33. Ghosh P, Bitsanis D, Ghebremeskel K, Crawford MA, Poston L (2001) Abnormal aortic fatty acid composition and small artery function in offspring of rats fed a high fat diet in pregnancy. J Physiol 533(Pt 3):815–822

34. Robb JL, Messa I, Lui E, Yeung D, Thacker J, Satvat E et al (2017) A maternal diet high in saturated fat impairs offspring hippocampal function in a sex-specific manner. Behav Brain Res 326:187–199

35. Tamashiro KL, Terrillion CE, Hyun J, Koenig JI, Moran TH (2009) Prenatal stress or high-fat diet increases susceptibility to diet-induced obesity in rat offspring. Diabetes 58 (5):1116–1125

36. Chen H, Simar D, Lambert K, Mercier J, Morris MJ (2008) Maternal and postnatal overnutrition differentially impact appetite regulators and fuel metabolism. Endocrinology 149 (11):5348–5356

37. Srinivasan M, Katewa SD, Palaniyappan A, Pandya JD, Patel MS (2006) Maternal high-fat diet consumption results in fetal malprogramming predisposing to the onset of metabolic syndrome-like phenotype in adulthood. Am J Physiol Endocrinol Metab 291(4): E792–E799

38. Koukkou E, Ghosh P, Lowy C, Poston L (1998) Offspring of normal and diabetic rats fed saturated fat in pregnancy demonstrate vascular dysfunction. Circulation 98 (25):2899–2904

39. Howie GJ, Sloboda DM, Kamal T, Vickers MH (2009) Maternal nutritional history predicts obesity in adult offspring independent of postnatal diet. J Physiol 587(Pt 4):905–915

40. Ferezou-Viala J, Roy AF, Serougne C, Gripois D, Parquet M, Bailleux V et al (2007) Long-term consequences of maternal high-fat feeding on hypothalamic leptin sensitivity and diet-induced obesity in the offspring. Am J Physiol Regul Integr Comp Physiol 293(3): R1056–R1062

41. Cerf ME, Muller CJ, Du Toit DF, Louw J, Wolfe-Coote SA (2006) Hyperglycaemia and reduced glucokinase expression in weanling offspring from dams maintained on a high-fat diet. Br J Nutr 95(2):391–396

42. Buckley AJ, Keseru B, Briody J, Thompson M, Ozanne SE, Thompson CH (2005) Altered body composition and metabolism in the male offspring of high fat-fed rats. Metabolism 54 (4):500–507

43. Cerf ME, Williams K, Nkomo XI, Muller CJ, Du Toit DF, Louw J et al (2005) Islet cell response in the neonatal rat after exposure to a high-fat diet during pregnancy. Am J Physiol Regul Integr Comp Physiol 288(5): R1122–R1128

44. Gregersen S, Dyrskog SE, Storlien LH, Hermansen K (2005) Comparison of a high saturated fat diet with a high carbohydrate diet during pregnancy and lactation: effects on insulin sensitivity in offspring of rats. Metabolism 54(10):1316–1322

45. Guo F, Jen KL (1995) High-fat feeding during pregnancy and lactation affects offspring metabolism in rats. Physiol Behav 57 (4):681–686

46. MacDonald KD, Moran AR, Scherman AJ, McEvoy CT, Platteau AS (2017) Maternal high-fat diet in mice leads to innate airway hyperresponsiveness in the adult offspring. Physiol Rep 5(5). pii: e13082. doi: https://doi.org/10.14814/phy2

47. Ito J, Nakagawa K, Kato S, Miyazawa T, Kimura F, Miyazawa T (2016) The combination of maternal and offspring high-fat diets causes marked oxidative stress and development of metabolic syndrome in mouse offspring. Life Sci 151:70–75

48. Taylor PD, Khan IY, Lakasing L, Dekou V, O'Brien-Coker I, Mallet AI et al (2003) Uterine artery function in pregnant rats fed a diet supplemented with animal lard. Exp Physiol 88 (3):389–398

49. Armitage JA, Khan IY, Taylor PD, Nathanielsz PW, Poston L (2004) Developmental programming of the metabolic syndrome by maternal nutritional imbalance: how strong is the evidence from experimental models in mammals? J Physiol 561(Pt 2):355–377

50. Taylor PD, Khan IY, Hanson MA, Poston L (2004) Impaired EDHF-mediated vasodilatation in adult offspring of rats exposed to a fat-rich diet in pregnancy. J Physiol 558 (Pt 3):943–951

51. Taylor PD, McConnell J, Khan IY, Holemans K, Lawrence KM, Asare-Anane H et al (2005) Impaired glucose homeostasis and mitochondrial abnormalities in offspring of rats fed a fat-rich diet in pregnancy. Am J Physiol Regul Integr Comp Physiol 288(1): R134–R139

52. Sanchez J, Priego T, Garcia AP, Llopis M, Palou M, Pico C et al (2012) Maternal supplementation with an excess of different fat sources during pregnancy and lactation differentially affects feeding behavior in offspring: putative role of the leptin system. Mol Nutr Food Res 56:1715–1728

53. Sun B, Purcell RH, Terrillion CE, Yan J, Moran TH, Tamashiro KL (2012) Maternal high-fat diet during gestation or suckling differentially affects offspring leptin sensitivity and obesity. Diabetes 61:2833–2841

54. Bayol SA, Farrington SJ, Stickland NC (2007) A maternal 'junk food' diet in pregnancy and lactation promotes an exacerbated taste for 'junk food' and a greater propensity for obesity in rat offspring. Br J Nutr 98(4):843–851

55. Sampey BP, Vanhoose AM, Winfield HM, Freemerman AJ, Muehlbauer MJ, Fueger PT et al (2011) Cafeteria diet is a robust model of human metabolic syndrome with liver and adipose inflammation: comparison to high-fat diet. Obesity (Silver Spring) 19:1109–1117

56. Akyol A, Langley-Evans SC, McMullen S (2009) Obesity induced by cafeteria feeding and pregnancy outcome in the rat. Br J Nutr 102(11):1601–1610

57. Morris MJ, Chen H, Watts R, Shulkes A, Cameron-Smith D (2008) Brain neuropeptide Y and CCK and peripheral adipokine receptors: temporal response in obesity induced by palatable diet. Int J Obes (Lond) 32:249–258

58. Pomar CA, van Nes R, Sanchez J, Pico C, Keijer J, Palou A (2017) Maternal consumption of a cafeteria diet during lactation in rats leads the offspring to a thin-outside-fat-inside phenotype. Int J Obes 41:1279. https://doi.org/10.1038/ijo.2017.42

59. Jacobs S, Teixeira DS, Guilherme C, da Rocha CF, Aranda BC, Reis AR et al (2014) The impact of maternal consumption of cafeteria diet on reproductive function in the offspring. Physiol Behav 129:280–286

60. Chen H, Morris MJ (2009) Differential responses of orexigenic neuropeptides to fasting in offspring of obese mothers. Obesity (Silver Spring) 17:1356–1362

61. Castro H, Pomar CA, Palou A, Pico C, Sanchez J (2017) Offspring predisposition to obesity due to maternal-diet-induced obesity in rats is preventable by dietary normalization before mating. Mol Nutr Food Res 61(3). https://doi.org/10.1002/mnfr.201600513

62. Akyol A, McMullen S, Langley-Evans SC (2012) Glucose intolerance associated with early-life exposure to maternal cafeteria feeding is dependent upon post-weaning diet. Br J Nutr 107(7):964–978

63. Mucellini AB, Goularte JF, de Araujo d, Cunha AC, Caceres RC, Noschang C et al (2014) Effects of exposure to a cafeteria diet during gestation and after weaning on the metabolism and body weight of adult male offspring in rats. Br J Nutr 111(8):1499–1506

64. Ong ZY, Muhlhausler BS (2014) Consuming a low-fat diet from weaning to adulthood reverses the programming of food preferences in male, but not female, offspring of 'junk food'-fed rat dams. Acta Physiol 210 (1):127–141

65. Bouanane S, Benkalfat NB, Baba Ahmed FZ, Merzouk H, Mokhtari NS, Merzouk SA et al (2009) Time course of changes in serum oxidant/antioxidant status in overfed obese rats and their offspring. Clin Sci (Lond) 116 (8):669–680

66. Sanchez-Blanco C, Amusquivar E, Bispo K, Herrera E (2016) Influence of cafeteria diet and fish oil in pregnancy and lactation on pups' body weight and fatty acid profiles in rats. Eur J Nutr 55(4):1741–1753

67. Eckburg PB, Bik EM, Bernstein CN, Purdom E, Dethlefsen L, Sargent M et al (2005) Diversity of the human intestinal microbial flora. Science 308 (5728):1635–1638

68. Flint HJ, Scott KP, Louis P, Duncan SH (2012) The role of the gut microbiota in nutrition and health. Nat Rev Gastroenterol Hepatol 9(10):577–589

69. Lepage P, Leclerc MC, Joossens M, Mondot S, Blottiere HM, Raes J et al (2013) A metagenomic insight into our gut's microbiome. Gut 62(1):146–158

70. Bajzer M, Seeley RJ (2006) Physiology: obesity and gut flora. Nature 444(7122):1009–1010

71. Dahlen HG, Downe S, Kennedy HP, Foureur M (2014) Is society being reshaped on a microbiological and epigenetic level by the way women give birth? Midwifery 30 (12):1149–1151

72. Dahlen HG, Kennedy HP, Anderson CM, Bell AF, Clark A, Foureur M et al (2013) The EPIIC hypothesis: intrapartum effects on the neonatal epigenome and consequent health outcomes. Med Hypotheses 80(5):656–662

73. Hyde MJ, Mostyn A, Modi N, Kemp PR (2012) The health implications of birth by caesarean section. Biol Rev Camb Philos Soc 87(1):229–243

74. Mueller NT, Whyatt R, Hoepner L, Oberfield S, Dominguez-Bello MG, Widen EM et al (2015) Prenatal exposure to antibiotics, cesarean section and risk of childhood obesity. Int J Obes 39(4):665–670

75. Vijay-Kumar M, Aitken JD, Carvalho FA, Cullender TC, Mwangi S, Srinivasan S et al (2010) Metabolic syndrome and altered gut microbiota in mice lacking Toll-like receptor 5. Science 328(5975):228–231

76. Wu GD, Chen J, Hoffmann C, Bittinger K, Chen YY, Keilbaugh SA et al (2011) Linking long-term dietary patterns with gut microbial enterotypes. Science 334(6052):105–108

77. Walker AW, Ince J, Duncan SH, Webster LM, Holtrop G, Ze X et al (2011) Dominant and diet-responsive groups of bacteria within the human colonic microbiota. ISME J 5 (2):220–230

78. David LA, Maurice CF, Carmody RN, Gootenberg DB, Button JE, Wolfe BE et al (2014) Diet rapidly and reproducibly alters the human gut microbiome. Nature 505 (7484):559–563

79. Armitage JA, Taylor PD, Poston L (2005) Experimental models of developmental programming; consequences of exposure to an energy rich diet during development. J Physiol 565(Pt 1):3–8

80. Erlanson-Albertsson C (2005) Appetite regulation and energy balance. Acta Paediatr Suppl 94(448):40–41

81. Zhang M, Balmadrid C, Kelley AE (2003) Nucleus accumbens opioid, GABAergic, and dopaminergic modulation of palatable food motivation: contrasting effects revealed by a progressive ratio study in the rat. Behav Neurosci 117(2):202–211

82. Ramsay JE, Greer I, Sattar N (2006) ABC of obesity. Obesity and reproduction. BMJ 333 (7579):1159–1162

83. Brand-Miller JC, Holt SH, Pawlak DB, Mcmillan J (2002) Glycemic index and obesity. Am J Clin Nutr 76:281S–285S

84. Schulze MB, Liu S, Rimm EB, Manson JE, Willett WC, Hu FB (2004) Glycemic index, glycemic load, and dietary fiber intake and incidence of type 2 diabetes in younger and middle-aged women. Am J Clin Nutr 80:348–356

85. Wang Jensen B, Nichols M, Allender S, de Silva-Sanigorski A, Millar L, Kremer P et al (2012) Consumption patterns of sweet drinks in a population of Australian children and adolescents (2003–2008). BMC Public Health 12:771

86. Nikpartow N, Danyliw AD, Whiting SJ, Lim H, Vatanparast H (2012) Fruit drink consumption is associated with overweight and obesity in Canadian women. Can J Public Health 103:178–182

87. Zhang C, Liu S, Solomon CG, Hu FB (2006) Dietary fiber intake, dietary glycemic load, and the risk for gestational diabetes mellitus. Diabetes Care 29:2223–2230

88. Walsh JM, McGowan CA, Mahony R, Foley ME, McAuliffe FM (2012) Low glycaemic index diet in pregnancy to prevent macrosomia (ROLO study): randomised control trial. BMJ 345:e5605

Chapter 10

Generation of the Maternal Low-Protein Rat Model for Studies of Metabolic Disorders

Dan Ma, Susan E. Ozanne, and Paul C. Guest

Abstract

Poor nutrition during pregnancy leads to an increased risk of metabolic disorders and other diseases in the offspring. This can be modelled in animals through manipulation of the maternal diet. One such model is the maternal low-protein rat which gives rise to offspring characterized by insulin resistance. This chapter gives a detailed protocol for generation of the maternal low-protein rat, which has been used in the study of several disorders including diabetes and psychiatric disorders.

Key words Low-protein rat, Metabolic programming, Insulin resistance, Diabetes, Psychiatric disorders, Biomarkers

1 Introduction

Suboptimal nutrition during development and early life can have lasting detrimental effects on long-term health. Undernutrition during pregnancy can lead to intrauterine growth restriction and a low birth weight, which has been associated with increased risk of developing metabolic pathologies such as type 2 diabetes mellitus, obesity, and cardiovascular disease [1–3]. There is now an increasing number of studies aimed at identifying the molecular mechanisms through which nutritional programming occurs [4, 5]. Such studies could lead to identification of potential novel biomarkers and drug targets and thereby pave the way for the development of newer and more effective intervention strategies which target individuals at risk of developing age-associated diseases.

Animal models are invaluable for studying the mechanisms linking early nutrition and the risk of developing specific diseases. The use of these models is often essential since the corresponding studies in humans are impractical, with the exception that epidemiological-based data may be available in some rare instances [6]. This includes the study of well-documented disasters such as the Dutch Hunger Winter, a severe famine which occurred at the

Paul C. Guest (ed.), *Investigations of Early Nutrition Effects on Long-Term Health: Methods and Applications*, Methods in Molecular Biology, vol. 1735, https://doi.org/10.1007/978-1-4939-7614-0_10, © Springer Science+Business Media, LLC 2018

end of World War II [7]. Individuals conceived at the peak of the famine showed an increased incidence of an adverse metabolic profile, with suboptimal glucose handling, higher body mass index (BMI), higher total and low-density lipoprotein cholesterol, and increased risk of psychiatric disorders. We and other researchers have attempted to mimic the effects of protein deficiency during pregnancy using a rat model of maternal protein restriction [8]. The offspring of these rats show similar phenotypes as above, including development of insulin resistance in skeletal muscle and adipose tissue and an age-dependent loss of glucose tolerance leading to development of type 2 diabetes [9, 10]. These animals may also develop a behavioral phenotype resembling certain aspects of psychiatric disorders such as a decrease in sensorimotor gating and a lower initial startle response [11], both of which are typical features of schizophrenia. In terms of molecular mechanisms, a multiplex immunoassay profiling study of maternal low-protein rat offspring found changes in the circulating levels of insulin, adiponectin, leptin, and a number of inflammation-related factors, consisted with effects seen in both metabolic and psychiatric disorders [12].

Here, we present a protocol for the preparation of 3-month-old offspring of rat dams fed a low-protein diet throughout pregnancy and lactation. This model can be used to study several diseases and conditions although the focus here was on measurement of basic organ weights, along with preparation of the serum, brain, and liver tissue for proteomic studies.

2 Materials

1. House female Wistar rats weighing 240–260 g individually at 22 °C on a 12 h/12 h light/dark cycle (see **Note 1**).

2. Control diet: 20% protein pellets.

3. Low-protein diet: 8% protein pellets (see **Note 2**).

4. S-Monovette 7.5 mL serum tubes (Sarstedt).

5. Standard dissection kit.

6. Blood glucose analyzer (HemoCue, Angelholm, Sweden).

7. Ultrasensitive Rat Insulin ELISA (Mercodia, Uppsala, Sweden).

8. Qproteome Cell Compartment Kit® (Qiagen).

9. Protease Inhibitor Mini Tablets, EDTA-free (Pierce).

10. ProteoExtract Protein Precipitation Kit® (Merck; Hull, UK).

11. Bio-Rad DC Protein Assay.

3   Methods

1. Allow rats to mate (*see* **Note 3**).

2. Feed dams ad libitum on either the control or low-protein diet from the day of detection of vaginal plugs (day 0) through to the 21st day of gestation (birth) and the 21st day of lactation (weaning) (Fig. 1).

3. Reduce each litter to eight pups after birth (*see* **Note 4**).

4. Wean all pups (from both the control and low-protein diet groups) onto the control diet and allow to feed ad libitum.

5. Remove access to chow at the end of the final day before tissue collection (on *postnatal* day 89) (*see* **Note 5**).

6. Measure and record body and organ weights (Table 1) (*see* **Note 6**).

7. Collect trunk blood and analyze one drop in duplicate from each animal using the blood glucose analyzer (*see* **Note 7**).

8. Collect the rest of the blood into 10 mL S-Monovette tubes and allow this to coagulate 90 min at room temperature (*see* **Note 8**).

9. Centrifuge 15 min at $1000 \times g$ to clarify the supernatant.

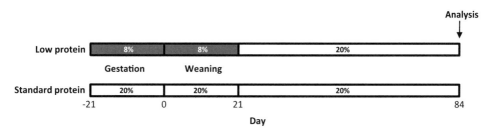

Fig. 1 Flow diagram showing generation of the maternal LP rat model

Table 1
Effects of low-protein maternal diet on fasting glucose levels and body and organ weights of 3-month-old offspring rats

	Control	Low protein	Ratio	*P*-value
Glucose (mM)	4.6 ± 0.6	4.8 ± 0.4	1.1	0.330
Body weight (g)	466 ± 14	334 ± 12	0.7	<0.001
Brain weight (g)	1.9 ± 0.1	1.8 ± 0.2	0.9	0.027
Liver weight (g)	19.0 ± 3.3	11.8 ± 2.6	0.6	<0.0001

$N = 8$ for both control and low-protein rats

10. Transfer the supernatant to new tubes and proceed immediately to analysis or store at −80 °C.

11. Analyze insulin levels using the rat insulin ELISA (data not shown) (*see* **Note 9**).

12. Dissect brain and analyze immediately or store at −80 °C prior to analysis (*see* **Note 10**).

13. Remove the liver and analyze immediately or store at −80 °C prior to analysis (*see* **Note 11**).

14. Chop 30–50 mg brain or liver tissue using a sterile scalpel to produce <0.5 mm^3 pieces on a chilled glass Petri dish.

15. Mix tissues 20 min on a rotary mixer at 4 °C in 1 mL ice-cold extraction buffer 1 (Qproteome Cell Compartment Kit) containing the recommended concentration of protease inhibitors.

16. Collect the supernatants after centrifugation for 20 min at 1000 × g at 4 °C.

17. Suspend the pellets in 0.5 mL of the same buffer, mix, and centrifuge as above.

18. Pool the two supernatants from above for each sample and designate this as fraction 1 (*see* **Note 12**).

19. Mix the pellets 20 min on the rotary mixer at 4 °C in 1 mL ice-cold buffer 2 (Qproteome Cell Compartment Kit) containing the recommended concentration of protease inhibitors.

20. Centrifuge 10 min at 6000 × g at 4 °C to obtain the supernatant.

21. Repeat the extraction of the pellet in 0.5 mL ice-cold buffer 2 plus protease inhibitors and centrifuge as above to obtain the supernatant.

22. Pool the supernatants from the two extractions in buffer 2 for each sample and designate this as fraction 2 (*see* **Note 13**).

23. Determine sample protein concentrations using the DC protein assay according to the manufacturer's instructions (*see* **Note 14**).

24. Precipitate the sample proteins to concentrate and/or remove any buffer contaminants as needed using the ProteoExtract Protein Precipitation Kit (*see* **Note 14**).

4   Notes

1. The researcher should make sure that all regulatory approvals have been completed before carrying out experiments.

2. Several types of low-protein diet are available commercially. It is important that the low-protein diet is matched as closely as

possible to the control diet used in terms of caloric content and all nutrients so that only the protein content differs. If this is not done, the true source of any resulting biomarker-related differences cannot be determined.

3. When setting numbers of animals for the mating stage of the maternal diet experiment, the researcher should be aware that the statistical unit is the mother. Thus, all power calculations carried out prior to the study should reflect the number of dams to be included in the study.

4. The litter size should be standardized and consistent to ensure the similar nourishment of offspring across different litters, and within the same low and control protein groups, during lactation. If possible, the same numbers of male and female pups should be included in the standardized litter in a given experiment.

5. Offspring can be studied at any age. Here we investigated the offspring at 12 weeks of age as we were interested in early effects seen in young adults.

6. Low-protein rat offspring will have significantly lower body weights compared to offspring of control protein dams.

7. We have shown previously that glucose levels increase significantly in low-protein compared to control rat offspring over a 17-month time course [12].

8. The researcher should make sure that the clotting time for serum preparation is standardized to minimize any potential confounding effects related to differences in blood coagulation.

9. In a previous study, we showed that the insulin/glucose ratio increased significantly over a 17-month time course [12]. This is consistent with reports showing that low-protein rat offspring have increased insulin resistance in later life compared to control rat offspring [13, 14].

10. The method used for dissection of the brain will depend on the brain region of interest and the amount of tissue generated will differ depending on which methodology is used.

11. The entire liver can be analyzed on specific portions. We usually analyze approximately 1 g taken from the bottom of the left lobe.

12. Using this kit, fraction 1 should be enriched in cytosolic proteins. If other fractionation is desired, many other kits or protocols can be used.

13. As above, other fractionation procedures can be applied as required by the user.

14. Other assay kits can be used, but the user should ensure compatibility with the reagents used in the extraction buffer.

15. Removal of detergents in buffers is normally required prior to mass spectrometry-based experiments.

References

1. Hales CN, Barker DJ, Clark PM, Cox LJ, Fall C, Osmond C et al (1991) Fetal and infant growth and impaired glucose tolerance at age 64. BMJ 303:1019–1022

2. Dorner G, Plagemann A (1994) Perinatal hyperinsulinism as possible predisposing factor for diabetes mellitus, obesity and enhanced cardiovascular risk in later life. Horm Metab Res 26:213–221

3. Eriksson JG, Forsen T, Tuomilehto J, Winter PD, Osmond C, Barker DJ (1999) Catch-up growth in childhood and death from coronary heart disease: longitudinal study. BMJ 318:427–431

4. McArdle HJ, Andersen HS, Jones H, Gambling L (2006) Fetal programming: causes and consequences as revealed by studies of dietary manipulation in rats—a review. Placenta 27(Suppl A):S56–S60

5. Tarry-Adkins JL, Ozanne SE (2016) Nutrition in early life and age-associated diseases. Ageing Res Rev pii: S1568–1637(16):30179–30179. https://doi.org/10.1016/j.arr.2016.08.003. Sep 1 [Epub ahead of print]

6. Martin-Gronert MS, Ozanne SE (2007) Experimental IUGR and later diabetes. J Intern Med 261:437–452

7. Tobi EW, Goeman JJ, Monajemi R, Gu H, Putter H, Zhang Y et al (2014) Nat Commun 5:5592. https://doi.org/10.1038/ncomms6592

8. Ozanne SE, Smith GD, Tikerpae J, Hales CN (1996) Altered regulation of hepatic glucose output in the male offspring of protein-malnourished rat dams. Am J Phys 270: E559–E564

9. Ozanne SE, Dorling MW, Wang CL, Nave BT (2001) Impaired PI 3-kinase activation in adipocytes from early growth-restricted male rats. Am J Physiol Endocrinol Metab 280: E534–E539

10. Ozanne SE, Olsen GS, Hansen LL, Tingey KJ, Nave BT, Wang CL et al (2003) Early growth restriction leads to down regulation of protein kinase C zeta and insulin resistance in skeletal muscle. J Endocrinol 177:235–241

11. Palmer AA, Printz DJ, Butler PD, Dulawa SC, Printz MP (2004) Prenatal protein deprivation in rats induces changes in prepulse inhibition and NMDA receptor binding. Brain Res 996:193–201

12. Guest PC, Urday S, Ma D, Stelzhammer V, Harris LW, Amess B et al (2012) Proteomic analysis of the maternal protein restriction rat model for schizophrenia: identification of translational changes in hormonal signaling pathways and glutamate neurotransmission. Proteomics 12:3580–3589

13. Petry CJ, Ozanne SE, Wang CL, Hales CN (2000) Effects of early protein restriction and adult obesity on rat pancreatic hormone content and glucose tolerance. Horm Metab Res 2:233–239

14. Fernandez-Twinn DS, Wayman A, Ekizoglou S, Martin MS, Hales CN, Ozanne SE (2005) Maternal protein restriction leads to hyperinsulinemia and reduced insulin-signaling protein expression in 21 month old female rat offspring. Am J Physiol Regul Integr Comp Physiol 288:R368–R373

Chapter 11

Investigation of Paternal Programming of Breast Cancer Risk in Female Offspring in Rodent Models

Camile Castilho Fontelles, Raquel Santana da Cruz, Leena Hilakivi-Clarke, Sonia de Assis, and Thomas Prates Ong

Abstract

Emerging experimental evidence show that fathers' experiences during preconception can influence their daughters' risk of developing breast cancer. Here we describe detailed protocols for investigation in rats and mice of paternally mediated breast cancer risk programming effects.

Key words Breast cancer, Paternal programming, Preconceptional diet, Female offspring, 7,12-Dimethylbenz[a]anthracene, Mammary gland development, Rodents

1 Introduction

Breast cancer is a major global public health problem. It is the most frequent cancer and the second most common cause of cancer death in women. Although several breast cancer risk factors have been identified, the exact causes of the disease remain largely unknown [1]. Reproductive (age at menarche and at first pregnancy, nulliparity, age at menopause), genetic (germ-line mutations in *BRCA1* and *BRCA2*), and environmental and lifestyle factors (alcohol intake, exposure to endocrine disruptors, pollution, sedentary lifestyle, obesity, and diet) have been shown to impact on female risk of developing breast cancer [2].

Timing of exposure to environmental and lifestyle factors has been identified as a key determinant in breast cancer susceptibility [3]. Among particularly vulnerable periods are in utero life and puberty when extensive mammary gland remodeling takes place [4–6]. In women, high birth weight is associated with increased risk of pre- and postmenopausal breast cancer in adulthood [7]. In addition, daughters of women that took the synthetic estrogen diethylstilbestrol during gestation to avoid miscarriage are also at increased risk of the disease [8] as are daughters of women having

Paul C. Guest (ed.), *Investigations of Early Nutrition Effects on Long-Term Health: Methods and Applications*, Methods in Molecular Biology, vol. 1735, https://doi.org/10.1007/978-1-4939-7614-0_11, © Springer Science+Business Media, LLC 2018

high levels of organochlorine dichlorodiphenyltrichloroethane (DDT) during pregnancy [9]. In rodent models, high-fat diet consumption during gestation and/or lactation has been found to increase susceptibility of female offspring to carcinogen-induced breast carcinogenesis [10, 11].

Increased interest has recently been directed toward the role of the father experiences on offspring health at the environmental level [12, 13]. Accumulating evidence shows that paternal experiences including obesity and malnutrition during preconception can increase metabolic diseases in the offspring [14] and epigenetic deregulation in sperm cells could represent potential underlying mechanism of these paternal metabolic programming effects [15]. However, few studies have been conducted to evaluate the influence of the fathers on female offspring breast cancer risk. In humans, paternal age and level of education have been shown to play a role [16, 17]. We have shown in mice and rats that paternal weight [18] and high-fat [19] or selenium-deficient [20] diets during preconception increased female offspring susceptibility to 7,12-dimethylbenz[α]anthracene (DMBA)-induced mammary carcinogenesis. These effects were associated with alterations in mammary gland development and paternal sperm and female offspring mammary gland microRNA profiles [18, 19].

In this chapter, we present protocols in rats and mice to investigate paternal programming of breast cancer risk in female offspring. Briefly, male rodents are submitted to different dietary interventions (i.e., high-fat or selenium-deficient diets) starting in prepuberty until sexual maturation and then mated with females receiving control diets. Paternal sperm cells are isolated to analyze molecular parameters (i.e., miRNA content, DNA methylation, histone modifications). Next, female offspring mammary glands are analyzed for morphological (number of terminal end buds [TEBs] and epithelial elongation) and molecular parameters (i.e., DNA methylation, histone modifications, miRNA content). Female offspring are further submitted to the classical breast carcinogenesis model based on DMBA initiation (Fig. 1) [6]. This is one of the most widely used animal models to study breast carcinogenesis, due to numerous similarities between rodent and human mammary gland development and carcinogenesis, as well as the rodent short life span, easy manipulation, limited genetic heterogeneity, and controlled handling environment [21]. DMBA is the most extensively used carcinogen in breast cancer studies due to its single dose regime in rats, which allows a distinction between the carcinogenic development steps (initiation, promotion, and progression), as well as reliability of tumor induction, organ site specificity, tumors of ductal histology (mainly carcinomas), and tumors of varying hormone responsiveness [22].

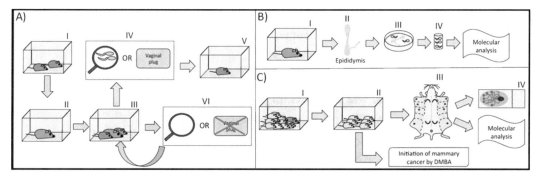

Fig. 1 Outline of protocols to investigate paternal programming of breast cancer risk in female offspring in rat and mouse models. Male rodents are submitted to different dietary interventions (i.e., high-fat or selenium-deficient diets) starting in prepuberty until sexual maturation and then mated with females receiving control diets (**a.III**). Fathers' sperm cells are isolated to analyze molecular parameters (i.e., miRNA content, DNA methylation, histone modifications) (**b.IV**). Female offspring mammary gland is analyzed for morphological (number of terminal end buds and epithelial elongation) and molecular parameters (i.e., DNA methylation, histone modifications, miRNA content) (**c.IV**). Female offspring are further submitted to the classical breast carcinogenesis model based on DMBA initiation (**c.II**)

2 Laboratory Rodents, Materials, and Reagents

2.1 Animals (See Note 1)

1. Sprague-Dawley rats.
2. C57BL/6 mice.

2.2 General Equipment

1. Biological upright microscope.
2. Standard laboratory CO_2 incubator.
3. High-speed refrigerated centrifuge and centrifuge tubes.
4. CO_2 chamber.
5. Electric shaver.
6. Sharp scissors.
7. Tweezers.
8. Glass microscope slides.
9. Conventional rodent cages.
10. Scale.
11. Swab.
12. Petri plate.
13. Class IIB2 hood.
14. 1 mL glass syringe.
15. Rigid dosing cannulae.
16. Caliper rule.
17. Tissue cassette.

2.3 Reagents

1. Ethanol (analytical grade).

2. Liquid nitrogen (*see* **Note 2**).

3. Deionized water.

4. Distilled water.

5. M2 mouse embryo culture medium.

6. Phosphate-buffered saline (PBS).

7. Somatic cells lysis buffer (SCLB): 0.1% sodium dodecyl sulfate (SDS), 0.5% triton-X 100.

8. 300 mL Carnoy solution (*see* **Note 3**): 60% ethanol, 30% chloroform, and 10% glacial acetic acid in distilled water.

9. 500 mL carmine (*see* **Note 4**): 0.2% carmine powder, 0.5% aluminum potassium sulfate in distilled water.

10. Xylene (*see* **Note 3**).

11. Permount mounting medium (*see* **Note 3**).

12. Corn oil.

13. DMBA: 0.5 mL 2% DMBA in corn oil for rats and 0.1 mL 1% DMBA in corn oil for mice (*see* **Note 5**).

14. 15% medroxyprogesterone acetate (for mouse protocol).

15. 100 mL buffered formaldehyde: 10% formaldehyde, 33 mM monosodium phosphate, 32 mM disodium phosphate (*see* **Note 6**).

16. DNA mini kit reagents (Qiagen; São Paulo, Brazil)

17. RNA miRNeasy Micro Kit (Qiagen).

18. QIAzol Lysis Reagent (Qiagen).

19. Histone Extraction Kit (Abcam; Cambridge, UK).

20. 10 mL RIPA buffer: 10 mL RIPA buffer, 1 phosphatase inhibitor tablet, 10 mM glycerophosphate, 1 mM sodium orthovanadate, 5 mM pyrophosphatase, 1 mM PMSF (*see* **Note 7**).

3 Methods

3.1 Paternal Dietary Intervention

1. Randomly distribute male rodents into control and experimental groups.

2. Maintain a maximum of two rats/cage or five mice/cage; but do not house animals singly in social isolation until 1 week prior to mating (*see* Subheading 3.2, **step 1**).

3. Start dietary interventions ad libitum for the defined length of time prior to mating (*see* **Note 8**).

4. Weigh the animals and the amount of food in each cage at the beginning of the experiment (*see* **Note 9**).

5. Weigh the animals and remaining food twice a week.

6. Calculate the average daily animal diet ingestion using the following formula:

(initial amount of diet)− (remaining amount of diet)/number of days between measurements × number of animals in the cage

7. On the same day of weighing, add a measured amount of diet to the cage.

3.2 Mating Control

1. One week prior mating, male rodents should be transferred to individual cages with the designed experimental diet and ad libitum water.

2. Introduce up to two female rodents per individual cage and allow them to stay together with one male overnight (Fig. 1a. III).

3. During mating, male and female rodents should only receive control diet (*see* **Notes 8**, **10**, and **11**).

4. On the next morning, confirm mating success by:

Rats: perform a vaginal smear on females and spread the contents on a glass slide in order to locate sperm cells on a microscope. Mice: assess the presence of a vaginal plug.

5. If sperm cells are present in the vaginal smear of rats or if there is a vaginal plug in mice (Fig. 1a.IV), separate the female in another cage with control diet and ad libitum water (Fig. 1a.V).

6. To avoid stress, keep a maximum of two female per cage until the second week of pregnancy (after this time, house one female per cage).

7. If sperm cells or vaginal plug are absent (Fig. 1a.VI), reintroduce the female at the male rodent's cage and repeat until sperm cells or a vaginal plug are found.

3.3 Sperm Cell Extraction and Purification

1. After euthanizing male rodents, remove the epididymis (Fig. 1b.II).

2. Transfer the epididymis to a petri plate with 1 mL of M2 mouse embryo culture medium (Fig. 1b.III).

3. Puncture the epididymis multiple times to allow sperm cells to swim out.

4. Incubate 1 h at 37 °C (*see* **Note 12**).

5. Remove the epididymis and transfer the medium with sperm cells to a centrifuge tube (Fig. 1b.IV).

6. Centrifuge at $3000 \times g$ for 5 min at 4 °C.

7. Remove the supernatant, and wash the pellet twice with PBS and once with deionized water.

8. Centrifuge at $3000 \times g$ for 5 min at 4 °C.

9. Remove supernatant (*see* **Note 13**).

10. Add 5 mL of SCLB and let it on ice for 1 h.

11. Centrifuge at 2000 × g for 5 min at 4 °C.

12. Remove supernatant, wash it twice with PBS.

13. Centrifuge at 3000 × g for 5 min at 4 °C.

14. Resuspend the pellet with the specific buffer according to further analysis (i.e., DNA, RNA, or protein analysis).

3.4 Gestational and Birth Weight Control

1. Weigh pregnant rodents once a week and assure that they have enough control diet and water, to avoid stress (*see* **Note 14**).

2. When offspring are born, wait 1 day before weighing pups.

3. Weigh pups once a week until they are 21 days old (*see* **Note 15**).

3.5 Litter Selection and Distribution

1. One day after birth, separate males from females.

2. A total of ten rat pups (eight females and two males) and eight mouse pups (six females and two males) should be left with the mother (Fig. 1c.I).

3. If there are not enough females, daughters from mothers from the same group can be used to compose the litter size (*see* **Note 16**).

4. Euthanize additional pups according to institutional guidelines.

3.6 Female Offspring Weight and Food Ingestion Control after Weaning

1. To control weight and food ingestion of the female offspring after weaning (22 days of birth onward), proceed as described for the fathers in Subheading 3.1.

2. When the animals reach 50 days of age (Fig.1c.II), six to eight of the female offspring should be euthanized in diestrus and have their mammary glands removed for morphological, immunohistochemistry, or molecular biology analysis.

3. The remaining animals should be submitted to the DMBA-based carcinogenesis model (avoid using sisters for the different analysis) (Fig. 1c.II).

3.7 Mammary Gland Extraction (See Note 17)

1. Euthanize 50-day-old female offspring according to institutional guidelines.

2. Shave abdominal fur.

3. Apply alcohol to the abdominal area.

4. Make a superficial incision with sharp scissors right above the vagina, without reaching the peritoneum.

5. Perform a superficial cut throughout the skin toward the head.

6. Separate the skin from the peritoneum using sharp scissors.

7. The cervical, thoracic, abdominal, and inguinal mammary glands should be visible (Fig. 1c.III).

8. Carefully separate the chosen mammary gland from the skin using small, sharp scissors.

9. Freeze the mammary glands that are going to be used in molecular analysis (i.e., DNA, RNA, proteins) in liquid nitrogen (Fig. 1c.IV).

10. Mark each slide to be able to identify the origins of the mammary glands. Use pencil for the marking, as pencil will not wash off during whole-mount preparation.

11. Stretch the mammary glands on a slide for whole-mount preparation that is going to be used in morphological analysis and submerge it in Carnoy solution (Fig. 1c.IV).

12. For immunohistochemistry analyses, place the tissue inside an immunohistochemistry cassette and submerge in buffered formaldehyde (Fig. 1c.IV).

3.8 Mammary Gland Whole-Mount Staining with Carmine Solution

1. Dilute carmine powder and aluminum potassium sulfate in distilled water.

2. Heat the solution until boiling process starts, and then immediately remove the solution from the heat.

3. Wait 20 min before refrigeration to approximately 8 °C.

4. Incubate mammary gland whole-mount in Carnoy overnight.

5. Perform three subsequent 15-min washes in 70, 50, and 30% ethanol, in that order.

6. Wash the mammary gland whole-mount in distilled water for 5 min.

7. Submerge the mammary gland whole-mount in carmine solution overnight.

8. Remove dye excess in distilled water.

9. Perform five subsequent 15-min washes in 70, 90, 95, 100, and 100% (free of water) ethanol.

10. Submerge the mammary gland whole-mount in xylene to remove fat (*see* **Note 18**).

11. Add Permount mounting medium (*see* **Note 19**).

12. Cover the slide and leave to dry (*see* **Note 20**).

3.9 Mammary Epithelial Elongation Measurement

1. Measure using a ruler, the distance between the lymph node and the end of the epithelial tree (Fig. 2a.I) (*see* **Note 21**).

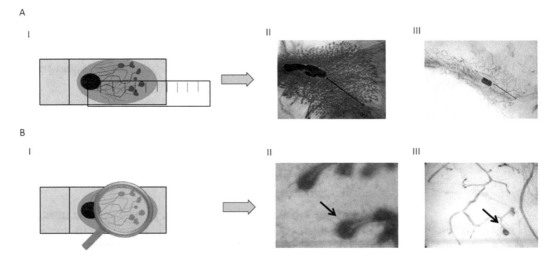

Fig. 2 Morphological analysis of mammary glands of 50-day-old rodent female offspring. Mammary epithelial elongation measurement (**a.I**) in rat (**a.II**) or mouse (**a.III**). Terminal end bud count (**b.I**) in mammary glands of rat (**b.II**) or mice (**b.III**). TEBs are located at the further edge of the epithelial tree from the nipple

3.10 TEB Assessment in Mammary Glands

1. Place slide in a light microscope.
2. Count all TEBs throughout the entire epithelial mammary tree (TEBs are located at the far end of the epithelial tree) (Fig. 2b.I) (*see* **Note 21**).

3.11 DMBA Preparation and Administration (See Notes 5 and 22)

1. Weigh inside a Class IIB2 hood the desired amount of DMBA in a Falcon tube covered with foil.
2. Dilute DMBA in corn oil at the following concentration: 20 mg/mL (rat experiment) and 10 mg/mL (mice experiment).
3. Mix the solution.
4. For female rats, administer by oral gavage 10 mg of DMBA per 50-day-old offspring.
5. For female mice, inject a subcutaneous dose of 15 mg/100 μL medroxyprogesterone acetate to 42-day-old offspring (*see* **Note 23**).
6. On the following week, administer by oral gavage 1 mg of DMBA per female mouse once a week, for 4 weeks (*see* **Note 24**).

3.12 Assessment of Tumor Development

1. Palpate, twice a week after the last DMBA administration, all animals to assess tumor appearance.
2. Record tumor appearance date and the affected mammary gland.

3. By the end of the experiment, calculate the following macroscopic parameters:

   (a) *Mammary tumor incidence through a Kaplan-Meier analysis*

   (b) *Tumor multiplicity (average number of tumors per animal) and weight*

   (c) *Latency of appearance of first tumor (average number of days post-DMBA initiation when the first tumor appears in an animal)*

4. Monitor tumor development for 20 weeks after the DMBA administration (if the tumor burden exceeds 10% of body weight, euthanize the animal prior to the end of the monitoring period).

3.13 Tumor Volume Measurement In Vivo

1. Using a caliper rule, measure the height (a), length (b), and width (c) of each tumor within each rat once a week.

2. The tumor volume is calculated by using the formula:
   $(1/6 \times 3.14) \times (a \times b \times c)$

3.14 Tumor Extraction

1. Euthanize female rodents according to institutional guidelines.

2. Shave abdominal fur.

3. Apply alcohol to the abdominal area.

4. Make a superficial incision with a sharp scissor right above the vagina, without reaching the peritoneum.

5. Perform a superficial cut throughout the skin toward the head.

6. Separate the skin from the peritoneum using a sharp scissor.

7. Visually locate tumor.

8. Carefully remove the tumor from the surrounding tissue.

9. Record weight of the tumor.

10. Freeze the tumors that are going to be used in molecular analysis in liquid nitrogen (i.e., DNA, RNA, proteins).

11. Place in cassettes samples of tumors for histopathological and immunohistochemical analysis and submerge in buffered formaldehyde.

3.15 Sperm and Mammary Gland DNA, miRNA, Protein, and Histone Extraction

1. Extract DNA using the DNA mini kit from sperm or mammary gland samples according to the manufacturer's instructions.

2. Add 700 μL of QIAzol Lysis Reagent to sperm or mammary gland samples, and extract miRNA using the miRNeasy Micro Kit according to manufacturer's instructions (*see* **Note 25**).

3. Extract proteins from sperm or mammary gland samples with RIPA buffer (add two volumes of RIPA buffer for each volume of tissue).

4. Extract histones from sperm or mammary gland samples using the Histone Extraction Kit according to the manufacturer's instructions.

5. Add pre-lysis buffer so that the final concentration is 200 mg/mL.

4 Notes

1. Animal experiments should be performed according to national and institutional regulations.

2. Liquid nitrogen should be handled with extreme caution since it is extremely cold ($-196\ ^{\circ}$C). Improper handling could result in severe frostbite or eye damage upon contact. Additionally, liquid nitrogen can expand by a factor of almost 700 and may cause an explosion in sealed containers as well as asphyxiation. When dealing with liquid nitrogen always wear safety equipment including heavy, loose-fitting leather or cryogenic gloves, as well as eye and face protection.

3. This should be carried out in a fume hood.

4. Carmine can be used multiple times. In addition, thymol can be added to increase the conservation period, given the solution is properly refrigerated ($-20\ ^{\circ}$C).

5. This is for oral gavage. DMBA preparation and administration (DMBA is a carcinogen and should be handled with care). Given its carcinogenic profile, DMBA should be handled inside a Class IIB2 hood with proper protective wear, including hair net, face mask, disposable gown, safety goggles, and gloves. Detailed procedures to safely handling DMBA can be found in Abel et al. [23]. Always follow your institutional safety guidelines. DMBA is also light-sensitive. Therefore, all DMBA solutions should be protected with foil paper and DMBA solutions should be made on the day that it is administered. Finally, all materials used in DMBA preparation and administration should be disposed of as hazardous waste.

6. First add both phosphates to water and then add formaldehyde to prevent crystallization of the phosphates.

7. RIPA buffer is typically 10 mM Tris–Cl (pH 8.0), 1 mM EDTA, 1% Triton X-100, 0.1% sodium deoxycholate, 0.1% SDS, and 140 mM NaCl.

8. During the male reproductive cycle, there are windows of susceptibility for environmental factors to induce alterations in the germ cell line [24]. These periods consist of an in utero stage, before puberty, during the reproductive cycle, and through postzygotic epigenetic reprogramming [24]. Although our research

groups have focused on the preconceptional period (from weaning to puberty) to investigate paternal breast cancer programming, further experimental interventions could be performed in any of these described periods. Prior to mating, male rodents should receive the experimental or control diets and water ad libitum, while the females (located in a separate cage) should receive control diet and water ad libitum (Fig. 1a.I, a.II). During mating, male rodents should have their experimental diets changed to control diets. This is necessary to avoid exposure of females to experimental diets and potential maternally mediated programming effects.

9. Experimental diets should be prepared to contain all the necessary nutrients of a balanced rodent food, such as those in the American Institute of Nutrition (AIN) Rodent Diets. The AIN-93G diet is used during rodent's growth, pregnancy, and lactation, and the AIN-93M is used for adult maintenance [25].

10. Mating is a delicate process. Therefore additional stress should be avoided. Verify if the circadian circle is being respected and ensure there are no excessive external noises. Also avoid entering the room during the dark period when animals are most active.

11. Both female and male mice should be kept together long enough in order to assure estrous cycle completion, since it is during the estrous phases that the female becomes receptive and able to successfully mate.

12. Under a microscope, check if there are enough sperm cells in the medium before removing the epididymis.

13. The solution containing the sperm can be frozen and kept at $-80\ ^{\circ}\text{C}$ for later use.

14. Choose only one person to handle pregnant rodents in order to avoid stressing the animals. In addition, reduce the number of experimental procedures that occur in the same room pregnant females are housed.

15. Before weighing newborn pups, rub your gloved hands in the cage shavings to acquire the cage smell. This decreases maternal stress levels and prevents cannibalism.

16. If there are more than eight rat or six mouse females in the litter, weigh all of them and choose those with a body weight close to the average litter weight, and remove pups that are noticeable lighter or heavier.

17. De Assiset et al. [26] described details of this method in a video publication.

18. Rat mammary gland can be left in xylene for several days (as long as the xylene levels are always above the slide) in

order to fully remove fat from the tissue. Approximately 3 weeks is sufficient to ensure that the slide is ready for the next step. In mice, xylene removes the fat faster.

19. Minimize bubble formation through slide coverage, since these will impair the visibility of the mammary epithelial tree during slide analysis.

20. The slide can be left to dry in a ventilated space for several days, since the mounting medium requires time to solidify.

21. This analysis should be conducted blinded and by two different researchers. TEBs are highly proliferative structures that stimulate duct elongation, as well as the development of the mammary gland. Furthermore, TEBs contain mammary gland stem cells, which are target sites of DMBA action and consequently of mammary tumor development.

22. The proposed protocol is the one adopted in our research laboratory. However, you can use the one from your institution.

23. Certain mouse strains are resistant to mammary carcinogenesis induction with DMBA or display a long latency period. In C5BL/6 mice, the use of MPA followed by DMBA increases proliferation rates and mutational risk in the target tissue, enabling mammary tumor appearance. Thus, MPA exposure induces the rapid development of mammary tumors and reduces the latency period.

24. Puberty onset in female rodents starts during weeks 4 to 5 and is completed on weeks 6 to 8 of age. This is the period when the susceptibility to breast cancer is highest, due to the higher number of TEBs.

25. It is advised to macerate the tissue with liquid nitrogen prior to molecular biology analyses.

Acknowledgments

C.C.F. was a recipient of a Ph.D. scholarship from the Brazilian National Council for Scientific and Technological Development (CNPq; Proc. 153478/2012-8). T.P.O. is the recipient of a researcher fellowship from CNPq (Proc. 307910/2016-4) and is supported by grants from CNPq (Proc. 448501/2014-7), the Food Research Center (FoRC), and the São Paulo State Research Funding Agency (Proc. 2013/07914-8). S.D.A. is supported by grants from the National Institutes of Health (K22CA178309-01A1) and the American Cancer Society (Research Scholar Grant).

References

1. Grover PL, Martin FL (2002) The initiation of breast and prostate cancer. Carcinogenesis 23:1095–1102

2. Barnard ME, Boeke CE, Tamimi RM (2015) Established breast cancer risk factors and risk of intrinsic tumor subtypes. Biochim Biophys Acta 1856:73–85

3. Hilakivi-Clarke L (2007) Nutritional modulation of terminal end buds: its relevance to breast cancer prevention. Curr Cancer Drug Targets 7:465–474

4. Trichopoulos D (1990) Is breast cancer initiated in utero? Epidemiology 1:95–96

5. Russo J, Russo IH (1980) Influence of differentiation and cell kinetics on the susceptibility of the rat mammary gland to carcinogenesis. Cancer Res 40:2677–2687

6. Russo J (2015) Significance of rat mammary tumors for human risk assessment. Toxicol Pathol 43:145–170

7. Hilakivi-Clarke L, de Assis S (2006) Fetal origins of breast cancer. Trends Endocrinol Metab 17:340–348

8. Palmer JR, Wise LA, Hatch EE, Troisi R, Titus-Ernstoff L, Strohsnitter W et al (2006) Prenatal diethylstilbestrol exposure and risk of breast cancer. Cancer Epidemiol Biomark Prev 15:1509–1514

9. Cohn BA, La Merrill M, Krigbaum NY, Yeh G, Park JS, Zimmermann L et al (2015) DDT exposure in utero and breast cancer. J Clin Endocrinol Metab 100:2865–2872

10. Hilakivi-Clarke L, Clarke R, Onojafe I, Raygada M, Cho E, Lippman M (1997) A maternal diet high in n - 6 polyunsaturated fats alters mammary gland development, puberty onset, and breast cancer risk among female rat offspring. Proc Natl Acad Sci U S A 94:9372–9377

11. de Oliveira Andrade F, Fontelles CC, Rosim MP, de Oliveira TF, de Melo Loureiro AP, Mancini-Filho J et al (2014) Exposure to lard-based high-fat diet during fetal and lactation periods modifies breast cancer susceptibility in adulthood in rats. J Nutr Biochem 25:613–622

12. Soubry A, Schildkraut JM, Murtha A, Wang F, Huang Z, Bernal A et al (2013) Paternal obesity is associated with IGF2 hypomethylation in newborns: results from a Newborn epigenetics study (NEST) cohort. BMC Med 11:29

13. Ng S-F, Lin RCY, Laybutt DR, Barres R, Owens JA, Morris MJ (2010) Chronic high-fat diet in fathers programs β-cell dysfunction in female rat offspring. Nature 467:963–966

14. Ferguson-Smith AC, Patti ME (2011) You are what your dad ate. Cell Metab 13:115–117

15. McPherson NO, Fullston T, Aitken RJ, Lane M (2014) Paternal obesity, interventions, and mechanistic pathways to impaired health in offspring. Ann Nutr Metab 64:231–238

16. Titus-Ernstoff L, Egan K, Newcomb P, Ding J, Trentham-Dietz A, Greenberg ER et al (2002) Early life factors in relation to breast cancer risk in postmenopausal women. Cancer Epidemiol Biomark Prev 11:207–210

17. Choi JY, Lee KM, Park SK, Noh DY, Ahn SH, Yoo KY et al (2005) Association of paternal age at birth and the risk of breast cancer in offspring: a case control study. BMC Cancer 5:143

18. Fontelles CC, Carney E, Clarke J, Nguyen NM, Yin C, Jin L et al (2016) Paternal overweight is associated with increased breast cancer risk in daughters in a mouse model. Sci Rep 6:28602

19. Fontelles CC, Guido LN, Rosim MP, Andrade Fde O, Jin L, Inchauspe J et al (2016) Paternal programming of breast cancer risk in daughters in a rat model: opposing effects of animal- and plant-based high-fat diets. Breast Cancer Res 18:71

20. Guido LN, Fontelles CC, Rosim MP, Pires VC, Cozzolino SM, Castro IA et al (2016) Paternal selenium deficiency but not supplementation during preconception alters mammary gland development and 7,12-dimethylbenz[a]anthracene-induced mammary carcinogenesis in female rat offspring. Int J Cancer 139:1873–1882

21. Líška J, Brtko J, Dubovický M, Macejová D, Kissová V, Polák Š et al (2016) Relationship between histology, development and tumorigenesis of mammary gland in female rat. Exp Anim 65:1–9

22. Russo J, Gusterson BA, Rogers AE, Russo IH, Wellings SR, van Zwieten MJ (1990) Comparative study of human and rat mammary tumorigenesis. Lab Investig 62:244–278

23. Abel EL, Angel JM, Kiguchi K, DiGiovanni J (2009) Multi-stage chemical carcinogenesis in

mouse skin: fundamentals and applications. Nat Protoc 4:1350–1362

24. Soubry A, Hoyo C, Jirtle RL, Murphy SK (2014) A paternal environmental legacy: evidence for epigenetic inheritance through the male germ line. BioEssays 36:359–371

25. Reeves PG, Nielsen FH, Fahey GCJ (1993) AIN-93 purified diets for laboratory rodents:

final report of the American Institute of Nutrition ad hoc writing committee on the reformulation of the AIN-76A rodent diet. J Nutr 123:1939–1195

26. de Assis S, Warri A, Cruz MI, Hilakivi-Clarke L (2010) Changes in mammary gland morphology and breast cancer risk in rats. J Vis Exp 16 (44):pii: 2260

Chapter 12

Studies of Isolated Peripheral Blood Cells as a Model of Immune Dysfunction

Hassan Rahmoune and Paul C. Guest

Abstract

Peripheral blood mononuclear cells (PBMCs) have been used as a surrogate model of immune function in studies of multiple medical areas, such as metabolic diseases and immune dysfunction. This chapter describes a standardized technique for blood draw and preparation of PBMCs from whole blood using density gradient centrifugation, followed by cell culture. The main focus is on collection of the PBMC culture media and extraction of cellular proteins in order to provide the materials for biomarker studies.

Key words Metabolic disorders, Immune dysfunction, Blood draw, Cell model, PBMCs

1 Introduction

Biomarker techniques have increased in use in recent years in the ongoing study of many diverse diseases. In clinical development, biomarker discovery, validation, and translation into diagnostic tests require reproducible sample collection, handling, and storage procedures. If possible, it is preferable that blood or blood cells are chosen for these studies since these are easily accessible from most individuals by standardized venipuncture methods. This is due to the fact that the use of blood can lead to the development of biomarker tests with greater clinical utility, and it also provides a readily accessible sample in emergency room situations [1]. This is because blood contains many factors which are produced by many tissue systems of the body and which act as messengers between these in both healthy and disease states [2, 3]. However, it is important that all procedures involved are standardized to enable reproducible and meaningful comparisons of results across different studies and in different clinics or laboratories.

Recent studies have highlighted the links between metabolism, inflammation, and the immune system in relation to nutritional programming and disease [4]. Also, studies have now begun to

Paul C. Guest (ed.), *Investigations of Early Nutrition Effects on Long-Term Health: Methods and Applications*, Methods in Molecular Biology, vol. 1735, https://doi.org/10.1007/978-1-4939-7614-0_12, © Springer Science+Business Media, LLC 2018

emerge which have investigated potential biomarkers that are capable of reflecting this important link. For example, a number of investigations have now been carried out using peripheral blood mononuclear cells (PBMCs) taken from patients and controls as a potential cellular model of various diseases marked by immune dysfunction, such as type 2 diabetes and obesity [5, 6]. PBMCs are also likely to be useful models for investigating drug effects and as potential tools for drug screening, considering that they contain many receptors and corresponding signaling pathways that are targets of several current medications [7–9]. As examples, one study used PBMCs to investigate the effects of metformin treatment on breast cancer and type 2 diabetes patients [10], another attempted to identify biomarkers of response to growth hormone therapy in PBMCs [11], and one used PBMCs to identify biomarkers of treatment response in multiple sclerosis patients [12].

Here we present a standard procedure for blood taking along with preparation of PBMCs from the blood and culture of these cells to provide the conditioned media and cellular extracts for use in biomarker studies. As examples, these procedures would be compatible with downstream analysis by immunoassay [13], multiplex immunoassay [14], two-dimensional gel electrophoresis [15], mass spectrometry-based techniques [16–18], and others.

2 Materials

2.1 Venipuncture

1. 14–20 G sterile blood draw needles.
2. Holder/adapter (see **Note 1**).
3. Tourniquet.
4. Alcohol wipes (70% isopropyl alcohol).
5. Latex, rubber, or vinyl gloves.
6. Evacuated 9 mL EDTA S-Monovette tubes (Sarstedt) (see **Note 2**).

2.2 Density Gradient Centrifugation

1. Dulbecco's phosphate-buffered saline (DPBS) (Invitrogen).
2. Ficoll-Paque Plus (density 1.077 g/mL) (GE Healthcare).
3. Sterile 15 and 50 mL Falcon tubes.
4. Centrifuge with swinging bucket rotor.
5. Cell storage solution: 10% DMSO/90% fetal bovine serum (FBS).
6. Thawing solution: RPMI 1640 (Sigma-Aldrich; Poole, UK), 10% fetal calf serum, 1% glutamine, 1% penicillin, 1% streptavidin, and 1% DNAse.

7. Stimulation solution: thawing solution, 1 mg/mL staphylococcal enterotoxin B (SEB), and 1 mg/mL CD28 (BD Bioscience) (without DNAse).

3 Methods

3.1 Venipuncture (See Note 1)

1. Before drawing blood, record all personal details, physiological status, and other metadata of each donor using a coded worksheet (Table 1) (*see* **Notes 3** and **4**).

2. Draw blood from the most appropriate arm vein of the donor (Fig. 1) (*see* **Note 5**).

3. Clean the arm of the donor with alcohol in a circular fashion, beginning at the target site.

4. Allow the area to air dry.

5. Insert the needle at a 15–20° of the vein attempting to avoid trauma and minimizing probing.

6. Take 8–10 mL of whole blood from the vein by drawing into the collection tube.

3.2 PBMC Preparation

1. Pre-warm the Ficoll-Paque to approximate room temperature, and ensure that the solution is mixed by repeated inversion so that it is homogeneous (*see* **Note 6**).

2. Add 15 mL Ficoll-Paque to a 50 mL tube (*see* **Note 7**).

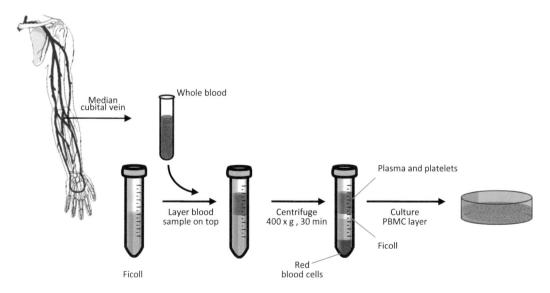

Fig. 1 Schematic showing experimental protocol for isolation of PBMCs from whole blood

3. Combine 8 mL blood with 8 mL DPBS (*see* **Note 8**).

4. Layer the blood/DPBS mixture gently on top of the Ficoll-Paque solution (*see* **Note 9**).

5. Centrifuge 30 min at $750 \times g$ at room temperature (*see* **Note 10**).

6. Remove the upper layer (plasma) using a clean pipette (*see* **Note 11**).

7. Collect PBMCs from the plasma/Ficoll-Paque interface using a sterile pipette (Fig. 1) (*see* **Note 12**).

8. Transfer the interface containing the PBMCs into a new sterile 50 mL Falcon tube.

9. Add 30 mL DPBS and centrifuge 5 min at $350 \times g$ (*see* **Note 13**).

10. Suspend the PBMC pellet gently using 10 mL DPBS, and add to a sterile 15 mL Falcon tube.

11. Centrifuge 5 min at $350 \times g$.

12. Remove the supernatant, and suspend the PBMC pellet gently using 10 mL DPBS.

13. Repeat **steps 9–12**, and either proceed immediately to the next step or freeze cells under liquid nitrogen in cell storage medium (*see* **Note 14**).

3.3 Cell Culture

1. Suspend cells in thawing solution and culture overnight at 37 °C in 5% CO_2/95% air (*see* **Note 15**).

2. After approximately 16 h, suspend 7×10^6 cells in thawing solution (without DNAse or stimulation solution (*see* **Note 16**).

3. Culture PBMCs in both thawing and stimulation solutions 72 h at 37 °C in 5% CO_2/95% air.

4. Collect cell supernatants by centrifugation of PBMCs 4 min at $1000 \times g$ at 4 °C.

5. Centrifuge the supernatants 4 min at $13,000 \times g$.

6. Collect the supernatants and store at −80 °C prior to biomarker analysis (*see* **Note 17**).

7. Wash the PBMC pellets from step 15 twice with ice-cold DPBS by centrifugation at $1000 \times g$.

8. Remove the media and store at −80 °C before use (*see* **Note 18**).

Table 1
Examples of metadata that may be collected for a biomarker study (*see* Note 19)

Samples	1. Date and time of collection (*see* **Notes 20** and **21**)
	2. Time in storage at the time of analysis
	3. Additives (*see* **Note 22**)
Personal data	1. Gender
	2. Ethnicity (*see* **Note 23**)
	3. Height (m) and weight (kg)
	4. Hip and waist measurement
	5. Fasting or fed
	6. Smoking (yes/no; number of cigarettes per day, number of years as asmoker)
	7. Alcohol consumption (number of units per week)
	8. If female, hormonal status (menstruation, menopause, hormone treatment)
	9. If female, pregnancy or breastfeeding status
	10. Age of disease onset and duration of disease
	11. Current medications and dosages
	12. Comorbidities (including disease duration and medications)
Physiological/ biochemical analysis	1. Systolic/diastolic blood pressure (mm hg)
	2. Clinical laboratory measurements (e.g., blood count, urea, creatinine, glucose, lipids)
	3. If plasma analyzed, total protein or albumin levels
	4. Sodium and potassium levels
	5. Liver enzymes in plasma
	6. Glucose tolerance test results
	7. C-reactive protein levels
Diet and lifestyle	1. If any, type of food and beverages ingested before sample collection
	2. If possible, participants should fast overnight before blood draw (*see* **Note 24**)
	3. If possible, donors should avoid heavy exercise, alcohol, and nicotine 12 h prior to sample collection
	4. Record use and levels of illicit drugs

4   Notes

1. It is important to always keep in mind when carrying out procedures with human bio-samples that safety comes first. All samples and materials should be processed and handled as if they are capable of transmitting infections or disease, and putting in place appropriate risk assessments prior to any experimentation is paramount. In addition, all materials should be disposed of with appropriate precautions, in accordance with local regulations at containment level two or three as required.

2. There are many evacuated systems available with syringe, one-draw or butterfly systems. The selection of choice should only be used after appropriate training in phlebotomy techniques.

3. It is important that consistency is maintained within and across experiments. Steps should be taken to ensure that all measurements are performed using the same types of materials, systems, and staff members, if possible. In addition, the same types of data should be collected from all donors to avoid the need to account for missing information. This is important so that statistical corrections can be made during data analysis [19, 20]. If these steps are not taken, inaccurate readings of biomarker levels are likely to occur. It is also important to maintain donor anonymity and data protection by recording all data in a secure database using codes for each subject (the codes and subject identity are recorded and kept separate so that all analyses are carried out blinded). In terms of materials, we normally use 10 mL Vacutainer K_2EDTA tubes from BD Bioscience. However, care should be taken to ensure that this is compatible with the downstream analytical platform.

4. If analyzing the plasma, the choice of tube should be made at the beginning of the study and maintained throughout. This is important for comparison of results within and across different studies. The chelating agent can be important as one study found significant differences in analyte levels after carrying out immunoassays of EDTA, heparin, and citrate plasma [21].

5. The median cubital vein is used commonly by phlebotomists due to its location and size (Fig. 1). If this is not easily accessible, other veins (such as those in the back of the hand) may be tested.

6. The Ficoll solution may settle heterogeneously during cold storage.

7. The amount of Ficoll used is related to the volume of blood.

8. Other solutions can be used here such as various tissue culture media containing FBS.

9. Care should be taken to minimize mixing of blood with the Ficoll-Paque layer as this will lead to uneven separation of blood components during centrifugation.

10. Note that the samples should be centrifuged with the brake turned off to minimize any mixing of the layers during deceleration.

11. While doing this step, care should be taken so that the lower layer is not disturbed to avoid contamination of the PBMCs.

12. The volume of collected cells should be 6–8 mL if using a 50 mL tube. The PBMC layer will be at the interface of the plasma and Ficoll layers. This is because PBMCs have an intermediate density between the plasma/DPBS mixture and Ficoll (Fig. 1). Note that the bottom layer contains red blood cells due to their higher density.

13. This is performed as a washing step to remove other materials such as the Ficoll from the PBMCs.

14. It is preferable to proceed immediately to the experiment (as opposed to freezing the cells) to ensure maximum cell survival in all subsequent steps.

15. Culturing the cells for several hours or overnight allows some recovery after the isolation or freeze/thawing procedures.

16. The aim of this experiment was to measure the effects of stimulation on the levels of various cytokines. This simulates the effects of inflammation.

17. The media contains several molecules such as cytokines that have been released from the cells under both stimulated and non-stimulated conditions. Thus some molecules may show differential release due to the inflammation. These can be identified by a method such as immunoassay [13] or multiplex immunoassay [14]. A concentration step may first be required to improve the sensitivity of detection.

18. Several proteomic platforms can be used to measure the effects of stimulation on the cellular proteins such as two-dimensional gel electrophoresis [15] and mass spectrometry-based techniques [16–18]. Previous studies have shown that cytokines [22] and several components of the glycolytic pathway [23, 24] are differentially regulated by stimulation of PBMCs from patients with various diseases compared to controls.

19. This can be used as a guide although the information required may vary in different studies in accordance with the objectives of the specific experiment.

20. As possible, all samples should be collected randomly from the test case and control individuals to minimize statistical biases.

21. Many environmental factors can have effects on the readings of some biomarkers. This includes seasonal effects and comorbid conditions, as well as the taking of some medications and other substances.

22. The addition of some substances such as preservatives or protease inhibitors can alter the results of a biomarker study.

23. Note that the distribution of the participants in each study should be equally represented across the test case and control subjects to maximize the chances of achieving reliable results that are not influenced by a sample bias.

24. Although fasting is usually desirable, this is not always possible for some participants at the time of blood draw. Any differences in fasting state should be recorded with information regarding the timing and composition of the last meal.

References

1. Singhal N, Saha A (2014) Bedside biomarkers in pediatric cardio renal injuries in emergency. Int J Crit Illn Inj Sci 4:238–246
2. Goldknopf IL (2008) Blood-based proteomics for personalized medicine: examples from neurodegenerative disease. Expert Rev Proteomics 5:1–8
3. Maggi E, Patterson NE, Montagna C (2016) Technological advances in precision medicine and drug development. Expert Rev Precis Med Drug Dev 1:331–343
4. Afman L, Milenkovic D, Roche HM (2014) Nutritional aspects of metabolic inflammation in relation to health–insights from transcriptomic biomarkers in PBMC of fatty acids and polyphenols. Mol Nutr Food Res 58:1708–1720
5. Cortez-Espinosa N, Cortés-Garcia JD, Martínez-Leija E, Rodríguez-Rivera JG, Barajas-López C, González-Amaro R et al (2015) D39 expression on Treg and Th17 cells is associated with metabolic factors in patients with type 2 diabetes. Hum Immunol 76:622–630
6. Mahmoud F, Al-Ozairi E, Haines D, Novotny L, Dashti A, Ibrahim B et al (2016) Effect of Diabetea tea ™ consumption on inflammatory cytokines and metabolic biomarkers in type 2 diabetes patients. J Ethnopharmacol 194:1069–1077
7. Jawa V, Cousens LP, Awwad M, Wakshull E, Kropshofer H, De Groot AS (2013) T-cell dependent immunogenicity of protein therapeutics: preclinical assessment and mitigation. Clin Immunol 149:534–555
8. Becker K, Schroecksnadel S, Gostner J, Zaknun C, Schennach H, Uberall F et al (2014) Comparison of in vitro tests for antioxidant and immunomodulatory capacities of compounds. Phytomedicine 21:164–171
9. Wesseling H, Guest PC, Lago SG, Bahn S (2014) Technological advances for deciphering the complexity of psychiatric disorders: merging proteomics with cell biology. Int J Neuropsychopharmacol 17:1327–1341
10. Damjanović A, Matić IZ, Đorić M, Đurović MN, Nikolić S, Roki K et al (2015) Metformin effects on malignant cells and healthy PBMC; the influence of metformin on the phenotype of breast cancer cells. Pathol Oncol Res 21:605–612
11. Welzel M, Appari M, Bramswig N, Riepe FG, Holterhus PM (2011) Transcriptional response of peripheral blood mononuclear cells to recombinant human growth hormone in a routine four-days IGF-I generation test. Growth Hormon IGF Res 21:336–342
12. Valenzuela RM, Kaufman M, Balashov KE, Ito K, Buyske S, Dhib-Jalbut S (2016) Predictive cytokine biomarkers of clinical response to glatiramer acetate therapy in multiple sclerosis. J Neuroimmunol 300:59–65
13. Keustermans GC, Hoeks SB, Meerding JM, Prakken BJ, de Jager W (2013) Cytokine assays: an assessment of the preparation and treatment of blood and tissue samples. Methods 61:10–17
14. Stephen L (2017) Multiplex immunoassay profiling. Methods Mol Biol 1546:169–176
15. Aquino A, Guest PC, Martins-de-Souza D (2017) Simultaneous two-dimensional difference gel electrophoresis (2D-DIGE) analysis of two distinct proteomes. Methods Mol Biol 1546:205–212
16. Garcia S, Baldasso PA, Guest PC, Martins-de-Souza D (2017) Depletion of highly abundant proteins of the human blood plasma: applications in proteomics studies of psychiatric disorders. Methods Mol Biol 1546:195–204
17. Faça VM (2017) Selective reaction monitoring for quantitation of cellular proteins. Methods Mol Biol 1546:213–221
18. Núñez EV, Domont GB, Nogueira FC (2017) iTRAQ-based shotgun proteomics approach for relative protein quantification. Methods Mol Biol 1546:267–274
19. Becan-McBride K (1999) Laboratory sampling: does the process affect the outcome? J Intraven Nurs 22:137–142
20. Bowen RA, Hortin GL, Csako G, Otañez OH, Remaley AT (2010) Impact of blood collection devices on clinical chemistry assays. Clin Biochem 43:4–25
21. Haab BB, Geierstanger BH, Michailidis G, Vitzthum F, Forrester S, Okon R et al (2005) Immunoassay and antibody microarray analysis of the HUPO plasma proteome project reference specimens: systematic variation between sample types and calibration of mass spectrometry data. Proteomics 5:3278–3291
22. Omodei D, Pucino V, Labruna G, Procaccini C, Galgani M, Perna F et al (2015) Immune-metabolic profiling of anorexic patients reveals an anti-oxidant and anti-inflammatory phenotype. Metabolism 64:396–405
23. Herberth M, Koethe D, Cheng TM, Krzyszton ND, Schoeffmann S, Guest PC et al (2011)

Impaired glycolytic response in peripheral blood mononuclear cells of first-onset antipsychotic-naive schizophrenia patients. Mol Psychiatry 16:848–859

24. Calton EK, Keane KN, Soares MJ, Rowlands J, Newsholme P (2016) Prevailing vitamin D status influences mitochondrial and glycolytic bioenergetics in peripheral blood mononuclear cells obtained from adults. Redox Biol 10:243–250

Chapter 13

Studies of a Neuronal Cell Line as a Model of Psychiatric Disorders

Keiko Iwata

Abstract

Mental disorders are generally characterized by a combination of abnormal thoughts, perceptions, emotions, behavior, and relationships with others. Although multiple risk factors, such as genetic and environmental factors and interaction of these factors, are suggested, the exact etiologies are not known. On the other hand, it has been strongly suggested that the dopaminergic system is impaired in a variety of mental disorders. In the described method, the SH-SY5Y neuroblastoma cell line is differentiated to neuronal cell which expresses NSE, neuronal marker, and dopamine transporter (DAT) by treatment with all-trans-retinoic acid. SH-SY5Y cells allow investigating neuronal phenotypes of mental disorders as an in vitro model of these disorders.

Key words Mental disorders, SH-SY5Y cell line, All-trans-retinoic acid, Differentiation, Dopamine transporter

1 Introduction

Mental disorders, such as depression, bipolar disorder, schizophrenia (and other psychoses), dementia, and developmental disorders including autism, are generally characterized by a combination of abnormal thoughts, perceptions, emotions, behavior, and relationships with others. Depression is characterized by sadness, loss of interest or pleasure, feelings of guilt or low self-worth, disturbed sleep or appetite, tiredness, and poor concentration. It is estimated that 300 million people are affected by depression [1]. Bipolar disorder affects about 60 million people worldwide [1]. It typically consists of both manic and depressive episodes separated by periods of normal mood. Manic episodes involve elevated or irritable moods, overactivity, pressure of speech, inflated self-esteem, and a decreased need for sleep. Schizophrenia affects about 21 million people worldwide [1]. Psychoses, including schizophrenia, are characterized by distortions in thinking, perception, emotions, language, sense of self, and behavior. Common psychotic

Paul C. Guest (ed.), *Investigations of Early Nutrition Effects on Long-Term Health: Methods and Applications*, Methods in Molecular Biology, vol. 1735, https://doi.org/10.1007/978-1-4939-7614-0_13, © Springer Science+Business Media, LLC 2018

experiences include hallucinations (hearing, seeing, or feeling things that are not there) and delusions (fixed false beliefs or suspicions that are firmly held even when there is evidence to the contrary). Dementia is usually of a chronic or progressive nature in which there is deterioration in cognitive function (i.e., the ability to process thoughts) beyond what might be expected from normal aging. It affects memory, thinking, orientation, comprehension, calculation, learning capacity, language, and judgment. It is estimated that 47.5 million people are affected by dementia [1]. Pervasive developmental disorders, such as autism, are characterized by impaired social behavior, communication and language, and a narrow range of interests and activities that are both unique to the individual and are carried out repetitively. It is estimated that 24.8 million people are affected by autism [2].

Multiple risk factors incorporating genetic and environmental susceptibility are associated with development of these disorders. Although many research groups have tried to elucidate etiologies of these mental disorders by epidemiological, genomic, molecular biological, neuropathological, and brain imaging approaches, the exact etiologies of most mental disorders are not known. On a neurobiological level, it has long been observed that basic neurotransmitter systems, such as the dopaminergic system, are impaired in a variety of mental disorders [3–6]. In addition, the dopaminergic system is also an important target for treatment of mental disorders [7].

The SH-SY5Y neuroblastoma cell line is the widely used in vitro model in research for mental disorders and has the advantage of low cost, ease of culture, and genetic modification and reproducibility. Despite its origin from a tumor, its neuroectodermal lineage allows investigating neuronal phenotypes mental disorders. The method described here is for a culture of the SH-SY5Y cell line and differentiation and measuring of differentiation status using quantitative real-time reverse transcription-polymerase chain reaction (qRT-PCR) measurement of a neuronal marker, neuron-specific enolase (*NSE*). In addition, measurement of dopamine transporter (*DAT*) is also described.

2 Materials

2.1 Equipment

1. CO_2 incubator at 37 °C (5% CO_2).
2. Laminar flow cabinet.
3. Inverted microscope.
4. NanoDrop ND-1000 spectrophotometer (Nanodrop Technologies).
5. Real-time PCR system (Applied Biosystems 7900HT real-time PCR system with software SDS 2.4; Applied Biosystems).

2.2 Reagents	1. Normal medium: Dulbecco's modified Eagle's media (DMEM), supplemented with 10% (v/v) fetal bovine serum (FBS), 2 mM L-glutamine, 100 U/mL penicillin, and 100 µg/mL streptomycin.

    2. Differentiation medium: normal medium containing 3% FBS and 10 µM all-trans-retinoic acid (ATRA) (Sigma-Aldrich) (*see* **Note 1**).

    3. Trypsin–EDTA (Thermo Fisher Scientific).

    4. Phosphate buffered saline (PBS) (pH 7.4) (*see* **Note 2**).

    5. 100 mm tissue culture-treated plates (Corning).

    6. 6-well tissue culture-treated plates (12 well can also be used) (Corning).

    7. Pasteur pipettes (Fisher Scientific).

    8. RNeasy Plus Mini Kit (Qiagen).

    9. β-mercaptoethanol.

    10. SuperScript IV first-strand synthesis system (Thermo Fisher Scientific).

    11. SsoAdvanced™ Universal SYBR® Green Supermix (Bio-Rad).

    12. Primers:

       ACTB

       Sense: 5′-ATTGGCAATGAGCGGTTC-3′.

       antisense: 5′-GGATGCCACAGGACTCCAT-3′.

       NSE

       Sense: 5′-AGGCCAGATCAAGACTGGTG-3′.

       antisense: 5′-CACAGCACACTGGGATTACG-3′;

       DAT

       Sense: 5′-CCATACTGCAAG GTGTGGGC-3′.

       antisense: 5′-CCAGGAGTTGTTGCAGTG GA-3′ [8].

    13. Statistical analysis software: SPSS (version 12.0 J; SPSS, Inc.).

3   Methods

3.1 Cell Culture	1. Ensure all media, solutions, and reagents added to cells are sterile and prewarmed to 37 °C prior to use.

    2. Culture SK-SY5Ycell line in tissue culture medium in 100 mm tissue culture-treated plates at 37 °C in 5% CO_2.

    3. Passage cells when they reach 85% confluence by aspirating the media and washing the cells once with PBS.

4. Add 1 mL trypsin–EDTA solution for 2–3 min to detach the cells.

5. Deactivate the trypsin by adding 0.3–0.4 mL of the cell suspension to 10 mL normal culture medium in 100 mm tissue culture-treated plate.

3.2 Differentiation

1. Day 0: plate cells in 6-well tissue culture-treated plates in normal medium.

2. Day 1: when cells reach approximately 50% confluence, replace normal medium with differentiation medium, and incubate at 37 °C in 5% CO_2 (*see* **Note 3**).

3. Day 3: replace medium with fresh differentiation medium.

4. Day 5: replace medium with fresh differentiation medium.

5. Day 7: confirm differentiation by quantitative polymerase chain reaction (qPCR) measurement of neuron-specific enolase (NES) and dopamine transporter (DAT).

3.3 Total RNA Preparation (See Note 4)

1. After 7-day differentiation, aspirate the medium and wash the cells three times with PBS.

2. Isolate total RNA from cell lines using the RNeasy Plus Mini Kit according to the manufacturer's instructions.

3. Add 350 µL RLT Plus buffer containing 1% β-mercaptoethanol to each cell culture, and homogenize by pipetting at least ten times.

4. To remove genomic DNA, transfer the homogenate to a gDNA eliminator spin column, and centrifuge 30 s at $10,000 \times g$.

5. Save the flow-through.

6. Add 350 µL 70% ethanol to the flow-through, and mix well by pipetting.

7. Transfer the sample to an RNeasy spin column, and centrifuge 15 s at $10,000 \times g$.

8. Discard the flow-through.

9. Add 700 µL RW1 buffer to the RNeasy spin column, and centrifuge 15 s at $10,000 \times g$.

10. Discard the flow-through.

11. Add 500 µL RPE buffer to the RNeasy spin column, and centrifuge 15 s at $10,000 \times g$ to wash the spin column membrane.

12. Discard the flow-through.

13. Add 500 µL RPE buffer to the RNeasy spin column, and centrifuge 2 min at $10,000 \times g$ to wash the spin column membrane.

14. Discard the flow-through.

15. Centrifuge 1 min at 10,000 × g to eliminate any possible carryover of RPE buffer.

16. Place the RNeasy spin column in a new 1.5 mL collection tube.

17. Add 30 μL RNase-free water to the spin column membrane, and centrifuge 1 min at 10,000 × g to elute RNA.

18. Determine RNA quantity and purity using NanoDrop ND-1000 spectrophotometer, and store at −80 °C prior to further use (*see* **Note 5**).

3.4 Reverse Transcription (RT) (See Note 4)

1. Convert total RNA to cDNA using the SuperScript IV first-strand synthesis system according to the manufacturer's instructions.

2. Combine the components of RNA primer mix in a PCR reaction tube (Table 1).

3. Mix and briefly centrifuge the components.

4. Heat the RNA primer mix at 65 °C for 5 min to anneal primer to template RNA.

5. Leave on ice at least 1 min.

6. Combine the components of RT reaction mix in a tube, mix, and briefly centrifuge the components (Table 2).

7. Add RT reaction mix to RNA primer mix.

Table 1
Components of RNA primer mix

Component	Volume (μL)
50 ng/μL random hexamers	1
10 mM dNTP mix (10 mM each)	1
Total RNA (10 pg–5 μg)	Up to 11
RNase-free water	To 13

Table 2
Components of RT reaction mix

Component	Volume (μL)
5× SSIV buffer	4
100 mM DTT	1
RNaseOUT™ Recombinant RNase Inhibitor	1
SuperScript® IV Reverse Transcriptase (200 U/μL)	1

8. Incubate the combined reaction mixture at 23 °C for 10 min.

9. Incubate the combined reaction mixture at 50 °C for 10 min.

10. Stop the reaction by incubating at 80 °C for 10 min.

11. Add 1 μL *E. coli* RNase H, and incubate at 37 °C for 20 min to remove RNA.

3.5 qPCR

1. qPCR is performed using the SsoAdvanced™ Universal SYBR® Green Supermix according to the manufacturer's instructions.

2. Prepare enough reaction mixture by combining the components except the cDNA.

3. Combine the components of the reaction mixture in a tube (Table 3).

4. Mix gently but thoroughly, and briefly centrifuge the components.

5. Add aliquots of the reaction mixture to each well of PCR plate.

6. Add cDNA to wells containing the reaction mixture.

7. Seal wells with optically transparent film, and mix gently but thoroughly by inverting the tube several times.

8. Program thermal cycling protocol on the real-time PCR instrument (Table 4).

Table 3
Components of reaction mixture for qRT-PCR

Component	Volume per 16 μL reaction
SsoAdvanced™ Universal SYBR® Green Supermix	8 μL
Forward and reverse primers (10 μM each)	1 μL each
cDNA (100 ng–100 fg)	4 μL
Nuclease-free water	3 μL

Table 4
Thermal cycling setting for qRT-PCR

Polymerase activation and DNA denaturation	95 °C	30 s	
Denaturation	95 °C	15 s	40 cycles
Annealing/extension (plate read)	60 °C	60 s	
Dissociation step	Use instrument default setting		

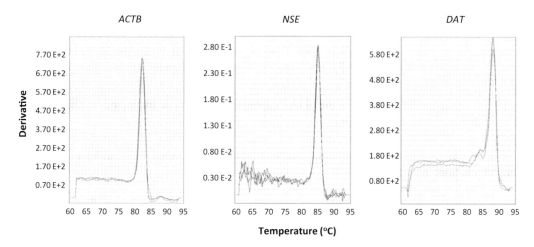

Fig. 1 Analysis of dissociation curves for *NSE*, *DAT*, and *ACTB*

9. Check each primer pair for potential primer dimer formation by analyzing dissociation curves (Fig. 1).

10. Calculate fold changes in the target mRNA levels (NSE and DAT) in the cells by normalizing the gene expression levels to those of ACTB ($2 - \Delta\Delta Ct$ method) [9] (*see* **Note 6**).

11. Calculate expression levels as percentages of the expression levels in control cells (*see* **Note 7**).

12. Analyze gene expression levels using two-tailed unpaired t-tests after no violation of the equal variance assumption is confirmed by F-test.

13. Consider values of $p < 0.05$ as statistically significant.

4   Notes

1. ATRA is light, heat, and air sensitive. A stock solution (10 mM in ethanol) should be stored in the dark at −70 °C for up to 2 weeks.

2. When you make the 10X stock of PBS, the pH should be approximately 6.8, and when diluted to 1x PBS, it should change to 7.4.

3. Final ethanol concentration from the ATRA stock should be less than 0.1% (v/v).

4. All materials must be RNase-free for all stages.

5. RNA should have an OD_{260}/OD_{280} ratio of between 1.7 and 2.1.

6. It is suggested that more than one gene should be used as a reference gene to obtain reliable results [10, 11]. Usually the author uses at least two internal controls, including ACTB, GAPDH, HRPT1, and B2M (depend on cell lines, treatment, etc.).

7. NSE and DAT expression should be increased in differentiated cells.

Acknowledgments

This work was supported in part by the Japan Society for the Promotion of Science (JSPS) Program for Advancing Strategic International Networks to Accelerate the Circulation of Talented Researchers, Grant No. S2603, and the Japan Foundation for Pediatric Research (to K.I.).

References

1. http://www.who.int/mediacentre/factsheets/fs396/en/

2. GBD 2015 Disease and Injury Incidence and Prevalence Collaborators (2016) Global, regional, and national incidence, prevalence, and years lived with disability for 310 diseases and injuries, 1990–2015: a systematic analysis for the Global Burden of Disease Study 2015. Lancet 388:1545–1602

3. Grace AA (2016) Dysregulation of the dopamine system in the pathophysiology of schizophrenia and depression. Nat Rev Neurosci 17:524–532

4. Nguyen M, Roth A, Kyzar EJ, Poudel MK, Wong K, Stewart AM et al (2014) Decoding the contribution of dopaminergic genes and pathways to autism spectrum disorder (ASD). Neurochem Int 66:15–26

5. Grace AA (2012) Dopamine system dysregulation by the hippocampus: implications for the pathophysiology and treatment of schizophrenia. Neuropharmacology 62:1342–1348

6. Hoenicka J, Aragues M, Ponce G, Rodriguez-Jimenez R, Jimenez-Arriero MA, Palomo T (2007) From dopaminergic genes to psychiatric disorders. Neurotox Res 11:61–72

7. Brandl EJ, Kennedy JL, Muller DJ (2014) Pharmacogenetics of antipsychotics. Can J Psychiatr 59:76–88

8. Green AL, Hossain MM, Tee SC, Zarbl H, Guo GL, Richardson JR (2015) Epigenetic regulation of dopamine transporter mRNA expression in human Neuroblastoma cells. Neurochem Res 40:1372–1378

9. Bookout AL, Mangelsdorf DJ (2003) Quantitative real-time PCR protocol for analysis of nuclear receptor signaling pathways. Nucl Recept Signal 1:e012. https://doi.org/10.1621/nrs.01012

10. Bustin SA (2002) Quantification of mRNA using real-time reverse transcription PCR (RT-PCR): trends and problems. J Mol Endocrinol 29:23–39

11. Vandesompele J, De Preter K, Pattyn F, Poppe B, Van Roy N, De Paepe A et al (2002) Accurate normalization of real-time quantitative RT-PCR data by geometric averaging of multiple internal control genes. Genome Biol 3. RESEARCH0034

Chapter 14

Assessment of Placental Transport Function in Studies of Disease Programming

Amanda N. Sferruzzi-Perri

Abstract

Environmental conditions during pregnancy affect fetal growth and development and program the off-spring for poor future health. These effects may be mediated by the placenta, which develops to transfer nutrients from the mother to the fetus for growth. The ability to measure the unidirectional maternofetal transfer of non-metabolizable radio-analogues of glucose and amino acid by the placenta in vivo has thus been invaluable to our understanding of the regulation of fetal growth, particularly in small animal models. Herein, I describe the method by which in vivo placental transfer function can be quantified in the mouse, an animal model widely used in studies of in utero disease programming.

Key words Placenta, Amino acid, Glucose, Transport, Nutrient allocation, Intrauterine growth restriction, Overgrowth

1 Introduction

In mammals, the main determinant of growth and development in utero is the placenta. The placenta constitutes the interface between the mother and her fetus which is responsible for controlling the amount of nutrients and oxygen transferred. In human pregnancies, abnormal birth weight is associated with changes in placental transporter capacity, which suggests that the placenta is a key mediator of alterations in fetal growth [1, 2]. Indeed, in animal models, placental supply capacity is modified in response to environmental and hormonal challenges in the mother and appears to link maternal perturbations to changes in fetal growth and offspring outcome [3–6]. Moreover, placental transport capacity adapts dynamically to both fetal signals of nutrient demand and maternal signals of nutrient availability to ensure appropriate allocation of available resources [7–11]. Therefore, by assessing placental transport capacity, we are able to better understand the regulation of fetal growth and identify programming mechanisms.

Paul C. Guest (ed.), *Investigations of Early Nutrition Effects on Long-Term Health: Methods and Applications*, Methods in Molecular Biology, vol. 1735, https://doi.org/10.1007/978-1-4939-7614-0_14, © Springer Science+Business Media, LLC 2018

This chapter aims to describe how the transfer of nutrients from the mother to the fetus by the placenta may be quantified in vivo. The technique involves injecting radiolabeled non-metabolizable substrates into the maternal circulation and then assessing the clearance rates across the placenta over time in relation to the accumulation of radiolabel in the fetus. The method has largely been used in studies of mouse, rat, and guinea pig pregnancy and has been most widely used to measure the placental transfer of glucose and amino acid, which are indicative of facilitated diffusion and active transport function, respectively [9, 12–32]. However, this method has also been used to examine the in vivo transplacental transport of radioactive sodium and calcium [33–36] and the passive permeability characteristics of the placenta for solute flux, using radioactive inert hydrophilic substrates [12, 37, 38]. Thus, by substituting the radiolabeled substrate or tracer used, this method can also be applied to study other transport systems in vivo [39]. Herein, the method for simultaneous measurement of the unidirectional placental transfer of ^{3}H-methyl-D glucose (^{3}H-MeG) and ^{14}C-aminoisobutyric acid (^{14}C-MeAIB), an amino acid analogue principally transferred by the System A transporters [40], is described for the mouse.

2 Materials

2.1 Equipment

1. Warming/heating pad to maintain the dam's body temperature.
2. Table lamp.
3. Incubator.
4. Liquid scintillation counter.
5. Liquid dispensing pump for scintillation fluid.
6. Refrigerated centrifuge.

2.2 Materials and Reagents

1. Small weigh boats. Stainless steel sterile scalpel blades (size #21).
2. Single-edge razor blades.
3. 27 gauge needles (0.4 mm × 20 mm).
4. 25 gauge needles (0.5 mm × 16 mm).
5. Polythene tubing (800/100/200, I.D. 0.58 mm, OD 0.96 mm).
6. 1 mL syringes.
7. Ethylenediaminetetraacetic acid (EDTA)-coated tubes.
8. 1.5 mL microcentrifuge tubes.
9. 15 mL screw cap tubes.

10. 5 mL plastic scintillation vials.

11. AS ScintLogic scintillation fluid (LabLogic).

12. ^{14}C-MeAIB (NEN NEC-671; specific activity 1.86 GBq/mmol).

13. ^{3}H-MeG (NEN NEC-377; specific activity 2.1 GBq/mmol).

14. Hypnorm (fentanyl citrate, 0.315 mg/mL, and fluanisone, 10 mg/mL).

15. Hypnovel (midazolam, 10 mg/2 mL).

16. Sterile water for injection.

17. Biosol (National Diagnostics; Atlanta, GA, USA).

18. Sterile physiological saline 0.9% NaCl.

19. 70% ethanol.

2.3 Instruments

1. Small curved, serrated forceps (for holding catheter in vessel).

2. Curved serrated forceps (for dissection).

3. Blunt dissecting scissors.

4. Dressing/operating sharp scissors.

5. Scalpel blade holder.

6. Three-sided, small needle file.

7. Pair of mosquito forceps.

8. Seeker needle.

3   Methods

3.1 Catheters Preparation

1. Stretch the polythene tubing to around 10 cm, leaving the first 1 cm un-stretched.

2. Using a single-edge razor blade, cut the stretched polythene tubing in such a way that both ends (about 0.5 cm) are un-stretched.

3. With the aid of the mosquito forceps, break off the hub of one 27 gauge needle.

4. To another 25 gauge needle, break off the pointed tip.

5. Using the needle file, file down the broken-off end of each needle (*see* **Note 1**).

6. With the aid of mosquito forceps, insert the broken-off edge of each needle into opposite ends of the polythene tubing (*see* **Note 2**).

7. Test the catheter with sterile water and ensure that the water is expelled afterward.

3.2 Radioactive Isotope (See Note 3)

1. Create a stock of ^{3}H-MeG (*see* **Note 4**) and ^{14}C-MeAIB, each at a concentration of 3.5 µCi/100 µL in sterile physiological saline (0.13 MBq).

2. Prepare a 1:1 mixture of 3.5 µCi of ^{3}H-MeG and 3.5 µCi of ^{14}C-MeAIB and store at −20 °C in a lockable freezer.

3. Prepare sufficient radioactive isotope for the batch of experiments to be performed (*see* **Note 5**).

3.3 Anesthetic Preparation

1. Prepare anesthetic in the following ratio: one part Hypnorm to one part Hypnovel and two parts sterile water (1:1:2) (*see* **Note 6**).

3.4 Time-Mating of Mice (See Note 7)

1. Order an adult mice from your preferred supplier and allow them to acclimatize in your animal facility for 1 week (*see* **Note 8**).

2. Time-mate female mice by placing one or two females in each stud male cage in the late afternoon (*see* **Note 9**).

3. The next day, between 0800 and 1000 h, check for the presence of a copulatory plug in the entrance to dam's genital tract (the day of plug indicates day 1 of pregnancy).

4. Females can either remain or be removed from the stud male cage (*see* **Note 10**).

3.5 Placental Transport Assay (See Notes 11 and 12)

1. On the day of pregnancy when placental transfer function will be assessed, weigh the dam.

2. Prepare the room by turning on the heating pad, covering the entire bench (and heating pad) with benchcote.

3. Label tubes and allow radioactive isotope mixture (containing 3.5 µCi of ^{3}H-MeG and 3.5 µCi of ^{14}C-MeAIB) to thaw at room temperature.

4. Draw up 200 µL of the radioactive isotope mixture into the catheter.

5. Induce anesthesia in the dam using IP injection of Hypnorm-Hypnovel solution (*see* **Note 13**).

6. Check the dam for reflexes by firmly squeezing her foot, and place her on her back, on the heat pad, to ensure body temperature is maintained.

7. Clean the neck area of the animal with a 70% ethanol.

8. Expose the maternal jugular vein (Fig. 1) using a scalpel blade fitted to a holder to make a 1.5–2 cm vertical incision in the skin of the neck, ~0.25–0.5 cm from the midline.

9. Then, using the blunt scissors and small curved serrated forceps, slowly use blunt dissection to clear the skin and expose the jugular vein.

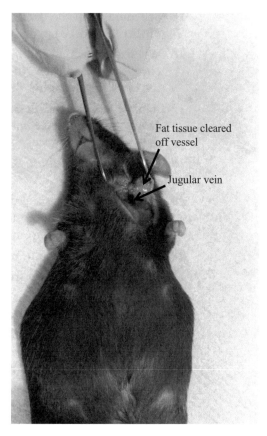

Fat tissue cleared
off vessel

Jugular vein

Fig. 1 Photo of an anaesthetized pregnant dam, showing the jugular vein and fat pad cleared away (work was conducted in the University of Cambridge Animal Facility abiding by the UK Home Office Animals (Scientific Procedures) Act 1986 and local ethics committee). Prior to incision, the neck area is normally cleaned with 70% ethanol

10. The fat pad needs to be slowly pushed/cleared away from the vessel using a cotton tip bud so that ~0.5–1 cm of the vessel can be easily observed (*see* **Note 14**).

11. Apply a little pressure on the distal end of the vessel using your finger or cotton tip bud to help it to bulge.

12. Then slowly insert the catheter into the vessel, keeping the insertion shallow/superficial to prevent injection through the vessel, and into the underlying interstitium.

13. Once in place, hold the catheter in the vessel using the small curved serrated forceps and slowly infuse the radioactive isotope into the dam over ~15 s (*see* **Notes 15** and **16**).

14. Slowly withdraw the catheter, and quickly place a cotton tip bud over the vessel entrance to stop the blood or isotope from leaking out, and immediately start the timer counting up.

15. Schedule 1 kill the dam at 1–4 min after tracer injection (*see* **Notes 17** and **18**).

16. When approaching the time to kill the dam, open the dam's chest cavity, and then cut through the top of the heart using the operating/dressing sharp scissors.

17. Rapidly collect the exsanguinated blood using the 1 mL syringe, and dispense into a labeled pre-chilled EDTA tube, shake, and keep on ice.

18. Turn dam over and use cervical dislocation to ensure death of the animal.

19. Open the peritoneal cavity, and count the number of viable and dead/resorbing conceptuses in each uterine horn before removing the uterus from the dam.

20. Separate each conceptus into its own small weigh boat.

21. Then, dissect each fetus from its placenta and fetal membranes.

22. After drying on tissue paper, weigh each fetus and placenta, and then move fetus into a new weigh boat and decapitate (*see* **Note 19**).

23. Mince the fetus in the weigh boat using a scalpel blade, and scrape the entire sample into a 15 mL screw cap tube.

24. To the 15 mL tubes containing minced fetuses, add 2 mL or 4 mL of Biosol for studies on days 15–16 and days 18–19 of pregnancy, respectively (*see* **Note 20**).

25. Manually shake the fetal samples to ensure the entire minced fetus is immersed in the Biosol.

26. Incubate samples at 55 °C for 1 week to ensure complete homogenization/solubilization of fetal tissue and release of radioactivity (*see* **Note 21**).

27. Centrifuge maternal blood at $3000 \times g$ for 10 min at 4 °C, and recover plasma into an Eppendorf tube (*see* **Note 22**).

28. Then determine counts in maternal plasma and add 198 μL Biosol, 2 μL plasma, and 4 mL scintillation fluid to a scintillation vial.

29. Cap the vial and then shake.

30. Prepare triplicates of each plasma sample, and also prepare a background sample which contains 200 μL Biosol and 4 mL scintillation fluid.

31. Allow all samples to sit in the dark to allow chemiluminescence to dissipate, and then determine ^{3}H and ^{14}C content using a liquid scintillation counter (*see* **Notes 23** and **24**).

32. To determine counts in homogenized fetuses, add 250 μL or 500 μL of fetal homogenate from days 15–16 to days 18–19 of pregnancy, respectively, to a scintillation vial.

33. Add 4 mL scintillation fluid, cap the vial, and then shake.

34. Prepare duplicates of each fetal homogenate, and prepare a background vial which replaces the volume of fetal homogenate with Biosol.

35. Allow samples to sit in the dark to allow chemiluminescence to dissipate, and then determine ^{3}H and ^{14}C content using a liquid scintillation counter (*see* **Notes 23–25**).

3.6 Calculations and Data Analysis

1. Subtract the background value from the mean maternal plasma and fetal counts.

2. Using the maternal plasma counts, create a one-phase exponential decay curve for ^{3}H and ^{14}C (*see* **Note 26** and Fig. 2).

3. Values should be in disintegrations per min (DPM) per μL maternal plasma.

4. Calculate the weight-specific clearance of each tracer (K_{mf}) by the placenta using the equation:
$$K_{mf} = N_X / W \int^X Cm\,(t)\mathrm{d}t.$$
where N_X is the radioactivity (DPM) in the fetus at the time of death (X),
W is the weight of the corresponding placenta in grams, and
$\int_O^X Cm\,(t)\mathrm{d}t$ is the integral (area under curve) of the first-order exponential decay curve of maternal plasma radioactivity (Cm, DPM per μL) with time, up to the time of death [12] (values should be presented as placental clearance of ^{3}H-MeG and ^{14}C-MeAIB expressed as μL/min/g placenta).

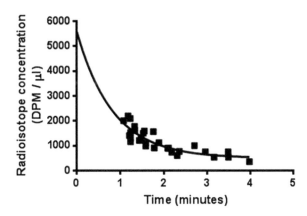

Fig. 2 Example of a one-phase exponential decay curve for a radioisotope. Each data point represents counts in a plasma sample from a single mouse dam. In larger species that have a greater blood volume, like the rat and guinea pig, repeated sampling of the dam is possible, and a radioactive isotope clearance curve can be generated for each animal

5. Use radioactive counts in each fetus to calculate the amount of radioisotope transferred per gram of fetus (also known as fetal accumulation) using the following formula:

$$[A \times (V_b + \text{fetal weight in grams})/V_f)]/\text{weight of fetus in g}$$

where

A = DPM in fetal sample

V_b = volume of Biosol used to homogenize sample (e.g., 2 mL for samples on day 16 of pregnancy)

V_f = volume of fetal homogenate counted (e.g., 0.25 mL for samples on day 16 of pregnancy)

(Values should be presented as fetal accumulation of ^{3}H-MeG and ^{14}C-MeAIB expressed as DPM/g fetus.) (*See* **Notes 27** and **28**.)

4   Notes

1. File at right angles to the needle and then around the edge of the break. Careful filing is essential to make sure there are no rough edges or splinters that can damage the polythene tubing (and make leaks in the catheter).

2. If necessary, use the seeker needle to widen the opening of the polythene tubing. Make sure that a seal is made between the needle and tubing and that the needle reaches down to the stretched part of the catheter.

3. Ensure all the regulatory procedures are in place, including appropriate training of individuals, to undertake ^{3}H and ^{14}C radioactive work in your workplace. Always abide by local, environmental, and institutional policies for working and disposing of radioactive substances.

4. The ^{3}H-MeG is often supplied in ethanol. If possible, purchase ^{3}H-MeG in a concentrated form, so that once it is diluted, there is only very little residual ethanol. The half-life of ^{3}H is 12.3 years. A decay correction may need to be applied when preparing new batches of ^{3}H-MeG.

5. This will eliminate introducing variation from preparing stocks on different days.

6. This anesthetic combination is most effective when prepared fresh on the day. Excess can be stored at 4 °C for use within the week.

7. Accurate timing of pregnancy is required as the nutrient transfer capacity of the placenta changes with gestational age.

8. Ordered females should be older than 7 weeks and males older than 10 weeks to ensure they are sufficiently reproductively mature.

9. To increase the chance of mating, females should be placed in the stud cage when they are in estrous.

10. Dams that are pregnant should start gaining weight from day 10 of pregnancy.

11. Make sure ethical approval for the proposed regulated procedures on mice (administration of substances) has been attained from the government and local committees.

12. This procedure has been performed on mice that are at day 15 or later in pregnancy.

13. Typically, 400 μL is given to mice at day 16 of pregnancy and 600 μL to those on day 19 of pregnancy. The Hypnorm-Hypnovel combination rapidly induces anesthesia with minimal cardiovascular depression and thus minimal changes in uterine blood flow.

14. The jugular will be surrounded by a fat pad (the size of the fat pad depends on maternal age and environmental manipulation).

15. If the first attempt does not work, apply pressure using cotton tip bud to the jugular vein to stop any bleeding. Then try the jugular vein on the other side of the animal.

16. You can practice performing the procedure by injecting physiological saline or a colored nontoxic substance.

17. Between 1 and 4 min, there is a minimal back flux of radioisotope from fetus to mother. Make sure that the average time for each experimental group is 2 min.

18. Weigh catheter before and after tracer injection to ensure that 200 μL has been successfully administered to the dam. Exclude animal from placental transfer analysis if volume injected varies by 10% or more than intended.

19. Yolk sacs or placentas can be taken for DNA extraction and sexing the fetuses, if required.

20. Placentas can also be minced and then digested in Biosol (2 mL) to determine radiolabel accumulation. Just note that due to large amounts of blood in the placenta, samples may have high levels of chemiluminescence which interferes with the discrimination of the ^{3}H and ^{14}C channels. Appropriate run programs should be developed in consultation with technical support for the scintillation counter.

21. To speed up the solubilization process, samples can be vortexed each day or placed in an incubator with a shaking element.

22. Plasma can be stored at $-20\,^{\circ}C$ or immediately analyzed for radioactivity content.

23. The length of time required to allow chemiluminescence to dissipate in the samples needs to be determined in consultation with technical support for the liquid scintillation counter.

24. During measurement of both maternal plasma and fetal radioactivity, simultaneously run quench-adjusted standards with samples to discern between the ^3H and ^{14}C channels. This should be done in consultation with technical support for the liquid scintillation counter.

25. If placental samples were homogenized in Biosol, add 250 µL of the homogenate to 4 mL scintillation fluid. Prepare duplicates of each sample, and determine ^3H and ^{14}C content using a liquid scintillation counter.

26. Prepare a radioisotope clearance curve for each tracer at each gestational age studied.

27. If the surface area of the placenta functioning in maternofetal exchange has been determined stereologically using the method described by Coan et al. [41], then transfer of each radiolabel per unit of surface area can be estimated yielding the estimated $µL/min/cm^2$ of placental exchange surface area [19].

28. If ^3H and ^{14}C were counted in placental samples, placental accumulation of ^3H-MeG and ^{14}C-MeAIB can also be determined using the formula:

$$[A \times (V_b + \text{placental weight in grams})/V_p)]/\text{weight of the placenta in grams}$$

where

$A = $ DPM in placental sample

$V_b = $ volume of Biosol used to homogenize sample (i.e., 2 mL)

$V_f = $ volume of placental homogenate counted (i.e., 0.25 mL)

(Values should be presented as placental accumulation of ^3H-MeG and ^{14}C-MeAIB expressed as DPM/g placenta.)

References

1. Sibley CP (2009) Understanding placental nutrient transfer - why bother? New biomarkers of fetal growth. J Physiol 587(14):3431–3440

2. Lager S, Powell TL (2012) Regulation of nutrient transport across the placenta. J Pregnancy 2012:179827. https://doi.org/10.1155/2012/179827

3. Dimasuay KG, Boeuf P, Powell TL, Jansson T (2016) Placental responses to changes in the maternal environment determine fetal growth. Front Physiol 7:12. https://doi.org/10.3389/fphys.2016.00012

4. Sferruzzi-Perri AN, Camm EJ (2016) The programming power of the placenta. Front Physiol 7:33. https://doi.org/10.3389/fphys.2016.00033

5. Sferruzzi-Perri AN, Sandovici I, Constancia M, Fowden AL (2017) Placental phenotype and the insulin-like growth factors: resource allocation to fetal growth. J Physiol 595:5057–5093

6. Lewis RM, Cleal JK, Hanson MA (2012) Review: placenta, evolution and lifelong health. Placenta 33(Suppl):S28–S32

7. Fowden AL, Sferruzzi-Perri AN, Coan PM, Constancia M, Burton GJ (2009) Placental efficiency and adaptation: endocrine regulation. J Physiol 587(14):3459–3472

8. Sibley CP, Brownbill P, Dilworth M, Glazier JD (2010) Review: adaptation in placental nutrient supply to meet fetal growth demand: implications for programming. Placenta 31 (Suppl):S70–S77

9. Sferruzzi-Perri AN, Lopez-Tello J, Fowden AL, Constancia M (2016) Maternal and fetal genomes interplay through phosphoinositol 3-kinase(PI3K)-p110a signalling to modify placental resource allocation. Proc Natl Acad Sci U S A 113(40):11255–11260

10. Zhang S, Regnault TR, Barker PL, Botting KJ, McMillen IC, McMillan CM et al (2015) Placental adaptations in growth restriction. Forum Nutr 7(1):360–389

11. Diaz P, Powell TL, Jansson T (2014) The role of placental nutrient sensing in maternal-fetal resource allocation. Biol Reprod 91(4):82. https://doi.org/10.1095/biolreprod.114.121798

12. Atkinson DE, Robinson NR, Sibley CP (1991) Development of passive permeability characteristics of rat placenta during the last third of gestation. Am J Phys 261(6 Pt 2): R1461–R1464

13. Sferruzzi-Perri AN, Owens JA, Standen P, Taylor RL, Heinemann GK, Robinson JS et al (2007) Early treatment of the pregnant guinea pig with IGFs promotes placental transport and nutrient partitioning near term. Am J Physiol Endocrinol Metab 292(3):E668–E676

14. Jansson N, Pettersson J, Haafiz A, Ericsson A, Palmberg I, Tranberg M et al (2006) Down-regulation of placental transport of amino acids precede the development of intrauterine growth restriction in rats fed a low protein diet. J Physiol 576(3):935–946

15. Sferruzzi-Perri AN, Owens JA, Standen P, Taylor RL, Robinson JS, Roberts CT (2007) Early pregnancy maternal endocrine IGF-I programs the placenta for increased functional capacity throughout gestation. Endocrinology 148 (9):4362–4370

16. Jones HN, Woollett LA, Barbour N, Prasad PD, Powell TL, Jansson T (2008) High-fat diet before and during pregnancy causes marked up-regulation of placental nutrient transport and fetal overgrowth in C57/BL6 mice. FASEB J 23(1):271–278

17. Constancia M, Angiolini E, Sandovici I, Smith P, Smith R, Kelsey G et al (2005) Adaptation of nutrient supply to fetal demand in the mouse involves interaction between the Igf2 gene and placental transporter systems. Proc Natl Acad Sci U S A 102(52):19219–19224

18. Sferruzzi-Perri AN, Vaughan OR, Coan PM, Suciu MC, Darbyshire R, Constancia M et al (2011) Placental-specific Igf2 deficiency alters developmental adaptations to undernutrition in mice. Endocrinology 152(8):3202–3212

19. Sferruzzi-Perri AN, Vaughan OR, Haro M, Cooper WN, Musial B, Charalambous M et al (2013) An obesogenic diet during mouse pregnancy modifies maternal nutrient partitioning and the fetal growth trajectory. FASEB J 27 (10):3928–3937

20. Higgins JS, Vaughan OR, de Liger EF, Fowden AL, Sferruzzi-Perri AN (2015) Placental phenotype and resource allocation to fetal growth are modified by the timing and degree of hypoxia during mouse pregnancy. J Physiol 594 (5):1341–1356

21. Coan PM, Angiolini E, Sandovici I, Burton GJ, Constância M, Fowden AL (2008) Adaptations in placental nutrient transfer capacity to meet fetal growth demands depend on placental size in mice. J Physiol 586(18):4567–4576

22. Coan PM, Vaughan OR, McCarthy J, Mactier C, Burton GJ, Constância M et al (2011) Dietary composition programmes placental phenotype in mice. J Physiol 589 (14):3659–3670

23. Coan PM, Vaughan OR, Sekita Y, Finn SL, Burton GJ, Constancia M et al (2010) Adaptations in placental phenotype support fetal growth during undernutrition of pregnant mice. J Physiol 588(3):527–538

24. Ganguly A, Collis L, Devaskar SU (2012) Placental glucose and amino acid transport in calorie-restricted wild-type and Glut3 null heterozygous mice. Endocrinology 153 (8):3995–4007

25. Varma DR, Ramakrishnan R (1991) Effects of protein-calorie malnutrition on transplacental kinetics of aminoisobutyric-acid in rats. Placenta 12(3):277–284

26. Wyrwoll CS, Seckl JR, Holmes MC (2009) Altered placental function of 11{beta}-hydroxysteroid dehydrogenase 2 knockout mice. Endocrinology 150(3):1287–1293

27. Vaughan OR, Fisher HM, Dionelis KN, Jefferies EC, Higgins JS, Musial B et al (2015) Corticosterone alters materno-fetal glucose partitioning and insulin signalling in pregnant mice. J Physiol 593(5):1307–1321

28. Vaughan OR, Sferruzzi-Perri AN, Coan PM, Fowden AL (2013) Adaptations in placental phenotype depend on route and timing of maternal dexamethasone administration in mice. Biol Reprod 89(4):1–12

29. Vaughan OR, Sferruzzi-Perri AN, Fowden AL (2012) Maternal corticosterone regulates nutrient allocation to fetal growth in mice. J Physiol 590(21):5529–5540

30. Jansson T, Persson E (1990) Placental transfer of glucose and amino acids in intrauterine growth retardation: studies with substrate analogs in the awake guinea pig. Pediatr Res 28 (3):203–208

31. Audette MC, Challis JR, Jones RL, Sibley CP, Matthews SG (2011) Antenatal dexamethasone treatment in midgestation reduces system A-mediated transport in the late-gestation murine placenta. Endocrinology 152 (9):3561–3570

32. Vaughan OR, Phillips HM, Everden AJ, Sferruzzi-Perri AN, Fowden AL (2015) Dexamethasone treatment of pregnant F0 mice leads to parent of origin-specific changes in placental function of the F2 generation. Reprod Fertil Dev 27(4):704–711

33. Dilworth MR, Kusinski LC, Cowley E, Ward BS, Husain SM, Constância M et al (2010) Placental-specific Igf2 knockout mice exhibit hypocalcemia and adaptive changes in placental calcium transport. Proc Natl Acad Sci U S A 107(8):3894–3899

34. Flexner LB, Pohl HA (1941) Transfer of radioactive sodium across the placenta of the guinea pig. J Physiology 132:594–606

35. Stulc J, Stulcova B (1986) Transport of calcium by the placenta of the rat. J Physiol 371:1–16

36. Bond H, Dilworth MR, Baker B, Cowley E, Requena Jimenez A, Boyd RD et al (2008) Increased maternofetal calcium flux in parathyroid hormone-related protein-null mice. J Physiol 586(7):2015–2025

37. Sibley CP, Coan PM, Ferguson-Smith AC, Dean W, Hughes J, Smith P et al (2004) Placental-specific insulin-like growth factor 2 (Igf2) regulates the diffusional exchange characteristics of the mouse placenta. Proc Natl Acad Sci U S A 101(21):8204–8208

38. Adams AK, Reid DL, Thornburg KL, Faber JJ (1988) In vivo placental permeability to hydrophilic solutes as a function of fetal weight in the guinea pig. Placenta 9(4):409–416

39. Stulc J (1997) Placental transfer of inorganic ions and water. Physiol Rev 77(3):805–836

40. Cramer S, Beveridge M, Kilberg M, Novak D (2002) Physiological importance of system A-mediated amino acid transport to rat fetal development. Am J Physiol Cell Physiol 282 (1):C153–C160

41. Coan PM, Ferguson-Smith AC, Burton GJ (2004) Developmental dynamics of the definitive mouse placenta assessed by stereology. Biol Reprod 70(6):1806–1813

Chapter 15

Assessment of Fatty Liver in Models of Disease Programming

Kimberley D. Bruce and Karen R. Jonscher

Abstract

Nonalcoholic fatty liver disease (NAFLD) is currently the most common cause of chronic liver disease worldwide and is present in a third of the general population and the majority of individuals with obesity and type 2 diabetes. Importantly, NAFLD can progress to severe nonalcoholic steatohepatitis (NASH), associated with liver failure and hepatocellular carcinoma. Recent research efforts have extensively focused on identifying factors contributing to the additional "hit" required to promote NALFD disease progression. The maternal diet, and in particular a high-fat diet (HFD), may be one such hit "priming" the development of severe fatty liver disease, a notion supported by the increasing incidence of NAFLD among children and adolescents in Westernized countries. In recent years, a plethora of key studies have used murine models of maternal obesity to identify fundamental mechanisms such as lipogenesis, mitochondrial function, inflammation, and fibrosis that may underlie the developmental priming of NAFLD. In this chapter, we will address key considerations for constructing experimental models and both conventional and advanced methods of quantifying NAFLD disease status.

Key words Fatty liver, NAFLD, NASH, Maternal obesity, High-fat diet, Histology, Fibrosis

1 Introduction

Nonalcoholic fatty liver disease (NAFLD) describes a spectrum of disorders characterized by the accumulation of ectopic fat accumulation in the liver without significant alcohol use. At one end of the spectrum is simple steatosis (NAFLD); however, 25% of patients with NAFLD can develop more severe steatosis with inflammation, termed nonalcoholic steatohepatitis (NASH) [1]. NASH can progress further still to fibrosis (26–37%) and cirrhosis, portal hypertension liver failure, and hepatocellular carcinoma (HCC) [2, 3]. Worryingly, an increasing number of younger individuals are being diagnosed with NAFLD [4, 5]. Recent estimates suggest that in Western societies, the number of children with NAFLD ranges from 3 to 10% in the general population and up to 70% in children that are considered obese [6]. The number of adolescents

Paul C. Guest (ed.), *Investigations of Early Nutrition Effects on Long-Term Health: Methods and Applications*, Methods in Molecular Biology, vol. 1735, https://doi.org/10.1007/978-1-4939-7614-0_15, © Springer Science+Business Media, LLC 2018

diagnosed with NAFLD has more than doubled in the last two decades [7]. Like adults, pediatric NAFLD can follow a severe disease progression to cirrhosis and end-stage liver disease; however, onset of severe disease may be accelerated as up to half of obese children with NAFLD and may have already progressed to NASH at time of diagnosis [8].

It is hypothesized that a multiple-"hit" mechanism is involved in the pathogenesis and progression of NAFLD. A preliminary hit may consist of suboptimal diets that promote hepatic lipid accumulation. Recent work suggests that there are, in fact, a multitude of factors that may act as the second hit to promote liver disease progression, including epigenetics, circadian rhythms, disturbances in intestinal microbiota, and developmental dietary exposures. There is now a significant body of research to support the notion that NAFLD may have developmental origins. During critical periods of development, the liver adapts to its environment, causing changes in metabolic function in response to suboptimal nutrition. These changes are often maladaptive and persist into adult life, increasing the susceptibility of developing metabolic disease, including NAFLD, in later life. For example, in a recent cross-sectional study of 543 children with biopsy-proven NAFLD, children with either high or low birth weight had >twofold increased incidence of NASH compared with those of normal birth weight and were more likely to have advanced fibrosis, even after adjusting for obesity [9].

The murine model of maternal obesity and high-fat feeding has been an invaluable tool in dissecting the mechanisms underlying the developmental priming of metabolic disease, including severe fatty liver (NASH). For example, offspring of obese/high-fat-fed mothers/dams develop a more severe fatty liver phenotype compared to offspring of lean/control-fed mothers/dams, involving mitochondrial dysfunction, disruption of sirtuin activity, oxidative stress, circadian disruption, altered epigenetic regulation, fibrosis, inflammation, elevated lipogenesis, and altered lipid composition [10–16]. This established model also provides an unrivaled platform in which to perform proof-of-concept and preclinical studies that test the efficacy of pharmacological and nutraceutical interventions with the potential to halt the development of severe fatty liver disease in this and future generations [17–21].

Histological assessment in both human and animal models is commonly used to stage fatty liver disease and assess efficacy of therapeutics. However, sample preparation for standard histopathological examination requires numerous steps that alter native conditions and destroy lipid droplets. Emerging technologies such as multiphoton optical microscopy allow characterization of intact biological tissues with minimum sample destruction [22, 23]. Also, coherent anti-Stokes Raman scattering (CARS) microscopy can be utilized to visualize lipid droplets [24–29], and second harmonic generation (SHG) microscopy can be used to visualize collagen fibril formation [29–40]. These techniques allow for three-dimensional optical sectioning and penetration of thick specimens for deep imaging and have subcellular spatial resolution [41].

SHG imaging uses a nonresonant process that leverages a coherent optical signal generated by light interacting with material containing non-centrosymmetric structures, such as elastin and fibrillary collagens, including types I, II, and III [42, 43]. Third harmonic generation (THG) signal increases when a laser beam focus spans an interface between optically different materials, such as a lipid droplet in the tissue [44]. Therefore, both SHG and THG have utility for NAFLD and NASH quantitative analysis, although SHG has been more extensively utilized. Liu et al. recently used SHG to develop an automated evaluation system to assess fibrosis in rodents [34]. Their scores favorably compared with those obtained using Ishak fibrosis scores, as well as measures of hydroxy-proline content and collagen proportionate area, demonstrating the utility of the SHG-based system for accurately and quantitatively staging liver fibrosis in several rodent models [34]. Pirhonen et al. analyzed 32 liver biopsies obtained from patients undergoing bariatric surgery using SHG and CARS [29]. Mean signal intensities from both imaging modalities correlated well with fibrosis stage and extent of steatosis. Also, a specific SHG signal correlated with fibrillary collagens I and III, and this was detected outside portal areas in samples classified as Stage 0. Both of these findings suggested that SHG may be a particularly sensitive technique for detecting nascent fibrosis. Indeed, we used SHG and found a small but significant increase in SHG signal in liver sections from mice flown aboard the Space Shuttle Atlantis for only 13 days [45, 46]. SHG imaging can be performed using either frozen tissue sections or tissue fixed with 2–4% paraformaldehyde. The use of fixed tissue is preferred to retain fibril conformation and tissue morphology.

Here we describe a detailed protocol in which to generate a model of developmentally primed NASH. Additionally, we provide detailed strategies on how to determine hepatic steatosis using standard histological methodologies and, in addition, emerging technologies such as second-generation (SGH) microscopy, which simultaneously allows quantification of steatosis and fibrosis.

2 Materials (*See* Note 1)

2.1 Animal Model

1. C57BL/6J mice (*see* **Note 2**).

2. 45 kCal% fat diet (DIO HFD, D12451; Research Diets, Inc) or a Western-style diet (TD.88137; Envigo), consisting of 42%

kcal from fat, 15% protein, 43% carbohydrate (34% sucrose by weight), and 0.2% cholesterol (4.5 kcal/g) (*see* **Note 3**).

3. Control diet 10 kCal% fat diet (DIO LFD D12450B, Research Diets, Inc).

4. Cryotubes.

2.2 Metabolic Phenotyping

1. Calorimetry equipment (Oxylet; Panlab SLU).

2. Closed modular indirect calorimetry system (Oxylet; Panlab SLU).

3. Computer-assisted data acquisition program (Metabolism; Panlab SLU).

4. Glucometer (Accu-Chek; Roche Diagnostics Ltd.).

2.3 Histological Analysis (H and E)

1. 10% buffered formalin (for fixed, paraffin-embedded sections).

2.4 Cryosectioning (Oil Red O)

1. Tissue-Tek OCT compound (Sakura).

2. Cryomold, standard or intermediate size (Sakura).

3. Cryostat.

4. Liquid nitrogen and isopentane or aluminum block.

5. Fisherbrand Superfrost Plus Microscope Slides ($25 \times 75 \times 1$ mm).

6. Pencil for marking slides.

7. Hydrophobic pen for isolating sections.

8. Oil Red O (Sigma-Aldrich).

9. Mounting medium (water soluble, Biocare).

10. Isopropyl alcohol.

11. Distilled water.

12. Light microscope with imaging software (AxioVision, Zeiss).

13. ImageJ http://rsbweb.nih.gov/ij/.

2.5 Second Harmonic Generation (SHG) Imaging

1. 16% paraformaldehyde (Electron Microscopy Sciences) diluted to 2–4% with HPLC-grade water for fixing tissue (*see* **Note 4**).

2. Corning #1.5 cover glass (*see* **Note 5**).

3. Phosphate-buffered saline (PBS).

4. Quick-dry clear nail polish.

5. Zeiss 780 microscope equipped with a titanium/sapphire Chameleon Ultra II laser (Coherent) (*see* **Note 6**).

6. Dichroic mirror transmitting two-photon autofluorescent signal at 425 nm (hq575/250 m-2p, Chroma Technology).

7. Zen 2012 software (Zeiss).

3 Methods

3.1 Animal Model

1. Carry out all animal procedures in accordance with the *Guide for the Care and Use of Laboratory Animals*, published by the National Research Council and institutional guidelines.

2. Maintain mice in standard 12 h light (subjective day) and 12 h dark (subjective night conditions), with a constant temperature 22 ± 2 °C with food and water available ad libitum, unless the study requires specific modifications.

3. Female mice should be randomly assigned to either the HFD/WD or LFD and fed a HFD/WD 6 weeks prior to mating, during pregnancy, and through lactation (*see* **Note 7**).

4. Pregnant dams are allowed to deliver the pups, and the litter size should be standardized (to six pups) to prevent nutritional biasing of litters.

5. Randomly assign pups from either HFD/WD or LFD dams to the experimental groups required, including diet switches, e.g., LFD > LFD, LFD > HFD, HFD > HFD, HFD > LFD, etc.

6. If the study includes a pharmacological or nutraceutical intervention, the timing should be considered carefully along with liver development (Fig. 1) (*see* **Note 8**).

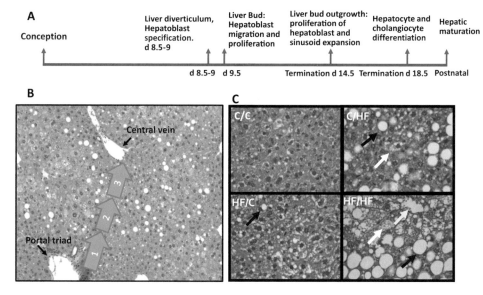

Fig. 1 Time line of liver development in the mouse and histological scoring of NASH. (**a**) Timeline of liver development in the mouse, adapted from [51], where "d" corresponds to embryonic day. (**b**) H and E stained liver section of a 6-month-old mouse showing zonal lipid accumulation associated with age. (**c**) Histology of livers from control-fed offspring from control-fed dams (C/C), high-fat-fed offspring from control dams (C/HF), control offspring from high-fat-fed dams (HF/C), and high-fat-fed offspring from high-fat-fed dams (HF/HF), showing macrovesicular steatosis (black arrows), microvesicular steatosis (white arrows), and ballooning degeneration (yellow arrows) (adapted from [16])

7. Allow pups to reach experimental age of phenotyping and terminal experiments [developmentally primed NAFLD will be clearly distinguishable by 15 weeks of age (12 weeks postweaning)].

8. Weigh mice at least weekly at the same time of the day each week.

9. At the end of the study, harvest liver tissue (this should be performed at the same time of the day since many genes/proteins associated with hepatic metabolism are under control of the endogenous molecular clock).

10. If female mice are included in the study, they should be analyzed at the same point within the estrous cycle to reduce variability (vaginal smears can be performed to determine the proportion of cells within the smear (*see* **Note 9**)) and cycle stage.

11. To measure changes in baseline metabolism, the mice should be fasted for at least 4 h prior to harvest (*see* **Note 10**).

12. Tissues can be either snap frozen in liquid nitrogen for molecular biology analysis, fixed in 10% buffer formalin for at least 24 h for paraffin embedding and histological analysis, or embedded in Tissue-Tek in an appropriately sized Cryomold and frozen for cryosectioning in liquid nitrogen-cooled isopentane (to prevent ice crystal formation) (*see* **Note 11**).

3.2 Metabolic Phenotyping (See Note 12)

1. Calorimetry should be determined in at least $n = 6$ per offspring group, in a closed modular indirect calorimetry system prior to terminal experiment.

2. Since the behavioral response to the calorimetry chamber can vary, each mouse should be allowed 1 day to habituate.

3. Metabolic measurements should commence from day 2.

4. Oxygen consumption (VO_2) and carbon dioxide production (VCO_2) should be recorded in 30 min–1 h bins using a computer-assisted data acquisition program over a 24 h period.

5. Calculate the animal's energy expenditure (EE, in kcal/day/body weight 0.75) from this data.

6. To measure change in glucose homeostasis, a 2 h intraperitoneal glucose tolerance test (Ip-GTT) is easy to perform and gives reproducible data (*see* **Note 13**).

7. Fast the mice for 4 h.

8. Cut the tail vein with a 23 G needle (a few microliters will be needed to determine fasting glucose concentration using a portable glucometer).

9. Inject sterile 20% D-glucose (2 g/kg mouse body weight) intraperitoneally, and measure blood glucose concentration using the glucometer at 15, 30, 60, and 120 min.

3.3 Histological Analysis

1. Stain paraffin-embedded sections with hematoxylin and eosin (H and E) to visualize tissue and nuclei.

2. Use standard light microscopy to visualize the hepatic architecture, and perform histological scoring based on steatosis and ballooning (*see* **Note 14**).

3. Mason trichrome staining can also be used to counter stain collagen, cytoplasm, and nuclei, allowing simultaneous semi-quantitative analysis of steatosis and fibrosis.

4. Picrosirius red can also be used to detect collagen fibrils.

5. Images of each sample should include a region around a portal triad (hepatic artery, portal vein, and bile duct) and a central vein so that zonal lipid accumulation can be assessed (*see* **Note 15**).

6. A modified version of the Kleiner scoring system can be used to generate a NAFLD activity score (NAS) by summing individual scores for necroinflammatory features [15, 16].

7. A minimum of ten images from each sample should be used for scoring each mouse.

8. Steatosis can be assessed by scoring the percentage of the liver parenchyma that contains steatotic regions: $< 5\% = 0$, $5–33\% = 1$, $33–66\% = 2$, and $> 66\% = 3$ (*see* **Note 16**).

9. Score lobular inflammation, characterized as infiltration of monocytes visible in the periportal regions, as no foci $= 0$, < 2 foci $= 1$, 2–4 foci $= 2$, and > 4 foci $= 3$ (a score of <3 correlates with mild NAFLD, a score of 3–4 correlates with moderate NAFLD, and a score of 5 or more correlates with NASH).

10. Six to ten mice from each offspring group should be imaged and scored, and the average score for each histological characteristic in each group should be summed to stage the group liver phenotype.

3.4 Cryosectioning (Oil Red O)

1. To determine the lipid content of the offspring liver, the cryopreserved livers can be sectioned using a cryostat for Oil Red O staining, which stains the neutral lipids (triglycerides) in the frozen liver sections.

2. Place the cryopreserved liver in the cryostat for 30 min to equilibrate.

3. Mount the liver on the chuck and cut the excess Tissue-Tek so that the sample takes a rhomboid shape with the smallest parallel side on the bottom edge (*see* **Note 17**).

4. Cut sections around 15 μm thick and store at $-80\,°C$ (*see* **Note 18**).

5. When ready to stain sections, take them out of the freezer and allow equilibration for 10 min.

6. Circle sections with a hydrophobic pen, add 200 μL of working Oil Red O solution to each section, and incubate at room temperature for 5 min.

7. Rinse slides for 30 min in running tap water.

8. Mount slides with mounting media, add coverslip, seal using clear nail polish around edges, and allow drying.

9. Image slides within 24 h using standard light microscopy and image analysis software.

10. Quantify lipids using ImageJ (*see* **Note 19**).

11. First, convert the RGB image to an 8 bit grayscale image and then use the Image → Adjust → Threshold tool to set the threshold according to your requirements.

12. Use the Analyze → Set measurement tool and activate the boxes for Limit to Threshold, Area and Integrated density.

13. Obtain results by running the Analyze → Measure tool (*see* **Note 20**).

14. Analyze data using Microsoft Excel or GraphPad Prism, considering that calculations of significance should be performed according to the study design.

3.5 Sample Preparation for Fresh/Fixed/Frozen Tissue Cryosections

1. Add OCT to a Cryomold, ensuring there are no bubbles.

2. Place the fresh or fixed liver tissue into the OCT and press gently with rounded forceps to flatten the tissue on the bottom of the Cryomold; be careful not to puncture the tissue (*see* **Note 21**).

3. Add OCT to cover the tissue and fill the mold, ensuring that no bubbles appear on top of the tissue.

4. Place the tissue on a level piece of dry ice or an aluminum block in a Styrofoam box that has been precooled with liquid N_2.

5. Allow to slowly freeze and store at −80 °C until use.

6. Using a cryostat set at −20 °C, cut ~6–12 μm-thick tissue sections from each OCT block and allow to adhere to microscope slide.

7. Ensure the tissue does not wrinkle (*see* **Notes 22** and **23**).

8. Allow slide to sit at room temperature for 20 min and then store at −20 °C (slides should be analyzed within 30 days).

9. Immediately prior to analysis, 1 mL ice-cold PBS is gently dripped (using a pipette) onto the OCT-covered areas and removed by gentle blotting or shaking.

10. This process is repeated twice; then excess PBS is removed.

11. A drop of PBS is placed on the tissue, and a coverslip is carefully applied and sealed with fast-drying nail polish.

12. Slides should be used immediately and disposed of in a sharps container after use.

3.6 Sample Preparation for Fixed, Paraffin-Embedded Tissue (See Note 24)

1. Fix the tissue at room temperature in 2–4% PFA for 4–12 h.
2. Remove the tissue to ice-cold PBS and store at 4 °C prior to paraffin embedding and sectioning (*see* **Notes 25** and **26**).
3. Prior to analysis, a drop of PBS is placed on the tissue and a coverslip carefully affixed and sealed with nail polish (slides may be reanalyzed for 2–3 months).

3.7 Data Acquisition

1. Use a 20× objective (0.8 NA), select the Chameleon laser, and set the laser power to 7% (*see* **Notes 27** and **28**).
2. Set the laser wavelength to 800 nm and the frequency to 80 mHz.
3. Set the two-photon autofluorescent (TPAF) signal to 450–700 nm and select non-descanned detectors.
4. Insert the microscope slide into the holder, with the coverslip facing the objective (*see* **Note 29**).
5. With the Oculars online set to "brightfield," adjust the height of the holder until the tissue is in focus (*see* **Note 30**).
6. Turn off the lights, cover the microscope and stage with a black cloth, and set the Oculars to "offline."
7. Set the resolution to 512 × 512 to quickly scan the sample and, when ready to acquire an image, change the resolution to a final setting of 1024 × 1024.
8. Acquire images from 15 random sites per sample using Zen software (*see* **Note 31**).

3.8 Image Analysis

1. Save images as .lsm files (Fig. 2).
2. In ImageJ or Fiji, select Image and then Color and Split Channels to separate the SHG channel from the autofluorescent channel.
3. In Set Measurements, select Area and Integrated Density.
4. Select the image with the SHG channel and click "Measure."
5. Copy results into the Excel or other programs for statistical analysis.

4 Notes

1. Many of the materials listed are dependent on the equipment in-house and the institutional core facilities available.
2. C57BL/6J mice have not only been repeatedly used to generate model of developmentally primed metabolic disease and NASH but are also the background strain for most transgenic and mutant mice and therefore should be carefully considered if future genetic studies are likely.

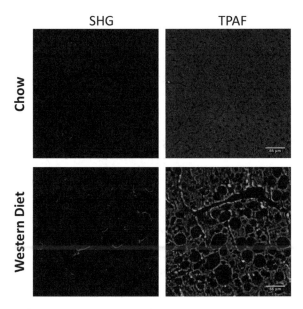

Fig. 2 SHG imaging may be used to assess fibrosis in the liver. Liver cryosections were obtained from offspring of mice fed either vivarium chow or a Western-style diet (WD; TD.88137, Envigo, Indianapolis, IN). Offspring were maintained on the maternal diet for 20 weeks and analyzed using SHG at 20× with a Zeiss LSM 780 laser scanning confocal/multiphoton excitation fluorescence microscope. Representative images are shown. SHG signal, displayed in red, shows the presence of abundant fibrillary collagen in offspring-fed WD that is absent in mice-fed vivarium chow. The autofluorescent channel (green signal) reveals large, non-fluorescing vacuoles in the WD-fed liver that are evidence of steatosis

3. These commercially available diets are very successful in generating NAFLD, but their usage should be carefully considered. Protein, sucrose, and the type of fat are matched. However, the fat source is soybean and lard (which is greater in the kCal% fat). Therefore, the HFD is only useful for studies investigating the specific effects of saturated fats. In addition, for studies where accurate measurements of food intake are important, custom formulation may be required to harden the diet and prevent excessive crumbling of the HFD, allowing accurate weight measurements and diet consumption. It is not advisable to use the very high-fat diet (65 kCal%) in transgenerational studies, since this profoundly affects reproductive capacity. The Western-style diet, comprised of milk fat, sucrose, and cholesterol, may accelerate risk for steatohepatitis.

4. Thick tissue slices may be placed in a 35 mm tissue culture dish with a high tolerance 1.5 coverslip and 14 mm glass diameter in ice-cold PBS. Another coverslip and a weight can be placed on top of the tissue to keep it from floating. If using a dish, be sure

to place it tightly in the holder to prevent spills. Either frozen cryosections or fixed sections may be used.

5. Coverslip thickness will depend on the specific microscope and objective used.

6. SHG signal is detected using a non-descanned detector (NDD).

7. This study design facilitates stratification of developmental (pre- and perinatal) vs. nondevelopmental (postnatal) diets. Further stratification may be required in order to delineate the specific effect of critical developmental windows (i.e., lactation). Since high-fat feeding causes a reproductive challenge for the dams, we recommend the use of proven breeders for these studies.

8. The timing of both the diet and the intervention should be carefully considered. If only the influence of maternal effects is of interest, then the diet/intervention should be provided during pregnancy and lactation, and offspring should be switched to control diet postweaning. If the intervention is for pre- or postconception, the diet timing should correlate.

9. Proestrus is characterized by predominantly nucleated epithelial cells in the vaginal smear and estrus, by anucleated, cornified cells. In metestrus, leukocyte, cornified, and nucleated epithelial cells can be seen. In diestrus, predominantly leucocytes are observed. We recommend performing all end point experiments/tissue harvests at metestrus when E2 levels are low and less likely to influence endocrine functions [47].

10. In order to determine changes in VLDL production by the liver, an overnight fast may be required.

11. The left and right lobes of the liver may harbor subtle molecular and biochemical differences. In addition, certain regions may be differentially vascularized, which influence lipid accumulation and hepatic microarchitecture. Therefore, the region harvested for each experiment should be recorded and be consistent. In addition, if SHG is going to be performed, tissues can be fixed in 4% PFA for 4–12 h and then stored in PBS and at 4 °C until embedded, rather than fixed in formalin.

12. To determine changes in phenotype that are associated with and may contribute to severe metabolic disease and fatty liver, we recommend performing quantitative measurements of energy expenditure and activity.

13. For the gold standard characterization of hepatic glucose homeostasis, including hepatic glucose output and insulin sensitivity, a hyperinsulinemic-euglycemic clamp study should be performed.

14. Paraffin embedding, cutting, and staining are performed by the local/institutional histology core.

15. Often, early-onset NAFLD, which occurs in pediatric liver disease, may be first observed in zone 1 hepatocytes that have a high mitochondrial density and are associated with the highest β-oxidation of fatty acids [48]. Importantly, in human histopathological studies, many of the pediatric cases appeared to have more zone 1 steatoses, more "periportal-only" fibroses, less ballooning, and rare Mallory's hyaline [49]. Zone 1 is the closest to the portal triad, including the hepatic artery, portal vein, and bile duct, and contains specialized hepatocytes associated with gluconeogenesis, β-oxidation of fatty acids, and cholesterol synthesis that contain elevated mitochondrial numbers. In contrast, zones 2 and 3 are found further from the portal triad, toward the central vein.

16. Ballooned hepatocytes are larger than average hepatocytes due to lipid accumulation and may appear to be undergoing necrosis/degeneration (scores are none = 0, few = 1, and prominent = 2).

17. This shape facilitates cutting. A reasonable amount of OCT around the sample allows for some curling under the no-roll plate without curling the sample.

18. Although thicker sections are easier to cut, layer of lipid staining causes low resolution and high variation in analysis.

19. The remaining steps in this section are adapted from [50].

20. The area value is the size of the selection within the set threshold. Integrated density gives two results: IntDen, which is the product of area and mean gray value, and RawIntDen, which represents the sum of all values of all pixels in the selection.

21. If using archived tissue, do not let tissue thaw during processing. Cut the tissue on a piece of glass placed on top of dry ice. If using freshly harvested tissue, ideally paddle forceps can be chilled in liquid N_2 and then used to freeze the liver prior to removal to minimize stress-related changes in metabolite abundances. Fixed tissue can be embedded in OCT and cryosectioned. If fixed tissue is used, fix in 2–4% paraformaldehyde for 2 h to overnight and then wash with a gradient of sucrose in PBS (5%, 10%) for 2 h and 25% overnight prior to embedding in OCT.

22. Fatty livers tend to be more difficult to section and tear more easily. It may be easier to cut thinner sections with the fattier livers. The caveat is that the thinner sections may heat more readily and the lipid droplets broaden when sliced; therefore, thicker sections provide better image quality.

23. Several sections can be placed on the same slide to either compare between groups or have multiple sections for improved statistics. Check with the microscope configuration to determine if there is room to allow for multiple sections and a 75 mm-long slide.

24. SHG can be performed on frozen, fixed, or fresh tissue sections. It is preferable to use 2–4% paraformaldehyde over formalin as the cross-linking induced by the formalin can interfere with the SHG signal. The fresh, unfixed tissue may be sectioned using a Vibratome for deeper imaging, if desired.

25. Time in fix should be consistent for all samples being compared, and only samples fixed under similar conditions may be compared.

26. Standard tissue sectioning, such as performed by a histopathology core facility, can be used, with 5–6 μm-thick sections of fixed tissue.

27. Laser power will need to be optimized for each system. If background is high, laser power can be increased while gain is decreased. Alternatively, try increasing the offset or increase signal averaging.

28. The $20\times$ objective allows for a larger section of the tissue to be visualized, albeit at lower resolution. A higher-resolution objective $(63\times)$ can be used in concert with a tile scan; however, the laser will scan the tissue more slowly, increasing the risk of damage by photobleaching or tissue heating.

29. Leave the slide slightly loose in the holder to protect the objective.

30. Once the sample is in focus, it is helpful to set a relative zero on the display in case the focus is accidentally changed.

31. When acquiring images for SHG, do not image near the edge of the tissue as collagen signal is more abundant near tissue edges and tears. Collagen will be more abundant around the central veins and portal triads; image acquisition in the interstitial zone, where bridging fibrils can be visualized, may be more informative for staging fibrosis than imaging at high magnification near veins.

Acknowledgments

The authors appreciate the help of Dr. Evgenia Dobrinskikh with the preparation of the manuscript and acquisition of SHG data. K.R.J. is grateful for funding from the NIH K25DK098615. The Zeiss LSM780 was funded by the NIH 1S10OD016257, and the Advanced Light Microscopy Core Facility is supported in part by the NIH/NCATS Colorado CTSI Grant Number UL1 TR001082, University of Colorado Anschutz Medical Campus, Aurora, CO, USA.

References

1. Marchesini G, Bugianesi E, Forlani G, Cerrelli F, Lenzi M, Manini R et al (2003) Nonalcoholic fatty liver, steatohepatitis, and the metabolic syndrome. Hepatology 37:917–923

2. Nelson R, Persky V, Davis F, Becker E (1997) Re: excess risk of primary liver cancer in patients with diabetes mellitus. J Natl Cancer Inst 89:327–328

3. Adami HO, Chow WH, Nyrén O, Berne C, Linet MS, Ekbom A et al (1996) Excess risk of primary liver cancer in patients with diabetes mellitus. J Natl Cancer Inst 88:1472–1477

4. Mencin AA, Lavine JE (2011) Advances in pediatric nonalcoholic fatty liver disease. Pediatr Clin N Am 58:1375–1392

5. Loomba R, Sirlin CB, Schwimmer JB, Lavine JE (2009) Advances in pediatric nonalcoholic fatty liver disease. Hepatology 50:1282–1293

6. Alisi A, Manco M, Vania A, Nobili V (2009) Pediatric nonalcoholic fatty liver disease in 2009. J Pediatr 155:469–474

7. Doycheva I, Watt KD, Rifai G, Abou Mrad R, Lopez R, Zein NN et al (2017) Increasing burden of chronic liver disease among adolescents and young adults in the USA: a silent epidemic. Dig Dis Sci 62:1373–1380

8. Goyal NP, Schwimmer JB (2016) The progression and natural history of pediatric nonalcoholic fatty liver disease. Clin Liver Dis 20:325–338

9. Newton KP, Feldman HS, Chambers CD, Wilson L, Behling C, Clark JM et al (2017) Low and high birth weights are risk factors for nonalcoholic fatty liver disease in children. J Pediatr 187:141–146.e1. https://doi.org/10.1016/j.jpeds.2017.03.007. pii: S0022–3476(17)30356-6. [Epub ahead of print]

10. Zhang J, Zhang F, Didelot X, Bruce KD, Cagampang FR, Vatish M et al (2009) Maternal high fat diet during pregnancy and lactation alters hepatic expression of insulin like growth factor-2 and key microRNAs in the adult offspring. BMC Genomics 10:478. https://doi.org/10.1186/1471-2164-10-478

11. Wankhade UD, Zhong Y, Kang P, Alfaro M, Chintapalli SV, Thakali KM et al (2017) Enhanced offspring predisposition to steatohepatitis with maternal high-fat diet is associated with epigenetic and microbiome alterations. PLoS One e0175675:12. https://doi.org/10.1371/journal.pone.0175675

12. Lemmens KJ, van de Wier B, Koek GH, Köhler E, Drittij MJ, van der Vijgh WJ et al (2015) The flavonoid monoHER promotes the adaption to oxidative stress during the onset of NAFLD. Biochem Biophys Res Commun 456:179–182

13. Kendrick AA, Choudhury M, Rahman SM, McCurdy CE, Friederich M, Van Hove JL et al (2011) Fatty liver is associated with reduced SIRT3 activity and mitochondrial protein hyperacetylation. Biochem J 433:505–514

14. Fujii S, Nishiura T, Nishikawa A, Miura R, Taniguchi N (1990) Structural heterogeneity of sugar chains in immunoglobulin G. Conformation of immunoglobulin G molecule and substrate specificities of glycosyltransferases. J Biol Chem 265:6009–6018

15. Bruce KD, Szczepankiewicz D, Sihota KK, Ravindraanandan M, Thomas H, Lillycrop KA et al (2016) Altered cellular redox status, sirtuin abundance and clock gene expression in a mouse model of developmentally primed NASH. Biochim Biophys Acta 1861:584–593

16. Bruce KD, Cagampang FR, Argenton M, Zhang J, Ethirajan PL, Burdge GC et al (2009) Maternal high-fat feeding primes steatohepatitis in adult mice offspring, involving mitochondrial dysfunction and altered lipogenesis gene expression. Hepatology 50:1796–1808

17. Tarry-Adkins JL, Fernandez-Twinn DS, Hargreaves IP, Neergheen V, Aiken CE, Martin-Gronert MS et al (2016) Coenzyme Q10 prevents hepatic fibrosis, inflammation, and oxidative stress in a male rat model of poor maternal nutrition and accelerated postnatal growth. Am J Clin Nutr 103:579–588

18. Ramaiyan B, Bettadahalli S, Talahalli RR (2016) Dietary omega-3 but not omega-6 fatty acids down-regulate maternal dyslipidemia induced oxidative stress: a three generation study in rats. Biochem Biophys Res Commun 477:887–894

19. Jonscher KR, Stewart MS, Alfonso-Garcia A, DeFelice BC, Wang XX, Luo Y et al (2017) Early PQQ supplementation has persistent long-term protective effects on developmental

programming of hepatic lipotoxicity and inflammation in obese mice. FASEB J 31:1434–1448

20. Chicco A, Creus A, Illesca P, Hein GJ, Rodriguez S, Fortino A (2016) Effects of post-suckling n-3 polyunsaturated fatty acids: prevention of dyslipidemia and liver steatosis induced in rats by a sucrose-rich diet during pre- and post-natal life. Food Funct 7:445–454

21. AlSharari SD, Al-Rejaie SS, Abuohashish HM, Ahmed MM, Hafez MM (2016) Rutin attenuates hepatotoxicity in high-cholesterol-diet-fed rats. Oxidative Med Cell Longev 2016:5436745. https://doi.org/10.1155/2016/5436745

22. Zipfel WR, Williams RM, Webb WW (2003) Nonlinear magic: multiphoton microscopy in the biosciences. Nat Biotechnol 21:1369–1377

23. Helmchen F, Denk W (2005) Deep tissue two-photon microscopy. Nat Methods 2:932–940

24. YM W, Chen HC, Chang WT, Jhan JW, Lin HL, Liau I (2009) Quantitative assessment of hepatic fat of intact liver tissues with coherent anti-stokes Raman scattering microscopy. Anal Chem 81:1496–1504

25. Brackmann C, Gabrielsson B, Svedberg F, Holmaang A, Sandberg AS, Enejder A (2010) Nonlinear microscopy of lipid storage and fibrosis in muscle and liver tissues of mice fed high-fat diets. J Biomed Opt 15:066008. https://doi.org/10.1117/1.3505024

26. Lin J, Lu F, Zheng W, Xu S, Tai D, Yu H et al (2011) Assessment of liver steatosis and fibrosis in rats using integrated coherent anti-stokes Raman scattering and multiphoton imaging technique. J Biomed Opt 16:116024. https://doi.org/10.1117/1.3655353

27. Le TT, Ziemba A, Urasaki Y, Brotman S, Pizzorno G (2012) Label-free evaluation of hepatic microvesicular steatosis with multimodal coherent anti-stokes Raman scattering microscopy. PLoS One 7:e51092. https://doi.org/10.1371/journal.pone.0051092

28. Urasaki Y, Fiscus RR, Le TT (2016) Molecular classification of fatty liver by high-throughput profiling of protein post-translational modifications. J Pathol 238:641–650

29. Pirhonen J, Arola J, Sädevirta S, Luukkonen P, Karppinen SM, Pihlajaniemi T et al (2016) Continuous grading of early fibrosis in NAFLD using label-free imaging: a proof-of-concept study. PLoS One 11:e0147804. https://doi.org/10.1371/journal.pone.0147804

30. Gailhouste L, Le Grand Y, Odin C, Guyader D, Turlin B, Ezan F et al (2010) Fibrillar collagen scoring by second harmonic microscopy: a new tool in the assessment of liver fibrosis. J Hepatol 52:398–406

31. Guilbert T, Odin C, Le Grand Y, Gailhouste L, Turlin B, Ezan F et al (2010) A robust collagen scoring method for human liver fibrosis by second harmonic microscopy. Opt Express 18:25794–25807

32. He Y, Kang CH, Xu S, Tuo X, Trasti S, Tai DC et al (2010) Toward surface quantification of liver fibrosis progression. J Biomed Opt 15:056007. https://doi.org/10.1117/1.3490414

33. Lee JH, Kim JC, Tae G, MK O, Ko DK (2013) Rapid diagnosis of liver fibrosis using multimodal multiphoton nonlinear optical microspectroscopy imaging. J Biomed Opt 18:076009. https://doi.org/10.1117/1.JBO.18.7.076009

34. Liu F, Chen L, Rao HY, Teng X, Ren YY, YQ L et al (2017) Automated evaluation of liver fibrosis in thioacetamide, carbon tetrachloride, and bile duct ligation rodent models using second-harmonic generation/two-photon excited fluorescence microscopy. Lab Investig 97:84–92

35. Stanciu SG, Xu S, Peng Q, Yan J, Stanciu GA, Welsch RE et al (2014) Experimenting liver fibrosis diagnostic by two photon excitation microscopy and bag-of-features image classification. Sci Rep 4:4636. https://doi.org/10.1038/srep04636

36. Sun TL, Liu Y, Sung MC, Chen HC, Yang CH, Hovhannisyan V et al (2010) Ex vivo imaging and quantification of liver fibrosis using second-harmonic generation microscopy. J Biomed Opt 15:036002. https://doi.org/10.1117/1.3427146

37. Sun W, Chang S, Tai DC, Tan N, Xiao G, Tang H et al (2008) Nonlinear optical microscopy: use of second harmonic generation and two-photon microscopy for automated quantitative liver fibrosis studies. J Biomed Opt 13:064010. https://doi.org/10.1117/1.3041159

38. Tai DC, Tan N, Xu S, Kang CH, Chia SM, Cheng C et al (2009) Fibro-C-index: comprehensive, morphology-based quantification of liver fibrosis using second harmonic generation and two-photon microscopy. J Biomed Opt 14:044013. https://doi.org/10.1117/1.3183811

39. Wu Q, Zhao X, You H (2017) Characteristics of liver fibrosis with different etiologies using a fully quantitative fibrosis assessment tool. Braz

J Med Biol Res 50:e5234. https://doi.org/10.1590/1414-431X20175234

40. Xu S, Kang CH, Gou X, Q P, Yan J, Zhuo S et al (2016) Quantification of liver fibrosis via second harmonic imaging of the Glisson's capsule from liver surface. J Biophotonics 9:3513–3563

41. Wang H, Liang X, Gravot G, Thorling CA, Crawford DH, ZP X et al (2017) Visualizing liver anatomy, physiology and pharmacology using multiphoton microscopy. J Biophotonics 10:46–60

42. Chen X, Nadiarynkh O, Plotnikov S, Campagnola PJ (2012) Second harmonic generation microscopy for quantitative analysis of collagen fibrillar structure. Nat Protoc 7:654–669

43. Ranjit S, Dobrinskikh E, Montford J, Dvornikov A, Lehman A, Orlicky DJ et al (2016) Label-free fluorescence lifetime and second harmonic generation imaging microscopy improves quantification of experimental renal fibrosis. Kidney Int 90:1123–1128

44. Debarre D, Supatto W, Pena AM, Fabre A, Tordjmann T, Combettes L et al (2006) Imaging lipid bodies in cells and tissues using third-harmonic generation microscopy. Nat Methods 3:47–53

45. Jonscher KR, Alfonso-Garcia A, Suhalim JL, Orlicky DJ, Potma EO, Ferguson VL et al (2016) Correction: spaceflight activates Lipotoxic pathways in mouse liver. PLoS One 11: e0155282. https://doi.org/10.1371/journal.pone.0155282

46. Jonscher KR, Alfonso-Garcia A, Suhalim JL, Orlicky DJ, Potma EO, Ferguson VL et al (2016) Spaceflight activates Lipotoxic pathways in mouse liver. PLoS One 11:e0152877

47. Caligioni CS (2009) Assessing reproductive status/stages in mice. Curr Protoc Neurosci Appendix 4:Appendix 4I. https://doi.org/10.1002/0471142301.nsa04is48

48. Patton HM, Lavine JE, Van Natta ML, Schwimmer JB, Kleiner D, Molleston J, Nonalcoholic steatohepatitis clinical research network (2008) Clinical correlates of histopathology in pediatric nonalcoholic steatohepatitis. Gastroenterology 135:1961–1971.e2. https://doi.org/10.1053/j.gastro.2008.08.050

49. Kleiner DE, Brunt EM, Van Natta M, Behling C, Contos MJ, Cummings OW et al (2005) Design and validation of a histological scoring system for nonalcoholic fatty liver disease. Hepatology 41:1313–1321

50. Mehlem A, Hagberg CE, Muhl L, Eriksson U, Falkevall A (2013) Imaging of neutral lipids by oil red O for analyzing the metabolic status in health and disease. Nat Protoc 8:1149–1154

51. Gordillo M, Evans T, Gouon-Evans V (2015) Orchestrating liver development. Development 142:2094–2108

Chapter 16

Capillary Blood Sampling from the Finger

Kieron Rooney

Abstract

As the development of point of care testing devices improves, the uptake of capillary blood sampling from the fingertip across consumer groups and health professionals is increasing. The method promises to be a relatively safe and efficient method for monitoring patient health and obtaining research data. However, if not performed well, this simple technique can result in unreliable data and unsafe practices with a biological hazard. In this chapter, notes from the experiences of training undergraduate coursework and postgraduate research students in the method of capillary blood sampling from the fingertip are described to inform those considering the implementation of this method in teaching or research environments.

Key words Biosampling, Blood draw, Capillary, Finger prick, Point of care, Analysis

---

1 Introduction

Capillary sampling from the fingertip provides a quick and relatively safe method for small volume blood sampling. There is currently an increasing market for point of care testing devices (POCDs) that are empowering individuals to monitor their own health. Of particular interest to the professional is that the method provides researchers an efficient tool for either the routine screening of participants or in some cases primary data collection. Evidence, for example, of the increasing significance of capillary sampling from the fingertip, and the use of POCDs, is inclusion of the practice as a graduate outcome (10.4.2) within the professional standards for accredited exercise physiologists [1].

In practice, however, the method for capillary blood sampling from the finger has proved to be a difficult task for the novice to master cleanly and with minimal pain or discomfort to the participant. Since 2007, I have been involved in the training of both undergraduate course work students and postgraduate research students for whom mastery of the finger-prick capillary blood sampling skill has been a key learning outcome.

Paul C. Guest (ed.), *Investigations of Early Nutrition Effects on Long-Term Health: Methods and Applications*, Methods in Molecular Biology, vol. 1735, https://doi.org/10.1007/978-1-4939-7614-0_16, © Springer Science+Business Media, LLC 2018

The methods and notes below have been constructed with these experiences in mind in aid of providing guidance to those that are seeking to develop and/or teach this skill.

2 Materials

2.1 General

1. Lancet gun and lancets (Fig. 1) (*see* **Note 1**).
2. Alcohol wipes (at least two per piercing).

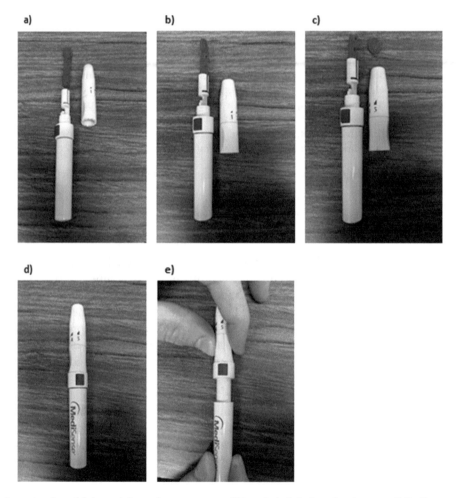

Fig. 1 Lancet setup. (**a**) Lancet "guns" may appear different, but their setup is essentially the same and follows the images. Step 1 requires unscrewing of the cap from just above the blue trigger button (use fingers to apply pressure on the indents as can be seen in step **e**). (**b**) A sterile, capped blue lancet is pushed into the barrel. This will require a little force as it is a tight fit. (**c**) Once the lancet is secured firmly in the barrel, twist off the safety cap. It is essential to twist off the cap as pulling may dislodge the lancet and/or increase risk of injury on the exposed needle. (**d**) Once the safety cap is removed, carefully return the gun cap over the lancet and screw into position, ensuring that the setting is on depth 5. Cock the trigger by extending the gun base until you hear a "click". (**e**) The blue trigger button should raise with a "click"

3. Tissues or gauze.

4. Gloves (box).

5. Bench coat or other contaminated waste protectors for the table.

6. Sharps bin and biohazard waste for contaminated materials.

7. General waste bin for non-contaminated waste.

8. Hot-water bottle and towel.

2.2 Optional
(See Note 2)

1. Point of care device (POCD) as required.

2. Test strips for POCD as required.

3. Control solutions for calibrating machines.

4. Capillary tube/Eppendorf (if the sample is to be stored for later analysis).

3   Method

1. 24 h prior to testing, check batteries in meters.

2. Check stocked supply of lancets and end point sample use (e.g., Eppendorf tubes for the purposes of sample storage or capillary tubes and POCD testing strips for immediate testing).

3. Check bins are emptied or have capacity for expected waste.

4. Print out hard copy of data collection sheets and leave these in the testing room.

5. Within 30 min of participant testing, boil water and prepare the hot-water bottle (*see* **Note 3**).

6. If using a POCD device, turn this on and calibrate against control solutions, and label any sample storage tubes.

7. On arrival of participant, outline brief procedure and obtain informed consent to proceed (*see* **Note 4**).

8. Seat the participant next to the table or bench on which the materials are set up.

9. Instruct the participants to rest their arm on the table with their hand placed on the hot-water bottle (*see* **Note 5**).

10. Put on the gloves, take the participant's hand and wipe the finger to be pierced with an alcohol wipe, and dry with tissues or gauze (*see* **Note 6**).

11. Angle the participant's hand so that the tip of the finger is facing the floor and the wrist is lower than that of the elbow (*see* **Note 7**).

12. Position the lancet lateral to the center of the finger pad (Fig. 2), and fire the piercing mechanism (*see* **Note 8**).

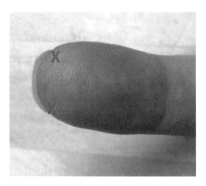

Fig. 2 Approximate position of piercing. Placement of the lancet should aim to avoid the absolute central tip of the finger

13. Place your index and middle fingers around the middle phalange of the finger you just pierced such that the distal phalange may be slightly squeezed over your finger utilizing the participant's knuckle as a hinge (*see* **Note 9**).

14. Discard the first drop of blood by wiping away with gauze or tissues.

15. Proceed to assist the formation of a second drop of blood by gently pushing the distal phalange over your index finger (this and all subsequent drops of blood can be used for sampling either directly onto a test strip or into a sampling tube for later processing) (*see* **Note 9**).

16. When all necessary blood samples have been collected, clean the participant's finger with an alcohol wipe and dry with either gauze or tissue (*see* **Note 10**).

17. Eject the lancet into the appropriate biohazard sharps bin and wipe down the lancet gun with an alcohol wipe.

18. Inform the participant of expected discomfort in the days to follow (*see* **Note 11**).

4   Notes

1. The length of lancet required will vary depending on the participant (adult, child, or infant) and the volume of blood required. The experiences informing these notes are from protocols requiring multiple blood sampling of 4–8 drops of blood from adults. We utilize a lancet gun (Fig. 1) and length of lancet to establish an approximate 2 mm piercing for sampling in accordance with [2]. With appropriate care as outlined below, this ensures patency for a period of time long enough to provide the required sample. Single-use lancets which

involve the blade being housed within a single use firing unit are also available.

2. Depending on whether or not the blood is to be sampled directly using a POCD with appropriate test strips or stored for later analysis.

3. Warming of the hands will enhance dilation and blood flow to the fingers. However, procedures are commonly conducted in cool, air-conditioned rooms. A safe and reliable way to ensure warming of the hands is to provide a hot-water bottle for participants to rest their hand on prior to sampling. A small hand towel or the like may be wrapped around the bottle to protect against burns from the potential leaking of hot water.

4. Information provided to participants may vary depending on institutional ethics committee requirements. The following is an example that has previously sufficed:

   The procedure of finger-prick blood collection may in some instances cause minor levels of stress and anxiety to individuals as well as acute levels of pain at the time of "prick." This pain will subside almost immediately, and investigators will attempt to reduce stress and anxiety as much as possible.

5. Understandably some participants may be anxious regarding the potential piercing, and this may result in the constriction of the arm muscles. This will impact the quality of sampling and the volume of blood sampled. It is important here to ensure that the participant is as relaxed as possible.

6. This is a good opportunity to assess how relaxed the participant's fingers may be. They should be pliable and easy to bend. If there is any tension, remind the participant to relax the fingers as any tension could restrict blood flow.

7. Having supervised many undergraduate students attempting this procedure for the first time, positioning of the hand is a commonly overlooked component of the procedure. You want to ensure blood flow to the fingertip is assisted by gravity, and when bleeding, a single drop is formed. Fingers that are pointing horizontally or even upward in some cases result in either very small if any drops of blood forming or more problematic is that drops do not form and rather blood ends up spreading over the finger and running down to the palm.

8. Incorrect positioning of the lancet can result in either very small if any drops of blood forming. Piercings on the absolute tip of the finger tend to be more painful to participants with enduring discomfort following the procedure. If one was to superimpose an analogue clock face over the pad of the finger, then ideal sites for piercing would be at approximately 01:00 or 11:00 positions (Fig. 2).

9. The key here is not to squeeze the finger of the participant such that you are attempting to drain blood but rather allow the fingertip to fill with blood (on account that it is facing the floor) and gently bend the distal phalange over your finger so that a blood drop may be eased out of the cut. You have to be mindful here that if you are squeezing the finger, you will be occluding any blood flow. As such, the key is to press the participant's finger over yours but not so hard as to block blood flow.

10. Some pressure may need to be applied to stop further bleeding. The participant may do this by pinching the gauze between the cut and thumb. This should only take about 10–15 s.

11. It is not uncommon for the fingertips to remain sensitive to the touch for the following 24–48 h. For this reason, it is important to consider which hand and fingers are to be sampled. Sampling from fingers of the non-dominant hand is recommended as participants are less likely to be using this hand for everyday tasks. The thumb is not recommended as the skin is commonly tough and difficult to sample from. Index fingers are also discouraged (particularly if the dominant hand is to be sampled) since the use of pens and cutlery tends to aggravate the fingertip. Of note, participants with a history of playing stringed instruments (e.g., guitar/banjo/violin) commonly present with callouses on the non-dominant hand which may also impede bleeding if a deep pierce is not achieved. Keyboard players have reported touching keys uncomfortable in the 12–24 h period following sampling.

References

1. Exercise and Sport Science Australia. Professional Standards September 2015. https://www.essa.org.au/wp-content/uploads/2016/12/AEP-Professional-Standards-with-coverpage_approved.pdf. Last accessed 20 July 2017

2. World Health Organisation (2010) WHO guidelines on drawing blood: best practices in phlebotomy. WHO Document Production Services, Geneva, Switzerland. ISBN 978 92 4 159922 1

Chapter 17

Physical Activity Assessment in Clinical Studies of Substance Use Disorder

Rhiannon Dowla, Bridin Murnion, Kia Roberts, Jonathan Freeston, and Kieron Rooney

Abstract

The therapeutic effect of exercise in promoting mental health is well known, and there is a growing body of evidence for incorporating physical activity-based interventions in the management of substance use disorders (SUD). A particular challenge in this area is a lack of standardized testing protocols between studies and clear descriptive statistics on the capacity of the SUD patient to perform exercise. Therefore, an essential starting point for new studies that seek to incorporate exercise into usual care therapy for SUD is an appropriate suite of baseline fitness assessments that include measures of aerobic capacity as well as muscular strength and/or endurance. We present here the methods and notes of our recent experiences in implementing baseline fitness testing of a patient population undergoing in-patient treatment for SUD. The tests described here have been adapted from freely available standardized tests that were developed for implementation with the general population. It is hoped that these experiences aid in the development of patient-specific physical activity programs that assist in the management of SUD.

Key words Fitness test, Endurance, Muscle, Drug, Patient, Cardiorespiratory

1 Introduction

Structured exercise and physical activity more generally are widely accepted to have positive effects on multiple parameters of health. However, the role of these in management of substance use disorder (SUD) is not well elucidated although it is an area of increasing interest within the community. For example, while a recent systematic review identified studies investigating the role of exercise in the therapy of alcohol dependence dating as far back as 1972 [1], no studies were identified that investigated the therapeutic effects of exercise in other drug-dependent patients prior to 1991, and only eight such studies were identified in the subsequent 20 years. Importantly, engagement with exercise-related activities has been reported to result in longer periods of abstinence during treatment

Paul C. Guest (ed.), *Investigations of Early Nutrition Effects on Long-Term Health: Methods and Applications*, Methods in Molecular Biology, vol. 1735, https://doi.org/10.1007/978-1-4939-7614-0_17, © Springer Science+Business Media, LLC 2018

for SUD [2] and in the improvement of psychological outcomes such as depression and anxiety across a number of studies [1].

A particular challenge for this area of growing interest is a lack of standardized testing protocols between studies and clear descriptive statistics on the capacity of the SUD patient to perform exercise. Furthermore, the likelihood that the SUD treatment seeker has at least one modifiable risk factor for cardiovascular disease is high [3] and, as such, exercise testing should be conducted following appropriate medical screening and with appropriate exercise professional supervision [4]. This presents another challenge in that, while anecdotally the number of exercise specialists within SUD contexts appears to be increasing, many clinics may not have adequately trained staff to perform exercise risk assessments for SUD patients on-site.

Interestingly, most studies investigating exercise in SUD therapy have incorporated aerobic exercise, while recent evidence identifies a high level of preference for resistance or strength training in this population [5]. As such, to enhance adherence, exercise programs should be constructed with the individual preferences, goals, and capacity of the patient in mind. For these reasons, an essential starting point for new studies that seek to incorporate exercise into usual care therapy for SUD is an appropriate suite of baseline fitness assessments that includes measures of both aerobic capacity and muscular strength and/or endurance.

Members of this team have been involved in two recent projects incorporating exercise testing in an in-patient setting for SUD. One of our recent projects on this is a pilot study for development of a large-scale in-patient program known as the Healthy Lifestyles Program (HeLP). We present here the protocols and notes for implementation of a range of muscular endurance and fitness assessments. The patients were seeking treatment for at least one of the following abuses: alcohol, benzodiazepines, opiates, or tetra-hydrocannabinol (THC). The purpose of these tests was to provide an assessment of the baseline level of fitness of treatment seekers to allow development of a physical activity program for implementation in the in-patient HeLP.

The second project laying the groundwork for this chapter was *A randomized controlled trial of daily aerobic exercise for inpatient cannabis withdrawal* (ACTRN12615000211561). We present here the protocol for baseline submaximal assessment of maximal cardiorespiratory fitness using a cycle ergometer, physical work capacity test protocol adapted from the American College of Sport Medicine Guidelines for Exercise Testing and Prescription [6]. A full account of how the data collected from this protocol can be analyzed for the prescription of exercise training intensity is available in the supplementary material of the full trial protocol [7].

Together, these protocols provide a comprehensive suite of baseline assessments of aerobic fitness and muscular endurance employed in an in-patient treatment facility for SUD.

2   Materials (*See* Note 1)

2.1 Muscular Endurance and Fitness Testing of Broad SUD (HeLP Pilot)

1. Exercise appropriate clothing (*see* **Note 2**).
2. 30 cm box or step.
3. Heart rate monitor (*see* **Note 3**).
4. Stopwatch.
5. Metronome.
6. Chair—approx. 17 in. tall.
7. Sphygmomanometer (*see* **Note 4**).
8. Measuring tape (*see* **Note 5**).
9. Markers for 6-min walk test (masking tape or chalk to mark corridor).
10. Borg Rating of Perceived Exertion (RPE) scale [8] (*see* **Notes 6 and 7**).

2.2 Physical Work Capacity (PWC) Test: Submaximal Assessment of Maximal Cardiorespiratory Fitness

1. Cycle ergometer preferably with adjustable power setting in W (*see* **Note 8**).
2. Heart rate monitor (*see* **Note 9**).
3. Stopwatch.
4. Scales for assessment of body weight.
5. Fan (optional).

3   Methods (*See* Note 10)

3.1 3-Minute Step Test (See Notes 11 and 12)

1. A chest strap heart rate monitor with wristwatch display is attached to the participant.
2. The participant stands in front of the 30 cm box/step.
3. The assessor sets metronome to the desired pace.
4. Instruct the participant to step up and down in time with the metronome (i.e., one step/beat to achieve the desired cadence) (*see* **Note 13**).
5. Run the test for 3 min.
6. Measure the pulse rate and rate of perceived exertion (RPE) at the end of min 1, 2, and 3.
7. Manually palpate radial pulse between min 3 and 4 as a recovery pulse.

3.2 Push-Up Test (Adapted from [6])

1. Male participants start in the standard "down" position (hands pointing forward and under the shoulder, back straight, head up, using the toes as the pivotal point).

2. Female participants start in the modified "knee push-up" position (legs together, lower leg in contact with floor with ankles plantar-flexed, back straight, hands shoulder width apart, head up, using the knees as the pivotal point).

3. Instruct the participant to raise the body by straightening the elbows and return to the "down" position until the chin touches the floor (the stomach should not touch the floor) (*see* **Note 14**).

4. Stop the test when the participant strains forcibly or is unable to maintain the appropriate technique within two repetitions.

5. Count the maximum number of push-ups performed consecutively without rest as the score (*see* **Note 15**).

3.3 30-Second Chair Stand Test (Text Below Adapted from [9]) (See Notes 16 and 17)

1. Have the participant sit in the middle of the chair.

2. Have the participant place their hands across the chest on opposite shoulders to avoid using armrests.

3. Have the participant place their feet flat on the ground.

4. Instruct the participant to keep their back straight and their arms across the chest while rising to a full standing position and then lowering to be seated again.

5. Continue the test for 30 s.

6. Score the number of rises in 30 s.

3.4 Curl-Up Test (Text Below Adapted from [6]) (See Notes 15 and 18)

1. Instruct the participant to lie on their back on the floor with knees at 90 degrees and arms across the chest.

2. Instruct the participant to squeeze/clench the stomach such that their lower back touches the floor.

3. Set a metronome to 50 beats/min and have the individual perform slow, controlled curl-ups to lift the shoulder blades off the floor in time with the metronome at a rate of 25/min.

4. Ensure that the trunk makes a 30-degree angle with the floor with no bending of the neck (monitor by the eye).

5. Conduct the test for 1 min to a maximum of 25 curl-ups.

6. Count the number of curl-ups completed without pausing as the score.

3.5 6-Minute Walk Test (Text Adapted from [10]) (See Note 19)

1. Mark a 20 m track in 5 m increments on the floor.

2. The participant should be wearing comfortable clothing and appropriate footwear.

3. Attach a chest strap heart rate monitor with wristwatch display to the participant to monitor heart rate throughout test.

4. Have the participant rest in a chair at the start point until heart rate returns to resting conditions.

5. Document heart rate and blood pressure immediately prior to commencing the test.

6. Initiate the stopwatch and advise the patient to commence walking (*see* **Note 20**).

7. At 3 min, collect heart rate and RPE readings without pausing the test.

8. At 6 min, advise the patient to cease walking and to sit.

9. Record the heart rate and blood pressure readings.

10. Record the distance walked.

11. Record the RPE.

3.6   Physical Work Capacity Test (Table 1) (See Note 21)

1. Weigh participant in light clothing.

2. Determine a heart rate target at which exercise will cease once reached or exceeded (critical heart rate target, e.g., 130, 150, or 170 bpm) (*see* **Note 22**).

3. Set seat height appropriate to the patient and ask the patient to mount the cycle ergometer (*see* **Note 23**).

4. Obtain resting heart rate readings.

5. Instruct the patient to commence cycling with no resistance set on the ergometer (*see* **Note 24**).

6. After 2–3 min of unloaded cycling, increase the resistance setting on the bike to achieve the desired target power output.

7. In this protocol, 25 W was selected as the first stage power target, and the participant should be instructed to maintain this power output for 4 min (*see* **Note 25**).

8. In the fourth min (3:00–4:00 min), record the heart rate every 10 s as in Table 2.

9. During this min, obtain a self-identified RPE from the participant.

10. At the completion of the fourth min, increase power output to the second stage target (*see* **Notes 26** and **27**).

11. Tabulate data as in Table 2.

12. For each power output, average the recorded heart rate for the final min.

13. Stop the exercise test if (1) the critical heart rate target has been reached or exceeded, (2) participants report a RPE of 18 (from a scale of 6 to 20), (3) participants request to stop, or (4) the predetermined maximal power output has been completed.

Table 1
Recorded information for physical work capacity test

Date:	Participant ID	Critical target heart rate _____ (bpm) Resting heart rate _____ (bpm)	
Power (W)	Time (min) 3:10 3:20 3:30 3:40 3:50 4:00	Heart rate (min)	RPE
Mean heart rate min 4 (bpm)			**RPE**
Power (W)	7:10 7:20 7:30 7:40 7:50 8:00		
Mean heart rate min 8 (bpm)			**RPE**
Power (W)	11:10 11:20 11:30 11:40 11:50 12:00		
Mean heart rate min 12 (bpm) Notes:			
Power (W):	15:10 15:20 15:30 15:40 15:50 16:00		**RPE**
Mean heart rate min 16 (bpm)			**RPE**
Power (W)	19:10 19:20 19:30 19:40 19:50 20:00		
Mean heart rate min 20 (bpm)			**RPE**
Power (W)	23:10 23:20 23:30 23:40 23:50 24:00		
Mean heart rate min 24 (bpm) Notes:			

Table 2
Heart rate and RPE recording format

1. Step test	HR	RPE
At rest		
Heart rate (at end of min 1)		
Heart rate (at end of min 2)		
Heart rate (at end of min 3)		
Total beats during min 3–4		
2. Push-ups (maximum count):		
3. Chair stand test (number of rises in 30 s)		
4. Curl-ups (total completed in 1 min):		
5. 6 min walk test		
At rest	Heart rate Blood pressure	RPE
At the end of min 3	Heart rate	RPE
At the end of min 6	Heart rate Blood pressure Distance walked	RPE

4 Notes

1. These protocols were approved by the Sydney Local Health District Human Research Ethics Committee—Concord Repatriation General Hospital (EC00118). All participants provided informed consent prior to implementation of the assessments. Potential participants underwent an initial medical screen by a doctor with all testing conducted under the supervision of an accredited exercise physiologist. The user must ensure similar standards and likewise adhere to the ethical requirements of their local institutions and ensure that all procedures undergo the necessary approvals.

2. Since patients attending an in-patient setting are not expecting to participate in an exercise program, many did not have appropriate footwear/clothing available. As such, where it was still safe to do so, several tests were completed barefoot or in suboptimal but acceptable footwear such as UGG boots or thongs. As such, clinics may consider advising patients prior to admission to bring adequate clothing if possible. In

addition, many participants were on regular PRN medication to assist with detox, and it was a common theme for participants to feel that they were shakier or unsteady on their feet due to these medications. Although not accounted for in this study, it would be advised to control for or consider PRN use in participants and the impact this would have on exercise capacity. Finally, it should be noted that none of the tests included in this protocol assess muscular strength. In the future, it would be beneficial to include some form of simple strength testing such as a grip strength test.

3. Heart rate can be measured by palpation of the wrist during exercise; however, for convenience, a wireless wearable heart rate monitor is best used.

4. Preferably a well-maintained and calibrated automated blood pressure monitor is available in the hospital setting. Alternatively a manual sphygmomanometer may be used for simple pre- and post-assessment.

5. Preferably, this should be made of steel for assessment of waist and hip circumference.

6. A standardized script was used to introduce and explain the RPE to the participants as follows (text adapted from [10]):

   Through some of the tests I will ask you to rate your perception of exertion. This feeling should reflect how heavy and strenuous the exercise feels to you, combining all sensations and feelings of physical stress, effort, and fatigue.

   Do not concern yourself with any one factor such as leg pain or shortness of breath, but try to focus on your total feeling of exertion.

   Look at the rating scale below while you are engaging in an activity; it ranges from 6 to 20, where 6 means "no exertion at all" and 20 means "maximal exertion." Choose the number from below that best describes your level of exertion.

   Try to appraise your feeling of exertion as honestly as possible, without thinking about what the actual physical load is. Your own feeling of effort and exertion is important, not how it compares to other people. Look at the scales and the expressions and then give a number.

7. The explanation of RPE sometimes needed to be simplified for participants to fully understand the scale. If participants didn't understand the scripted explanation, a simplified explanation was provided including relevant examples.

8. For this protocol we used the Keiser M3 Indoor Cycle Ergometer Magnetic Spin Bike. The in-built electronic display provides real-time heart rate and power output.

9. It is preferable that the heart rate monitor syncs wirelessly to the cycle ergometer power output display as mentioned above. However, if this is not available, then either a wearable heart rate monitor can be separately purchased, or heart rate can be determined by palpation of the wrist during exercise.

10. In the muscular endurance and fitness testing of broad SUD (HeLP Pilot), assessments can be carried out faster in an assessment room located within the hospital ward except for the 6-minute walk test which can be completed in a flat, covered corridor outside the ward. Assessments are completed as a single testing session. The assessment room should have sufficient space for the participant to complete a push-up and a firm stable surface to place a step on. Data can be recorded using Table 2. The purpose of these assessments is to gather a range of fitness outcomes across specific tasks that can be used to inform the development of an activity-based intervention. Due to heart rate being the dependent factor of the 3-minute step test, this test should be completed first to avoid elevated heart rate due to prior exercise testing influencing the data. The remaining tests should be ordered to maximize recovery time between tests utilizing the same muscle groups. The order we used was: (1) 3-minute step, (2) push-up, (3) 30-second chair stand test, (4) partial curl-up, and (5) 6-minute walk test. For all tests, it is preferable that the participant has not consumed food for at least 3 h prior to testing to minimize the impact of acute food consumption on performance and reduce the likelihood of non-exercise abdominal discomfort as patients approach higher intensities (in the case of the physical work capacity test or while performing the curl-up test).

11. The step test protocol completed is adapted from the YMCA step test protocol to better match the expected physical capacity of the population. The YMCA step test protocol can usually be completed if the participant can reach a workload of nine metabolic equivalents (METS) [11]. 15% of the general population assessed in that study however were unable to complete the YMCA step test. Since the SUD population reported at least one cardiac risk factor, we assumed that the number of participants unable to complete the original YMCA protocol would be higher than that of general population. For this reason, the step test protocol was adapted for the SUD population, with the cadence being decreased from 24 rises to 15 rises per min.

12. It was common for participants to lose timing with the metronome and often needed additional cueing from the investigator to maintain the pace of the metronome when stepping. This involved the vocal instruction of "up, up, down, down" in time with the tone.

13. Discontinue testing if the participant is unable to maintain the prescribed pace, experiences chest pain or dizziness, or requests to stop.

14. For both men and women, the participant's back must be straight at all times, and the participant must push up to a straight arm position.

15. Push-up and curl-up test results were compared to the American College of Sports Medicine (ACSM) normative values [6].

16. In the chair stand test, it is preferable that the chair does not have armrests. This is to minimize any tendency of the participants to use their arms to assist. While completing this test, if the participant uses the handrests or moves them from the chest for stabilization, the test is stopped.

17. When completing the chair stand test, if the participant is halfway to a full stand at the final count (30 s), then that half raise is counted as a full rise.

18. The partial curl-up test was well received by most participants as they were familiar with this form of exercise. The results for this test then were better than expected with many participants achieving the maximum score despite poor performance in other tests. While it is currently not clear why this is the case, alternate abdominal endurance tests may be preferable when working with this population to more clearly identify any differences in abdominal endurance across the population.

19. Prior to the 6-minute walk test, participants were given the following instructions:

    The goal of this test is to walk as many laps as possible in the 6 minutes. You can slow down or stop at any time. However I want you to continue walking as soon as possible. The goal is to walk as quickly as possible, but do not run or jog. I will tell you at the end of each minute. At 3 minutes and 6 minutes I will ask you for your heart rate and RPE. If you experience any discomfort including chest pain or dizziness tell me immediately. Do you have any questions?

20. Terminate the test if the participant develops chest pain and intolerable dyspnea or becomes acutely unwell.

21. The purpose of this test is to gain an accurate assessment of maximal aerobic capacity using extrapolation of data collected submaximally. This allows us to recruit normal population individuals and gain an indicator of their maximal capacity without exposing them to the risk of working at maximum. Further, this data allows us to control for individual variations in fitness that will be evident in free-living humans.

22. For the purposes of the trial reported here, we identified a critical heart rate target of 75% of age predicted heart rate maximum (i.e., $0.75 \times [(206.9 - (0.67 \times age)])$).

23. Seat height should be adjusted such that either the leg is straight when the patient is seated with their heel placed on the pedal or there is a slight bend in the knee when the patient is seated with the ball of their foot placed on the pedal. A third condition is commonly used that requires the patient to stand next to the bike and the seat is aligned with the greater trochanter of the femur. However, in practice this was difficult to correctly identify in all patient body types and through clothing.

24. We did not prescribe a fixed cadence to be maintained by participants. Rather we adjusted the resistance setting in accordance with cadence to ensure controlled power output. Participants were informed that the faster they pedaled, the less resistance would be required to be placed on the wheel. Many participants then self-selected a cadence of around 75–80 revolutions/min.

25. Typically the test starts at a power output of either 25 W (for a mostly sedentary individual) or 50 W (for a sedentary but mostly active individual). For the purposes of this study and assumed fitness of the population, all participants commenced at 25 W. The participant cycles at this intensity for 4 min at which time power output is increased by 25 W. Power output is increased every 4 min until the target heart rate is achieved. For the most accurate submaximal assessment of VO_2 max, a minimum of three completed stages is preferred.

26. This protocol required participants to cycle for 4 min at up to 6 submaximal work rates (i.e., 25, 50, 75, 100, 125, and 150 W).

27. Arguably the most consistent complaint received during this test was in regard to the seat. The bike seat is small and hard and is quite uncomfortable after a period of about 10 min or so. We often then wrapped a towel around the seat to increase padding. Also, participants commonly reported that their performance was limited by soreness localized to their thighs. This is most likely a result of not being accustomed to cycling exercise. Walking-based tests could be encouraged for future studies although an appropriate assessment of potential risk of falls will need to be completed.

References

1. Zschucke E, Heinz A, Strohle A Exercise and physical activity in the therapy of substance use disorders. ScientificWorldJournal 2012, 2012:901741. https://doi.org/10.1100/2012/901741

2. Weinstock J, Barry D, Petry NM (2008) Exercise-related activities are associated with positive outcome in contingency management treatment for substance use disorders. Addict Behav 33:1072–1075

3. Kelly PJ, Baker AL, Deane FP, Kay-Lambkin FJ, Bonevski B, Tregarthen J (2012) Prevalence of smoking and other health risk factors in people attending residential substance abuse treatment. Drug Alcohol Rev 31:638–644

4. Riebe D, Franklin BA, Thompson PD, Ewing Garber C, Whifield GP, Magal M et al (2015) Updating ACSM's recommendations for exercise participation health screening. Med Sci Sports Exerc 47:2473–2479

5. Abrantes AM, Battle CL, Strong DR, Ing E, Dubreuil ME, Gordon A et al (2011) Exercise preferences of patients in substance abuse treatment. Ment Health Phys Act 4:79–87

6. ACSM (2009) ACSM's guidelines for exercise testing and prescription, 8th edn. Lippincott Williams and Wilkins, Philadelphia, PA. - ISBN-10: 0781769027

7. Allsop DJ, Rooney K, Arnold J, Bhardwaj AK, Bruno R, Bartlett DJ et al (2017) Randomised controlled trial (RCT) of daily aerobic exercise for inpatient cannabis withdrawal: a study protocol. Mental Health and Physical Activity (In Press, Available online June 27, 2017 https://doi.org/10.1016/j.mhpa.2017.06.002)

8. Borg G (1982) Psychophysiological bases of perceived exertion. Med Sci Sports Exerc 14:377–381

9. Centers for Disease Control and Prevention National Center for Injury Prevention and Control. The 30-Second Chair Stand Test. Protocol last accessed online July 11, 2017 (https://www.cdc.gov/steadi/pdf/30_second_chair_stand_test-a.pdf)

10. Queensland Cardiorespiratory physiotherapy network. Technical Standard for Functional Exercise Testing–6 Minute Walk Test. Protocol accessed July 11, 2017 (https://www.health.qld.gov.au/data/assets/pdf_file/0018/426204/qcpn_6mwt.pdf)

11. Beutner F, Ubrich R, Zachariae S, Engel C, Sandri M, Teren A et al (2015) Validation of a brief step-test protocol for estimation of peak oxygen uptake. Eur J Prev Cardiol 22:503–512

Chapter 18

Cardiopulmonary Exercise Testing

Derek Tran

Abstract

Cardiopulmonary exercise testing (CPET) is an objective assessment of exercise capacity. It has become increasingly popular in clinical, research, and athletic performance settings. CPET allows for investigation of the cardiovascular, pulmonary, and skeletal muscle systems during exercise-induced stress. The main variable of maximal oxygen uptake (VO_2max) reflects the gold standard measure of exercise capacity. This chapter will describe the method of performing a graded maximal CPET with the Vmax 229 Cardiopulmonary Exercise Testing Instrument and CardioSoft program.

Key words Cardiopulmonary exercise testing, Exercise capacity, Exercise, Functional capacity, Physical activity, CPET, CPX, Aerobic power

1 Introduction

Determining exercise capacity is a valuable measure which when interpreted correctly is used in clinical decision-making [1, 2]. It is also useful for evaluating outcomes of clinical studies. However, defining exercise capacity can be challenging in clinical and general populations. Reproducible objective results are difficult to obtain. The ability to perform physical work is associated with the body's ability to deliver and extract oxygen [3]. Recruitment of skeletal muscles to induce movement requires energy derived from adenosine triphosphate (ATP). During exercise, ATP hydrolysis provides the energy required for myocyte cross bridging and maintaining sodium–potassium pump function. However, muscular stores of ATP are limited and maintaining physical exertion requires the resynthesis of ATP. Resynthesis is commonly described as being derived from anaerobic and aerobic pathways. The primary source of ATP resynthesis during sustained exercise occurs through aerobic pathways. Although systems do not operate in isolation, the exchange of oxygen and production of carbon dioxide varies and is distinct between energy systems. Inferences can be made by recognizing the typical patterns of gas exchange which can identify the

Paul C. Guest (ed.), *Investigations of Early Nutrition Effects on Long-Term Health: Methods and Applications*, Methods in Molecular Biology, vol. 1735, https://doi.org/10.1007/978-1-4939-7614-0_18, © Springer Science+Business Media, LLC 2018

contribution of the predominant energy systems providing the source of ATP. During steady-state exercise, exhaled gases are equivalent to the output of internal respiration allowing for the investigation of metabolic demand [3]. The analysis of these exhaled gases can be performed during cardiopulmonary exercise testing (CPET).

CPET is considered as a gold standard noninvasive measure of assessing cardiorespiratory fitness and exercise capacity. It can evaluate the limits of oxygen delivery and uptake for ATP resynthesis. Abnormalities which are not apparent at rest can be revealed with CPET [3, 4]. The value of objectively evaluating exercise capacity promotes the clinical application of CPET (Table 1). Accurate methods for assessing cardiorespiratory fitness are necessary as it has important implications on prognosis [5] and all-cause mortality [6]. Estimation equations vary in accuracy and often overestimate an individual's exercise capacity. This can lead to poor clinical decisions or misinterpretation of the effectiveness of clinical studies. The most common method of performing CPET involves a maximal exercise test on a cycle ergometer or treadmill while collecting expired gases for analysis. The test begins at low work rates which progressively increase until indications for termination [2] are observed or volitional exhaustion. Individualized workloads should be selected with the aim to reach peak exercise within 8–12 min [7]. Breath-by-breath gas analysis can provide us with an understanding of limitations to exercise capacity. During CPET breath-by-breath measures of oxygen uptake (VO_2), ventilation (VE), and carbon dioxide output (VCO_2) can be obtained. Recently the value of CPET extends beyond the measure of VO_2. Technological advancements provide the clinician with various variables which can assist in exploring the physiology of the individual.

Table 1
Indications for cardiopulmonary exercise testing (Reproduced from Stickland et al. [8])

(a) Assessment of unexplained dyspnea
(b) Evaluation of disease severity
(c) Development of an exercise prescription for pulmonary rehabilitation
(d) Identification of gas exchange abnormalities
(e) Preoperative assessment
• Lung cancer surgery
• Lung volume reduction surgery
• Heart or lung transplantation
(f) Evaluation for lung/heart transplantation
(g) Objective evaluation of exercise capacity

The most well-known variable attained from CPET is maximal oxygen uptake (VO_2max) reflecting aerobic power. However, many individuals do not meet the traditional criteria for VO_2max where a plateau of VO_2 is observed. As a result, peak VO_2 is often used and defined as the highest VO_2 attained. In healthy subjects peak VO_2 and VO_2max are often the same [3, 9]. Other variables collected during CPET can help determine whether or not sufficient effort has been made for a valid result (*see* below). Impairment of systemic or peripheral delivery and extraction of oxygenated blood will limit VO_2 values. Our understanding of VO_2 can be derived from the Fick equation:

$$VO_2 = (CO) \times (CaO_2 - CvO_2)$$

The Fick equation derives VO_2 from the product of cardiac output (CO) and the difference between arterial oxygen content (CaO_2) and mixed venous oxygen content (CvO_2). CO can be determined by the product of heart rate (HR) and stroke volume. The Fick equation demonstrates how the coupling of the cardiovascular, pulmonary, and skeletal muscle systems is associated with exercise capacity. VO_2 can be expressed as relative (mL/kg/min) or absolute (L/min). In addition, most laboratories will also report values as percent predicted to permit intersubject comparisons. The most common normative reference values proposed are the Jones et al. [10] and the Hansen et al. [11] equations.

Peak HR can assist the technician in determining whether sufficient effort has been made for a valid result. Achieving >85% of the subjects age-predicted maximal HR is used [7]. Failure to achieve >85% of age-predicted maximal HR may indicate insufficient effort or chronotropic incompetence in the absence of HR-limiting drugs [12]. In some individuals, stroke volume may plateau as exercise intensity increases, although this response is not consistent in all individuals [13, 14]. Therefore, increasing HR is the only method to continue to augment cardiac output to meet metabolic demands during high-intensity exercise in some individuals and clinical populations [15]. This makes HR a valuable measure when assessing for exercise intolerance. Heart rate recovery (HRR) can also be used as an indicator of autonomic function. An attenuated HRR has been associated with cardiac autonomic dysfunction [16, 17] and with poor prognosis in many disease states [18–20].

Another useful measure which can be determined from CPET is the ventilatory anaerobic threshold (VAT). The VAT is associated with metabolic acidosis and is a widely used index for assessing submaximal exercise capacity independent of the patient's motivation. It identifies the point where oxygen delivery is unable to meet metabolic demands and the onset of increased anaerobic pathways for ATP resynthesis. Metabolic acidosis is commonly and inappropriately described in the literature as the result of lactic acidosis

[21]. In fact, lactate production inhibits acidosis [22]. The production of lactate is necessary for performing high-intensity exercise not only for the purpose of inhibiting acidosis but because it also removes pyruvate from the cytosol and provides a source of NAD+ for glycolysis [22]. Muscular acidosis is attributed to increased nonmitochondrial ATP turnover [21]. However, the rate of lactate production is still associated with metabolic acidosis and can be advantageous as an indirect marker [21]. There are various methods of calculating VAT that have been proposed. The V-slope method is the most common and is described as "the departure of VO_2 from a line of identity drawn through a plot of VCO_2 versus VO_2" [7] and is suggested as a reproducible method [1, 23].

The respiratory exchange ratio (RER) is the ratio of VCO_2/VO_2. During high-intensity exercise, the buffering of protons increases VCO_2 output at a greater rate than VO_2 resulting in a higher RER. In some individuals, using HR alone to determine effort may be insufficient. Peak RER is another variable used to determine adequate effort during a test with ≥ 1.10 used as an indication of good effort [7]. An RER of <1.00 suggests poor effort from the individual and results should be interpreted with caution. During steady-state exercise, RER is equivalent to the respiratory quotient (RQ) [1]. Therefore, RQ may reflect the primary source of fuel utilization during steady-state exercise [1, 24].

The ventilatory equivalent for CO_2 (VE/VCO_2) slope is an index of ventilatory efficiency [7]. In healthy individuals, a normal VE/VCO_2 slope is <30 [25]. An elevated VE/VCO_2 slope is an indicator of the severity of systemic disease and is likely attributed to ventilation-perfusion mismatching [7]. Increased physiological dead space can also contribute to an elevated slope in some conditions. An increased slope is a hallmark feature of clinical conditions such as pulmonary hypertension and heart failure [26–28].

The addition of an electrocardiogram (ECG) enables detailed assessment of exercise tolerance. It can attribute or exclude various arrhythmias which limit exercise performance. An ECG also allows the test to be terminated if ischemia is observed and can be evaluated by ST segment depression.

The value of CPET extends beyond the variables described above. This chapter will describe the methods to perform a CPET ramp protocol using the Vmax 229 Cardiopulmonary Exercise Testing Instrument in conjunction with a 12-lead ECG and simultaneous blood pressure measurements.

2   Materials

1. Electronically braked cycle ergometer (Lode Corival; Lode BV, Groningen, The Netherlands).

2. Vmax 229 Cardiopulmonary Testing Instrument includes (a) mass flow sensor, (b) sensor cable, (c) sample line, (d) directional sense line [DIR], (e) span 1 gas cylinder (16% O_2, 4% CO_2, balance N_2), (f) span 2 gas cylinder (26% O_2, balance nitrogen), (g) 3 L calibration syringe (SensorMedics Corporation).

3. 12-lead ECG resting and exercise stress testing system (Cardio-Soft, version 6.51, GE Healthcare, Waukesha, Wisconsin).

4. ECG tabs.

5. Face mask and head strap.

6. Alcohol wipes.

7. Disposable razor.

8. Sphygmomanometer.

9. Stethoscope.

10. Oximeter with forehead probe (Radical, Massimo Corp, Irvine, USA).

11. Head band.

3 Methods

3.1 Calibration and Warm-Up

1. Turn on the Vmax unit at least 30 min prior to beginning.
2. Turn on the PC.
3. Select and enter the Vmax software.

3.2 Flow Calibration

1. Prepare the mass flow sensor by attaching the flow meter sensor cable, sample line (white tip), and DIR line (*see* **Note 1**).
2. Insert the mass flow sensor into the syringe.
3. Select "Flow Sensor Calibration [1]" on the home screen.
4. Select F1 and when prompted withdraw (inspiratory stroke) and pump (expiratory stroke) the syringe completely twice to purge.
4. Press the space bar on the key board and wait 15 s.
5. Withdraw and pump the complete volume of the syringe at different flow rates corresponding to the yellow bars (bars will turn green when successful) (*see* **Note 2**).
6. When successful a verification screen will appear.
7. Perform five full inspiratory and expiratory strokes.
8. Four of the five strokes should be as follows: the first (inspiratory and expiratory) strokes should correspond to the lowest dotted line; the second (inspiratory and expiratory) strokes should correspond to the highest dotted line; the third (inspiratory and expiratory) strokes should correspond to the midpoint between

the lowest and highest dotted lines; and the fourth (inspiratory and expiratory) strokes should represent peak flows.

9. If successful, verification is automatic.

10. Select F3 to store calibration and exit.

3.3 Gas Calibration

1. Select calibrate O_2 and CO_2 (located on the top toolbar).

2. Remove the sample line (white tip) from the mass flow sensor and insert it into calibration fitting in the Vmax unit.

3. Turn on gas cylinders and regulators (*see* **Note 3**).

4. Select F1 for gas calibration.

5. When successful a green box will appear.

6. Return the sample line (white tip) to the mass flow sensor.

7. Turn off gas cylinders.

8. Select F3 to store and exit back to the home screen.

3.4 Subject Setup

1. Take the subjects height (cm) and weight (kg) (*see* **Note 4**).

2. Record other relevant information (i.e., medications).

3. For ECG, prepare the skin by cleaning the surface with alcohol wipes (shave parts of the chest where electrodes will be located if contact surface is impeded by chest hair).

4. Place electrodes for LA and RA (this should be placed on the midclavicular line below the clavicle).

5. Place electrodes for LL (midclavicular line two rib spaces below V5) and RL on the apex of the right ribs.

6. Palpate the fourth intercostal space and position the electrode to the right (V1) and left (V2) of the sternum.

7. Palpate the fifth intercostal space and position an electrode (V4) in line with the midclavicular line on the left rib cage (*see* **Note 5**).

8. V3 should be placed on the surface of the fifth rib between V2 and V4.

9. Place V5 on the left anterior axillary line horizontal to V4.

10. Place V6 on the left midaxillary line horizontal to V5.

11. Attach the cables to the corresponding electrode (*see* **Note 6**).

12. Tighten the CardioSoft unit around the waist of the subject.

3.5 Bike Setup

1. Lower the seat completely and instruct the subject to sit down and place their feet into the pedals.

2. Ask the subject to stand on one pedal and adjust the seat height to achieve a 30° angle of knee flexion or to the comfort of the subject.

3. Adjust the handles accordingly so the subject is comfortable.

3.6   Oximeter

1. Attach headband firmly on the subject's head.

2. Attach oximeter sensor on the forehead of the subject (*see* **Note 7**).

3. Turn on the oximeter.

4. Ensure a perfusion index of >1 (*see* **Note 8**).

3.7   Inserting Subject Details and Selecting Ramp Protocol (Work Rate Estimation)

1. Select "New study" from the home screen.

2. Insert subject demographics as appropriate.

3. Select F3 to save.

4. Select "Exercise/metabolic tests [4]" from the main screen.

5. Select test protocol as "ECG Ergometer Study" from the test protocol selection.

6. Select F10 "protocol settings" and set measurement modes as in Table 2.

7. Select F3 to save.

8. Select F2 to view subject reference values and predicted peak power output determined from the Jones equations [10].

9. Divide the predicted peak power output by 10 and round appropriately to determine work rate (e.g., if predicted peak power output is 150 watts (W), then the ramp increment would be 150 W/10 min $=$ 15 W/min).

10. Click ESC to return to "Metabolic Study" screen.

11. Click "Start Test (F1)" and two screens will appear.

12. Adjust work rate increment in CardioSoft (right screen) by selecting the icon with the dials.

13. Change to the "protocol editor" tab.

14. Amend the ramp using the drop down box.

15. Select ok.

3.8   Exercise Test

1. Position the mask to cover the mouth and nose of the subject (*see* **Note 9**).

2. Tighten straps to ensure a tight seal and no gas leakage (*see* **Note 10**).

3. Firmly insert the mass flow sensor.

4. Once the subject is set up and all former steps are completed, select "pre-test" on the CardioSoft program (right screen).

5. Take baseline blood pressure and RPE (rate of perceived exertion). Blood pressure can be entered in the CardioSoft program (right screen) ("F12 enter BP").

6. Collect ~2 min of resting data (*see* **Note 11**).

7. Select "Exercise" on the CardioSoft program.

Table 2
Ergometer study measurement modes

(a) Breath-by-breath exercise
(b) Bike—external
(c) ECG—CardioSoft
(d) Left graph—single plot
(e) Txt—CPX tabular display
(f) Signal out—analog group
(g) Analog in—1—off

8. Instruct the subject to initiate pedalling at a speed of 65–75 rpm and to continue until exhaustion (*see* **Note 12**).

9. Blood pressure should be taken and recorded every 2 min.

10. Monitor CPET variables for indications to terminate.

11. Terminate the test by selecting "recovery" when the subject reaches volitional fatigue or when clinically indicated.

12. Instruct the subject to continue pedalling for 2–3 min to aid recovery.

13. Monitory CPET variables until they return to baseline and collect recovery data for ~10 min.

14. Click "test end" on the CardioSoft program once variables return to baseline measures.

15. Remove the face mask, forehead probe, and ECG tabs/cables from the subject.

16. Exit CardioSoft program (last icon on the right).

17. Select exit on the Vmax program (tool bar).

18. Select "y to end test."

19. Input relevant information (i.e., main reason for termination and RPE).

20. Click F3 to save results.

21. Results are automatically generated by the Vmax program.

22. Amend results as appropriate (*see* **Note 13**).

23. Select F3 to save results.

24. Exit program.

4 Notes

1. Cable ends are different sizes and will only fit into the correct corresponding hole in the mass flow sensor.

2. The first few strokes will not register and you will see nothing on the screen. This is normal.

3. The regulators should be set at 50–60 PSI.

4. Ensure that the subject has removed their jacket/shoes and emptied their pockets prior to taking height and weight. This is important as prediction equations use height and weight to calculate reference values.

5. For females, place V3-V6 directly under the bra line.

6. Tape down the wires attached to the electrode. This will aid in reducing noise produced during exercise.

7. A finger oximeter may also be used, but this will be less accurate when contrasted to a forehead probe. In addition, in patients with poor peripheral circulation and/or connective tissue disease such as systemic scleroderma, it may be difficult to detect a signal.

8. Ensure you place the forehead probe on the temporal artery. This will provide you with an adequate perfusion index.

9. Select an appropriate mask size for the subject. This will ensure minimal gas leakage.

10. Apply petroleum jelly around the mask to aid in forming a tight seal. Place your hand over the mass flow sensor hole and ask the subject to exhale. If there is evident air escape, readjust the mask and/or select a different size mask.

11. Observe the resting gas exchange variables. VCO_2 (red dots) and VO_2 (blue dots) should be steady and be grouped closely together. This indicates minimal/no gas leakage. If gas leakage is apparent, readjust the face mask.

12. Inform the subject that the rpm reading is available for him/-her to view on the bike. In addition, provide verbal feedback on rpm to aid the subject in cycling at the correct speed.

13. It is important to review the results generated by the Vmax program. A skilled technician should determine the AT using the V-slope method.

References

1. Albouaini K, Egred M, Alahmar A, Wright DJ (2007) Cardiopulmonary exercise testing and its application. Postgrad Med J 83 (985):675–682

2. American Thoracic Society; American College of Chest Physicians (2003) ATS/ACCP statement on cardiopulmonary exercise testing. Am J Respir Crit Care Med 167(2):211–277

3. Wasserman K, Hansen JE, Sue DY, Stringer WW, Sietsema KE, Sun X-G et al(2011) Principles of exercise testing and interpretation: including pathophysiology and clinical applications, 5th Revised edn. Lippincott Williams and Wilkins (Philadelphia, PA); . ISBN-10: 1609138996

4. Wasserman K, Whipp BJ (1975) Exercise physiology in health and disease. Am Rev Respir Dis 112(2):219–249

5. Stelken AM, Younis LT, Jennison SH, Douglas Miller D, Miller LW, Shaw LJ et al (1996) Prognostic value of cardiopulmonary exercise testing using percent achieved of predicted peak oxygen uptake for patients with ischemic and dilated cardiomyopathy. J Am Coll Cardiol 27(2):345–352

6. Kodama S, Saito K, Tanaka S, Maki M, Yachi Y, Asumi M et al (2009) Cardiorespiratory fitness as a quantitative predictor of all-cause mortality and cardiovascular events in healthy men and women: a meta-analysis. JAMA 301 (19):2024–2035

7. Balady GJ, Arena R, Sietsema K, Myers J, Coke L, Fletcher GF et al (2010) Clinician's guide to cardiopulmonary exercise testing in adults: a scientific statement from the American heart association. Circulation 122(2):191–225

8. Stickland MK, Butcher SJ, Marciniuk DD, Bhutani M (2012) Assessing exercise limitation using cardiopulmonary exercise testing. Pulm Med 2012:824091. https://doi.org/10.1155/2012/824091

9. Day JR, Rossiter HB, Coats EM, Skasick A, Whipp BJ (2003) The maximally attainable VO2 during exercise in humans: the peak vs. maximum issue. J Appl Physiol 95 (5):1901–1907

10. Jones NL, Makrides L, Hitchcock C, Chypchar T, McCartney N (1985) Normal standards for an incremental progressive cycle ergometer test. Am Rev Respir Dis 131 (5):700–708

11. Hansen JE, Sue DY, Wasserman K (1984) Predicted values for clinical exercise testing. Am Rev Respir Dis 129(2P2):S49–S55. https://doi.org/10.1164/arrd.1984.129.2P2.S49

12. Brubaker PH, Kitzman DW (2011) Chronotropic incompetence. causes, consequences, and management. Circulation 123 (9):1010–1020

13. Vella CA, Robergs RA (2005) A review of the stroke volume response to upright exercise in healthy subjects. Br J Sports Med 39 (4):190–195

14. Smith DL, Fernhall B (2011) Advanced cardiovascular exercise physiology. Human Kinetics

Ltd, Champaign, IL. 1 Feb. 2011. ISBN-10: 0736073922

15. Holverda S, Gan CT, Marcus JT, Postmus PE, Boonstra A, Vonk-Noordegraaf A (2006) Impaired stroke volume response to exercise in pulmonary arterial hypertension. J Am Coll Cardiol 47(8):1732–1733

16. Imai K, Sato H, Hori M, Kusuoka H, Ozaki H, Yokoyama H et al (1994) Vagally mediated heart rate recovery after exercise is accelerated in athletes but blunted in patients with chronic heart failure. J Am Coll Cardiol 24 (6):1529–1535

17. Arai Y, Saul JP, Albrecht P, Hartley LH, Lilly LS, Cohen RJ et al (1989) Modulation of cardiac autonomic activity during and immediately after exercise. Am J Phys 256(1): H132–H141

18. Cole CR, Blackstone EH, Pashkow FJ, Snader CE, Lauer MS (1999) Heart-rate recovery immediately after exercise as a predictor of mortality. N Engl J Med 341(18):1351–1357

19. Watanabe J, Thamilarasan M, Blackstone EH, Thomas JD, Lauer MS (2001) Heart rate recovery immediately after treadmill exercise and left ventricular systolic dysfunction as predictors of mortality. the case of stress echocardiography. Circulation 104(16):1911–1916

20. Vivekananthan DP, Blackstone EH, Pothier CE, Lauer MS (2003) Heart rate recovery after exercise is apredictor of mortality, independent of the angiographic severity of coronary disease. J Am Coll Cardiol 42(5):831–838

21. Robergs RA, Ghiasvand F, Parker D (2004) Biochemistry of exercise-induced metabolic acidosis. Am J Physiol Regul Integr Comp Physiol 287(3):R502–R516. https://doi.org/10.1152/ajpregu.00114.2004

22. Baker JS, McCormick MC, Robergs RA (2010) Interaction among skeletal muscle metabolic energy systems during intense exercise. J Nutr Metab 2010:905612. https://doi.org/10.1155/2010/905612

23. Milani RV, Lavie CJ, Mehra MR, Ventura HO (2006) Understanding the basics of cardiopulmonary exercise testing. Mayo Clin Proc 81 (12):1603–1611

24. Brooks GA (1998) Mammalian fuel utilization during sustained exercise. Mammalian fuel utilization during sustained exercise. Comp Biochem Physiol B Biochem Mol Biol 120 (1):89–107

25. Sun X-G, Hansen JE, Garatachea N, Storer TW, Wasserman K (2002) Ventilatory efficiency during exercise in healthy subjects. Am J Respir Crit Care Med 166 (11):1443–1448

26. Tumminello G, Guazzi M, Lancellotti P, Piérard LA (2007) Exercise ventilation inefficiency in heart failure: pathophysiological and clinical significance. Eur Heart J 28(6):673–678

27. Coats AJS (2005) Why ventilatory inefficiency matters in chronic heart failure. Eur Heart J 26 (5):426–427

28. Voelkel NF, Schranz D (2015) The right ventricle in health and disease (respiratory medicine). Humana, New York. (Springer; New York, NY, USA); 2015 edition (29 Oct. 2014). ASIN: B00S15C7R0.

Chapter 19

Cardiovascular Assessment in Human Research

Marjan Mosalman Haghighi and Julian Ayer

Abstract

A number of noninvasive tests exist for assessing cardiovascular structure and function. This chapter describes protocols for flow-mediated dilation (FMD) of the brachial artery and pulse wave velocity (PWV) and pulse wave analysis (PWA) for measurement of arterial stiffness. The chapter also describes the different methodological approaches involved in applying these techniques for optimizing their validity, comparability, and potential uses as clinical and physiological research tools.

Key words Cardiovascular disease, Arterial stiffness, Flow-mediated dilation, Carotid intima-media thickness, B-mode ultrasonography, Pulse wave velocity

1 Introduction

Noninvasive tests of cardiovascular structure and function have been widely used in biological research. For example, flow-mediated dilation (FMD) of the brachial artery has been used to test endothelial function, and structural changes in the arterial wall may be assessed by measurement of carotid intima-media thickness (CIMT). Arterial stiffness may also be assessed by techniques such as pulse wave velocity (PWV) and pulse wave analysis (PWA). The noninvasive nature of these techniques allows repeated measurements over time, such that the effect of environmental exposures and/or interventions on vascular structure and function may be assessed. However, it is important to be aware of different methodological approaches which limit their validity, comparability, and potential uses as clinical and physiological research tools.

The endothelium is the thin layer of cells that line the interior surface of blood vessels, forming an interface between circulating blood in the lumen and the rest of the vessel wall. One of the functions of the endothelium is to control vasomotor tone, via production of vasodilator and vasoconstrictor substances (nitric oxide is just one of these substances and is a key for vasoactive integrity). Endothelial dysfunction is thought to be an important factor in the development of hypertension, atherosclerosis, heart failure, and other vascular diseases. Endothelial function may be assessed by B-mode ultrasound through FMD. Over the last 30 years, a large body of evidence has linked reduced FMD to risk factors for cardiovascular disease [1]. Furthermore, reduced FMD has been found to be present in patients with incident cardiovascular disease [2]. FMD is stimulated by a sheer stress induced by hyperemic blood flow through medium-sized arteries. This sheer stress provokes the endothelium to release nitric oxide with subsequent vasodilation that can be imaged and quantitated as an index of vasomotor function [3–5]. Changes in FMD may provide important prognostic clinical information in individuals with or at risk of vascular diseases. Despite the fact that FMD is noninvasive, inexpensive and safe to be repeated, it is challenging to measure reliably. Variability is often attributed to sonographer technique, measurement error, and environmental conditions. In addition, variations may be caused by fluctuations in biological parameters such as glucose, insulin, and triglyceride levels. These can affect the accuracy and sensitivity of this test [6].

The arterial wall is composed of three layers termed the intima, media, and adventitia. CIMT is a noninvasive technique that once again uses B-mode ultrasonography to measure the intima-media complex. It has been shown to correlate with the histological intima-medic complex [7]. IMT is increased in subjects with risk factor for cardiovascular disease including hypercholesterolemia and hypertension. Furthermore IMT may undergo progressive reduction with cholesterol-lowering treatment, suggesting that it may be useful in assessment of the early phase of atherosclerosis [8]. CIMT is associated with cardiovascular events such as myocardial infarction or stroke [9–12]. Benefits of IMT assessment of vascular structure include that it is safe, noninvasive, inexpensive, and easy to repeat. Interventions where change in CIMT is a primary outcome may need to last for at least 6 months to detect meaningful change. Protocols for CIMT assessment are variable and may include measurement of external, internal, and/or common carotid arteries but need to be uniform across a particular study. However, reader variation and lack of reproducibility can be reduced by implementation of carotid intima border edge detection software programs [13].

Central arterial blood pressure may be noninvasively estimated from radial and brachial arterial waveforms through the development of transfer functions. These transfer functions and the consequent estimation of central blood pressure have been validated by invasive pressure monitoring in adults. In addition to central blood pressure, other parameters of the central arterial waveform may be estimated by this technique such as augmentation index (AIx = augmentation pressure/pulse pressure × 100%). Higher AIx is independently associated with adverse cardiovascular events and all-cause mortality, after adjusting for other cardiovascular risk factors.

The arterial system is accompanied by structural changes brought about by factors such as aging or risk factors for cardiovascular disease. These structural changes include fragmentation and degeneration of elastase, increases in collagen, thickening of the arterial wall, and progressive dilation of the arteries. These changes result in a gradual stiffening of the vasculature and an increase in velocity of the pressure waveform as it travels from central to peripheral arteries. In a normal elastic aorta, the pressure wave reflects from the periphery and returns to the heart during diastole. This reflected wave helps augment pressure during diastole. As the aorta stiffens, the velocity of the pressure wave increases, and the reflected pressure wave eventually reaches the heart at systole instead of diastole, causing augmentation of the systolic blood pressure (SBP) and increased cardiac afterload. Arterial stiffness is a main contributor to cardiovascular diseases, and it is important to diagnose this early [14–16]. PWV is the difference in transit time of the pulse waveform between two points in the arterial system divided by the distance between those two points [15]. For PWV the points of measurement of the waveforms are often the radial, carotid, femoral, or dorsalis pedis arteries. The two pulse waveforms may be measured sequentially with electrocardiogram gating or contemporaneously.

This chapter describes detailed protocols for and FMD, CIMT PWV, and PWA analyses.

2   Materials

2.1   FMD

1. Ultrasound machine, rapid pressure cuff, air-source machine, arm border, ECG electrodes, gel and brachial artery analyzer (Medical Imaging Applications LLC; Coralville, IA, USA).

2.2   CIMT

1. Ultrasound machine, ECG electrodes, gel and carotid artery analyzer (Medical Imaging Applications LLC; Coralville, IA, USA).

2.3   PWV and PWA

1. SphygmoCor XCEL (AtCor Medical Pty Ltd.; West Ryde, Australia), blood pressure cuffs (brachial and thigh cuffs), tonometer, and tape measurement.

3   Methods

3.1   FMD

1. Ensure that subjects are fasted and have avoided exercise, caffeine, drugs, alcohol, stimulants, and medications for at least 6 h (*see* **Note 1**).

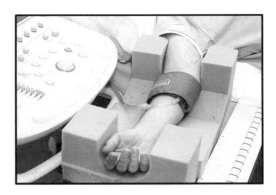

Fig. 1 FMD procedure: position of blood pressure cuff around the forearm immediately below the elbow

2. Conduct the test in a quiet and dark room with a controlled temperature (21 °C) after the subject has been resting quietly.

3. Explain the procedure to subjects and ensure that they are happy to continue with the test.

4. Ensure that subjects remain in the supine position for approximately 10 min before testing.

5. Give a pillow to the subjects to ensure their comfort with the shoulders off the pillow.

6. Place three ECG lead electrodes: one below the left clavicle, one below the right clavicle, and one approximately 5 cm left of the umbilicus.

7. Position the subject's right arm comfortably at 90° to the body on an arm board or a small table.

8. Place a blood pressure cuff around the forearm immediately below the elbow (1–2 cm distal to the medial epicondyle) (Fig. 1).

9. With a vascular ultrasound probe, locate a straight segment of the brachial artery (Fig. 2) (*see* **Note 2**).

10. Perform correction movements as necessary such as a slight increase of tilting angle or fine movements in the medial-lateral or proximal-distal axis to obtain the best image (*see* **Note 3**).

11. When a satisfactory image is obtained, secure the ultrasound probe in position using a probe holder, if available, or by holding the probe in place over the vessel (*see* **Note 4**).

12. Make a Doppler assessment of blood flow using pulse wave Doppler with angle correction to align the measurement in the direction of blood flow.

13. Baseline diameter must be examined before cuff inflation for a period of at least 1 min, or continuous capture of the data may be undertaken from baseline to test end (*see* **Note 5**).

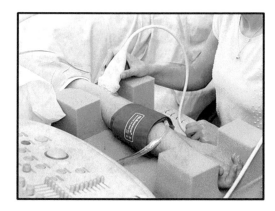

Fig. 2 FMD procedure: positioning of ultrasound probe over a straight segment of the brachial artery

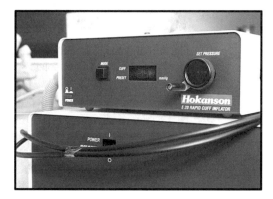

Fig. 3 FMD procedure: inflation of forearm blood pressure cuff to induce ischemia

14. Rapidly inflate the forearm blood pressure cuff to induce fore-arm ischemia, with an automated cuff deflator with the cuff inflated to 200 mm Hg for children and 250 mm Hg for adults (Fig. 3) (*see* **Note 6**).

15. Inflate the cuff for 5 min and then note the time of deflation.

16. Immediately after cuff deflation, assess the PW Doppler measurement of brachial blood flow to determine the peak Doppler velocity (the peak Doppler flow pattern should be stored).

17. Capture the brachial artery diameter in a location that is similar to the baseline assessment for a period of at least 2 min (*see* **Note 7**).

18. Perform FMD analysis using the brachial artery analyzer using a region of interest (ROI) between two points on the near and far wall of the brachial artery at baseline and then post-cuff deflation (Fig. 4) (*see* **Note 8**).

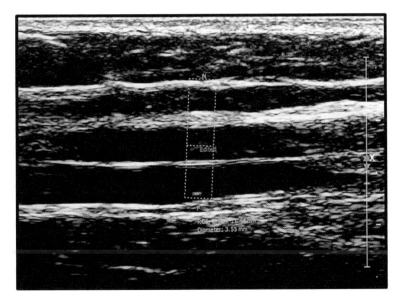

Fig. 4 FMD procedure: analysis using the brachial artery analyzer of a region of interest (ROI) between two points on the near and far wall of the brachial artery

19. Record baseline and peak brachial artery diameter for FMD calculation using the formula: $FMD = (peak\,diameter - baseline\,diameter)/baseline\,diameter$ (*see* **Note 9**).

3.2 CIMT

1. Allow subjects to rest in a supine position for at least 10 min (do not use a pillow for this test).
2. Explain the procedure to the subjects (*see* **Note 10**).
3. Place three ECG lead electrodes as above.
4. Using the ultrasound probe, locate the carotid artery just medial to the body of the sternocleidomastoid muscle with the neck turned slightly away from the side being imaged (Fig. 5).
5. Optimize images of the common carotid artery for depth, and gain with a focus on the far artery wall without image zoom (Fig. 6) (*see* **Note 11**).
6. Obtain a straight arterial segment of at least 10 mm proximal to the carotid bulb, including at least three loops of both right and left common carotid arteries in the protocol (Fig. 7) (*see* **Note 12**).
7. Measurement of far wall IMT at least 5 mm proximal to the carotid bulb should be performed using dedicated software.
8. Record end diastolic mean, maximum, and minimum IMT.

3.3 PWV and PWA

1. After the subject has rested in the supine position for 10–15 min, use a standard brachial cuff (e.g., SphygmoCor

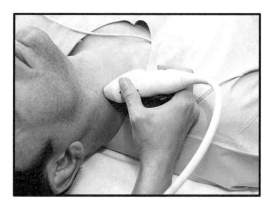

Fig. 5 CIMT procedure: positioning of ultrasound probe over the carotid artery just medial to the body of the sternocleidomastoid muscle

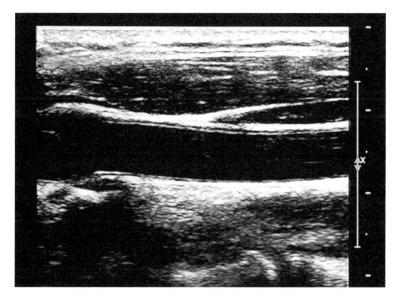

Fig. 6 CIMT procedure: image of the common carotid artery to optimize for depth and gain with a focus on the far artery wall

    XCEL) to measure brachial systolic and diastolic pressures, and capture a brachial waveform (Fig. 8) (*see* **Note 13**).

2. Analyze the brachial waveform using the SphygmoCor XCEL to provide a central aortic waveform (Fig. 9).

3. For measurement of PWV by carotid tonometer and femoral cuff, a thigh cuff should be placed around subject's upper thigh (Fig. 10) (*see* **Note 14**).

4. Measure the distance from the carotid to the top of thigh cuff and between top of cuff and the femoral artery (measurement

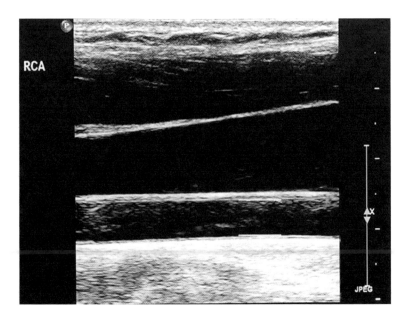

Fig. 7 CIMT procedure: image of a straight arterial segment of at least 10 mm proximal to the carotid bulb

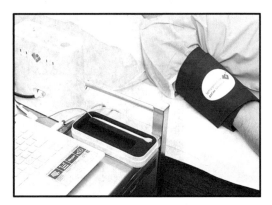

Fig. 8 PWV and PWA procedures: SphygmoCor XCEL brachial cuff used to measure systolic and diastolic pressures

preferably is performed on the right femoral artery) (Fig. 11) (*see* **Note 15**).

5. Place a tonometer over the right carotid artery (a sensor in the tonometer and the cuff will detect the pressure waveforms) (Fig. 12).

6. Data capture should be undertaken with good-quality waveforms with clear evidence of the upstroke of the pressure waves, obtaining at least three measurements (automated commercial software quality control may be available) (Fig. 13) (*see* **Note 16**).

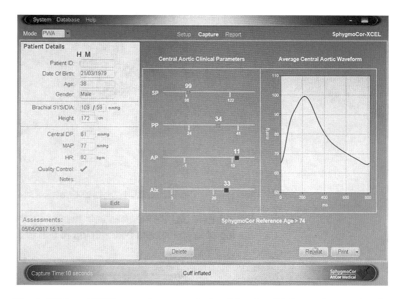

Fig. 9 PWV and PWA procedures: analysis of the brachial waveform using the SphygmoCor XCEL to provide a central aortic waveform

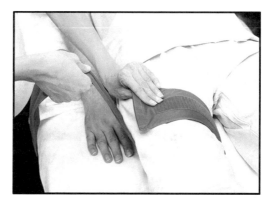

Fig. 10 PWV and PWA procedures: positioning of thigh cuff around the subject's upper thigh for measurement of PWV by carotid tonometer and femoral cuff

4   Notes

1. FMD can be acutely influenced by food intake, exercise, smoking, drinking caffeine, alcohol, and medications. Thus, it is recommended that when assessing FMD, subjects are fasted and have avoided exercise, caffeine, drugs, alcohol, stimulants, and medications for at least 6 h. In the case where medications cannot be avoided, tests should be conducted after a standardized time unless this is contraindicated by the subject's

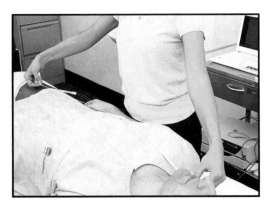

Fig. 11 PWV and PWA procedures: measuring the distance from the carotid to the top of thigh cuff and between top of cuff and the femoral artery

Fig. 12 PWV and PWA procedures: placement of tonometer over the right carotid artery

medical condition. Premenopausal women should be assessed in a standardized phase of the menstrual cycle (ideally days 1–7).

2. To improve the image quality, identify a straight non-branching segment with equal contrast on anterior and posterior walls. FMD may be studied in radial, axillary, and superficial femoral arteries. Measuring arteries with diameter smaller than 2.5 mm is difficult. Conversely, vasodilation is generally less difficult to perceive in vessels larger than 5.0 mm in diameter.

3. The arterial segment needs to be horizontally positioned and not at an angle for proper analysis. To confirm that the identified vessel is an artery and not a vein, color Doppler and pulse wave (PW) Doppler may be used.

4. A probe holder can be used especially in continuous captures. The probe holder positioning device allows the ultrasound

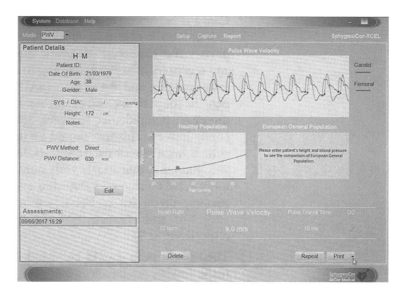

Fig. 13 PWV and PWA procedures: waveforms with clear evidence of the upstroke of the pressure waves

probe to be stabilized on a subjects arm, so that the brachial artery can be analyzed longitudinally. Control of the probe during scanning and acquisition is maintained using the fine adjustment screw.

5. The FMD response is characteristically presented as a change in arterial diameter from baseline diameter.

6. The time of cuff inflation should be noted. Subjects should be warned that they may feel a slight discomfort or tingling sensation during cuff inflation or even after deflation. Typically, the cuff is inflated to at least 50 mm Hg above systolic pressure to occlude arterial inflow for a standardized length of time. This causes ischemia and consequent dilation of downstream resistance vessels via autoregulatory mechanisms [17]. It has been shown that the duration of ischemia is an important determinant of the mechanisms responsible for the subsequent vasodilator responses [18, 19].

7. If baseline data is subsequently not analyzable, then the last 30 s of post-cuff deflation diameter may be used as the baseline data.

8. The intima-media or media-adventitia boundaries may be used as long as the positioning is consistent between baseline and post-deflation. Auto-contouring is possible with the software but manual adjustment may be required.

9. With continuous data capture, the area under the diameter-time curve may provide additional information about flow-mediated dilatation.

10. CIMT is becoming more important because it is a noninvasive tool for early detection in children and young adults with obesity and metabolic syndrome due to inactivity, poor diet, and sedentary lifestyle. According to the Society of Atherosclerosis Imaging and Prevention (SAIP) guideline, this test has benefits for patients with prediabetes, diabetes, metabolic syndrome, hyperlipidemia, smokers, age > 40 years, and having family history. This test should be repeated annually for individuals with abnormal results (CIMT >1 mm) and every 2–5 years for individuals with normal results. In women, a mean CIMT value of 1.069 mm and 1.153 mm in men was strongly predictive of CAD (sensitivity $= 79.1\%$, specificity $=89.7\%$) and (sensitivity $=66.4\%$, specificity $=74.2\%$), respectively [13].

11. The carotid bulb should be included in the image as a landmark. Images can be selected at the optimal angle of interrogation, an anatomic landmark, and at 60, 90, 120, 150, and 180° marked for the right side and 300, 270, 240, 210, and 180° for the left side in adult research. According to standardized protocols, right and left carotid arteries should be scanned.

12. Protocols may also include imaging of the external carotid artery and with varying angles.

13. PWV is associated with CHD, stroke, and CV mortality. Independent of age, gender, race, and SBP, individuals with aortic PWV values >641 cm/s for men and 627 cm/s for women (the 25th percentile of the aortic PWV distribution) have a more than twofold increase in the risk of CVD, a two- to threefold increase in stroke, and a more than 50% increase in CHD events compared with those with values below this level [20].

14. The cuff may be applied through light clothing. High-fidelity tonometers have been introduced for accurate, noninvasive measurement of arterial pulse contour, and there is now a better understanding of arterial hemodynamic and appreciation of disease and aging effects in humans. It is now possible to record the pulse wave accurately in the radial artery [21] (Fig. 14), to synthesize the ascending aortic pulse waveform, to identify systolic and diastolic periods, and to generate indices of ventricular-vascular interaction previously only possible with invasive arterial catheterization. After 10–15 min of rest in the spine position, a tonometer should be held on the radial artery and captured symmetrical waves for 10 s (one full page). The radial waveform could be analyzed by mathematical techniques by the SphygmoCor, CVMS software (V9) to provide a central aortic waveform.

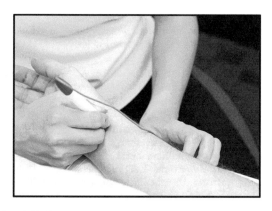

Fig. 14 PWV and PWA procedures: measuring PWA by brachial artery cuff

15. There is another method to measure PWV which is called brachial-ankle PWV (baPWV) which assesses stiffness of peripheral arteries. It is a simpler method than other noninvasive autonomic devices to measure with using pressure cuffs wrapped on the brachium and ankle. The technical simplicity and short sampling time of this method make it more feasible for screening a large population than other methods. However, up to the year 2002 [22], there was not sufficient evidence about validity and reproducibility of baPWV. PWV can be measured in radial, carotid, femoral, or dorsalis pedis arteries. The two pulse waveforms may be measured sequentially with ECG gating or contemporaneously.

16. Most softwares can perform automated calculations of PWV.

References

1. Celermajer DS, Sorensen KE, Gooch VM, Spiegelhalter DJ, Miller OI, Sullivan ID et al (1992) Non-invasive detection of endothelial dysfunction in children and adults at risk of atherosclerosis. Lancet 340(8828):1111–1115

2. Cox DA, Vita JA, Treasure CB, Fish RD, Alexander RW, Ganz P et al (1989) Atherosclerosis impairs flow-mediated dilation of coronary arteries in humans. Circulation 80:458–465

3. Yeboah J, Crouse JR, Hsu FC, Burke GL, Herrington DM (2007) Brachial flow-mediated dilation predicts incident cardiovascular events in older adults. Circulation 115:2390–2397

4. Thijssen DH, Black MA, Pyke KE, Padilla J, Atkinson G, Harris RA et al (2011) Assessment of flow-mediated dilation in humans: a methodological and physiological guideline. Am J Physiol Heart Circ Physiol 300:H2–H12. https://doi.org/10.1152/ajpheart.00471. 2010

5. Inaba Y, Chen JA, Bergmann SR (2012) Carotid plaque, compared with carotid intima-media thickness, more accurately predicts coronary artery disease events: a meta-analysis. Atherosclerosis 220:128–133

6. West S et al (2004) Biological correlates of day-to-day variation in flow-mediated dilation in individuals with type 2 diabetes: a study of test–retest reliability. Diabetologia 47:1625–1631

7. Den Ruijter HM, Peters SA, Anderson TJ, Britton AR, Dekker JM, Eijkemans MJ et al (2012) Common carotid intima-media thickness measurements in cardiovascular risk prediction: a meta-analysis. JAMA 308:796–803

8. Sorof JM, Alexandrov AV, Cardwell G, Portman RJ (2003) Carotid artery intimal-medial thickness and left ventricular hypertrophy in children with elevated blood pressure. Pediatrics 111:61–66

9. Salonen JT, Salonen R (1991) Ultrasonographically assessed carotid morphology and the risk of coronary heart disease. Arterioscler Thromb 1:1245–1249

10. Bots ML, Hoes AW, Koudstaal PJ, Hofman A, Grobbee DE (1997) Common carotid intima-media thickness and risk of stroke and myocardial infarction. Circulation 96:1432–1437

11. Chambless LE, Heiss G, Folsom AR, Rosamond W, Szklo M, Sharrett AR et al (1997) Association of coronary heart disease incidence with carotid arterial wall thickness and major risk factors: the atherosclerosis risk in communities (ARIC) study, 1987–1993. Am J Epidemiol 146:483–494

12. O'leary DH, Polak JF, Kronmal RA, Manolio TA, Burke GL, Jr Wolfson SK (1999) Carotid-artery intima and media thickness as a risk factor for myocardial infarction and stroke in older adults. N Engl J Med 340:14–22

13. Cobble M, Bale B (2010) Carotid intima-media thickness: knowledge and application to everyday practice. Postgrad Med 122:10–18

14. Lakatta EG (2003) Arterial and cardiac aging: major shareholders in cardiovascular disease enterprises. Circulation 107:490–497

15. Blacher J et al (1999) Aortic pulse wave velocity as a marker of cardiovascular risk in hypertensive patients. Hypertension 33:1111–1117

16. Sugawara J, Hayashi K, Yokoi T, Cortez-Cooper MY, DeVan AE, Anton MA et al (2005) Brachial–ankle pulse wave velocity: an index of central arterial stiffness? J Hum Hypertens 19:401–406

17. Corretti MC, Anderson TJ, Benjamin EJ, Celermajer D, Charbonneau F, Creager MA et al (2002) Guidelines for the ultrasound assessment of endothelial-dependent flow-mediated vasodilation of the brachial artery: a report of the International Brachial Artery Reactivity Task Force. J Am Coll Cardiol 39:257–265

18. Joannides R, Haefeli WE, Linder L, Richard V, Bakkali EH, Thuillez C et al (1995) Nitric oxide is responsible for flow-dependent dilatation of human peripheral conduit arteries in vivo. Circulation 91:1314–1319

19. Mullen MJ, Kharbanda RK, Cross J, Donald AE, Taylor M, Vallance P et al (2001) Heterogenous nature of flow-mediated dilatation in human conduit arteries in vivo. Circ Res 88:145–151

20. Sutton-Tyrrell K, Najjar SS, Boudreau RM, Venkitachalam L, Kupelian V, Simonsick EM et al (2005) Elevated aortic pulse wave velocity, a marker of arterial stiffness, predicts cardiovascular events in well-functioning older adults. Circulation 111:3384–3390

21. Wilkinson IB, Cockcroft JR, Webb DJ (1997) Pulse wave analysis and arterial stiffness. J Cardiovasc Pharmacol 32:S33–S37

22. Yamashina A, Tomiyama H, Takeda K, Tsuda H, Arai T, Hirose K et al (2002) Validity, reproducibility, and clinical significance of non-invasive brachial-ankle pulse wave velocity measurement. Hypertens Res 25:359–364

Chapter 20

Nutritional Programming Effects on the Immune System

Donald B. Palmer

Abstract

The relationship between patterns of early growth and age-associated diseases such as type 2 diabetes and cardiovascular disease is well established. There is also strong evidence from both human and animal studies that early environmental factors such as maternal nutrition may influence lifespan. Interestingly, more recent studies have demonstrated that nutritional programming in early life effects immunity, such that altered lifespan can also lead to programmed changes in immune function. Here we describe the use of immunohistology and flow cytometry techniques to study two key immune lymphoid organs: one that is involved in developing immune cells (thymus) and another which is the site of immune activation (spleen).

Key words Thymus, Spleen, Antibodies, Flow cytometry, Immunoperoxidase, Immunofluorescence

1 Introduction

It is now emerging that adverse conditions experienced in early life can alter the susceptibility to diseases in adulthood. This concept, which was originally put forward by Barker and colleagues, is now termed "the developmental origins of health and disease hypothesis" [1]. It proposes that environmental factors, such as nutrition, during gestation and early life can program permanent structural, physiological, and metabolic changes due to developmental adaptation, which consequently predisposes individuals to cardiovascular, metabolic, and endocrine diseases if exposed to a different environment postnatally [2, 3]. This implies that "events" which occur during critical periods of developmental plasticity can permanently alter the physiological and metabolic phenotype of an organism.

Animal models have contributed substantially to our knowledge and understanding of developmental programming and have provided direct evidence that early growth patterns can have profound long-term effects on later health [4]. One such animal model has shown that alteration of nutrition during fetal and early postnatal life, as well as affecting long-term metabolism of the offspring,

Paul C. Guest (ed.), *Investigations of Early Nutrition Effects on Long-Term Health: Methods and Applications*, Methods in Molecular Biology, vol. 1735, https://doi.org/10.1007/978-1-4939-7614-0_20, © Springer Science+Business Media, LLC 2018

also influences lifespan in rodents [5, 6]. Mice (and rats) exposed to a maternal low-protein diet during pregnancy are born small but when cross-fostered after birth to control-fed dams undergo rapid "catch-up" growth and have a substantially reduced lifespan (termed recuperated offspring), compared to control offspring when animals are weaned onto a standard diet and fed ad libitum. In contrast, offspring born to control-fed animals but nursed by low-protein-fed dams (termed postnatal low-protein offspring, PLP) grow slowly and have an increased longevity.

A functional immune system is considered a vital necessity for the host's continual survival against the daily onslaught of foreign organisms and pathogens. The development of the immune system involves a complex series of events that occurs during fetal and postnatal periods of mammalian growth. Genetic disorders that result in the disruption of this process can lead to the host becoming immunocompromised [7]. As the fetal and early postnatal period is such a critical time for development of a functional immune system [7], it may be vulnerable to the effects of a suboptimal environment during this period. Indeed, human studies have suggested that the early nutritional environment and growth patterns can have an impact on long-term immune function [8, 9].

In humans as well as in many other species, it is generally recognized that the immune system undergoes age-associated alterations, termed immunosenescence, which culminate in a progressive deterioration in its ability to respond to infection and vaccination [10]. Changes in cellular-mediated immunity have been attributed as the main cause of immunosenescence. In particular, this can involve the decline in naïve T-cell output which is intimately linked to regression of the thymus, the sole organ responsible for the generation of T cells [10]. Indeed, age-associated thymic involution is considered one of the most universally recognized changes of the ageing immune system [11].

With the demonstration that maternal diet can influence lifespan, this observation implies that the various pathways controlling cellular senescence and ageing are responsive to environmental signals during development. It therefore seems plausible that immunosenescence can also be influenced by early life events. Indeed, histological and flow cytometric examination of two key lymphoid organs thymus (primary) and spleen (secondary) in nutritionally induced short-lived and long-lived mice reveal alteration of these organs implying that immunosenescence can also be influenced by early life events [12, 13].

This chapter describes the histological and flow cytometric protocols needed to examine the thymus and spleen from nutritionally induced short-lived and long-lived mice. Specifically, using a low-protein model of nutritionally induced fetal growth restriction followed by postnatal catch-up growth (recuperated mice) which results in reduced lifespan, and nutritionally induced early postnatal growth restriction, resulting in extend lifespan (postnatal low protein mice) [8, 9]. We and others have described age-associated changes to the thymus and spleen using such techniques which represent important tools in deciphering immune function [14–16].

2   Materials (*See* Note 1)

2.1   Preparing Tissue for Cryosectioning for Immunohistology

1. Dissecting kit.
2. Liquid nitrogen.
3. Small flask.
4. Optimal cutting temperature (OCT) embedding medium.
5. Cryo molds.
6. Forceps.
7. Freezer −80 °C.
8. Multi-spot polytetrafluoroethylene (PTFE)-coated microscope slides.
9. Cryostat microtome.
10. Acetone.

2.2   Preparing Cells from Tissue for Flow Cytometric Analysis

1. Petri dish.
2. Phosphate buffered saline (PBS).
3. 0.85% ammonium chloride.
4. 21 G needles.
5. 20 mL-capacity universal tubes.
6. Centrifuge.
7. Multicolor flow cytometry.
8. Neubauer chamber.
9. Trypan blue solution.
10. Nylon cell strainer.

2.3   Immunohistology (Peroxidase/ Fluorescence)

1. Primary antibody.
2. Secondary conjugated antibody [horse radish peroxidase (HRP)/fluorochrome].
3. Mouse serum.
4. Staining rack to lay slides during staining procedure.
5. 3,3′-diaminobenzidine (DAB) peroxidase (HRP) substrate kit.
6. Mountant.
7. Coverslip.
8. Glass Coplin staining jar.
9. Meyer's hematoxylin solution.

2.4 Flow Cytometric Staining

1. Fluorescence conjugated primary antibodies.
2. Phosphate buffer solution (cold).
3. 4 mL-capacity (FACS) tubes.
4. Ice bucket.
5. 1% cold paraformaldehyde.
6. 0.85% ammonium chloride.
7. Centrifuge.
8. Neubauer chamber.
9. Multicolor flow cytometer.

3 Methods

3.1 Slide Preparation

1. Remove tissue and mount in OCT embedding medium in a cryogenic mold (*see* **Note 2**).
2. Snap-freeze in liquid nitrogen and then store molds at −80 °C.
3. Prepare serial 6–7 μm sections of tissue using a cryostat microtome.
4. Place each section onto multi-spot PTFE-coated microscope slides and allow to air-dry overnight at room temperature (RT).
5. Fix slides by immersion in 100% acetone for 10 min (*see* **Note 3**).
6. Allow slides to dry and then cover with foil and store at −20 °C (*see* **Note 4**).

3.2 Immunohistology (Immunoperoxidase)

1. Remove slide(s) from freezer (keeping wrapped in foil) and allow to reach RT (5–10 min) (*see* **Note 5**).
2. Label slides appropriately and add inoculate 100 μL diluted antibody in PBS and incubate for 1 h at RT (*see* **Note 6**).
3. Wash 2× using PBS in the staining jar at 5 min intervals.
4. Wipe excess PBS, being careful not to damage tissue section, and quickly add 100 μL of diluted secondary reagent in PBS (*see* **Note 7**).
5. Incubate secondary reagent for 1 h at RT in the dark.
6. Wash 2× using PBS in the staining jar at 5 min intervals.
7. Add DAB substrate (around 300 μL per section enough to cover the tissue) and leave until brown reaction appears (30 s to 1 min).
8. Wash 1× using PBS in the staining jar.
9. Add counterstain using Meyer's hematoxylin solution and leave for 30 s to 1 min.

10. Wash in the staining jar in running tap water for 2 min.

11. Wipe excess water and mount sections with a mountant and coverslip.

12. View under a microscope the following day.

3.3 Immunohistology (Immunofluorescence)

1. Remove slide(s) from freezer (keeping wrapped in foil) and allow to reach RT (*see* **Note 8**).

2. Add 100 μL diluted antibody in PBS and leave for 1 h at RT in the dark (*see* **Note 9**).

3. Wash 2× using PBS in the staining jar at 5 min intervals.

4. If directly conjugated primary antibody was used, wipe excess PBS and mount sections with a mountant and coverslip.

5. View under a fluorescence microscope the following day (*see* **Note 10**).

6. If primary antibody is unconjugated or biotinylated, wipe excess PBS, being careful not to damage tissue section, and quickly add 100 μL diluted secondary reagent in PBS (*see* **Note 7**).

7. Incubate secondary reagent for 1 h at RT in the dark.

8. Wash 2× using PBS in the staining jar at 5 min intervals.

9. After washing in PBS, wipe excess PBS from slide and mount sections with a mountant and coverslip.

10. View under a fluorescence microscope the following day (*see* **Notes 10** and **11**).

11. Add a total volume of 100 μL of the two antibodies diluted in PBS and leave for 1 h at RT in the dark.

12. Wash 2× using PBS in the staining jar at 5 min intervals.

13. Wipe excess PBS, being careful not to damage tissue section, and quickly add 100 μL diluted secondary reagent in PBS (*see* **Note 12**).

14. Incubate secondary reagent for 1 h at RT in the dark.

15. After washing in PBS, wipe excess PBS from slide and mount sections with a mountant and coverslip.

16. View under a fluorescence microscope the following day (*see* **Note 10**).

3.4 Flow Cytometry: Thymus

1. Remove thymus and place in small (3 cm) Petri dish in 2 mL PBS.

2. Tease thymus using a fine needle (*see* **Note 13**).

3. Recover thymic suspension, make up to 10 mL PBS, and centrifuge for 6 min at $1600 \times g$.

4. Resuspend in 20 mL PBS and determine cell count using a Neubauer chamber and trypan blue exclusion.

5. Use 1×10^6 cells per FACS staining.

3.5 Flow Cytometry: Spleen

1. Remove spleen and place in small Petri dish (3 cm) in 2 mL PBS.

2. Tease spleen using a needle (may need to filter through a nylon cell strainer) (*see* **Note 14**).

3. Recover splenic suspension, make up to 10 mL PBS, and centrifuge as above.

4. Resuspend in 5 mL of 0.85% ammonium chloride (removes red blood cells) for 5 min at RT (*see* **Note 15**).

5. Add 15 mL PBS and centrifuge for 6 min as above.

6. Resuspend in 20 mL PBS and determine cell count using a Neubauer chamber and trypan blue exclusion.

7. Use 1×10^6 cells per FACS staining.

3.6 Staining Flow Cytometry of Antibody Surface-Stained Cells

1. Add 1×10^6 cells per stain per FACS tube.

2. Add PBS to each FAC tube to bring to an appropriate volume and centrifuge for 6 min as above.

3. Remove supernatant and resuspend in 100 μL anti-mouse labeled-conjugated antibody diluted in PBS (*see* **Note 16**).

4. Incubate for 1 h on ice in the dark.

5. Wash cells $2\times$ using 2 mL cold PBS via centrifugation for 6 min as above.

6. Resuspend cells in 0.5 mL containing 1% cold paraformaldehyde for 15 min, in the dark.

7. Wash once in cold PBS and resuspend in 0.4 mL cold PBS and store covered in foil at 4 °C (*see* **Note 17**).

8. Analyze cells in a multicolor flow cytometer.

4   Notes

1. Nutritionally induced short-lived and long-lived mice are housed in appropriate facilities. Mice are scarified using schedule 1 method and tissues removed and used accordingly.

2. Freeze tissue as soon as possible. The thymus is a two-lobed tissue, so it is best to snap-freeze each lobe separately. The mouse spleen is approximately 1 cm in length and again can be cut into 2–3 pieces and snap-frozen separately.

3. In most instances we use fresh acetone rather than recycled acetone as fixation efficacy is reduced in the latter. A large

number of antibodies that we and others have used in immu-
nohistological studies are able to stain frozen sections. However, it is worthwhile checking the application notes of any
antibodies that you have purchased. Additionally, you may
wish to freeze un-fixed slides that can be used in other studies.
For instance, the identification of senescent cells by measuring
senescence-associated β-galactosidase activity requires the use
of unfixed sections (*see* ref. 14).

4. Slides are wrapped in two in foil. For long-term storage (i.e.,
>1 year), store at −80 °C.

5. Staining is performed in humid conditions. We normally do
staining in a slide box with a base containing moist tissue to
maintain a humid environment.

6. The nature of the primary antibody can be in various forms:
monoclonal (normally rat anti-mouse antigens), polyclonal
(often rabbit/goat/sheep anti-mouse antigens), unconjugated, or biotinylated (mostly monoclonal antibodies). This
need would dictate the type of secondary reagent that would
be used to detect binding activity. These would include HRP
conjugated antibodies which recognized the primary antibody
or streptavidin conjugated to HRP in order to detect biotinylated antibodies. We use primary antibodies at a range from
1:20 to 1:100, but the user should check manufacturer's
recommended usage.

7. When adding the secondary reagent, this needs to be carried
out quickly as tissue will dry out. We use the secondary reagent
at a concentration of 1:100–200 but check manufacturer's
recommended usage. Be aware that some secondary reagent
may show some background staining. This is often removed by
adding mouse serum at 2.5% with the PBS diluted secondary
reagent.

8. Staining is performed in humid and dark conditions. We normally do staining in a slide box with a base containing moist
tissue to maintain a humid environment. We also cover the
glass Coplin staining jar in foil.

9. In some instances, the primary antibody used in immunofluorescence is directly labeled to a conjugate fluorochrome [e.g.,
fluorescein isothiocyanate (FITC)] or alternatively it is either
unconjugated or biotinylated. In this regard you would use an
appropriately conjugated fluorochrome secondary reagent.

10. It is essential that you view immunofluorescence within 24 h as
the intensity of the fluorochrome will decrease over time.
Examining the slides immediately after the staining may
prove difficult due to Brownian motion of the mountant. Slides
should be stored in a dark humid environment in a sealed box
at 4 °C.

11. This method is for single immunofluorescence staining. In some instance double immunofluorescence staining may be required, for example, to simultaneously examine the appearance of T and B cells in the spleen (*see* Ref. 16). In this procedure one of the antibodies would be directly conjugated to a fluorochrome (e.g., FITC); the second antibody would preferably be biotinylated.

12. The secondary reagent should be conjugated to a fluorochrome [e.g., tetramethylrhodamine (TRITC)] which is different to the fluorochrome conjugated to the primary antibody.

13. When teasing the thymus, do this quickly, but also be careful as they can die by apoptosis. If you do see >15% cell death, you may wish to tease in 2% fetal calf serum/PBS.

14. It might be difficult to make single splenic cell suspension, so this may require filtration through a nylon cell strainer.

15. Try not to keep incubation of splenocytes in 0.85% ammonium chloride to less than 5 min, as longer incubation times can lead to lysis of lymphocytes.

16. In general, flow cytometric analysis of thymocytes and splenocytes often uses three to four multicolor analysis, involving anti-mouse antibody conjugated to different fluorochromes [FITC, R-phycoerythrin, allophycocyanin (APC)] being used simultaneously (*see* ref. 15). When performing such analyses, the user should prepare single staining of the individual fluorochrome in order to run compensation mechanisms due to fluorescence emission spectra overlap.

17. Cells can be analyzed almost immediately, but preferably within 2–3 days.

References

1. Barker DJ (2004) The developmental origins of chronic adult disease. Acta Paediatr 93:26–23
2. Hanson MA, Gluckman PD (2008) Developmental origins of health and disease: new insights. Basic Clin Pharmacol Toxicol 102:90–93
3. Singhel A (2006) Early nutrition and long-term cardiovascular health. Nutr Rev 64:544–549
4. Cottrell EC, Ozanne SE (2004) Early life programming of obesity and metabolic disease. Physiol Behav 94:17–28
5. Ozanne SE, Hales CN (2004) Lifespan: catch-up growth and obesity in male mice. Nature 427:411–412
6. Ozanne SE, Hales CN (2005) Poor fetal growth followed by rapid postnatal catch-up growth leads to premature death. Mech Aging Dev 126:852–854
7. Fischer A, Le Deist F, Hacein-Bey-Abina S, Andre-Schmutz I, BasileGde S, de Villartay JP et al (2005) Severe combined immunodeficiency. A model disease for molecular immunology and therapy. Immunol Rev 203:98–109
8. McDade TW, Beck MA, Kuzawa CW, Adair LS (2001) Prenatal undernutrition and postnatal growth are associated with adolescent thymic function. J Nutr 131:1225–1231
9. Raqib R, Alam DS, Sarker P, Ahmad SM, Ara G, Yunus M et al (2007) Low birth weight is associated with altered immune function in rural Bangladeshi children: a birth cohort study. Am J Clin Nutr 85:845–852

10. Aw D, Silva AB, Palmer DB (2007) Immunosenescence: emerging challenges for an ageing population. Immunology 120:435–446

11. Shanley DP, Aw D, Manley NR, Palmer DB (2009) An evolutionary perspective on the mechanisms of immunosenescence. Trends Immunol 30:374–381

12. Chen JH, Tarry-Adkins JL, Heppolette CA, Palmer DB, Ozanne SE (2009) Early-life nutrition influences thymic growth in male mice that may be related to the regulation of longevity. Clin Sci (Lond) 118:429–438

13. Heppolette CA, Chen JH, Carr SK, Palmer DB, Ozanne SE (2016) The effects of aging and maternal protein restriction during lactation on thymic involution and peripheral immunosenescence in adult mice. Oncotarget 7:6398–6409

14. Aw D, Silva AB, Maddick M, von Zglinicki T, Palmer DB (2008) Architectural changes in the thymus of aging mice. Aging Cell 7:158–167

15. Aw D, Silva AB, Palmer DB (2010) The effect of age on the phenotype and function of developing thymocytes. J Comp Pathol 142(Suppl 1):S45–S59

16. Aw D, Hilliard L, Nishikawa Y, Cadman ET, Lawrence RA, Palmer DB (2016) Disorganization of the splenic microanatomy in ageing mice. Immunology 148:92–101

Chapter 21

Small RNA Sequencing: A Technique for miRNA Profiling

Lucas Carminatti Pantaleão and Susan E. Ozanne

Abstract

Identifying microRNA (miRNA) signatures in animal tissues is an essential first step in studies assessing post-transcriptional regulation of gene expression in health or disease. Small RNA sequencing (sRNA-Seq) is a next-generation sequencing-based technology that is currently considered the most powerful and versatile tool for miRNA profiling. Here, we describe a sRNA-Seq protocol including RNA purification from mammalian tissues, library preparation, and raw data analysis.

Key words sRNA-Seq, MicroRNAs, Programming, Transcriptional changes, Epigenetics

1 Introduction

Parental programming of offspring health involves a substantial number of developmental and complex epigenetic changes that ultimately leads to permanent transcriptomic alterations [1, 2]. High-throughput gene expression studies during the last decade have identified programmed changes in mRNA expression as a consequence of changes in maternal behavior, maternal diet, or the maternal environment which correlate with offspring phenotype [3–5]. However, only more recently has the importance of small noncoding RNAs such as miRNAs become apparent in a programming context. Initial studies demonstrated effects of the maternal environment on offspring miRNA levels [6–8]; however, it is now clear that the miRNA signature in sperm of male mammals can contribute to programming through the paternal lineage [9].

MiRNAs are short noncoding RNA species that downregulate gene expression in metazoans by incomplete complementarity to target sequences usually found in the $3'UTR$ of specific mRNAs [10]. MiRNAs act to reduce levels of corresponding target gene proteins by decreasing mRNA stability and/or by blocking translation. Studies in rodent models demonstrate that early changes in miRNA expression can lead to differential expression of key target genes that could directly affect the overall health of the offspring by

Paul C. Guest (ed.), *Investigations of Early Nutrition Effects on Long-Term Health: Methods and Applications*, Methods in Molecular Biology, vol. 1735, https://doi.org/10.1007/978-1-4939-7614-0_21, © Springer Science+Business Media, LLC 2018

contributing to conditions as diverse as insulin resistance [6], increased susceptibility to ischemic-sensitive brain injury [11], and adipose tissue inflammation [12]. Detecting dysregulated miRNAs in key tissues, organs, or cell types throughout the life course of animals that have been exposed to a suboptimal environment in early life can therefore aid in our understanding of the mechanisms that lead to parental programming of chronic diseases in the offspring.

The first step in studying the role of miRNAs in programming models is to define the miRNA signature in tissues of programmed subjects when compared to control ones. In early studies, this was accomplished by microarray profiling using a probe-based system. However, this presented a few limitations, such as missing recently annotated miRNAs, not predicting novel ones, and poor validation rates [13]. Recent advances in next-generation sequencing (NGS) have offered a more versatile and reliable way of profiling miRNAs. Small RNA sequencing (sRNA-Seq) is a NGS technology that involves sequencing of cDNA synthesized from the miRNA fraction of a tissue sample, following insertion of adapters on both 3′ and 5′ ends and amplification by PCR. The resulting raw data consists of unique reads that are later mapped to sequences of mature miRNAs available on miRbase or to the annotated genome. Due to the flexibility in the post-sequencing analysis, prediction of novel miRNAs is now possible, and final data can be updated by remapping raw reads to the latest annotation on miRbase [13, 14].

In this chapter, we describe in detail methodology for miRNA profiling by NGS including all the procedures required from RNA isolation and purification from animal tissues to RNA integrity testing, library preparation, and post-sequencing analysis.

2 Materials (*See* Note 1)

2.1 Dissection Material and RNA Extraction Reagents

1. Clean surgery material (*see* **Note 2**).
2. 2 mL microcentrifuge tubes (*see* **Note 2**).
3. 1.5 mL microcentrifuge tubes (*see* **Note 2**).
4. Qiazol (Qiagen) (*see* **Note 3**).
5. Chloroform.
6. RNeasy mini kit (Qiagen).
7. RNeasy MinElute cleanup kit (Qiagen) (for small RNA purification only).
8. Nuclease-free water.
9. RNase-free DNase set (Qiagen) (for total RNA purification only).
10. 70% ethanol (for small RNA purification only).
11. 80% ethanol (for small RNA purification only).

12. 100% ethanol.

13. Tissue disruptor (*see* **Note 4**).

14. Microcentrifuge.

15. Nanodrop.

2.2 RNA Integrity Check: On-Chip Electrophoresis

1. Agilent RNA 6000 Nano kit (Agilent).

2. RNaseZAP® (Ambion).

3. 2100 Bioanalyzer System (Agilent).

4. Chip priming station (Agilent).

5. IKA vortex mixer (Agilent).

2.3 RNA Integrity Check: Denaturing Agarose Gel Electrophoresis

1. 10× MOPS buffer: 0.2 M MOPS (pH 7), 50 mM sodium acetate, 10 mM EDTA.

2. 37% formaldehyde.

3. Agarose.

4. 1 kb HyperLadder (Bioline).

5. 2× RNA loading dye (New England Biolabs).

6. Electrophoresis system.

2.4 Library Prep Material

1. TruSeq® Small RNA Library Preparation kit (Illumina; San Diego, CA, USA).

2. T4 RNA ligase 2, deletion mutant (Epicentre; Madison, WI, USA).

3. 200 µL nuclease-free tubes.

4. High sensitivity DNA kit (Agilent).

5. 5 µm filter tubes (IST Engineering; Milpitas, CA, USA).

6. Gel breaker tubes (IST Engineering) (*see* **Note 5**).

7. 6% 10-well TBE gels (Novex).

8. 5× TBE running buffer (Novex).

9. Sterile razor blade.

10. SYBR Safe (Invitrogen).

11. 3 M NaOAc, pH 5.2.

12. 10 mM Tris–HCl, pH 8.5.

13. 80% ethanol.

14. Glycogen.

15. 10 mM Tris–HCl, pH 8.5.

16. Tween 20.

17. Thermocycler.

18. XCell SureLock™ Mini-Cell Electrophoresis System (Thermo Fisher Scientific).

19. Unit 2100 Bioanalyzer System (Agilent).

20. Chip priming station (Agilent).

21. IKA vortex mixer (Agilent).

2.5 Denaturation and Dilution of Libraries for Sequencing

1. MiSeq reagent kit (Illumina).

2. 1.0 N NaOH, molecular biology grade.

3. 10 mM Tris–HCl (pH 8.5), 0.1% Tween 20.

4. Hybex Microsample Incubator (SciGene; Sunnyvale, CA, USA).

5. Thermal block for 1.5 mL microcentrifuge tubes.

6. MiSeq® System (Illumina).

2.6 Analysis of sRNA-Seq Data

1. Microsoft Office Excel.

2. sRNAbench (http://bioinfo5.ugr.es/sRNAbench/sRNAbench.php) implemented in Java (https://www.java.com/en/).

3. R (https://www.r-project.org) with the package EdgeR (http://www.bioconductor.org) installed.

3 Methods

3.1 RNA Extraction from Tissue Samples

1. Quickly dissect and snap freeze tissue, organ, or area of interest using clean instruments and 2 mL microcentrifuge tubes (*see* **Note 2**).

2. Tissues can be processed straight after collection or alternatively placed in −80 °C freezer for long-term storage (*see* **Notes 6** and **7**).

3. For hepatic tissue, place 30 mg of powdered sample in a 2 mL microcentrifuge tube containing 700 μL Qiazol or another equivalent phenol-based reagent, and homogenize using a tissue disruptor at full speed (*see* **Notes 4** and **8**).

4. Leave the tube at room temperature for 5 min.

5. Add 140 μL chloroform to the lysate and mix by vortexing for 15 s.

6. Incubate the lysate for 3 min.

7. Centrifuge samples at $12,000 \times g$ at 4 °C for 15 min.

8. Transfer 300 μL of the upper aqueous phase to a 1.5 mL centrifuge tube, and follow the complete protocol for total RNA or small RNA purification provided in the Qiagen RNeasy mini kit handbook (https://www.qiagen.com).

3.2 RNA Purification: Total RNA Purification (See Note 9)

1. Add 450 mL absolute ethanol and mix by inverting the tube six times.

2. Transfer 700 μL of the sample into an RNeasy mini spin column attached to a 2 mL collection tube.

3. Follow the remaining steps recommended by the manufacturer, including the optional on-column DNase digestion using RNase-free DNase set as described in Appendix B of the Qiagen RNeasy mini kit handbook.

4. After digestion, proceed with the remaining washing steps, including the extra spin with the column placed in a new collection tube, and with the RNA elution using 50 μl nuclease-free water.

5. Measure RNA concentration in the extracts (e.g., a nanodrop system).

3.3 RNA Purification: Small RNA Purification (See Note 10)

1. Add 300 μL 70% ethanol to the aqueous solution collected during **step 8**, Subheading 3.1, and mix it by vortexing.

2. Transfer the sample into an RNeasy mini spin column attached to a 2 mL collection tube.

3. Centrifuge 30 s at $8000 \times g$ at room temperature.

4. Transfer the flow-through to a new 2 mL microcentrifuge tube, and add 390 μl 100% ethanol.

5. Proceed with the small RNA isolation as described in Appendix A of Qiagen RNeasy mini kit handbook, including a column wash with 80% ethanol and an extra spin with the column placed onto a new collection tube.

6. After the washing steps, place the column into a 1.5 mL microcentrifuge tube.

7. Add 14 μL nuclease-free water to the column membrane and centrifuge 1 min at $8000 \times g$.

3.4 RNA Integrity Check: On-Chip Capillary Electrophoresis (See Notes 11 and 12)

1. Proceed with cleaning of the 2100 Bioanalyzer System electrode before starting.

2. Fill an electrode cleaner provided by the manufacturer with 350 μL RNase ZAP and place it on the electrode.

3. Close the lid for at least 1 min and remove it.

4. Load another electrode cleaner containing 350 μL RNase-free water on the electrode and close the lid briefly.

5. Remove the electrode cleaner and wait 10 s before using it.

6. Dilute an aliquot of each RNA extract to approximately 200 ng/μL.

7. Follow the instructions provided in the Agilent RNA 6000 Nano kit quick start guide (http://www.agilent.com) to set

up the gel-dye containing chip, to prepare the ladders, to load the samples into the gel matrix, and to run the electrophoresis in the 2100 Bioanalyzer System (*see* **Note 13**).

8. Analyze the output data and graphs (*see* **Note 14**).

3.5 RNA Integrity Check: Agarose Gel Electrophoresis

1. Weigh 1 g agarose and add it to 87 mL of nuclease-free water on a 500 mL laboratory glass bottle.

2. Melt it by mixing and heating in a microwave at medium power until agarose is completely dissolved.

3. Wait until agarose solution cools to approximately 60 °C.

4. While still warm, add 10 mL 10× MOPS buffer and 3 mL 37% formaldehyde to the agarose solution in a fume hood.

5. Add 10 μL 10,000× SYBR Safe to the agarose/MOPS/form-aldehyde solution.

6. Pour the gel on an electrophoresis tank and let it set for at least 30 min in a fume hood.

7. While waiting for the gel to cast, add 10 μL 2× RNA sample buffer to a 10 μL 200–1000 ng/μL RNA sample on a 200 μL microtube, and heat it to 55 °C for 15 min using a thermo-cycler or a thermal block.

8. Quickly transfer the tube to ice.

9. Fill the electrophoresis tank with 1× MOPS buffer, and pipet 10 μL 1 kb HyperLadder and 20 μL samples in the gel lanes.

10. Electrophorese at 60–100 V for 1 h, and analyze the bands using a transilluminator UV system (*see* **Note 15**).

3.6 Library Preparation and Sequencing (See Note 16)

1. Insert adapters on both ends of small RNAs by using T4 RNA ligase 2 deletion mutant.

2. Reverse transcribe using adapter-specific primers and reverse transcriptase provided.

3. Amplify cDNA through PCR by using primers to adapters provided (indexing will be accomplished in this step by using a different reverse prime for each library).

4. Validate libraries by on-chip electrophoresis:

   (a) Set up a gel-dye containing chip following the instructions described in the Agilent High Sensitivity DNA Kit Guide (http://www.agilent.com).

   (b) Load 1 μL 5× diluted sample from each library and 1 μL ladder in the gel-dye containing chip.

   (c) Analyze the output data and graphs.

5. Pool up to 12 samples by mixing 4 μL each library and add 12 μL 5× gel loading dye.

6. Load 50 μL of the pooled samples into two lanes of a 10-well 6% TBE gel along with the ladders provided, and run the electrophoresis at 145 V for 1 h.

7. Follow the instructions provided by the manufacturer, and excise the piece of gel corresponding to the correct molecular weight using a sterile razor blade.

8. Proceed with cDNA isolation from gel and the optional concentration step as described by the manufacturer.

9. Quality check the libraries by loading 1 μL samples and 1 μL ladder in a High Sensitivity DNA gel-dye containing chip (*see* **Note 17**).

10. Dilute the library pool to 2 nM.

11. Repeat **step 9** in Subheading 3.6 to check the final normality of the library pool.

12. Add Tween 20 to a final concentration of 0.1% and store libraries in a −20 °C freezer.

13. Follow the MiSeq System Denature and Dilute Libraries Guide to denature and dilute 5 μL of the 2 nM library pool, using freshly made 0.2 N NaOH and the Hybridization buffer provided in the MiSeq reagent kit.

14. Follow the MiSeq System User Guide to load the library pool into the sequencer and to set up the sequencing protocol.

3.7 Data Analysis (See Note 18)

1. Download the sRNAbench database (sRNAbenchDB) and the most recent version for replacement of the file sRNAbench.jar.

2. Download the mature and hairpin miRNA sequences for the studied species from the miRbase website (http://www.mirbase.org), and extract these into the libs folder of the sRNAbench database.

3. Follow the protocol described in the sRNAbench manual for library mapping using the default parameters and for adapter trimming and preprocessing.

4. Proceed with the differential expression analysis using either sRNAbench or RStudio and the package EdgeR (*see* **Note 19**).

4   Notes

1. Make up all solutions and dilutions with ultrapure water unless another reagent is required.

2. We encourage material autoclaving prior to tissue collection.

3. Other phenol-based reagents such as TRIzol and TRI Reagent can be used as an alternative.

4. A bead-milling based system can be used instead. Follow the manufacturer protocol for the tissue to be processed.

5. Breaker tubes can be replaced by autoclaved 600 μL tubes containing handmade pores using a small needle.

6. Never leave unprocessed tissue at room temperature or in ice as this will lead to RNA degradation.

7. Whenever working with large tissues (i.e., liver), we recommend powdering and homogenizing prior to extraction.

8. Time of homogenization and amount of initial sample might vary according to tissue and equipment used.

9. Either total RNA or small RNA fraction can be used as input for library preparation.

10. A small RNA enrichment protocol involves removal of RNA molecules longer than 200 nucleotides and can be used as an alternative to total RNA isolation. This procedure will provide clearer bands in the gel isolation step.

11. If using total RNA as an input, it is necessary to test RNA integrity by checking the ribosomal fractions prior to library preparation.

12. We recommend the use of Agilent RNA 6000 Nano kit for RNA integrity and concentration analysis. This method is based on an on-chip micro-capillary electrophoresis separation of nucleic acids according to size. It also offers an unbiased and automated RNA quality scoring system through identification and quantification of the two main ribosomal RNA fractions. In case of technical restrictions such as nonavailability of a bioanalyzer, a denaturing agarose gel can be used instead.

13. If the RNA yield after extraction is low, use the Agilent 6000 Pico kit.

14. Output data from the 2100 Bioanalyzer System will be comprised of an electropherogram highlighting two ribosomal RNA fractions, a quantitation of total RNA concentration and an integrity score (RNA Integrity Number—RIN) automatically calculated. For sRNA-Seq, we recommend an RIN larger than 9.

15. An agarose gel lane loaded with decent quality mouse RNA will present a 28S rRNA band at 4.7 kb that is twice the intensity of an 18S rRNA band at 1.9 kb.

16. Follow carefully all the procedures in the library prep protocol as described by the manufacturer of the TruSeq Small RNA Library Prep kit (https://www.illumina.com). When using total RNA as an input, dilute RNA extracts to 200 ng/μL in a total 5 μL solution. If using purified small RNA, a 2–10 ng/μL solution should be used instead.

17. This analysis will provide you a graphical qualitative representation and the molarity of the library pool.

18. After sequencing, the raw data collected (fastq or fastq.gz format) will have to be trimmed, aligned, and statistically analyzed by using sRNAbench, R, Bioconductor, and EdgeR.

19. To avoid noise introduced by miRNAs of extremely low or no expression, set the parameter "minimum read count" (minRC) during differential expression analysis.

Acknowledgments

LCP was the recipient of a CNPq Science Without Borders Post-Doctoral Fellowship (National Council of Technological and Scientific Development—CNPq—Brazil—PDE/204416/2014-0). SEO is funded by the UK Medical Research Council (MC_UU_12012/4).

References

1. Ong TP, Ozanne SE (2015) Developmental programming of type 2 diabetes: early nutrition and epigenetic mechanisms. Curr Opin Clin Nutr Metab Care 18:354–360

2. Godfrey KM, Costello PM, Lillycrop KA (2016) Development, epigenetics and metabolic programming. Nestle Nutr Inst Workshop Ser 85:71–80

3. Weaver ICG, Meaney MJ, Szy M (2006) Maternal care effects on the hippocampal transcriptome and anxiety-mediated behaviours in the offspring that are reversible in adulthood. Proc Natl Acad Sci U S A 103:3480–3485

4. Mortensen OH, Olsen HL, Frandsen L, Nielsen PE, Nielsen FC, Grunnet N et al (2010) Gestational protein restriction in mice has pronounced effects on gene expression in newborn offspring's liver and skeletal muscle; protective effect of taurine. Pediatr Res 67:47–53

5. Mukhopadhyay P, Horn KH, Greene RM, Michele Pisano M (2010) Prenatal exposure to environmental tobacco smoke alters gene expression in the developing murine hippocampus. Reprod Toxicol l29:164–175

6. Fernandez-Twinn DS, Alfaradhi MZ, Martin-Gronert MS, Duque-Guimaraes DE, Piekarz A, Ferland-McCollough D et al (2014) Downregulation of IRS-1 in adipose tissue of offspring of obese mice is programmed cell-autonomously through post-transcriptional mechanisms. Mol Metab 3:325–333

7. Zhang CR, Ho MF, Vega MC, Burne TH, Chong S (2015) Prenatal ethanol exposure alters adult hippocampal VGLUT2 expression with concomitant changes in promoter DNA methylation, H3K4 trimethylation and miR-467b-5p levels. Epigenetics Chromatin 8:40. https://doi.org/10.1186/s13072-015-0032-6

8. Ferland-McCollough D, Fernandez-Twinn DS, Cannell IG, David H, Warner M, Vaag AA et al (2012) Programming of adipose tissue miR-483-3p and GDF-3 expression by maternal diet in type 2 diabetes. Cell Death Differ 19:1003–1012

9. Rodgers AB, Morgan CP, Leu NA, Bale TL (2015) Transgenerational epigenetic programming via sperm microRNA recapitulates effects of paternal stress. Proc Natl Acad Sci U S A 112:13699–13704

10. Wilczynska A, Bushell M (2016) The complexity of miRNA-mediated repression. Cell Death Differ 22:22–33

11. Wang L, Ke J, Li Y, Ma Q, Dasgupta C, Huang X et al (2017) Inhibition of miRNA-210 reverses nicotine-induced brain hypoxic-ischemic injury in neonatal rats. Int J Biol Sci 13:76–84

12. Alfaradhi MZ, Kusinski LC, Fernandez-Twinn DS, Pantaleão LC, Carr SK, Ferland-McCollough D et al (2016) Maternal obesity in pregnancy developmentally programs adipose tissue inflammation in young, lean male

mice offspring. Endocrinology 157:4246–4256

13. Git A, Dvinge H, Salmon-Divon M, Osborne M, Kutter C, Hadfield J et al (2010) Systematic comparison of microarray profiling, real-time PCR, and next-generation sequencing technologies for measuring differential microRNA expression. RNA 16 (5):991–1006

14. Kang W, Friedländer MR (2015) Computational prediction of miRNA genes from small RNA sequencing data. Front Bioeng Biotechnol 3:7

Chapter 22

Pulse-Chase Biosynthetic Radiolabeling of Pancreatic Islets to Measure Beta Cell Function

Paul C. Guest

Abstract

Pulse-chase radiolabeling of cells with radioactive amino acids is a common method for studying the biosynthesis of proteins. The labeled proteins can then be immunoprecipitated and analyzed by electrophoresis and gel imaging techniques. This chapter presents a protocol for the biosynthetic labeling and immunoprecipitation of pancreatic islet proteins which are known to be affected in disorders such as diabetes, obesity, and metabolic syndrome.

Key words Metabolic disorder, Diabetes, Obesity, Pancreatic islets, Insulin, Pulse-chase radiolabeling, Immunoprecipitation, Electrophoresis

1 Introduction

Metabolic disorders such as obesity and type 2 diabetes mellitus affect a high proportion of people in the world today [1]. Almost two billion people (approximately 27% of the population) have been classified as either overweight (body mass index [BMI] = 25–30 kg/m^2) or obese (BMI > 30 kg/m^2) [2], with obesity being responsible for approximately 5% of all deaths [3]. The global prevalence of diabetes is now around 8.5% of the adult population, having nearly doubled since 1980 [4]. This is reflected by the increase in the associated risk factors of being overweight or obese. Therefore, these diseases impair quality of life and physical and mental well-being, and they can have significant knock-on effects on the workforce, the health services, society in general, as well as the global economy [5].

Recent studies have been focused on the effect of undernutrition or overnutrition during important time periods of fetal development as potential risk factors for the development of metabolic diseases and other medical disorders in later life [6, 7]. Several mechanisms have been linked to this programming effect on health, including epigenetics and perturbations in oxidative stress [8, 9]

Paul C. Guest (ed.), *Investigations of Early Nutrition Effects on Long-Term Health: Methods and Applications*, Methods in Molecular Biology, vol. 1735, https://doi.org/10.1007/978-1-4939-7614-0_22, © Springer Science+Business Media, LLC 2018

and insulin signaling pathways [10, 11]. Investigations of the mechanisms involved are needed to help lay the groundwork for identification of novel biomarkers [12, 13] and for development of new therapeutic strategies to help delay or minimize the effects of poor antenatal programming [14, 15].

Type 2 diabetes is characterized by fasting hyperglycemia due to defects in pancreatic islets and through insulin resistance in target tissues such as the liver, muscle, fat, and brain [16]. Insulin is synthesized in the islet β-cells and stored in secretory granules. Following a rise in the blood glucose concentration after a meal or intake of food, glucose is taken up into the cells and metabolized. This leads to depolarization of the membrane, resulting in activation of voltage-dependent calcium channels and increased concentrations of calcium in the cytoplasm [17]. The increased calcium levels activate fusion of the plasma and secretory granule membranes leading to secretion of insulin and other soluble granule contents through the fusion pore (Fig. 1).

Within the secretory granules, proteolytic conversion of proinsulin to insulin requires two Ca^{2+}-dependent endoproteases called prohormone convertase (PC)1 and PC2, which cleave the precursor on the carboxy-terminal side of residues Arg^{31}-Arg^{32} and Lys^{64}-Arg^{65} [18–20] and the Zn^{2+}-dependent carboxypeptidase H

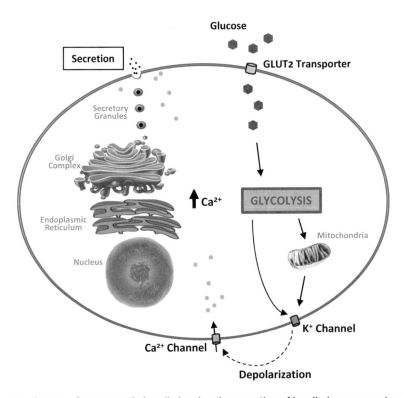

Fig. 1 Schematic diagram of a pancreatic b-cell showing the secretion of insulin in response to a rise in blood glucose

(CPH), which removes the exposed basic residues [21]. Thus Ca^{2+} is not only required for secretion of the insulin secretory granule contents, but also for proteolytic activation of insulin and other proproteins within the secretory pathway [18].

This chapter describes a protocol for looking at the effects of Ca^{2+} depletion on proteolytic processing of insulin and other secretory granule proteins (CPH, PC1, and PC2) in isolated rat islets. It describes the preparation of immunoadsorbents, pulse-chase radiolabeling of islets, the immunoprecipitation of insulin-, CPH-, PC1-, and PC2-related proteins, as well as the subsequent gel-based analyses. The same protocols can also be used to investigate other isolated neuroendocrine cells, providing that antibodies are available for immunoprecipitation of the target proteins.

2   Materials

2.1   Immuno-adsorbent Preparation

1. Purified monoclonal antibody for proinsulin (*see* **Note 1**).
2. Polyclonal antiserum against proCPH (*see* **Note 2**).
3. Polyclonal antiserum against proPC1 (*see* **Note 2**).
4. Polyclonal antiserum against proPC2 (*see* **Note 2**).
5. CNBr-activated Sepharose 4 (GE Healthcare).
6. Activation solution: 1 mM HCl.
7. Coupling solution: 100 mM $NaHCO_3$ (pH 8.3), 500 mM NaCl.
8. Quench solution: 100 mM Tris–HCl (pH 8.0).
9. Wash solution 1: 100 mM Tris–HCl (pH 8.0), 500 mM NaCl.
10. Wash solution 2: 100 mM NaH_3CO_2 (pH 4.0), 500 mM NaCl.
11. Storage solution: 10 mM NaH_2PO_4, 2 mM K_2HPO_4, 137 mM NaCl, 2.7 mM KCl.
12. 20 mL-capacity centrifuge tubes.
13. 50 mL-capacity centrifuge tubes.
14. UV spectrophotometer.

2.2   Islet Incubations

1. 600 large rat pancreatic islets (*see* **Note 3**).
2. Islet recovery medium: Dulbecco's modified Eagle's medium (DMEM), containing 10% newborn calf serum and 5.5 mM glucose.
3. High-glucose incubation medium: 25 mM $NaHCO_3$ (pH 7.4), 115 mM NaCl, 5.9 mM KCl, 1.2 mM $MgCl_2$, 1.2 mM NaH_2PO_4, 1.2 mM Na_2SO_4, 2.5 mM $CaCl_2$, 0.1% bovine serum albumin (BSA), 16.7 mM glucose.

4. High-glucose incubation medium minus $CaCl_2$: 25 mM $NaHCO_3$ (pH 7.4), 115 mM NaCl, 5.9 mM KCl, 1.2 mM $MgCl_2$, 1.2 mM NaH_2PO_4, 1.2 mM Na_2SO_4, 0.1% BSA, 16.7 mM glucose, 1 mM EGTA.

5. Low-glucose chase media: 25 mM $NaHCO_3$ (pH 7.4), 115 mM NaCl, 5.9 mM KCl, 1.2 mM $MgCl_2$, 1.2 mM NaH_2PO_4, 1.2 mM Na_2SO_4, 2.5 mM $CaCl_2$, 0.1% BSA, 3.3 mM glucose.

6. Low-glucose chase media minus $CaCl_2$: 25 mM $NaHCO_3$ (pH 7.4), 115 mM NaCl, 5.9 mM KCl, 1.2 mM $MgCl_2$, 1.2 mM NaH_2PO_4, 1.2 mM Na_2SO_4, 0.1% BSA, 3.3 mM glucose, 1 mM EGTA.

7. 1.5 mL-capacity microcentrifuge tubes.

8. 1.8 mL-capacity screw cap, round bottom cryogenic vials (Sigma-Aldrich) (see **Note 4**).

9. ^{35}S–methionine (PerkinElmer).

2.3 Immuno-precipitation, Electrophoresis, and Fluorography

1. Pancreatic islet lysis buffer: 25 mM $Na_2B_4O_7$ (pH 9), 3% BSA, 1% Tween-20, 1 mM phenylmethanesulfonyl fluoride, 0.1 mM E-64, 1 mM EDTA, 0.1% NaN_3 (see **Note 5**).

2. Immunoprecipitation preclearing reagent: 100 mg/mL suspension of Cowan-strain *Staphylococcus aureus* cells.

3. Immunoprecipitation wash buffer: 50 mM Tris–HCl (pH 7.5), 150 mM NaCl, 1% Triton X-100, 1% deoxycholate, 0.1% SDS, 5 mM EDTA.

4. Insulin elution buffer: 25% acetic acid (see **Note 6**).

5. Protein A Sepharose (see **Note 7**).

6. Protein A Sepharose rehydration buffer: 20 mM NaH_2PO_4 (pH 8.0), 150 mM NaCl.

7. Protein A Sepharose elution buffer: 20 mM HCl.

8. 1.5 mL-capacity microcentrifuge tubes.

9. Alkaline-urea electrophoresis gel: polymerized from 7.5% acrylamide, 0.2% NN'-methylenebisacrylamides, 12.5 mM Tris/80 mM glycine (pH 8.6), 8 M urea.

10. Alkaline-urea gel electrophoresis buffer: 12.5 mM Tris/80 mM glycine (pH 8.6).

11. Alkaline-urea gel loading buffer: 2.5 mM Tris–HCl (pH 8.6), 8 M urea, 0.001% Bromophenol Blue.

12. SDS-polyacrylamide electrophoresis (PAGE): gels polymerized from 15% acrylamide and 0.08% NN'-methylenebisacrylamide in a Tris–glycine buffer system using the buffer system of Laemmli [14].

13. SDS-PAGE loading buffer: 125 mM Tris–HCl (pH 6.8), 2% SDS, 0.25 M sucrose, 5 mM EDTA, 65 mM dithiothreitol, 0.005% Bromophenol Blue.

14. Fluorography reagent: 20% 2,5-diphenyloxazole in acetic acid.

15. MSE sonifier and microprobe (Crawley, UK) (*see* **Note 8**).

16. Cronex 4 X-ray film (Dupont) (*see* **Note 9**).

3   Methods

3.1 Preparation of Insulin Immunoadsorbent

1. Dialyze the antibody using three changes of coupling solution over 6 h total time at 4 °C (*see* **Note 10**).

2. Measure the absorbance of the final antibody solution at 280 nm in the UV spectrophotometer and calculate the concentration (*see* **Note 11**).

3. Add 20 mL ice-cold activation solution to 1 g dried resin and gently mix for 2 h at 4 °C in a 20 mL centrifuge tube (*see* **Note 12**).

4. Centrifuge the resin for 5 min at $1000 \times g$ in a swinging bucket rotor and discard the supernatant.

5. Add the dialyzed antibody to the resin at 10 mg antibody/5 mL swollen resin and mix overnight at 4 °C.

6. Centrifuge the resin for 5 min at $1000 \times g$ and keep the supernatant to test the coupling efficiency (*see* **Note 13**).

7. Add 20 mL coupling solution to the resin and mix for 30 min with low agitation at room temperature.

8. Centrifuge for 5 min at $1000 \times g$ and discard the supernatant.

9. Add 20 mL quench solution and mix for 2 h as above at room temperature.

10. Centrifuge for 5 min at $1000 \times g$ and discard the supernatant.

11. Add 20 mL wash solution 1 to the resin, mix gently, centrifuge for 5 min at $1000 \times g$, and discard the supernatant.

12. Add 20 mL wash buffer 2, centrifuge for 5 min at $1000 \times g$, and discard the supernatant.

13. Repeat **steps 11** and **12** for a total of three times and add 20 mL storage solution to the resin.

14. Centrifuge for 5 min at $1000 \times g$, discard the supernatant, and add 20 mL storage buffer to the final immunoadsorbent.

15. Store at 4 °C for up to 1 month if no preservatives are used (*see* **Note 14**).

3.2 Preparation of CPH, PC1, and PC2 Immunoadsorbents

1. Add 20 mL of rehydration solution to 4 g of dry Protein A Sepharose in three separate 50 mL centrifuge tubes and mix for 30 min gently for complete hydration.

2. Centrifuge for 5 min at $1000 \times g$, remove the supernatant, and suspend the resin in 20 mL of the same buffer.

3. Repeat the centrifugation and washing steps twice and store in 16 mL of rehydration buffer at 4 °C for up to 1 month if no preservatives are used (*see* **Note 14**).

4. Add 200 μL of suspended Protein A Sepharose to microcentrifuge tubes containing 15 μL of each antiserum, and mix gently overnight at 4 °C (*see* **Note 15**).

5. Centrifuge for 5 min at $1000 \times g$, remove supernatant and wash three times by centrifugation and resuspension in 500 μL of the rehydration buffer.

6. Centrifuge for 5 min at $1000 \times g$ and remove the supernatant, leaving a packed gel of approximately 50 μL for immediate use.

3.3 Biosynthetic Radiolabeling of Pancreatic Islets

1. Preincubate 100 isolated islets for 1 h in 300 μL of Kreb's bicarbonate buffer containing high-glucose incubation medium with and without $CaCl_2$ at 37 °C in 1.8 mL cryovials under 95% O_2/5% CO_2 (Fig. 2) (*see* **Note 16**).

2. Recover the islets gently by centrifugation at $100 \times g$ for 10 s and suspend the loose pellets in 100 μL of the same pre-warmed (37 °C) medium containing 150 μCi of ^{35}S-methionine.

3. Incubate for 15 min at 37 °C under 95% O_2/5% CO_2.

4. Recover the islets by centrifugation at $100 \times g$ for 10 s, carefully remove, and retain the radioactive supernatant.

5. Gently resuspend the islets in 300 μL of nonradioactive low-glucose chase media with and without $CaCl_2$.

6. Incubate for 105 min at 37 °C under 95% O_2/5% CO_2 (*see* **Note 17**).

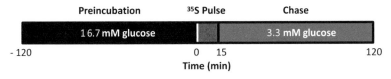

Fig. 2 Time course diagram showing the ^{35}S-methionine pulse-chase biosynthetic labeling protocol. The preincubation and labeling periods are carried out in the presence of high glucose (16.7 mM) to maximize incorporation of ^{35}S-methionine into the newly synthesized proteins. The chase period is carried out in the presence of low glucose (3.3 mM) so that the proteins reach their final intracellular destination without being secreted

7. Recover the islets by centrifugation at $100 \times g$ for 10 s, carefully remove and retain supernatant (*see* **Note 18**).

8. Add 200 µL lysis buffer to the pellets and sonicate for 15 s at approximately ¼ power (*see* **Note 19**).

9. Centrifuge the lysates at $13,000 \times g$ for 5 min and retain the supernatants for immunoprecipitation.

3.4 Immuno-precipitation of Insulin and Other Pancreatic Islet Proteins

1. Incubate islet lysates for 1 h at room temperature in 1.5 mL microcentrifuge tubes containing 50 µL of preclearing reagent (*see* **Note 20**).

2. Centrifuge the samples for 5 min at $13,000 \times g$, and retain the supernatants.

3. Immunoprecipitate the insulin-related molecules by adding the lysates to the 50 µL packed gel of 3B7 immunoadsorbent and incubate overnight at 4 °C.

4. Centrifuge for 1 min in a swinging bucket rotor at $500 \times g$ and retain the supernatant for the CPH, PC1, and PC2 immunoprecipitations (*see* **Note 21**).

5. Wash the anti-insulin immunoadsorbent by repeated centrifugation and resuspension in 4×1 mL lysis solution, 2×1 mL immunoadsorbent wash solution and 2×1 mL distilled water.

6. Elute the insulin-related peptides with 2×250 µL insulin elution buffer, freeze-dry and reconstitute in 50 µL of alkaline-urea gel loading buffer.

7. Incubate the supernatant obtained after immunoprecipitation of the insulin-related molecules overnight at 4 °C with 50 µL packed gel of CPH immunoadsorbent.

8. Wash the CPH immunoadsorbent as above, and retain the supernatant for immunoprecipitation of the other proteins.

9. Elute the CPH-related peptides with 2×100 µL Protein A Sepharose elution buffer.

10. Combine the two eluates, freeze-dry, and reconstitute in 50 µL of SDS-PAGE loading buffer.

11. Repeat **steps 7–10** of Subheading 3.4 for immunoprecipitation of the PC1- and PC2-related peptides.

3.5 Electrophoresis and Fluorography of ^{35}S-labeled Immunoprecipitates

1. Pre-run alkaline-urea gels in tank buffer for 600 Vh, replace the upper tank buffer, load the insulin immunoprecipitates in alkaline-urea gel loading buffer, and subject to electrophoresis for 1000 Vh (*see* **Note 22**)

2. Disassemble the gel plates and rock the gels gently for 2×5 min in acetic acid and 2 h in fluorography solution and then leave for 30 min under running cold water (*see* **Note 23**).

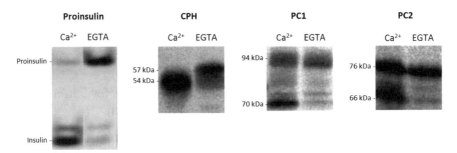

Fig. 3 Islet content of ^{35}S-labeled proinsulin, proCPH, proPC1, and proPC2 forms in pulse-chase-radiolabeled islets as determined by immunoprecipitation, PAGE, and fluorography. The islets were incubated throughout either in 2.5 mM CaCl$_2$ or in Ca^{2+}-free medium containing 1 mM EGTA to measure the effect of Ca^{2+} depletion on proprotein conversion [15] (*see* **Note 23**). The migration of the precursor and mature forms of each protein is indicated on the left side of each image

3. Vacuum dry and expose the gel to X-ray film for 6–72 h (*see* **Note 24**).

4. For the PC2 immunoprecipitates, heat the samples for 5 min at 95 °C, and electrophorese the samples to the point where the dye front just reaches the bottom of the gel.

5. Perform fluorography as above for 2–20 days (Fig. 3) (*see* **Note 25**).

4   Notes

1. We used the 3B7 clone which recognizes epitopes in human, rat, and mouse proinsulin [18].

2. Antisera were raised in rabbits against glutathione S-transferase fusion proteins incorporating amino acids 111–137, 162–388, and 42–454 of rat PC1, PC2, and CPH, respectively. The fusion proteins were produced using a bacterial expression vector, as described by Bennett et al. [20].

3. We used isolated rat islets in this study prepared as described [22] although rat pancreatic beta cell lines (and other cells as appropriate) can also be used. If this is the case, 1.5×10^6 cells would be approximately equivalent to 600 islets since each islet contains approximately 5000 cells.

4. These tubes are used for the islet incubations as they allow free movement of large numbers of islets. In addition, the screw caps enable a tight seal so that incubations can be carried out under a blanket of humidified 95% O$_2$/5% CO$_2$.

5. BSA is added as a carrier to minimize protein loss on tube walls during immunoprecipitation. In the case of insulin, very high levels of BSA are required due to the "sticky" nature of this molecule.

6. The elution of insulin requires strong acid conditions due to the high affinity of the 3B7 antibody and the relatively low solubility of insulin in neutral pH solutions.

7. Protein A Sepharose is used commonly to capture polyclonal antibodies in indirect immunoprecipitation protocols (as is the case for CPH, PC1, and PC2 in this study). One should keep in mind that the bond is non-covalent and elution will occur for both the antibody and protein. In contrast, the direct method used for immunoprecipitation of the proinsulin-related molecules does not result in elution of the antibody, since this is covalently bound to the resin.

8. Other sonication devices can be used. The probe size is important as this should fit inside a 1.5 mL-capacity microcentrifuge tube down to the narrowest part of the bottom.

9. We use Dupont Cronex 4 X-ray film (Stevenage, Herts, UK) although other such films can be used.

10. The Tris buffer should be removed completely as this contains primary amines and therefore will also react with the activated resin.

11. We use 2 mg antibody/mL activated resin (swollen volume) to achieve a final immunoadsorbent with a high binding capacity.

12. 1 g of dried resin will swell to an approximate volume of 4 mL.

13. The coupling efficiency of the antibody to the resin can be measured by reading the final optical density at 280 nm in a UV spectrophotometer compared to the starting optical density (before coupling). It is desirable to achieve an efficiency >80%.

14. The immunoadsorbent can be stored for long periods of time. However, preservatives such as NaN_3 should be used. Note that these should be removed prior to use in an immunoprecipitation procedure by washing the desired amount of resin.

15. In this step, immunoglobulins in serum bind to the Protein A moieties on the resin.

16. The incubations should be carried out in at least triplicates for each condition. However, the overall numbers should not be too high to minimize the number of animals used in the study.

17. In this study, we used islets isolated from rat pancreas prepared as described previously [22]. However, pancreatic beta cell lines (or other neuroendocrine cells) can also be used. As a general rule, 5×10^5 cells would be approximately equivalent to 100 islets as each islet contains approximately 5000 cells.

18. This may contain some radiolabeled proteins even though the chase is carried out under non-stimulatory (low-glucose) conditions. Use appropriate precautions when handling and

disposing the radioactive materials. The addition of ice-cold medium containing nonradioactive methionine stops the uptake of ^{35}S-methionine and halts metabolic activity of the islet cells.

19. Adjust the power setting using experimentally determined analyses if using other sonication and probe devices.

20. This step removes immunoglobulin-like molecules in the lysates that could interfere with the immunoprecipitation step.

21. These immunoprecipitations are carried out sequentially to minimize the number of islet incubations and therefore the number of animals used. If using cells, it would be preferable to carry out separate incubations for each protein targeted for immunoprecipitation.

22. This is equivalent to approximately 1.75 lengths of the migration of the dye front. This can be measured and carried out by stopping electrophoresis after the dye has run 1gel length, adding new dye to a blank well and restarting electrophoresis for a further run of 0.75 gel length.

23. The gel turns white during this step as a sign that it is completely impregnated by the scintillant.

24. Longer exposure periods may be required as the levels of the islet accessory proteins are lower compared to that of the insulin-related molecules. Obtaining the best exposure may require a few attempts and adjusting the times accordingly. However, it is recommended to minimize the exposure period to less than 30 days (the half-life of ^{35}S is 87 days).

25. The image shows that intracellular conversion of all molecules was reduced by extracellular depletion of calcium. This is consistent with previous studies [23].

References

1. http://www.who.int/nmh/publications/ncd_report_chapter1.pdf

2. http://www.who.int/mediacentre/factsheets/fs311/en/

3. http://www.mckinsey.com/industries/healthcare-systems-and-services/our-insights/how-the-world-could-better-fight-obesity

4. http://apps.who.int/iris/bitstream/10665/204871/1/9789241565257_eng.pdf

5. Piot P, Caldwell A, Lamptey P, Nyrirenda M, Mehra S, Cahill K et al (2016) J Glob Health 6:010304. https://doi.org/10.7189/jogh.06.010304

6. Carolan-Olah M, Duarte-Gardea M, Lechuga J (2015) A critical review: early life nutrition and prenatal programming for adult disease. J Clin Nurs 24:3716–3729

7. Lopes GA, Ribeiro VL, Barbisan LF, Marchesan Rodrigues MA (2016) Fetal developmental programing: insights from human studies and experimental models. J Matern Fetal Neonatal Med 23:1–7

8. Lee HS (2015) Impact of maternal diet on the epigenome during in utero life and the developmental programming of diseases in childhood and adulthood. Forum Nutr 7:9492–9507

9. Tarry-Adkins JL, Ozanne SE (2017) Nutrition in early life and age-associated diseases. Ageing Res Rev 39:96–105. pii: S1568-1637(16)

30179-9. https://doi.org/10.1016/j.arr.2016.08.003

10. Alfaradhi MZ, Ozanne SE (2011) Developmental programming in response to maternal overnutrition. Front Genet 2:27–39

11. Fernandez-Twinn DS, Alfaradhi MZ, Martin-Gronert MS, Duque-Guimaraes DE, Piekarz A, Ferland-McCollough D et al (2014) Downregulation of IRS-1 in adipose tissue of offspring of obese mice is programmed cell-autonomously through post-transcriptional mechanisms. Mol Metab 3:325–333

12. Camm EJ, Martin-Gronert MS, Wright NL, Hansell JA, Ozanne SE, Giussani DA (2011) Prenatal hypoxia independent of undernutrition promotes molecular markers of insulin resistance in adult offspring. FASEB J 25:420–427

13. Martínez JA, Cordero P, Campión J, Milagro FI (2012) Interplay of early-life nutritional programming on obesity, inflammation and epigenetic outcomes. Proc Nutr Soc 71:276–283

14. Dixon JB (2009) Obesity and diabetes: the impact of bariatric surgery on type-2 diabetes. World J Surg 33:2014–2021

15. Khavandi K, Brownrigg J, Hankir M, Sood H, Younis N, Worth J et al (2014) Interrupting the natural history of diabetes mellitus: lifestyle, pharmacological and surgical strategies targeting disease progression. Curr Vasc Pharmacol 12:155–167

16. Porte D Jr, Kahn SE (1991) Mechanisms for hyperglycemia in type II diabetes mellitus: therapeutic implications for sulfonylurea treatment—an update. Am J Med 90(6A):8S–14S

17. Takahashi N (2015) Imaging analysis of insulin secretion with two-photon microscopy. Biol Pharm Bull 38:656–662

18. Davidson HW, Rhodes CJ, Hutton JC (1988) Intraorganellar calcium and pH control proinsulin cleavage in the pancreatic beta cell via two distinct site-specific endopeptidases. Nature (London) 333:93–96

19. Bailyes E, Shennan KIJ, Seal AJ, Smeekens SP, Steiner DF, Hutton JC et al (1992) A member of the eukaryotic subtilisin family (PC3) has the enzymatic properties of the type 1 proinsulin-converting endopeptidase. Biochem J 285:391–394

20. Bennett DL, Bailyes EM, Nielsen E, Guest PC, Rutherford NG, Arden SD et al (1992) Identification of the type 2 proinsulin processing endopeptidase as PC2, a member of the eukaryote subtilisin family. J Biol Chem 267:15229–15236

21. Davidson HW, Hutton JC (1987) The insulin-secretory-granule carboxypeptidase H. Purification and demonstration of involvement in proinsulin processing. Biochem J 245:575–582

22. Guest PC, Rhodes CJ, Hutton JC (1989) Regulation of the biosynthesis of insulin secretory granule proteins: co-ordinate translational control is exerted on some, but not all, granule matrix constituents. Biochem J 257:432–437

23. Guest PC, Bailyes EM, Hutton JC (1997) Endoplasmic reticulum Ca2+ is important for the proteolytic processing and intracellular transport of proinsulin in the pancreatic beta-cell. Biochem J 323:445–450

Chapter 23

Rapid and Easy Protocol for Quantification of Next-Generation Sequencing Libraries

Steve F. C. Hawkins and Paul C. Guest

Abstract

The emergence of next-generation sequencing (NGS) over the last 10 years has increased the efficiency of DNA sequencing in terms of speed, ease, and price. However, the exact quantification of a NGS library is crucial in order to obtain good data on sequencing platforms developed by the current market leader Illumina. Different approaches for DNA quantification are available currently and the most commonly used are based on analysis of the physical properties of the DNA through spectrophotometric or fluorometric methods. Although these methods are technically simple, they do not allow exact quantification as can be achieved using a real-time quantitative PCR (qPCR) approach. A qPCR protocol for DNA quantification with applications in NGS library preparation studies is presented here. This can be applied in various fields of study such as medical disorders resulting from nutritional programming disturbances.

Key words Next-generation sequencing, qPCR, RNA, Primers, Template

1 Introduction

Significant advances in next-generation sequencing (NGS) and high-throughput transcriptomic profiling methods have brought us one step closer toward the understanding of complex metabolic diseases such as diabetes at the molecular level [1, 2]. An increasing number of life science researchers are now performing NGS sequencing approaches, including DNA sequencing, RNA sequencing, resequencing, microsatellite analysis, single-nucleotide polymorphism (SNP) genotyping, gene regulation analysis, cytogenetic analysis, DNA-protein interaction analysis (ChIP-Seq), sequencing-based methylation analysis and metagenomics [3]. The typical library preparation protocol takes over 3 h and can contain several hands-on pipetting and clean-up steps throughout the protocol. Not only are each of these steps time consuming and require significant manual effort, they can also introduce biases or lead to sample loss. This can be compounded by the fact that all library preparation methods produce some DNA molecules that do not contain ligated adapters

Paul C. Guest (ed.), *Investigations of Early Nutrition Effects on Long-Term Health: Methods and Applications*, Methods in Molecular Biology, vol. 1735, https://doi.org/10.1007/978-1-4939-7614-0_23, © Springer Science+Business Media, LLC 2018

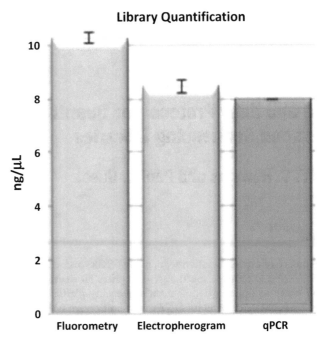

Fig. 1 Comparison of library concentrations determined using either fluorometry, electrophoretogram, or qPCR (JetSeq; Bioline; London, UK) techniques

at both ends and therefore cannot be sequenced. This is crucial for PCR-based protocols since these incompletely ligated products cannot be amplified.

In addition, standard methods for quantifying NGS libraries have a number of disadvantages [4]. For example, electrophoresis and spectrophotometry techniques only measure total nucleic acid concentrations (Fig. 1) without providing appropriate concentrations of PCR-amplifiable DNA molecules [5]. These methods also have low sensitivity, use nanograms of sometimes precious samples, and are not suitable for high-throughput workflows. Therefore, it is important to know exactly how much of sequenceable library has been produced to maximize loading onto the flow cell. qPCR has been recognized as a highly sensitive method for DNA quantification for many uses [6], including NGS library preparation [7]. To achieve this level of performance, the qPCR reaction needs to be specific, efficient, and reliable. qPCR allows the measurement of only those DNA molecules that contain the appropriate adapter sequences, since the primers are designed to match the specific adapters for the sequencing platform being used. The qPCR quantification determines library concentrations as a function of the PCR cycle number to yield a quantitative estimate of the initial template concentration based on comparison with a known DNA standard concentration. It is important that a constant quantity of the control template is available, since each round of library

Fig. 2 Structure of an Illumina library showing position of the primers (Q-primer 1 and Q-primer 2) for qPCR. These primers bind to the P5 and P7 regions of the adaptors

quantification is used to adjust the loading concentration of constructed libraries for sequencing. Even a slight variation of the standard can lead to fluctuation of cluster density, which either compromises the data quality or wastes the sequencing capacity. This is particularly important since reduced cluster density variability eliminates the need for titrations across multiple libraries.

This chapter presents a protocol for use of the Bioline JetSeq Library Quantification Kit [8], which provides a fast and reliable solution to determine the library concentration with a direct estimate cluster density on Illumina sequencing platforms. This qPCR-based method saves time and resources by providing the JetSeq FAST Lo-ROX Mix (a SYBR®-based qPCR mix), six pre-diluted DNA standards (ranging from 10 pM to 100 aM to minimize pipetting errors), primer mix (using primers designed for the universal P5 and P7 sequences in the library adapters, to ensure reproducible and precise qPCR results; Fig. 2), and a dilution buffer (optimized for NGS library samples).

2 Materials (*See* Notes 1 and 2)

1. Pre-diluted library (*see* **Notes 3** and **4**).

2. Tenfold serial dilution DNA standards (1–6) (*see* **Note 5**).

3. Primer mix: 400 nM oligonucleotide Q-primers 1 and 2 (*see* **Note 6**).

4. 2× JetSeq Fast Mix: hot-start DNA polymerase, SYBR Green 1 dye, dNTPs, stabilizers, and enhancers (*see* **Note 7**).

5. Dilution buffer (*see* **Note 8**).

6. qPCR thermocycler (*see* **Notes 9** and **10**).

3   Methods (*See* Note 11)

3.1   PCR

1. Prepare a 1:10,000 diluted sample of the library using the JetSeq dilution buffer including additional dilution samples of each library (such as 1:100,000 and/or 1:1,000,000) to improve the accuracy of the quantification (*see* **Notes 3** and **4**).

2. Prepare a qPCR master mix based on a standard 20 μL final reaction volume containing primers, template (diluted library or DNA standards), and probe mix (*see* **Note 12**).

3. Suggested thermal cycling conditions are 1 cycle at 95 °C for 2 min for polymerase activation and then 35 cycles at 95 °C for 5 s for denaturation and 60 °C for 45 s for annealing/extension (*see* **Notes 13** and **14**).

3.2   Analysis and Quantification

1. Generate a standard curve from the standard samples.

2. Correlate the concentration values of the different standard dilutions against their respective Ct (averaged from each triplicate) using data generated by qPCR.

3. Use this data to calculate the efficiency by plotting the average Ct value against the log concentration using the formula below where a is the slope of the standard curve (*see* **Note 15**).

$$\text{Efficiency (\%)} = \left(10^{\left(\frac{-1}{a}\right)} - 1\right) \times 100$$

4. Check that the reaction efficiency is between 90% and 110% for the DNA standards (*see* **Note 16**).

5. From the Ct values obtained for the library sample dilutions, calculate the library concentration (L) using the standard curve following the formula below (a and b represent the standard curve slope and y-intercept, respectively) (Fig. 3) (*see* **Note 17**).

$$\text{Library concentration (pM)} = 10^{\left[\left(\frac{Ct-b}{a}\right) \times \left(\frac{342}{\text{average fragment length}}\right) \times \text{dilution factor}\right]}$$

4   Notes

1. The DNA standard in this protocol consists of a 342 bp fragment of linear DNA fragment, flanked by additional stabilizing DNA. The DNA standards should only be used in conjunction with the other reagents in this kit for the quantification of DNA libraries prior to use in Illumina-based next-generation sequencing.

2. To help prevent any carry-over DNA contamination, we recommend that separate areas are used for library preparation,

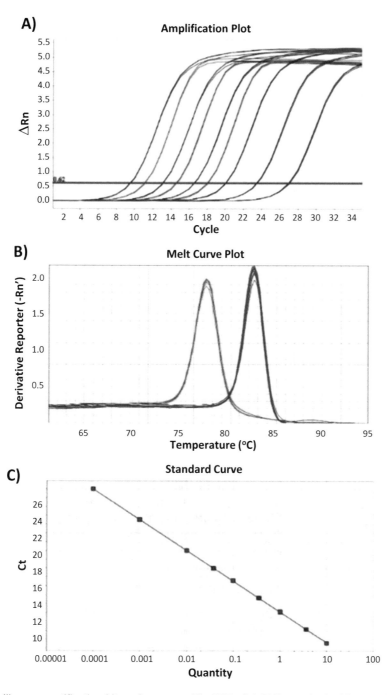

Fig. 3 JetSeq library quantification kit performance. (**A**) qPCR of 6 DNA standards (blue) and tenfold serial dilutions of the Illumina NGS library (red) were carried out with supplied primer sets in triplicate. (**B**) A melt curve was created to verify amplification of a single specific product for the standards and the library at all concentrations. (**C**) Using the standard curve, a size-adjusted library concentration was then determined. The amplification plots demonstrate a limit of detection of 0.0001 pM

reaction setup, PCR amplification, and any downstream analysis. It is essential that any tubes containing amplified PCR product are not opened in the PCR setup area.

3. We recommend that a no template control (NTC) is included in each assay to detect any contamination introduced during reaction setup. A NTC reaction should give a Ct value that is at least 3.5 cycles later than the average Ct value for standard 6.

4. A 1:10,000 diluted sample of the library should be prepared using the JetSeq dilution buffer. In order to improve the accuracy of the quantification, we suggest including additional dilution samples of each library (such as 1:100,000 and 1:1,000,000). Using two or more different dilutions of the library will ensure that at least one dilution falls within the dynamic range of the generated standard curve. This may be useful for high-concentration libraries. In order to improve the accuracy and reproducibility of the quantification, we recommend that each standard and library sample is assayed in triplicate. In addition, the use of a master mix in setting up the reactions will help to reduce pipetting errors and increase reproducibility.

5. The DNA standards supplied with the JetSeq kit range from 10 pM to 100 aM (0.0001 pM), which is equivalent to 2.75–0.0000275 pg/µL (or approximately 6,000,000–60 double stranded DNA molecules/µL). This broad range ensures that the diluted library under analysis falls within the dynamic concentration range of the qPCR assay, regardless of starting library DNA quantity.

6. Independent of the preparation method, submitted libraries for running on Illumina sequencers will consist of a sequence of interest flanked on either side by adapter constructs. These adapter constructs have flow cell binding sites at each end designated P5 and P7, which allow the library fragment to attach to the flow cell surface (Fig. 1).

Sequence of the primers:

Q-primer 1 (forward): 5′-AAT GAT ACG GCG ACC ACC GA-3′.

Q-primer 2 (reverse): 5′-CAA GCA GAA GAC GGC ATA CGA-3′.

7. The use of the SYBR Green 1-based mix allows a melt curve to be run to provide information about the specificity of the reaction. The melt curves for the DNA standard and library will have a characteristic double peak since the standards are a specific length, whereas the library is a heterogeneous population of molecules, with a range of lengths and differing guanine-cytosine (G-C) compositions. In addition, the melt

curve analysis is useful for the identification of carry-over adapter dimers that may have formed during library construction. This is important as these dimers will lead to an overestimation of library concentration if present in significant quantities.

8. A supplied dilution buffer is recommended, as it reduces potential variation and increases consistency in performance.

9. Many instruments can be used including the 7500 FAST, 7900HT FAST, ViiA7™, and StepOne™ from Applied Biosystems (Waltham, Massachusetts, USA), the Mx4000™ from Stratagene/Agilent (Santa Clara, California, USA), the iCycler™ and MyiQ5™ from Bio-Rad (Hercules, California, USA), the LightCycler® from Roche (Basel, Switzerland), the RotorGene™ from Qiagen (Hilden, Germany), and the MIC from Bio Molecular Systems (Upper Coomera, QLD, Australia). Other instruments can be used, but it is important to check with the manufacturer for compatibility, as described above.

10. Some of the older instruments require the use of a passive reference such as ROX or fluorescein, to normalize expression levels between wells of reaction plates. If normalization is required with these instruments, it will reduce the selection of fluorescent dyes that can be used for the samples.

11. The high sensitivity of qPCR necessitates prevention of any carry-over DNA as a potential contaminant. Separate areas should be maintained for reaction setup, PCR amplification, and any post-PCR gel analysis. It is also essential that any tubes containing amplified PCR product are not opened in the PCR setup area. One of the biggest causes of contamination and increased background noise is from using the same pipettes for extraction, PCR setup, and post-run analysis. Even if aerosol resistant tips are used all the time, this should be avoided. Instead, the user should have a dedicated set of pipettes for each stage. In addition, there should be a different dedicated location for each step. It is also important to detect the presence of any contaminating DNA that may affect the reliability of the data by including the NTC reaction, substituting PCR-grade water for the template.

12. Always set up a NTC, and these should give Ct values that are at least 3.5 cycles later than the average Ct value for Standard 6.

13. If the average library fragment size is larger than 500 bp, increase the annealing/extension time to 60 s.

14. After the reaction has reached completion, the user should refer to the instrument instructions for the option of melt-profile analysis.

15. Many qPCR machines generate this type of curve using the software provided by the manufacturer.

16. Optimal analysis settings, such as baseline and threshold parameters, are required for accurate quantification data. The efficiency of amplification is 90–110% (slope $= -3.6$ to -3.1) and R^2 is >0.99. An efficiency below 90% and a R^2 value below 0.99 indicate that the PCR conditions are suboptimal and pipettes should be recalibrated. A PCR efficiency that is above 110% indicates that the template contains inhibitors.

17. This formula is an example using a fragment length for the standards of 342 bp.

References

1. Cui H, Dhroso A, Johnson N, Korkin D (2015) The variation game: cracking complex genetic disorders with NGS and omics data. Methods 79-80:18–31

2. Han Y, He X (2016) Integrating epigenomics into the understanding of biomedical insight. Bioinform Biol Insights 10:267–289

3. Su Z, Ning B, Fang H, Hong H, Perkins R, Tong W et al (2011) Next-generation sequencing and its applications in molecular diagnostics. Expert Rev Mol Diagn 11:333–343

4. Meyer M et al (2008) From micrograms to picograms: quantitative PCR reduces the material demands of high-throughput sequencing. Nucleic Acids Res 36:e5. https://doi.org/10.1093/nar/gkm1095

5. Robin JD, Ludlow AT, LaRanger R, Wright WE, Shay JW (2016) Comparison of DNA quantification methods for next generation sequencing. Sci Rep 6:24067. https://doi.org/10.1038/srep24067

6. Tipu HN, Shabbir A (2015) Evolution of DNA sequencing. J Coll Physicians Surg Pak 25:210–215

7. Calcagno A-M et al.. Poster: a rapid and easy protocol for quantification of Illumina-NGS libraries. Presented at the 4thqPCR & Digital PCR Congress, London, UK, 20–21 October 2016

8. http://www.bioline.com/uk/

Chapter 24

Telomere Length Analysis: A Tool for Dissecting Aging Mechanisms in Developmental Programming

Jane L. Tarry-Adkins and Susan E. Ozanne

Abstract

Accelerated cellular aging is known to play an important role in the etiology of phenotypes associated with developmental programming, such as cardiovascular disease and type 2 diabetes. Telomere length analysis is a powerful tool to quantify cellular aging. Here we describe a telomere length methodology, refined to quantify discrete telomere length fragments. We have shown this method to be more sensitive in detecting small changes in telomere length than the traditional average telomere length comparisons.

Key words Developmental programming, Telomere length analysis, Pulsed field gel electrophoresis, Southern blotting

1 Introduction

In the early 1990s, Hales and Barker conducted several seminal studies, in which they demonstrated strong associations between individuals with a low birth weight and the increased incidence of impaired glucose tolerance [1], type 2 diabetes (T2D), hypertension and hyperlipidemia [2], and cardiovascular disease (CVD) [3] in later life. They proposed the "thrifty phenotype hypothesis" to explain these observations. The hypothesis proposed that in response to in utero undernutrition, the fetus permanently alters its organ structure and adapts its metabolism to ensure immediate survival of the organism. This occurs through the "sparing" of certain vital organs, especially the brain, at the expense of other organs, including the heart, pancreas, liver, kidney, and skeletal muscle. In addition, metabolic "programming" was proposed to occur, favoring nutrient storage. Support for the thrifty phenotype hypothesis was obtained from studies in both humans and animal models (reviewed in ref. 4). Subsequent studies in humans and animal models demonstrated that a broad range of suboptimal early environments including maternal protein restriction, maternal hypoxia, placental insufficiency, maternal caloric restriction,

Paul C. Guest (ed.), *Investigations of Early Nutrition Effects on Long-Term Health: Methods and Applications*, Methods in Molecular Biology, vol. 1735, https://doi.org/10.1007/978-1-4939-7614-0_24, © Springer Science+Business Media, LLC 2018

maternal stress, and maternal obesity were associated with increased risk of the development of age-associated disease, including insulin resistance, dysregulated glucose homeostasis, CVD, renal failure, and nonalcoholic fatty liver disease (reviewed in ref. 5). This has been termed the developmental origins of health and disease hypothesis.

Underlying mechanisms which are responsible for developmental programming are still emerging. However, it is apparent that most of the conditions associated with a suboptimal early environment are those traditionally associated with aging. Therefore, accelerated cellular aging has been implicated as a contributing mechanism (reviewed in ref. 5). For this reason, attention has been given to processes such as telomere shortening that are associated with aging of cells. Telomeres are hexameric repeat DNA sequences of $TAGGG^n$ found at the ends of chromosomes which are essential to preserve genomic integrity and stability [6]. Telomeric DNA is largely double stranded; however, the extreme terminus is a single-stranded G-rich $3'$ overhang that serves as a template for elongation and forms a telomeric "T-loop." This loop is stabilized by telomere binding proteins, notably telomere repeat binding factor-1 (TRF1) and telomere repeat binding factor-2 (TRF2). In normal somatic cells (i.e., those without the expression of the telomere elongation enzyme, telomerase), eukaryotic telomeres shorten with each cellular division. This is due to the end replication problem [7], in which full extension of a linear chromosome is not possible due to DNA polymerase requiring a labile primer to initiate unidirectional $5'$-$3'$ synthesis. Therefore, some bases at the $3'$ end of each template strand are not copied. More recently it has been demonstrated that telomeres also shorten in response to oxidative stress [8].

When telomeres reach a critically short length, irreversible cell senescence can be triggered via the activation of the retinoblastoma (Rb) and/or P53 proteins and expression of their regulators such as $P16^{INK4a}$, P21, and ARF [9] and/or apoptosis, via activation of the caspase proteins [9]. Consequently, many studies have sought to investigate the effect of telomere length dynamics upon cellular aging and disease pathogenesis. Several of these investigations have shown that telomere shortening is implicitly associated with cellular aging [10, 11] and disease pathogenesis [12–15], strongly suggesting that telomere length assessment is a robust marker of cell aging and a potential predictor of long-term health and, in some cases, of longevity [16].

There are currently five main methods used to measure telomere length: (1) the qPCR assay, (2) Q-FISH/FLOW-FISH, (3) single telomere length analysis, (4) dot blot, and (5) the telomere fragment length (TFL) assay using Southern blot [17]. These methodologies all have their own strengths and weaknesses. Although the TFL assay is a labor-intensive technique, which

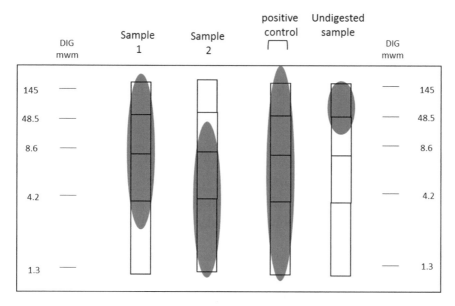

Fig. 1 Schematic telomere smear blot, representing analysis of specific telomere repeat factor lengths

requires a high amount of starting material, its strengths include the ability to provide telomere length distributions and reliable absolute telomere length measurement [17]. Over the last 15 years, we have developed a modified TRF assay in our laboratory in which the analysis of the resultant telomere smear provides information on telomere length distribution [18–22]. This varies from the traditional TRF analysis which defines only the mean telomere length. We have shown our methodology can detect differences in telomere length distribution that are missed if only mean length is calculated [18] (Fig. 1). This is particularly important as there is increasing evidence suggesting that regardless of mean telomere length, one critical short telomere may cause a cell to enter senescence [23].

In this chapter, we describe our detailed methodology.

2   Materials

2.1   High-Molecular-Weight DNA Extraction

1. 100 mM Tris-HCl (pH 8.0).

2. DNA lysis solution: 10 mM Tris-HCl (pH 8.0), 100 mM NaCl, 0.1 mM EDTA, 0.1% Triton-X 100.

3. Proteinase K solution (Promega).

4. Phenol solution and equilibration buffer (Sigma-Aldrich).

5. Chloroform.

6. Isoamylalcohol.

7. 7.5 M ammonium acetate.

8. 70% cold ethanol.

9. 100% cold ethanol.

10. Nuclease free water.

2.2
Spectrophotometric
Analysis

1. Nano-drop spectrophotometer (ThermoFisher Scientific).

2. DNA samples.

2.3 Agarose Gel
Electrophoresis

1. Powdered agarose (Bioline, UK, Ltd.; London, UK).

2. 10× Tris/borate/EDTA buffer (TBE).

3. SYBR safe dye (Invitrogen).

4. High-range-molecular-weight ladder (Bioline).

2.4 Restriction
Digestion

1. Digestion buffer (Telo TAGGG kit; Sigma-Aldrich).

2. 40 U/µL *Hinf*1 (Telo TAGGG kit).

3. 40 U/µL *Rsa*1 (Telo TAGGG kit).

4. DNA positive control (Telo TAGGG kit).

5. 5× loading buffer (Telo TAGGG kit).

2.5 Pulsed Field Gel
Electrophoresis (PFGE)

1. Running buffer: 1× TBE.

2. PFGE grade agarose (Bio-Rad; Hercules, CA, USA).

3. Gel cassette for the Chef-DR III PFGE unit (Bio-Rad).

4. Chef-DR III unit, cooling unit, and pump (Bio-Rad).

5. Dioxgenin (DIG)-labeled molecular weight ladder (Telo TAGGG kit).

6. Lambda full range molecular weight ladder (New England Biolabs).

2.6 Gel Visualization

1. Gene Snap transilluminator and software (Syngene; Cambridge, UK).

2.7 Southern Blotting

1. Acid nicking solution: 0.25 M HCl.

2. Denaturing solution: 1.5 M NaCl, 0.5 M NaOH.

3. Neutralizing solution: 0.5 M Tris base (pH 8.0), 1.5 M NaCl.

4. Southern blot vacuum transfer apparatus (Bio-Rad).

5. 10× SSC buffer: 300 mM sodium citrate (pH 7.0), 3 M NaCl.

6. Nylon membrane, cut to a size of 14.5 cm × 21 cm (Merck Millipore).

7. Whatman filter paper (3 mm), cut to a size of 14.5 cm × 21 cm.

8. Stratalinker cross-linker (Agilent; Santa Clara, CA, USA).

**2.8 Dioxgenin (DIG)
Telomere Detection
(All Reagents from
the Telo TAGGG Kit,
Roche)**

1. Hybridization granules (*see* **Note 1**).
2. DIG probe (Telo TAGGG kit).
3. Stringent wash I solution: 2× SSC, 0.1% SDS.
4. Stringent wash II solution: 0.2× SSC, 0.1% SDS.
5. 1 L 1× wash buffer.
6. 1× maleic acid buffer.
7. 150 mL 2× blocking buffer, 0.8× maleic acid buffer.
8. 100 mL 1× blocking buffer, 0.9× maleic acid buffer.
9. 1:10,000 anti-DIG antibody in 100 mL 1× blocking buffer.
10. 150 mL 1× detection buffer.
11. Substrate solution (*see* **Note 2**).

**2.9 Telomere Blot
Processing
and Telomere Length
Analysis**

1. Xomat film processor (Eastman Kodak Company).
2. Dark processing cassette.
3. Alpha Ease densitometry program (Alpha Innotech; Exeter, UK).

3 Methods

3.1 DNA Extraction

1. On dry ice, add 50–100 mg tissue (dependent upon tissue type) to a 2 mL Eppendorf tube.
2. Homogenize tissue with 650 μL DNA lysis buffer using an electric homogenizer.
3. Add 5 μL proteinase K solution per tube, and mix thoroughly using a vortex.
4. Incubate samples at 55 °C overnight in a water bath.
5. Vortex samples after removal from the water bath.
6. Add an equal volume (650 μL) of phenol/chloroform/isoamylalcohol (PCI) (25:24:1) per sample (*see* **Note 1**).
7. Mix the contents by hand until an emulsion forms.
8. Centrifuge the mixture at 30,000 × g for 3 min in a microcentrifuge at room temperature.
9. Remove the aqueous, DNA-containing phase into a fresh microcentrifuge tube and repeat the PCI extraction.
10. Remove the aqueous DNA-containing phase into a fresh microcentrifuge tube.
11. Precipitate the DNA by adding 500 μL of a mixture of 7.5 M ammonium acetate and 100% ethanol (0.5:2) (volume-volume) to the DNA solution.

12. Centrifuge the DNA at 30,000 × *g* for 3 min in a microcentrifuge.

13. Remove and discard the supernatant.

14. Add 500 μL 70% ethanol to the DNA pellet and repeat centrifuge as above.

15. Remove and discard the supernatant.

16. Add 500 μL 100% ethanol to the DNA pellet and repeat centrifuge as above.

17. Remove and discard the supernatant.

18. Dissolve DNA in 50 μL water (37 °C for 60 min) and store at −20 °C.

3.2 DNA Quantification by Nano-drop (See Note 2)

1. Take 1.3 μL DNA solution per sample to quantify using the nano-drop spectrophotometer.

2. Ensure the DNA has a 260/280 nm ratio of 1.8 or higher and a 260/230 nm ratio of approximately 2.0.

3. Calculate the volumes of DNA and water required for 1.2 μg starting material for the restriction digest.

3.3 DNA Integrity Check Using Agarose Gel Electrophoresis

1. Prepare a 1% agarose gel (containing 1:10,000 dilution of SYBR safe stain) in a gel cassette.

2. Add a gel comb and leave to set for 30 min.

3. Place in a horizontal electrophoresis tank and add 1× TBE running buffer.

4. Load samples containing 5 μL DNA, 5 μL water, and 4 μL 5× loading buffer into the set gel.

5. Add 10 μL high-molecular-weight ladder onto gel.

6. Run the gel for approximately 1 h at 130 V until adequate separation has been achieved.

7. Visualize the DNA integrity on a gel transilluminator, ensuring that the DNA is of a high molecular weight.

3.4 Restriction Digest (See Note 3)

1. Add the appropriate volume of DNA and water for 1.2 μg DNA per sample (to a final volume of 17 μL per sample).

2. To each sample, add 2 μL digestion buffer and 1 μL *Hinf1/ Rsa1* (1:1) restriction enzyme mix (*see* **Note 4**).

3. Prepare the control DNA sample as a positive control by omitting the restriction enzymes.

4. Incubate the samples at 37 °C for 2 h.

5. Quench the samples using 4 μL of 5× SDS loading buffer.

6. Centrifuge the samples and store at −20 °C until use.

3.5 Pulsed Field Gel Electrophoresis (PFGE)

1. Before beginning, set up the PFGE gel cassette by placing the clean black base gel plate onto the clear gel casting stand with two white edges.

2. Attach the two white side pieces and fix with four white screws.

3. Drain the PFGE tank of the 2 L of standing water (to protect the PFGE).

4. Release clamp on the drain tubing and disconnect the cooling unit tube from the tank unit.

5. Switch on the drive pump and hold the cooling unit tubing up to enable liquid to drain out of the tubes.

6. Switch off drive pump and tip the unit at an angle to ease emptying.

7. Level the unit using a leveling bubble centrally on the electrophoresis unit.

8. Reconnect the cooling unit tubing into the electrophoresis unit, and tighten the clamp on the drain tubing.

9. Pour the running buffer into the PFGE electrophoresis unit, close the lid, and switch on the cooling unit.

10. Ensure the temperature on the cooling unit is set to 14 °C.

11. Prepare a 0.8% PFGE grade agarose gel in 200 mL TBE.

12. Mix the contents of the bottle by swirling, and heat in a microwave for approximately 3 min at full power, until the agarose mixture is fully dissolved.

13. Leave the mixture to cool at room temperature for 10 min before adding 1:10,000 dilution of SYBR safe stain.

14. Pour the stained agarose gel into the pre-setup gel cassette and add the comb.

15. Leave the gel to set for 30 min, removing any air bubbles as they appear.

16. Once the agarose gel is set, remove it from the frame, keeping the gel on the black base plate.

17. Remove any extraneous gel from the underside of the base plate to avoid blocking the PFGE.

18. Lift the lid on the PFGE unit and place the gel into the electrophoresis tank.

3.6 Loading Samples onto the PFGE Gel

1. Ensure that the cooling unit and pump are both switched off while loading the gel, to prevent sample disturbance.

2. Load the gel lambda molecular weight ladder into the appropriate well by pushing up the syringe containing the molecular weight marker and removing a segment of the gel with a scalpel blade.

3. Prepare the dioxygenin (DIG)-labeled molecular weight ladder by adding 8 μL DIG marker, 4 μL water, and 4 μL loading buffer.

4. Heat for 5 min at 65 °C to denature the marker.

5. Load the full load volumes of each restricted DNA sample and undigested DNA control into the appropriate lanes, ensuring that the samples and control have previously been centrifuged.

6. Add 8 μL of the DIG-labeled molecular weight marker in two lanes on either side of the gel.

7. Set the power pack to the following parameters: initial switch time, 1 s; final switch time, 30 s; V/cm, 6; included angle, 120°; for rat samples, set the run time for 7.5 h; for human samples, set the run time for 8 h.

3.7 Gel Visualization of the Restricted DNA Samples

1. After the gel has run, place the gel into a container and restain the gel with SYBR stain in TBE buffer and shake for 30 min at room temperature.

2. 5 min before the visualization step will be completed, melt a 1% solution of agarose in the microwave (3 min at full power) and allow to cool.

3. Visualize the gel using the transilluminator and software package from Gene Snap and ensure that all telomeric DNA samples have been digested (*see* **Note 5**).

4. Before taking a picture, ensure a small ruler is placed on the gel in order to aid with high-molecular-weight quantification later, as the high-molecular-weight gel marker is not DIG-labeled.

3.8 Pre-Southern Blotting

1. Remove the SYBR safe stain solution and fill the empty loading wells of the gel with the previously prepared melted agarose gel solution.

2. Place the filled, stained gel into a container and depurinate in 1 L acid nicking solution and incubate for 10 min on a shaker at room temperature.

3. Rinse the gel twice with water.

4. Denature the gel in 1 L denaturation solution and incubate for 30 min on a shaker at room temperature.

5. Rinse the gel twice with water.

6. Add 1 L neutralizing solution and incubate for 30 min on a shaker at room temperature.

7. Transfer the gel to a container with 10× SSC, and begin the transfer procedure.

3.9 Southern Blotting and DNA Cross-Linking (See Note 6)

1. During the previous DNA neutralization step, ensure that the vacuum pump to the Southern blot vacuum apparatus has been switched on and that the porous vacuum base plate has been thoroughly soaked in water.

2. Wet the precut nylon membrane in a container containing water by slowly lowering the membrane at a 45° angle into the water.

3. Then lower both the membrane and the filter paper into a container with 10× SSC buffer.

4. Place the presoaked porous vacuum base plate in the base of the vacuum stage.

5. Place the pre-wetted filter paper on the porous vacuum base plate, ensuring that it is placed centrally (*see* **Note 7**).

6. Place the pre-wetted membrane on top of the filter paper and remove any air bubbles by rolling a 10 mL plastic pipette over the membrane.

7. Wet the reservoir O-ring seal with water.

8. Place the green precut window gasket on top of the membrane/filter paper stack and make sure the window gasket covers the entire O-ring seal on the vacuum stage.

9. At the same time, ensure that the membrane/filter paper stack is overlapping the window gasket and realign as necessary.

10. Gently place the gel, well side up, on top of the window gasket (*see* **Note 8**).

11. Remove any air bubbles using the 10 mL pipette (*see* **Note 9**).

12. Place the white sealing frame on top of the vacuum stage and lock the sealing frame into the four latch posts (*see* **Note 10**).

13. Unscrew the vacuum regulator bleed value to prevent strong initial vacuum.

14. Connect the waste flask to the vacuum pump and connect the other tubing from the flask to the vacuum stage.

15. Start the vacuum source and slowly turn the bleeder valve to 5 mmHg.

16. With a finger, apply gentle pressure on top of the gel along the window border (*see* **Note 11**).

17. Gently pour 1–1.5 L 10× SSC buffer into the upper reservoir (*see* **Note 12**).

18. Run the Southern transfer for 90 min.

19. After the vacuum Southern transfer, UV cross-link the DNA on the membrane by washing the membrane with 2× SSC buffer, then place the membrane (DNA side up) into the Stratalinker DNA cross-linker and cross-link at 12,000 J.

20. The cross-linked DNA membrane may be air-dried and stored at 4 °C until required for detection.

3.10 DIG Telomere Detection (See Notes 13 and 14)

1. Pre-warm 130 mL pre-dissolved pre-hybridization solution at 42 °C in a hybridization oven (*see* **Note 15**).

2. Place Stringent Wash II buffer in a shaking water bath at 50 °C until required.

3. Place the dried DNA membrane, DNA side inward, into a clean large hybridization tube.

4. Pour 60 mL of the pre-warmed pre-hybridization solution quickly into the hybridization tube.

5. Seal the lid and place the tube on the rotor in the hybridization oven for 45 min at 42 °C.

6. Just before the pre-hybridization step ends, put a further 60 mL of the pre-hybridization solution into a bottle, and add 8.4 μL DIG telomere probe.

7. Remove the pre-hybridization solution (*see* **Note 16**) and add the pre-made hybridization solution to the tube containing the membrane.

8. Incubate in the hybridization oven for 3 h at 42 °C.

9. Remove the hybridization solution (*see* **Note 16**) and place the membrane (DNA side up) into a container with 100 mL Stringent Wash I and incubate 20 min at room temperature on a rocker.

10. Repeat the above step with a further 100 mL Stringent Wash I.

11. Discard the Stringent Wash I and add 100 mL pre-warmed (at 50 °C) Stringent Wash II to the membrane.

12. Seal up the container thoroughly with parcel tape and place the box in the pre-heated 50 °C shaking water bath for 20 min.

13. Discard the Stringent Wash II solution and repeat the above step with a further 100 mL 50 °C Stringent Wash II.

14. Discard the Stringent Wash II buffer and add approximately 200 mL 1× wash buffer and incubate at room temperature for 30 min, with gentle agitation.

15. Discard the 1× wash buffer and add 150 mL freshly prepared 2× blocking buffer.

16. Incubate at room temperature for 1 h, with gentle agitation (*see* **Note 13**).

17. During the last 10 min, prepare the anti-DIG antibody by spinning down the antibody at $13,000 \times g$ for 8 min at 4 °C.

18. Take a 10 μL aliquot of the antibody and add it to 100 mL 1× blocking buffer, freshly prepared (*see* **Note 13**).

19. Incubate at room temperature for 30 min with gentle agitation.

20. Rinse the membrane three times for 15 min each time, with approximately 200 mL 1× wash buffer per rinse.

21. Incubate the membrane in 150 mL 1× detection buffer (*see* **Notes 13** and **14**) for 3–5 min at room temperature, with gentle agitation.

22. Remove membrane from the detection buffer and place DNA side up on a piece of clean filter paper to remove excess detection reagent (*see* **Note 17**).

23. Immediately place the wet membrane DNA side up onto an opened, clean hybridization bag and quickly apply an even covering of substrate solution onto the membrane.

24. Immediately cover the membrane with the top layer of the hybridization bag and spread the substrate solution homogenously using tissue, over the membrane, avoiding air bubbles.

25. Leave for 5 min before developing the blot using an automatic developer and chemiluminescent films.

3.11 Analysis of Telomere Blots

1. Scan the telomere blot and save the image as a jpeg file.

2. Use a densitometry package such as Alpha Ease; draw four distinct regions corresponding to the following molecular weight ranges: 145–48.5 kb, 48.5–8.6 kb, 8.6–4.2 kb and 4.2–1.3 kb (*see* Fig. 1).

3. Place these grid squares in an area which represents the background intensity of the film.

4. Then place these grid squares to encompass each of the discreet telomere length fragment sizes, as quantified the molecular weight markers.

5. In order to quantify the percentage of different size telomere lengths in samples, the intensity (as expressed as photo-stimulated luminescence, PSL) needs to be quantified (*see* Note 18).

6. Calculate the percentage of PSL (% PSL) in each molecular weight range using the following formula: % PSL = intensity of a defined region ± background × 100/total lane intensity ± background (*see* **Note 19**).

4 Notes

1. Dissolve in 64 mL deionized water at 37 °C overnight.

2. Use fresh.

3. 24 h prior to first use, equilibrate the phenol solution to pH 7.9 ± 0.2, by adding the entire contents of the equilibration buffer provided to the large bottle of phenol. Mix gently and allow the phases to separate before use.

4. Ensure that all of these steps are carried out on ice.

5. Therefore, leave no visible DNA smear. The control DNA should produce a molecular weight smear of DNA, and the high-molecular-weight marker should also be clearly visible.

6. Prepare a stock 1:1 mixture of *Rsa1* and *Hinf1* restriction enzymes for use in the restriction digest.

7. This should be in an area where the cut window gasket will go, to ensure a tight seal.

8. The gel must overlap the window gasket.

9. As a final check, ensure that the edges of the gel overlap the window gasket by at least 0.5 cm.

10. The spring handle of the sealing frame has a precut region in the middle. Press down on this exposed area of the sealing frame with your thumbs until the latches snap onto all four latch posts.

11. This helps to form a tight seal between the gel and the window gasket.

12. Check to see whether the gel has been displaced. The gel should stick to the window gasket. If the gel floats, simply disassemble the sealing frame to drain the buffer and repeat the appropriate steps.

13. During the assembly of the Southern blot, work quickly to ensure that the membrane does not dry out.

14. All buffers for the DIG telomere detection can be made several hours in advance, except the $2 \times$ blocking buffer, the $1 \times$ blocking buffer with DIG antibody, and the $1 \times$ detection buffer.

15. Overnight is ideal.

16. During all steps in the DIG telomere detection process, the membrane must always be DNA side upward. You can reuse both the pre-hybridization and hybridization solutions at least three times, as long as it is frozen at $-20\ ^\circ\text{C}$ after use.

17. Do not allow to dry.

18. Each telomeric sample must be divided into specific grid squares according to molecular size ranges as detailed.

19. Any gel where the % PSL in any of the four telomeric regions analyzed is >1.5 SD from the mean needs to be discarded from the analysis and that gel repeated.

Acknowledgments

The authors are members of the University of Cambridge MRC Metabolic Disease Unit and are funded by the UK Medical Research Council (MC UU12012/04).

References

1. Hales CN, Barker DJ, Clark PM, Cox LJ, Fall C, Osmond C et al (1991) Fetal and infant growth and impaired glucose tolerance at age 64. BMJ 303:1019–1022

2. Barker DJ, Hales CN, Fall CH, Osmond C, Phipps K, Clark PM (1993) Type 2 - (non-insulin dependent) diabetes mellitus, hypertension and hyperlipidaemia (syndrome X): relation to fetal growth. Diabetes 36:62–67

3. Fall CHD, Osmond C, Barker DJP, Clark PMS, Hales CN, Stirling Y et al (1995) Fetal and infant growth and cardiovascular risk factors in women. BMJ 310:428–432

4. Tarry-Adkins JL, Ozanne SE (2011) Mechanisms of early life programming: current knowledge and future directions. Am J Clin Nutr 94:1765S–1771S

5. Tarry-Adkins JL, Ozanne SE (2017) Nutrition in early life and age-associated diseases. Ageing Res Rev 39:96–105

6. Blackburn EH (1984) The molecular structure of centromeres and telomeres. Annu Rev Biochem 53:163–194

7. Olovnikov AM (1996) Telomeres, telomerase and aging: origin of the theory. Exp Gerontol 31:443–448

8. von Zglinicki T (2000) Role of oxidative stress in telomere length regulation and replicative senescence. Ann N Y Acad Sci 908:99–110

9. Sharpless NE, de Pinto RA (2004) Telomeres, stem cells, senescence and cancer. J Clin Invest 113:160–168

10. Lenart P, Krejci L (2016) DNA, the central molecule of aging. Mutat Res 786:1–7

11. Jaskelioff M, Muller FL, Paik JH, Thomas E, Jiang S, Adams AC et al (2011) Telomerase reactivation reverses tissue degeneration in aged telomerase-deficient mice. Nature 469 (7328):102–106

12. Blasco MA (2005) Telomeres and human disease: ageing, cancer and beyond. Nat Rev Genet 13:611–622

13. Haycock PC, Heydon EE, Kaptoge S, Butterworth AS, Thompson A, Willeit P (2014) Leucocyte telomere length and risk of cardiovascular disease: systemic review and meta-analysis. BMJ 349:g4227. https://doi.org/10.1136/bmj.g4227

14. Zhao J, Miao K, Wang H, Ding H, Wang DW (2013) Association between telomere length and type 2 diabetes mellitus: a meta-analysis. PLoS One 8:e79993. https://doi.org/10.1371/journal.pone.0079993

15. Cawthorn RM, Smith KR, O'Brien E, Sivatchenko A, Kerber RA (2003) Association between telomere length in blood and mortality in people aged 60 years or older. Lancet 361 (9355):393–395

16. Heidinger BJ, Blount JD, Boner W, Griffiths K, Metcalfe NB, Monaghan P (2012) Telomere length in early life predicts lifespan. Proc Natl Acad Sci U S A 109:1743–1748

17. Nussey DH, Baird D, Barrett E, Boner W, Fairlie J, Gemmell N et al (2014) Measuring telomere length and telomere dynamics in evolutionary biology and ecology. Methods Ecol Evol 5:299–310

18. Cherif C, Tarry JL, Ozanne SE, Hales CN (2003) Ageing and telomeres: a study into organ- and gender-specific telomere shortening. Nucleic Acid Res 31:1576–1583

19. Tarry-Adkins JL, Joles JA, Chen JH, Martin-Gronert MS, van der Giezen DM, Goldschmeding R et al (2007) Protein restriction in lactation confers nephroprotective effects in the male rat and is associated with increased antioxidant expression. Am J Physiol Regul Integr Comp Physiol 293:R1259–R1266

20. Tarry-Adkins JL, Martin-Gronert MS, Chen JH, Cripps RL, Ozanne SE (2008) Maternal diet influences DNA damage, aortic telomere length, oxidative stress and antioxidant capacity in rats. FASEB J 22:2037–2044

21. Tarry-Adkins JL, Chen JH, Smith NS, Jones RH, Cherif H, Ozanne SE (2009) Poor maternal nutrition followed by accelerated catch up growth leads to telomere shortening and increased markers of cell senescence in rat islets. FASEB J 23:1521–1528

22. Aiken CE, Tarry-Adkins JL, Ozanne SE (2013) Suboptimal nutrition in utero causes DNA damage and accelerated aging of the female reproductive tract. FASEB J 27:3959–3965

23. Hemann MT, Strong MA, Hao LY, Greider CW (2001) The shortest telomere, not average telomere length, is critical for cell viability and chromosome stability. Cell 107:67–77

Chapter 25

Qualitative and Quantitative NMR Approaches in Blood Serum Lipidomics

Banny Silva Barbosa, Lucas Gelain Martins, Tássia B. B. C. Costa, Guilherme Cruz, and Ljubica Tasic

Abstract

Nuclear magnetic resonance (NMR) spectroscopy in combination with chemometrics can be applied in the analysis of complex biological samples in many ways. For example, we can analyze lipids, elucidate their structures, determine their nutritional values, and determine their distribution in blood serum. As lipids are not soluble in water, they are transported in blood as lipid-rich self-assembled particles, divided into different density assemblies from high- to very-low-density lipoproteins (HDL to VLDL), or by combining with serum proteins, such as albumins (human serum albumins (HSA)). Therefore, serum lipids can be analyzed as they are using only a 1:1 (v/v) dilution with a buffer or deuterated water prior to analysis by applying ^{1}H NMR or ^{1}H NMR edited-by-diffusion techniques. Alternatively, lipids can be extracted from the serum using liquid partition equilibrium and then analyzed using liquid-state NMR techniques. Our chapter describes protocols that are used for extraction of blood serum lipids and their quantitative ^{1}H NMR (^{1}H qNMR) analysis in lipid extracts as well as ^{1}H NMR edited by diffusion for direct blood serum lipid analysis.

Key words Blood serum lipids, Lipidomics, ^{1}H NMR, ^{1}H qNMR, ^{1}H NMR edited by diffusion

1 Introduction

Lipidomics is an emerging area of research that refers to the determination of the composition of lipids and their quantitative and nutritional analyses and also to the elucidation of their roles in biological matrixes [1, 2]. Due to its origin, lipidomics is intrinsically linked to metabolomics and can be considered as a subfield [3].

Lipids are hydrophobic molecules that perform diverse functions in the human system [4]. They are composed of structurally nonuniform building blocks and also show hundreds to thousands of possible permutations of fatty acyls or other moieties which are functionally or biosynthetically related [5–7]. Many lipids are building blocks of living cells, some act as signaling molecules,

Paul C. Guest (ed.), *Investigations of Early Nutrition Effects on Long-Term Health: Methods and Applications*, Methods in Molecular Biology, vol. 1735, https://doi.org/10.1007/978-1-4939-7614-0_25, © Springer Science+Business Media, LLC 2018

but almost all are critical in human health. For example, polyunsaturated fatty acids (PUFAs) such as the omega-3 and omega-6 types, which are present in organisms mainly as triacylglycerols and phospholipids, are essential. Among these PUFAs, the most known acids are docosahexaenoic acid (DHA – 22:6, $\Delta^{4,7,10,13,16,19}$), cis-eicosapentaenoic acid (EPA – 20:5, $\Delta^{5,8,11,14,17}$), and eicosatetraenoic acid (ARA – 20:4, $\Delta^{5,8,11,14}$) [8]. In the form of phospholipids, especially in glycerophosphocholines and glycerophosphoserines, differential metabolism of PUFAs has been linked to many disorders of the central nervous system [9]. On the other hand, triacylglycerols (TAGs) are energy storage molecules that are transported from the liver to the blood and thereafter to adipocytes. After long starvation periods and energy deprivation, TAGs are decomposed by lipolysis to fatty acids (FAs) which are returned to the blood to be distributed to the tissues that use them, such as the muscles [10, 11].

The World Health Organization [12] recommendation for a healthy diet cites that up to 30% of the energy in the diet should be obtained from fat. The fat intake should be composed of up to 10% saturated fatty acids (SFAs) and 6–10% PUFAs. Additionally, the ratio of omega-6 to omega-3 should be 4–5:1 to avoid hypolipidemic and hypocholesterolemic effects. Long-lasting effects of a wrong diet on human health have been well studied, and most of the analyses used in monitoring health problems include determination of lipid profiles in blood. Additionally, a change in concentration of certain lipids in blood and tissues has been observed in several human diseases such as diabetes, obesity, atherosclerosis, and Alzheimer's disease [13].

Composition of lipids and their distribution in human blood serum can be assessed by applying nuclear magnetic resonance (NMR) spectroscopy techniques, through qualitative and quantitative ^{1}H NMR experiments [14, 15]. Since it is a nondestructive and reproducible technique that can be performed with often simple or no sample preparation, NMR has been used for metabolomics of body fluids or tissues, in order to obtain biomarkers for numerous diseases or to monitor responses to pharmaceutical treatments [13, 16, 17]. The NMR methodology to be used depends on the molecules that are expected to be monitored and/or observed. When the strategy consists of studying human serum lipids, the extraction process eliminates other molecules, such as proteins and metabolites, keeping only the lipids in the sample. Thus, blood serum samples can be analyzed [18] without the interference of other molecules (Fig. 1). ^{1}H NMR spectra of isolated lipids can lead to insights on lipid composition and distribution, and FAs, cholesterol, TAGs, phospholipids, and quantities of omega-3 and -6 polyunsaturated fatty acids could be identified and quantified [19, 20]. The assignments of the most common proton signals

SAMPLE PREPARATION

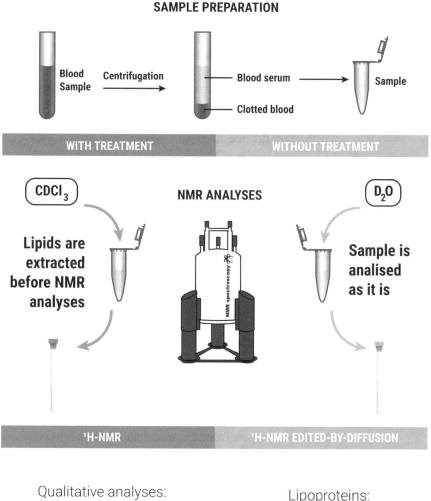

Fig. 1 Schematic representation of key steps in lipidomics by NMR. Blood sample is obtained by centrifugation of clothed blood where supernatant is collected and then analyzed. Lipids can be extracted (green) using $CHCl_3$ and CH_3OH liquid/liquid extraction procedure and dissolved in $CDCl_3$ prior to analysis or analyzed (blue) directly by adding 1/1 (v/v) of D_2O. NMR analyses are performed using ^1H NMR

related to fats can be found in the literature and in the lipid library [21].

However, when the research strategy is based on analysis of lipids as they appear in blood in their natural environment and without prior sample treatment, self-assembly suspension particles known as lipoproteins are observed. In this case, metabolites and proteins are not physically eliminated from the sample, making lipid evaluation impossible. These lipoproteins can be observed by applying a simple ^1H NMR pulse sequence [22]. Alternatively, it is possible to use ^1H NMR pulse sequences with spin filters, which involve the phenomena of relaxation and diffusion, both based on spin-echo pulse sequence block. Pulse sequences based on the spin relaxation, such as ^1H NMR edited-by-T_2-filter or CPMG pulse sequence, can filter the protein signals present in the human blood serum but do not filter the signals that refer to the metabolites [23, 24]. On the other hand, pulse sequences based on the diffusion, ^1H NMR edited by diffusion, permit detection of assemblies of lipids (HDL, LDL, and others), can eliminate all signals of the proteins and the low molecular weight metabolites present in human blood serum, but artificially introduce losses in the signal-to-noise ratio [25–27]. Therefore, the nutritional quality of lipids can be measured directly from serum, without previous extraction by applying ^1H NMR edited-by-diffusion experiments, in which the levels of lipoproteins such as HDL and LDL might be obtained (Fig. 1). But, this last methodology must not be used for quantification of lipids, since many of these species are not free but rather bound to or interact with serum proteins. Also, all ^1H NMR pulse sequence experiments used to evaluate diluted human blood serum must contain a water-removing signal block, and, in most cases, this block consists of the WATERGATE pulse sequence [28, 29].

Interpretation of the large amount of NMR data is facilitated by the application of multivariate statistical analysis. After exclusion of certain signals, for example, water and internal reference, the blood serum ^1H NMR spectra contain a few thousand points. Thus, in order to reduce the data matrix, the spectra are usually divided into segments of equal width (e.g., 0.001 ppm [30], 0.01 ppm [31, 32], 0.03 ppm [33, 34], 0.04 and 0.05 ppm) called bins or buckets [35].

In order to explore and discover the overall structure of the data, find trends, and groupings of samples, which can indicate the separation between the groups of interest, the exploratory analysis is applied. This initial overview of the dataset is made by unsupervised methods which do not assume any prior knowledge about classification of the samples. The most broadly used method is principal component analysis (PCA), which searches for common patterns by compressing the data into principal components that maximize description of the variance [16, 35–38]. The scores are the projection of the data into a coordinate system built from principal components, which provide visualization of the grouping of the samples. In addition, the loadings show the contribution of each original variable to the principal component. Thus, it is possible to determine the regions of the NMR spectra responsible for group separations [36, 39].

Partial least squares discriminant analysis (PLS-DA), one of the most widely used supervised method, is employed to better discriminate the samples of the different groups. It is a classification method aimed at grouping the objects into one of the classes, compressing the data into linear combinations of the original variables in the direction that maximizes the separation between classes (named latent variables) [16, 35, 36]. PLS-DA can be applied to improve the separation between samples of subjects with different lipid profiles due to nutritional deficits or disease presence [33, 34].

This chapter presents experimental procedures used for evaluation of nutritional quality of human blood serum and aims to describe sample preparation, NMR data recording, and their interpretation.

2 Materials

2.1 Sample Extraction

1. Serum samples (*see* **Notes 1** and **2**).
2. Biofreeze at −80 °C (*see* **Note 3**).
3. 99.8% chloroform.
4. 99.8% methanol.
5. 0.15 mM sodium chloride.
6. 99.8% deuterated chloroform, 0.03% tetramethylsilane (TMS) (*see* **Note 4**).
7. 5 mg/mL 99.86% reference standard – 1,2,4,5-tetrachloro-3-nitrobenzene (the solvent used for preparation of this solution is deuterated chloroform) (*see* **Note 5**).
8. 5 mm NMR tubes (*see* **Note 6**).

2.2 Serum Sample for Direct NMR Analyses

1. 99.96% deuterium oxide.

2.3 NMR Spectra Acquisition (Fig. 1)

1. Bruker Avance III 600 MHz spectrometer (*see* **Note 7**).
2. Triple-resonance broadband observed (TBO) probe (*see* **Note 8**).
3. Acquisition and data processing software Bruker TopSpin version 3.7 (*see* **Note 9**).

2.4 Chemometrics

1. Spreadsheet software: Microsoft Office Excel (*see* **Note 10**).
2. Access to chemometrics online platform MetaboAnalyst (*see* **Note 11**).

2.5 Nutritional Quality Evaluation of Lipids

1. Spreadsheet software: Microsoft Office Excel (*see* **Note 10**).

3 Methods

3.1 Extraction of Lipids (See Note 12)

1. Put 250 µL human serum sample into a 30 mL falcon tube (*see* **Note 13**).

2. Add 5 mL methanol and 10 mL chloroform and vortex 45 s (*see* **Note 14**).

3. Add 15 mL of sodium chloride solution and vortex 45 s (*see* **Note 15**).

4. Centrifuge 20 min at 2200 × *g* (*see* **Note 16**).

5. Lipids are dissolved in the chloroform, and thus collect the chloroform phase, and so use rota-evaporation to remove the solvent.

6. Transfer lipids to a flask with a known weight, and after determination of the weight of extracted lipids, store at −80 °C.

7. Add 100 µL of the standard solution into the lipid extract (*see* **Note 17**).

8. Add 500 µL of deuterated chloroform.

9. Transfer the solution into a NMR tube.

3.2 Blood Sample for Direct Analysis by NMR

1. To 250 µL of the human serum sample, add 250 µL of deuterium oxide.

2. Transfer the solution into a NMR tube.

3.3 Qualitative and Quantitative ^{1}H NMR Spectra Acquiring

1. Acquisition of qualitative ^{1}H NMR spectra is performed using parameters presented in Table 1 (*see* **Note 18**).

2. Process qualitative spectra, reference the chemical shifts to the solvent signal (1H, s, 7.26 ppm, residual proton in CDCl$_3$),

Table 1
Steps to 1D spectra acquisition

A	Solvent (CDCl$_3$)
B	Lock (automatic)
C	Probe match/tune (automatic correction)
D	Shimm (automatic)
E	Acquisition pars. Pulse sequence zg30, fid size 32,768, number of scans 56, spectral width 20.5533 ppm, acquisition time 1.33, receiver gain 203, delays (d1) 2 s
F	Prosol pars. Use the same number of scans utilized in acquisition parameters
G	Receiver gain (automatic)
H	Start acquisition

Table 2
Steps to quantitative 1D spectra acquisition

A	Solvent (CDCl$_3$)
B	Lock (automatic)
C	Probe match/tune (automatic correction)
D	Shimm (automatic)
E	Acquisition pars. Pulse sequence zg, size of fid 65,536, number of scans 56, spectral width 10.0115 ppm, acquisition time 8.18, receiver gain 203, delays (d1) 25 s (*see* **Note 19**)
F	Prosol pars. Use the same number of scans utilized in acquisition parameters.
G	Receiver gain (automatic)
H	Start acquisition

and export as American Standard Code for Information Interchange (ASCII) files.

3. Acquisition of quantitative ^{1}H NMR spectra is performed using parameters presented in Table 2 (*see* **Note 19**).

4. Process quantitative spectra, reference the chemical shifts to TMS signal (12H, s, 0.00 ppm, TMS), integrate all signals, calibrate ^{1}H to the standard signal, and export integrals to a data table (*see* **Notes 20** and **21**).

3.4 ^{1}H NMR Experiment Edited by Diffusion Using Stimulated Echo (See Note 22)

1. Acquisition of a ^{1}H NMR spectra edited by diffusion is qualitative (*see* **Note 23**), and the acquisition parameters are shown in Table 3 (*see* **Notes 24–26**).

2. The processing is to be performed equal to a common ^{1}H NMR spectra recording (*see* **Note 27**).

3.5 Lipid NMR Data Analysis by Chemometrics

1. Access the online platform MetaboAnalyst (www.metaboanalyst.ca), Statistical Analysis option and upload the data (*see* **Note 28**).

2. Choose preprocessing: sample normalization and data scaling (*see* **Note 29**).

3. Perform principal component analysis (PCA) and evaluate the sample's natural tendency to group into the classes and identify eventual outliers (*see* **Note 30**).

4. Remove outliers if observed and perform another PCA.

Table 3
Steps for acquisition of ^{1}H NMR spectra edited by diffusion

A	Solvent (D$_2$O)
B	Lock (automatic, *see* **Note 24**)
C	Probe match/tune (automatic correction)
D	Shimm (automatic)
E	Acquisition pars. Pulse sequence stebpgp1s191d, size of fid 65,536, number of scans 512, spectral width 10.0115 ppm, delays (d1) 5 s, diffusion time (d20) 400 ms (*see* **Notes 25** and **26**)
F	Prosol pars. Use the same number of scans utilized in acquisition parameters
G	Receiver gain (automatic)
H	Start acquisition

5. Observe the PCA scores and identify which principal components (PCs) caused separation of samples into the groups of interest.

6. Analyze loadings and identify spectral regions with high loadings in previously chosen PCs (these signals are related to the lipids that show different distribution in analyzed groups) (*see* **Note 31**).

7. If the PCA is not indicative for sample grouping, execute partial least squares discriminant analysis (PLS-DA), and evaluate the scores and loadings graphs the same as for the PCA data analysis (*see* **Note 32**).

8. Assign lipid signals using NMR data such as chemical shifts, coupling constants, and multiplicity of picks, by comparing data with the NMR scientific literature information (*see* **Note 33**).

3.6 ^{1}H NMR Spectral Data in the Evaluation of Lipid's Nutritional Quality

1. Transport integral values to a spreadsheet software (*see* **Note 34**).

2. Calculate concentration of each lipid type following Eq. 1 (*see* **Note 35**).

3. Calculate concentrations (average and standard deviation) of each lipid type.

$$P_{Sample} = (I_{Analyte}/I_{CRM}) \times (N_{CRM}/N_{Analyte}) \times (M_{Analyte}/M_{CRM}) \times (m_{CRM}/m_{Sample}) \times P_{CRM}$$

P_{Sample} = Purity of sample as mass fraction (%)
P_{CRM} = Purity of Certified Reference Material as mass fraction (%)
$I_{Analyte}$ = Integral of the analyte signal
I_{CRM} = Integral of the Certified Reference Material signal
$N_{Analyte}$ = Number of analyte protons
N_{CRM} = Number of Certified Reference Material protons
$M_{Analyte}$ = Molecular mass of the analyte (g/mol)
M_{CRM} = Molecular mass of the Certified Reference Material (g/mol)
m_{Sample} = Mass of sample (mg)
m_{CRM} = Mass of Certified Reference Material (mg)

Equation 1 ^{1}H qNMR in evaluation of lipids nutritional quality [40]

4 Notes

1. While handling human serum, wear a lab coat, eyewear, and gloves for self-protection during all processes, up to the point of sample storage after NMR analysis (*see* **Notes 24–26**)..

2. It is necessary to obtain authorization of local ethics committee and a document with experiment details, such as start and final dates and procedures to be included in the research.

3. Sample freezing and thawing procedures should be minimized as much as possible to avoid subtle changes in the spectroscopic profiles of lipids that could occur by degradation via enzymatic reactions and lead to a systematic error among the samples.

4. The deuterated solvent may contain reference compounds or not. The most common reference compound is TMS (tetra-methylsilane), which has a chemical shift at 0.00 ppm.

5. It is important to use specific standards for NMR analysis with certified purity. This standard is indicated for quantification of lipids because its signal is a singlet with chemical shift in the region of aromatic compounds, avoiding overlap with that of lipids.

6. For quantification, it is necessary to use specific NMR tubes of high quality.

7. Equipment such as 400 MHz or 500 MHz can be used for higher resolution.

8. Other types of probes can be used, such as TBI or BBI.

9. In addition to Topspin software, Icon software can be used as acquisition software. Data can be processed by other software (for instance, MestreNova).

10. Other software can be used such as Origin.

11. The chemometrics analyses can be made using many options of software, such as Pirouette, Matlab, Statistica, and the Unscrambler. We recommend the online platform MetaboAnalyst, which is free of charge and user-friendly [40].

12. Lipids are soluble in organic solvents and can be extracted from serum sample applying liquid/liquid extraction method in a similar manner to that described by Tukiainen et al. in 2008 [18].

13. At this step, a larger vessel is used so that there is room for stirring with the addition of the extractive solvents.

14. The stirring step turn the lipid extraction possible due to the miscibility of extraction solvents and serum, therefore the low polarity solvent, i.e. chloroform is allowed to be in contact with lipids from serum. However, it is not miscible with aqueous solutions like the serum, and so it needs the insertion of another solvent such as methanol.

15. At this step, the aqueous saline solution acts by increasing the coefficient of solubility of the polar phase, separating the solution in phases.

16. Centrifugation is a key step and should not be skipped since it ensures complete separation of the phases (lipids are contained in a single phase – the chloroform phase).

17. This step is essential for the quantification experiment since it refers to a known concentration of analyte.

18. The entire acquisition procedure occurs with greater efficiency in multicomponent analysis if automation is used. For this, Icon was used since it minimizes errors.

19. The value of d1 can be varied in groups of samples. To determine the value of d1, the experiment of determining T1 to calculate d1 must be performed, which will be used as parameter as five times T1.

20. The data should follow the following presentation: chemical shift in the first column and intensity in other columns.

21. If the data matrix has a large number of variables, it is advisable to apply a binning procedure. To choose the width of the bin, it is important to keep in mind that the smaller the bin, the less resolution is lost.

22. Analyzing lipids in human serum presents a difficulty due to the need to separate the lipids from the other components of the serum. In metabolomics, when it is necessary to eliminate large

molecules, such as proteins and self-assembly structures (as lipids in serum), a transverse relaxation filter can be used (T_2 filter using Carr-Purcell-Meiboom-Gill sequence). Although it is possible to use the CPMG sequence with parameters that eliminate the proteins and keep the lipids in the spectrum, the metabolites will also be maintained and will hinder the lipid analysis. However, the use of ^{1}H NMR edited by diffusion allows both, metabolites and proteins, to be eliminated. Metabolites are eliminated after the coding and decoding processes of the nuclear spins, intermediated by the diffusion time. Proteins are eliminated by a side effect of the diffusion time. During the diffusion time, the nuclear spins lose coherence by relaxation. As the proteins relax rapidly, they are eliminated depending on the diffusion time used.

23. As noted above, nuclear spins relax during the diffusion time. The loss of magnetization due to relaxation is different to each spin group and to each diffusion time chosen. For this reason, the ^{1}H NMR spectra edited by diffusion is a qualitative analysis. As a methodology used to observe lipids in serum just as they are, the sample preparation for the ^{1}H NMR experiment edited by diffusion is straightforward. This methodology is the only one able to promote a virtual elimination of signals of metabolites and proteins from the sample and enables observation of signals of lipid-rich self-assembled particles as illustrated in Fig. 2.

24. For high magnetic field equipment, the lock phase is easily lost after automatic shimming. To work around this problem, it is necessary to make an automatic adjustment of the lock phase after shimming is performed. For diffusion experiments, a precise lock phase should be used.

25. For good metabolite elimination, 200 ms of diffusion time is enough. However, this time is too short to promote protein relaxation. The choice of 400 ms for diffusion time is suitable to eliminate both metabolites and proteins, but as a consequence the signal-to-noise ratio is impaired because lipids also undergo spin relaxation during the diffusion time (although this is less than for proteins). In order to obtain a proper signal-to-noise ratio, the number of scans should be high.

26. Biological samples contain a large amount of water. Thus, residual H (HDO) signal needs to be suppressed in the spectra so that the remaining components in solution can be observed. There are several ways to suppress water, but the stebpgp1s191d pulse sequence, used to obtain the ^{1}H NMR spectra edited by diffusion, has a WATERGATE block to this role. In the WATERGATE block, a null is created over the

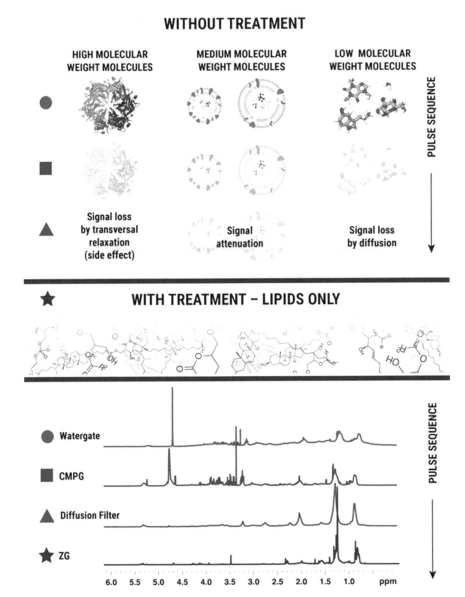

Fig. 2 Illustration of a composition of a blood serum that contains molecules with low up to high molecular weight species (upper panel). Lipids belong primarily to a group of medium molecular weight compounds (i.e., self-assembled lipid particles) but might also be associated to many serum proteins and thus occur in a form bound to a carrier protein. After extraction, lipids are solubilized in $CDCl_3$ previous to analysis (purple star). When analyzing lipids by NMR using different methodologies, we can observe different species (lower panel) by (1) WATERGATE suppression (orange circle) where all species are observed in water (metabolites, lipids, proteins); (2) by applying CPMG technique (blue square) or T_2-edited ^1H NMR (protein signals are being filtered and thus disappear from the spectra); (3) by applying ^1H NMR (green triangle) edited-by-diffusion method (all picks from proteins and molecules with low molecular weight disappear from the spectra with just lipids being observed); and (4) when analyzing extracts from the blood serum (purple star); only free lipids are observed

water signal in spectra. It is necessary to calculate the distance of the next null. This is the delay for binomial water suppression (d19), and it can be calculated by $1/2*d$ (d is the distance in Hz). This parameter is primordial for obtaining a good spectrum.

27. Lipids are found in human serum, such as in any aqueous solution, as self-assembly structures (VLDL, LDL, IDL, HDL). For this reason, the signals have bad definition, and the spectrum can be processed using a high lb value (line broadening; lb = 3).

28. The data is the table of peak integrals (described in **Note 20**) added to the class information. Depending on the software used, the type of file extension required may vary. For MetaboAnalyst, it is necessary that the data are in the .csv extension. For Pirouette, it is necessary that it is saved in Excel 93. If MestreNova is used to process the ^{1}H NMR spectra, there is an option to save the data as a .csv file.

29. As the concentration of the samples is known, there is no need for normalization. For PCA and PLS-DA analysis, it is recommended to apply mean center for variable preprocessing or data scaling, and the choice depends on the data. Sometimes autoscaling is the best option for the dataset, although this should be tested.

30. Outliers are samples that behave differently in the dataset. That may be a consequence of experimental errors or specific characteristics of a subject and does not inform about the group studied.

31. Some programs (for instance, MetaboAnalyst) can cross the data of scores and loadings through biplots, facilitating interpretation of data.

32. PLS-DA is a technique that improves the separation between the classes. Since it is a classification method, the samples are classified into the groups, and it is possible to obtain the accuracy of the model constructed by cross-validation.

33. Lipids have characteristic profiles that are known by NMR. However, because the blood shows multicomponent distribution of lipids, there is a lot of overlapping of signals. Still, blood serum NMR spectra show the most important lipid picks such as those of H of glycerol, H of methylenes, $-CH_2$ of the homologous series and C-2 present in fatty acids, and H of cholesterol methyl groups.

34. In a spreadsheet, data should be presented as chemical shift in the first column and the absolute integrals in the other columns.

35. Using molecular mass (M) representative for each lipid class, nutritional values can be determined for: (1) triacylglycerols (4.19 ppm, M from TAG 18:0/20:5 (5Z,8Z,11Z,14Z,17Z)/22:1(11Z) compound), (2) phospholipids (3.94 ppm, M from palmitoyloleoylphosphatidylcholine), (3) cardiolipins (3.85 ppm, M from tetralinoleoylcardiolipin), and (4) sterols (0.68 ppm, M from cholesterol). Also, the following compounds can be quantified by ^{1}H qNMR: fatty acids (FAs) (1.29 ppm, M from palmitic acid), monounsaturated FAs (2.03 ppm, M from oleic acid), other polyunsaturated FAs (except linoleic, 2.83 ppm, M from adrenic acid), linoleic acid (2.76 ppm), arachidonic acid (1.68 ppm), EPA (0.98 ppm), and DHA (2.38 ppm).

References

1. Watson AD (2006) Thematic review series: systems biology approaches to metabolic and cardiovascular disorders. Lipidomics: a global approach to lipid analysis in biological systems. J Lipid Res 47:2101–2111

2. Wenk MR (2005) The emerging field of lipidomics. Nat Rev Drug Discov 4:594–610

3. Navas-Iglesias N, Carrasco-Pancorbo A, Cuadros-Rodríguez L (2009) From lipids analysis towards lipidomics, a new challenge for the analytical chemistry of the 21st century. Part II: analytical lipidomics. TrAC Trends Anal Chem 28:393–403

4. Rai S, Bhatnagar S (2017) Novel lipidomic biomarkers in hyperlipidemia and cardiovascular diseases: an integrative biology analysis. OMICS 21:132–142

5. Vaz FM, Pras-Raves M, Bootsma AH, van Kampen AHC (2015) Principles and practice of lipidomics. J Inherit Metab Dis 38:41–52

6. Gurr MI, Harwood JL, Frayn KN (2002) Lipid biochemistry: an introduction, 5th edn. Blackwell, Oxford

7. Ståhlman M, Borén J, Ekroos K (2012) High-throughput molecular lipidomics. In: Lipidomics. Wiley-VCH, Weinheim

8. Gurr MI (1999) Lipids in nutrition and health: a reappraisal, 1st edn. Woodhead, Cambridge

9. Chaung H-C, Chang C-D, Chen P-H, Chang CJ, Liu SH, Chen CC (2013) Docosahexaenoic acid and phosphatidylserine improves the antioxidant activities in vitro and in vivo and cognitive functions of the developing brain. Food Chem 138:342–347

10. Hyötyläinen T, Bondia-Pons I, Orešič M (2013) Lipidomics in nutrition and food research. Mol Nutr Food Res 57:1306–1318

11. Kaur N, Chugh V, Gupta AK (2014) Essential fatty acids as functional components of foods – a review. J Food Sci Technol 51:2289–2303

12. World Health Organization (2015) Healthy diet. World Health Organization, Geneva

13. Hu C, van der Heijden R, Wang M, van der Greef J, Hankemeier T, Xu G (2009) Analytical strategies in lipidomics and applications in disease biomarker discovery. J Chromatogr B 877:2836–2846

14. Stark RE, Gaede HC (2001) NMR of a phospholipid: modules for advanced laboratory courses. J Chem Educ 78:1248

15. Kaffarnik S, Ehlers I, Gröbner G, Schleucher J, Vetter W (2013) Two-dimensional 31 P, 1 H NMR spectroscopic profiling of phospholipids in cheese and fish. J Agric Food Chem 61:7061–7069

16. Yu Y, Vidalino L, Anesi A, Macchi P, Guella G (2014) A lipidomics investigation of the induced hypoxia stress on HeLa cells by using MS and NMR techniques. Mol BioSyst 10:878

17. Gross RW, Han X (2011) Lipidomics at the interface of structure and function in systems biology. Chem Biol 18:284–291

18. Tukiainen T, Tynkkynen T, Mäkinen V-P, Jylänki P, Kangas A, Hokkanen J et al (2008) A multi-metabolite analysis of serum by 1H NMR spectroscopy: early systemic signs of Alzheimer's disease. Biochem Biophys Res Commun 375:356–361

19. Nuzzo G, Gallo C, D'Ippolito G, Cutignano A, Sardo A, Fontana A (2013) Composition and quantitation of microalgal lipids by ERETIC 1H NMR method. Mar Drugs 11:3742–3753

20. Siciliano C, Belsito E, De Marco R, Di Gioia ML, Leggio A, Liguori A (2013) Quantitative determination of fatty acid chain composition in pork meat products by high resolution 1H NMR spectroscopy. Food Chem 136:546–554

21. Ala-Korpela M, Korhonen A, Keisala J, Hörkkö S, Korpi P, Ingman LP et al (1994) 1H NMR-based absolute quantitation of human lipoproteins and their lipid contents directly from plasma. J Lipid Res 35:2292–2304

22. Meiboom S, Gill D (1958) Modified spin-Echo method for measuring nuclear relaxation times. Rev Sci Instrum 29:688–691

23. Carr HY, Purcell EM (1954) Effects of diffusion on free precession in nuclear magnetic resonance experiments. Phys Rev 94:630–638

24. Lopes TIB, Geloneze B, Pareja JC, Calixto AR, Ferreira MM, Marsaioli AJ (2015) Blood metabolome changes before and after bariatric surgery: a 1 H NMR-based clinical investigation. OMICS 19:318–327

25. Dyrby M, Petersen M, Whittaker AK, Lambertc L, Nørgaarda L, Bro R et al (2005) Analysis of lipoproteins using 2D diffusion-edited NMR spectroscopy and multi-way chemometrics. Anal Chim Acta 531:209–216

26. Price WS, Elwinger F, Vigouroux C, Stilbs P (2002) PGSE-WATERGATE, a new tool for NMR diffusion-based studies of ligand-macromolecule binding. Magn Reson Chem 40:391–395

27. Liu M, Mao X, Ye C, Huang H, Nicholson J, Lindon JC (1998) Improved WATERGATE pulse sequences for solvent suppression in NMR spectroscopy. J Magn Reson 132:125–129

28. Lee GSH, Wilson MA, Young BR (1998) The application of the "WATERGATE" suppression technique for analyzing humic substances by nuclear magnetic resonance. Org Geochem 28:549–559

29. Tranchida F, Shintu L, Rakotoniaina Z, Tchiakpe L, Deyris V, Hiol A et al (2015) Metabolomic and lipidomic analysis of serum samples following Curcuma longa extract supplementation in high-fructose and saturated fat fed rats. PLoS One 10:e0135948

30. Fernando H, Kondraganti S, Bhopale KK, Volk DE, Neerathilingam M, Kaphalia BS et al (2010) 1H and 31P NMR lipidome of ethanol-induced fatty liver. Alcohol Clin Exp Res 34:1937–1947

31. Fernando H, Bhopale KK, Kondraganti S, Kaphalia BS, Shakeel Ansari GA (2011) Lipidomic changes in rat liver after long-term exposure to ethanol. Toxicol Appl Pharmacol 255:127–137

32. Kostara CE, Tsimihodimos V, Elisaf MS, Bairaktari ET (2017) NMR-based lipid profiling of high density lipoprotein particles in healthy subjects with low, normal, and elevated HDL-cholesterol. J Proteome Res 16:1605–1616

33. Kostara CE, Papathanasiou A, Psychogios N, Cung MT, Elisaf MS, Goudevenos J et al (2014) NMR-based lipidomic analysis of blood lipoproteins differentiates the progression of coronary heart disease. J Proteome Res 13:2585–2598

34. Smolinska A, Blanchet L, Buydens LMC, Wijmenga SS (2012) NMR and pattern recognition methods in metabolomics: from data acquisition to biomarker discovery: a review. Anal Chim Acta 750:82–97

35. Savorani F, Rasmussen MA, Mikkelsen MS, Engelsen SB (2013) A primer to nutritional metabolomics by NMR spectroscopy and chemometrics. Food Res Int 54:1131–1145

36. Beger RD, Schnackenberg LK, Holland RD, Li D, Dragan Y (2006) Metabonomic models of human pancreatic cancer using 1D proton NMR spectra of lipids in plasma. Metabolomics 2:125–134

37. Ouldamer L, Nadal-Desbarats L, Chevalier S, Body G, Goupille C, Bougnoux P (2016) NMR-based lipidomic approach to evaluate controlled dietary intake of lipids in adipose tissue of a rat mammary tumor model. J Proteome Res 15:868–878

38. Bro R, Smilde AK (2014) Principal component analysis. Anal Methods 6:2812

39. Weber M, Hellriegel C, Rueck A et al (2014) Using high-performance 1H NMR (HP-qNMR®) for the certification of organic reference materials under accreditation guidelines – describing the overall process with focus on homogeneity and stability assessment. J Pharm Biomed Anal 93:102–110

40. Xia J, Wishart DS (2016) Using MetaboAnalyst 3.0 for comprehensive metabolomics data analysis. In: Curr Protoc Bioinforma. Wiley, Hoboken, NJ, pp 14.10.1–14.10.91

Chapter 26

In Vivo Electrical Stimulation for the Assessment of Skeletal Muscle Contractile Function in Murine Models

Kaio F. Vitzel, Marco A. Fortes, Gabriel Nasri Marzuca-Nassr, Maria V. M. Scervino, Carlos H. Pinheiro, Leonardo R. Silveira, and Rui Curi

Abstract

Skeletal muscle electrical stimulation is commonly used for clinical purposes, assisting recovery, preservation, or even improvement of muscle mass and function in healthy and pathological conditions. Additionally, it is a useful research tool for evaluation of skeletal muscle contractile function. It may be applied in vitro, using cell culture or isolated fibers/muscles, and in vivo, using human subjects or animal models (neuromuscular electrical stimulation – NMES). This chapter focuses on the electrical stimulation of the sciatic nerve as a research method for evaluation of the contractile properties of murine hind limb muscles. Variations of this protocol allow for the assessment of muscle force, fatigue resistance, contraction and relaxation times, and can be used as a model of contraction-induced muscle injury, reactive oxygen species production, and muscle adaptation to contractile activity.

Key words Murine muscle contraction, Relaxation, Fatigue, NMES, Contractile parameters, Electrostimulation

1 Introduction

Neuromuscular electrical stimulation (NMES) consists in the delivery of electrical stimuli to motor nerves or superficial skeletal muscles, inducing muscle contraction [1]. It is used as a common therapeutic strategy for several disorders affecting the skeletal muscle. Nonetheless, it is a similarly effective research tool for in vivo assessment of neuromuscular function in different conditions (e.g., normal, fatigued, or injured muscle) and evaluation of muscle contraction properties in a standardized way (e.g., force-frequency relationship, fatigue during constant stimulation, comparison between maximal voluntary contraction and NMES-induced contraction) [1].

Muscle contraction may be "partial," as a muscle fasciculation, or "total," as a tetanic contraction, depending on the manipulations

Paul C. Guest (ed.), *Investigations of Early Nutrition Effects on Long-Term Health: Methods and Applications*, Methods in Molecular Biology, vol. 1735, https://doi.org/10.1007/978-1-4939-7614-0_26, © Springer Science+Business Media, LLC 2018

of NMES parameters [2]. Important parameters that must be considered are pulse amplitude, duration, frequency, and the location for delivery of the stimuli (nerve or muscle) [3]. For example, electrical stimulation directly to the nerve produces repetitive depolarization of axons in contact with the electrodes that will reach the neuromuscular junction and elicit muscle contraction. If pulse voltage is increased, axonal depolarization and recruitment of motor units are enhanced (spatial summation), producing more intense muscle contraction [3].

Frequency selection is also crucial and depends on the aim of the evaluation, as it influences contraction pattern, recruitment of motor units, and fiber types elicited. Usually, frequency is considered low when below 20 Hz and high when above 30 Hz. If above 80 Hz, they are regarded as tetanic frequencies, when complete muscle relaxation between stimuli is prevented and temporal summation occurs until a fused tetanus producing maximal force is achieved. Isotonic muscle contractions (twitch) are elicited at low frequencies (1–10 Hz), and maximal tetanic muscle contractions are achieved at high frequencies (~100 Hz) [4–6]. Low-frequency stimulation activates fewer and smaller motor units, with predominantly oxidative/slow/type I fibers, and high-frequency stimulation activates more and larger motor units, with glycolytic/fast/type II muscle fibers. For example, rat soleus muscle, composed mainly by oxidative/slow/type I muscle fibers (83% type I, 17% type IIa, 0% type IIb) [7], presents a median motor unit firing frequency of 19–26 Hz [8]. The extensor digitorum longus (EDL) muscle, composed mainly by glycolytic/fast/type II muscle fibers (3% type I, 54% type IIa, 43% type IIb) [7], presents a median motor unit firing frequency of ~30 Hz [9]. During volitional contractions of increasing intensity, smaller motor units are recruited first, followed by larger ones, a pattern that is reversed during NMES [10].

Electrical stimulation protocols performed in muscle cell culture and isolated fibers, both in vitro models, provide a highly controlled environment for muscle studies, and the potential of manipulating the culture conditions and perfusion media confers additional experimental advantage. These are powerful approaches for investigation of specific mechanisms underlying muscle function. However, in situ and in vivo models take into account whole-body dynamics, including cardiovascular, respiratory, neural, and endocrine features. They represent the "final effect" of the mechanisms investigated by in vitro models and molecular studies on muscle contractile function. Therefore, in vivo models are a necessary complement to in vitro studies. They are also useful tools on investigation of muscle function during pathological conditions, such as diabetic myopathy, disuse-induced atrophy, muscular dystrophy, cachexia, several neuromuscular disorders, and other situations, including ageing, muscle regeneration, hypertrophy, stem cell treatment, and nutritional supplementation [11–16].

The method for murine NMES described here is based on Wojtaszewski et al. [17] and Silveira et al. [18] with modifications by Pinheiro et al. [12] and Fortes et al. [15]. It provides a direct assessment of basic parameters of neuromuscular excitability (rheobase and chronaxie) and skeletal muscle contractile function (force and shortening/relaxation properties). Specific contraction protocols can be developed according to the purposes of the study. This experimental setup supports short- or long-duration protocols, as well as the development of force-frequency curves (Fig. 1), providing models for analysis of muscle strength, fatigue, contraction-induced injury, reactive oxygen species production, recruitment of specific motor units/fiber types, and muscle adaptation to contractile activity.

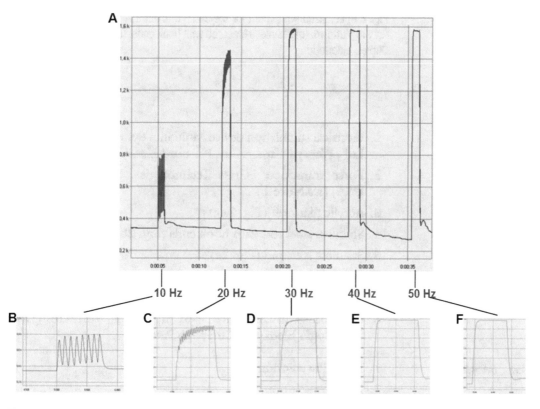

Fig. 1 Screenshots of myograms ("force × time" graphs) from rat triceps surae muscle group at increasing frequencies for development of a force-frequency curve. Pulse duration: 500 μs. Voltage: 10–15 V. Force is presented in gram-force (1gf = 9.806 × 10^{-3} N). Time is presented in s. (**a**) Overall image showing summation effect and progressive force increase at 10 Hz (**b**), 20 Hz (**c**), and 30 Hz (**d**), until a fused tetanic contraction and force plateau are achieved at 40 Hz (**e**) and 50 Hz (**f**). Images acquired from the data analysis software AqAnalysis® (Lynx Tecnologia Eletronica Ltda, Sao Paulo, Brazil)

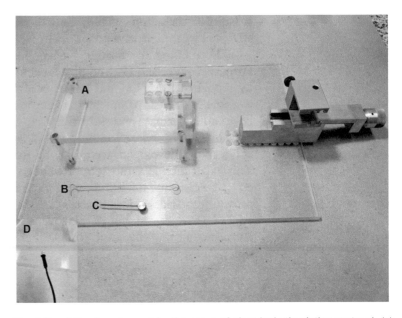

Fig. 2 Specialized equipment for the setup of electrical stimulation protocol. (**a**) Acrylic platform. (**b**) Stainless steel rod with hook ends. (**c**) Stainless steel pin. (**d**) Curved electrode

2   Materials

2.1   Equipment

1. Electrical stimulation device MultiStim System D330 (Digitimer Ltd., Welwyn Garden City, Hertfordshire, UK) (*see* **Note 1**).
2. Force transducer (Grass Technologies, West Warwick, RI, USA) (*see* **Note 1**).
3. Acrylic platform (Fig. 2a).
4. Stainless steel rod with hook ends (Fig. 2b).
5. Stainless steel pin (Fig. 2c).
6. Curved electrodes (Fig. 2d).
7. Heating lamp.
8. Computer with AqDados® and AqAnalysis® softwares (Lynx Tecnologia Eletronica Ltda, Sao Paulo, Brazil).

2.2   Reagents

1. Ketamine solution.
2. Xylazine solution.

3   Methods (*see* Note 2)

3.1   Anesthesia

1. Anesthetize rats or mice by intraperitoneal injection of ketamine and xylazine 90/10 mg/kg (*see* **Note 3**).

2. Maintain consistency in terms of time of day, and use a heating lamp and/or heated platform for maintenance of regular body and muscle temperatures during the protocol (*see* **Note 4**).

3.2 Sciatic Nerve Access and Electrode Placement

1. Using a pair of scissors and tweezers, retract the skin covering the anesthetized rat/mouse hind limb, exposing the thigh and calf muscles.

2. Using a craniolateral approach, palpate the femur and estimate the incision site at about 1/3 to 1/2 of the proximal length of the thigh (the interval between superficial gluteal muscles and biceps femoris is identified as the access point) (*see* **Note 5**).

3. Place a pair of scissors orthogonally to the muscle surface to pierce the fascia and displace the muscle fibers, by inserting the closed scissors, slightly opening and retracting it while still partially open.

4. Perform this movement along the length of the boundary between the gluteal and biceps femoris muscles.

5. Use a pair of tweezers to hold the 1-cm-long incision open, and move the muscle layer, until detection of the sciatic nerve underneath (Fig. 3a).

6. Place the curved electrodes on the nerve, which is still above the trifurcation level, avoiding pulling and straining motions (Fig. 3b).

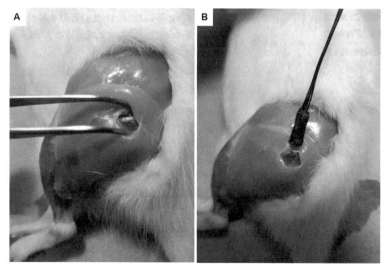

Fig. 3 Sciatic nerve access and electrode placement. (**a**) Access point between gluteal and biceps femoris muscles, where the sciatic nerve is detected beneath the superficial muscle tissue. (**b**) The curved electrode is placed avoiding straining of the nerve

3.3 Tenotomy

1. To evaluate the contractile function of a single muscle (soleus, EDL, or gastrocnemius) or muscle group (triceps surae, plantaris + gastrocnemius + soleus), tenotomize corresponding synergistic and antagonist muscles and retract their fascia.

2. Prevent excessive blood loss if damaging main blood vessels, such as the saphenous artery (some loss cannot be avoided).

3. Alternatively, the contractile function of the whole calf may be assessed without the need of tenotomy (*see* **Note 6**).

3.4 Animal Setup on Acrylic Platform

1. Place the animal in supine position on the acrylic platform while the forelimbs and the remaining hind limb are immobilized and taped to the platform (the hip joint should be positioned at approximately 130° angle).

2. Insert a stainless steel pin across the knee of the hind limb to be stimulated, beneath the distal portion of the quadriceps femoris muscle, as close as possible to its insertion point in the knee.

3. Attach the pin to the platform, stabilizing and fixing the joint, also at an approximate angle of 130°.

4. Place one hook end of the stainless steel rod under the corresponding distal tendon of the target muscle, and attach the other to the force transducer (Fig. 4) (*see* **Note 7**).

3.5 Adjustments of Electrical Stimulation Parameters and Pretest Assessment

1. Perform NMES using symmetrical monophasic square waves produced by an electrical stimulation device connected to the electrode placed at the sciatic nerve (*see* **Note 8**).

2. For pretest assessments, use an electrical stimulus to produce twitch contractions consisting of 500 μs pulse duration and 1 Hz frequency.

3. Apply increasing voltage until maximum force production is reached (usually around 10–15 V) (*see* **Note 9**).

4. Measure the electrophysiological parameters of excitability rheobase and chronaxie (or strength-duration time constant) before starting the protocol (*see* **Note 10**).

3.6 Protocol Execution and Examples

1. Apply electrical stimulation impulses fixed at 500 μs pulse duration at a voltage found to elicit maximal force production (usually around 10–15 V) [1] (*see* **Note 11**).

2. Manipulate frequency to produce twitch or tetanic contractions at 1 and 100 Hz, respectively [1] (Fig. 5).

3. Normalize the obtained absolute twitch and tetanic forces to muscle wet mass, dry mass, or cross-sectional area (specific force).

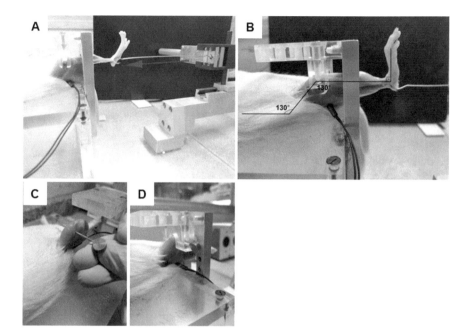

Fig. 4 Animal setup for electrical stimulation. (**a**) Rod placement under the EDL muscle tendon, connecting the muscle to the force transducer. The arrow shows the direction the rod is pulled during contractions, transmitting the force that will be measured by the transducer and recorded by the software. (**b**) Axis showing hip and knee articulations positioned at approximately 130° angle and ankle joint at 90° angle. The electrode is already connected to the sciatic nerve and the stainless steel rod is placed under the calcaneal (Achilles) tendon for electrical stimulation of the triceps surae muscle group. (**c**) Detail of pin placement under quadriceps femoris muscle and across the knee for joint stabilization. (**d**) Detail of pin placement showing how it fixes the knee joint to the acrylic platform

3.7 Twitch Parameters

1. Use a series of 5–10 twitch contractions at 1 Hz for estimation of maximal absolute twitch force, time to peak (TTP – time between the onset of force development until peak tension), rate of force development (RFD – amount of force generated per time unit during muscle shortening), half relaxation time (HRT – time of muscle relaxation halfway from peak tension), and late relaxation time (LRT – time of muscle relaxation between 50% and 25% of peak tension) [13] (Fig. 6).

2. After at least 1 min recovery period, other protocols may be performed.

3.8 Long-Duration Low-Frequency Fatigue Protocol

1. Elicit a series of twitch contractions at 1 Hz over 60 min of electrical stimulation, and measure decrease of force production and alteration of shortening and relaxation properties over time [11, 12] (*see* Fig. 7) (*see* **Note 12**).

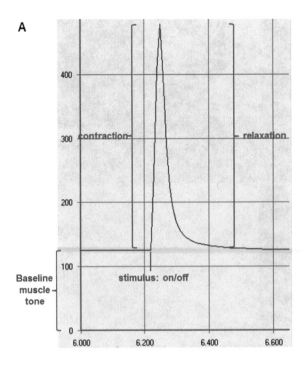

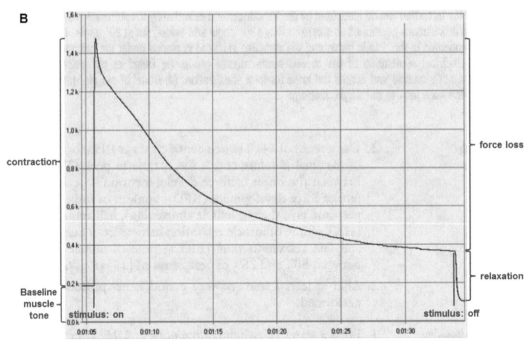

Fig. 5 Screenshots of myograms depicting twitch and tetanic contractions from rat triceps surae muscle group. Force is presented in gram-force (1gf $= 9.806 \times 10^{-3}$ N). (**a**) Twitch contraction and force increase upon electrical stimulation, followed by muscle relaxation and force decrease when the stimulus was ceased. Pulse duration: 500 μs. Frequency: 1 Hz. Voltage: 10–15 V. Time is presented in s. (**b**) Tetanic contraction, force loss during persistent stimulation for approximately 30 s, and relaxation after stimulus was ceased. Pulse duration: 500 μs. Frequency: 100 Hz. Voltage: 10–15 V. Time is presented as min:s. Images acquired from the data analysis software AqAnalysis

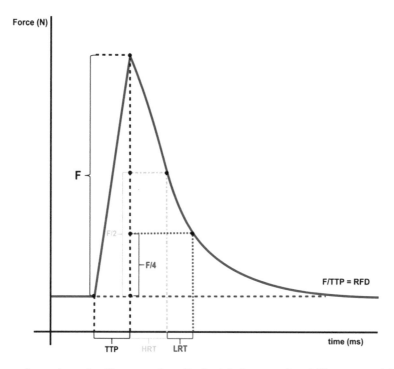

Fig. 6 Summary of muscle contractile parameters. F, absolute force produced (N): measured from baseline muscle tone to achieved peak tension and may be normalized to wet muscle mass, dry muscle mass, or cross-sectional area of the muscle for calculation of specific muscle force (N/g or N/mm²). TTP, time to peak (ms): period between the onset of force development until peak tension. RFD, rate of force development (N/ms): amount of force generated per time unit (F/TTP) during muscle contraction. HRT, half relaxation time (ms): time of muscle relaxation halfway from peak tension. LRT, late relaxation time (ms): time of muscle relaxation between 50% and 25% of peak tension

3.9 Short-Duration High-Frequency Fatigue Protocol

1. Elicit five to ten successive tetanic contractions at 100 Hz, with 1 s duration/each and 10 s of recovery between them, and measure decrease of force production over the course of the protocol [13, 14, 16] (Fig. 8) (*see* **Note 13**).

3.10 Long-Duration High-Frequency Fatigue Protocol

1. Elicit 6 series of 14 tetanic contractions with 2 s duration/each, with an interval of 20 s between contractions and 5 min between series.

2. Measure decrease of force production during successive contractions and force recovery between series as the protocol progresses (Fig. 9) (*see* **Note 14**).

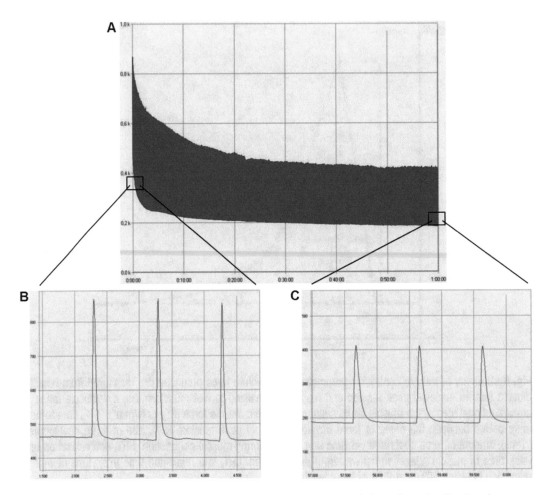

Fig. 7 Screenshots of myograms from rat triceps surae muscle group during a long-duration low-frequency fatigue protocol, consisted of one maximal twitch contraction per second during 60 min. Pulse duration: 500 μs. Frequency: 1 Hz. Voltage: 10–15 V. Force is presented in gram-force (1gf $= 9.806 \times 10^{-3}$ N). (**a**) Overall image of the protocol showing twitch force decrease as the protocol progresses. Time is presented as h:min. (**b**) Zoomed-in image of **a**, showing twitch contractions in detail at the beginning of the protocol. Time is presented in s. (**c**) Zoomed-in image of **a**, showing that twitch contractions at the end of the protocol presented reduced force production (peak force decrease) and impaired shortening and relaxation (less steep shortening and relaxation curves) when compared to the initial contractions. Time is presented in s. Images acquired from the data analysis software AqAnalysis

---

4   Notes

1. The electrical stimulation device and force transducer models and manufacturers may vary without significantly influencing the mentioned protocols.

2. All procedures should be approved by institutional ethical committees and performed accordingly to international and local principles of care and use of laboratory animals.

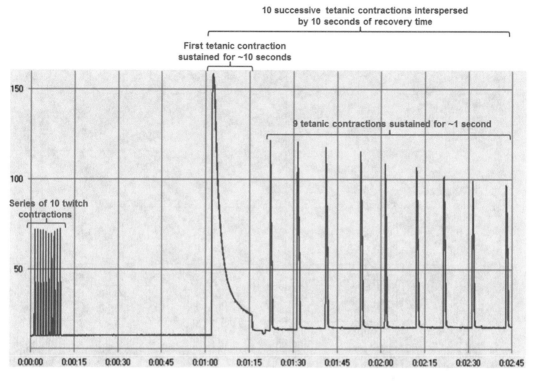

Fig. 8 Screenshot of a myogram from rat soleus muscle during a short-duration high-frequency fatigue protocol. It consisted of a series of ten twitch contractions, a recovery period of 1 min, and ten successive tetanic contractions interspersed by 10 s of recovery time. The first contraction was sustained for 10 s, and the remaining ones were sustained for 1 s. Tetanic contractions had pulse duration of 500 μs, frequency of 100 Hz, and voltage of 10–15 V. Twitch contractions had the same parameters, except for frequency (1 Hz). Force is presented in gram-force (1gf $= 9.806 \times 10^{-3}$ N). Time is presented as min:s. Image acquired from the data analysis software AqAnalysis

3. Alternative anesthetic regimens may be used [19], such as intraperitoneal injection of barbiturates, the addition of ace-promazine (3 mg/kg b.w.) to the ketamine/xylazine solution for longer surgical procedures, or the use of inhalation anesthesia.

4. Besides the regular parameters for assessment of anesthesia effectiveness, such as rate and depth of respiration and response to painful stimuli [19], this experimental setup provides a very sensitive monitoring of muscular tone and contraction. Any minimal alteration of baseline muscle tension or unintended muscle contraction can be detected by the force transducing software, and anesthesia can be promptly adjusted. The time of day for experimentation and preexperimental routines should be standardized, as muscle contractile function and anesthesia susceptibility may be affected by stress and circadian cycle. The anesthesia causes a rapid decrease in body temperature and may

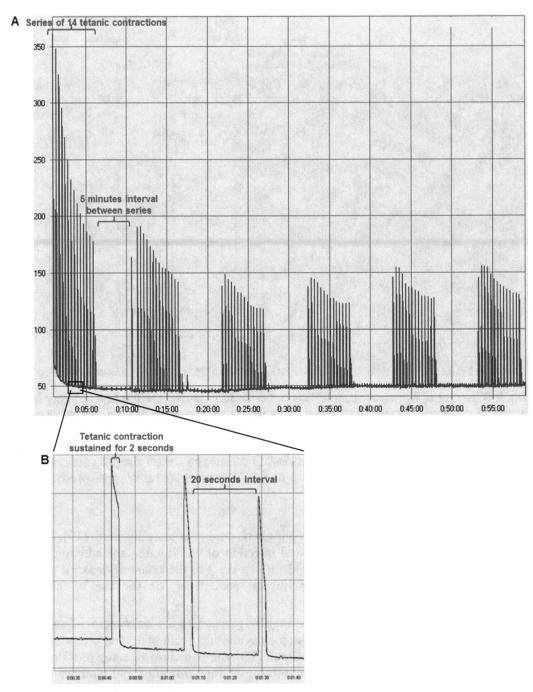

Fig. 9 Screenshots of myograms from mice triceps surae muscle group during a long-duration high-frequency fatigue protocol. Pulse duration of 500 μs. Frequency of 100 Hz. Voltage of 10–15 V. Force is presented in gram-force (1gf $= 9.806 \times 10^{-3}$ N). (**a**) Overall image of the protocol, showing the tetanic force decrease during 6 series of 14 tetanic contractions separated by 5 min of recovery between series. Time is presented in min. (**b**) Zoomed-in image of the protocol showing further details about the contraction series, such as the tetanic contraction duration of 2 s, interspersed by 20 s of interval between contractions. Time is presented as min:s. Images acquired from the data analysis software AqAnalysis

lead to hypothermia [19]. It also affects muscle temperature, potentially interfering with muscle force, contraction, relaxation, and fatigue [20]. Thus, the use of a heating lamp and/or heated platform is necessary for the maintenance of regular body and muscle temperatures throughout the protocol.

5. The surgical procedures for accessing the sciatic nerve are usually performed while flexing the hip and knee joints of the animal.

6. Activation of both plantar flexors and extensor muscles by sciatic nerve stimulation results in plantar flexion, as the former is stronger than the latter [4]. This stimulation results in concentric contraction (shortening) of the plantaris, gastrocnemius, and soleus muscles and eccentric contraction (lengthening) of the tibialis anterior and EDL muscles [21, 22]. Thus, it does not require muscle tenotomy and fascia removal, as it will consider the resulting action of all synergistic and antagonist muscles on plantar flexion.

7. By manipulating the traction of the rod coupled to the tendon and the force transducer, the resting length (L0) of the muscle and the ankle joint angle are adjusted, until reaching approximately 90°, enough to enable achievement of maximum tension upon stimulation.

8. The induced contraction tractions the rod placed between the distal tendon of the muscle and the force transducer. The resulting force is measured by the force transducer as the numerical and graphical information is transferred to the computer software and displayed on the screen.

9. Small adjustments to muscle length and ankle joint angle might support the achievement of maximum force.

10. The rheobase is the minimal voltage necessary on a stimulus with infinitely long duration to produce excitation, and chronaxie is the duration of the electrical stimulus to elicit a detectable muscle contraction, using twice the voltage of rheobase [23]. The first one can be estimated by setting pulse duration at the highest value allowed by the device and keeping frequency at 1 Hz while gradually increasing voltage from zero until occurrence of a minimal contraction. The second can be acquired by doubling the rheobase voltage and keeping the frequency at 1 Hz while gradually increasing pulse duration until a muscle contraction is detected.

11. The described electrical stimulation parameters and protocols design were presented as examples. They may be altered according to the evaluated muscle and aims of the study and still be suitable for execution using the proposed experimental setup.

12. This protocol also promotes contraction-induced muscle injury [11, 12] and oxidative stress [12].

13. The first tetanic contraction is used for evaluation of maximal tetanic force production [15].

14. The initial contraction of the first series is used for assessment of maximal tetanic force. This protocol is able to activate signaling pathways and induce expression of genes associated to exercise adaptation.

Acknowledgments

The authors are grateful to Adhemar Pettri Filho (in memoriam) for the excellent technical assistance and constant support for the establishment and improvement of the method.

References

1. Maffiuletti NA (2010) Physiological and methodological considerations for the use of neuromuscular electrical stimulation. Eur J Appl Physiol 110:223–234

2. Collins DF (2007) Central contributions to contractions evoked by tetanic neuromuscular electrical stimulation. Exerc Sport Sci Rev 35:102–109

3. Bergquist AJ, Clair JM, Lagerquist O, Mang CS, Okuma Y, Collins DF (2011) Neuromuscular electrical stimulation: implications of the electrically evoked sensory volley. Eur J Appl Physiol 111:2409–2426

4. Baar K, Esser K (1996) Phosphorylation of p70 (S6k) correlates with increased skeletal muscle mass following resistance exercise. Am J Phys 276:C120–C127

5. Brown MD, Cotter MA, Hudlicka O, Vrbova G (1976) The effects of different patterns of muscle activity on capillary density, mechanical properties and structure of slow and fast rabbit muscles. Pflugers Arch 361:241–250

6. Donselaar Y, Eerbeek O, Kernell D, Verhey BA (1987) Fibre sizes and histochemical staining characteristics in normal and chronically stimulated fast muscle of cat. J Physiol 382:237–254

7. Laughlin MH, Armstrong RB (1983) Rat muscle blood flows as a function of time during prolonged slow treadmill exercise. Am J Phys 244:H814–H824

8. Eken T, Elder GC, Lomo T (2008) Development of tonic firing behavior in rat soleus muscle. J Neurophysiol 99:1899–1905

9. Eken T (1998) Spontaneous electromyographic activity in adult rat soleus muscle. J Neurophysiol 80:365–376

10. Binder-Macleod SA, Snyder-Mackler L (1993) Muscle fatigue: clinical implications for fatigue assessment and neuromuscular electrical stimulation. Phys Ther 73:902–910

11. Bassit RA, Pinheiro CH, Vitzel KF, Sproesser AJ, Silveira LR, Curi R (2010) Effect of short-term creatine supplementation on markers of skeletal muscle damage after strenuous contractile activity. Eur J Appl Physiol 108:945–955

12. Pinheiro CH, Vitzel KF, Curi R (2012) Effect of N-acetylcysteine on markers of skeletal muscle injury after fatiguing contractile activity. Scand J Med Sci Sports 22:24–33

13. Pinheiro CH, Gerlinger-Romero F, Guimaraes-Ferreira L, de Souza AL Jr, Vitzel KF, Nachbar RT et al (2012) Metabolic and functional effects of beta-hydroxy-beta-methylbutyrate (HMB) supplementation in skeletal muscle. Eur J Appl Physiol 112:2531–2537

14. Pinheiro CH, de Queiroz JC, Guimaraes-Ferreira L, Vitzel KF, Nachbar RT, de Sousa LG et al (2012) Local injections of adipose-derived mesenchymal stem cells modulate inflammation and increase angiogenesis ameliorating the dystrophic phenotype in dystrophin-deficient skeletal muscle. Stem Cell Rev 8:363–374

15. Fortes MA, Pinheiro CH, Guimaraes-Ferreira-L, Vitzel KF, Vasconcelos DA, Curi R. (2015) Overload-induced skeletal muscle hypertrophy is not impaired in STZ-diabetic rats. Physiol Rep 3(7): pii: e12457. doi:10.14814/phy2.12457.

16. Abreu P, Pinheiro CH, Vitzel KF, Vasconcelos DA, Torres RP, Fortes MS et al (2016) Contractile function recovery in severely injured gastrocnemius muscle of rats treated with either oleic or linoleic acid. Exp Physiol 101:1392–1405

17. Wojtaszewski JF, Hansen BF, Urso B, Richter EA (1996) Wortmannin inhibits both insulin- and contraction-stimulated glucose uptake and transport in rat skeletal muscle. J Appl Physiol (1985) 81:1501–1509

18. Silveira L, Hirabara SM, Alberici LC, Lambertucci RH, Peres CM, Takahashi HK et al (2007) Effect of lipid infusion on metabolism and force of rat skeletal muscles during intense contractions. Cell Physiol Biochem 20:213–226

19. Gargiulo S, Greco A, Gramanzini M, Esposito S, Affuso A, Brunetti A et al (2012) Mice anesthesia, analgesia, and care, part I: anesthetic considerations in preclinical research. ILAR J 53:E55–E69

20. Allen DG, Lamb GD, Westerblad H (2008) Skeletal muscle fatigue: cellular mechanisms. Physiol Rev 88:287–332

21. Funai K, Parkington JD, Carambula S, Fielding RA (2006) Age-associated decrease in contraction-induced activation of downstream targets of Akt/mTor signaling in skeletal muscle. Am J Physiol Regul Integr Comp Physiol 290:R1080–R1086

22. Parkington JD, LeBrasseur NK, Siebert AP, Fielding RA (2004) Contraction-mediated mTOR, p70S6k, and ERK1/2 phosphorylation in aged skeletal muscle. J Appl Physiol (1985) 97:243–248

23. Nodera H, Kaji R (2006) Nerve excitability testing and its clinical application to neuromuscular diseases. Clin Neurophysiol 117:1902–1916

Chapter 27

Experimental Model of Skeletal Muscle Laceration in Rats

Phablo Abreu, Gabriel Nasri Marzuca-Nassr, Sandro Massao Hirabara, and Rui Curi

Abstract

This is a modified experimental model previously developed in mouse to study skeletal muscle laceration in rats. All experimental procedures are performed during the light period, including anesthesia and surgery. The animals are randomly distributed into control and injured groups prior to the procedure. This experimental model can be used to investigate skeletal muscle laceration repair.

Key words Skeletal muscle injury, Repair, Muscle regeneration, Damage, Trauma

1 Introduction

Muscle laceration occurs when the muscle is exposed to a crushing force or to a sharp object, leading to destruction of tissue structures (e.g., cell membrane, myofibrils, and innervation). There is accumulation of extracellular matrices during recovery (fibrosis) that impairs muscle regeneration and function recovery [1, 2]. Several experimental models have been used to study muscle laceration in order to improve treatment efficacy and muscle repair after severe injury in humans [1–19].These reports evaluated managements that promoted muscle regeneration. However, the main limitation of these works is the diversity of procedures used to reproduce the muscle laceration in animals, making the reproduction of procedures used in the studies difficult.

A list of experimental models applied to hind limbs skeletal muscle injury by laceration in laboratory animals is given in Table 1. The gastrocnemius muscle, due location and mixture of fiber types, is the most commonly used one in the animal models. The tibialis anterior muscle models, for example, have a predominance of type II muscle fibers [6]. These models listed in Table 1 have been described in mice, rats, and dogs and employ different muscles such as gastrocnemius, tibialis anterior, and quadriceps. An

Paul C. Guest (ed.), *Investigations of Early Nutrition Effects on Long-Term Health: Methods and Applications*, Methods in Molecular Biology, vol. 1735, https://doi.org/10.1007/978-1-4939-7614-0_27, © Springer Science+Business Media, LLC 2018

Table 1
Experimental models of muscle injury by laceration

Reference	Muscle	Animal
Negishi et al. (2005) [6]	Tibialis anterior	Mice
Menetrey et al. (1999) [1] Menetrey et al. (2000) [4] Li and Huard (2002) [7] Li et al. (2004) [8] Shen et al. (2005) [9] Kaar et al. (2008) [10] Chan et al. (2010) [11] Park et al. (2012) [12] Hwang et al. (2013) [13]	Gastrocnemius	Mice
Foster et al. (2003) [14] Hwang et al. (2006) [15] Freitas et al. (2007) [16]	Gastrocnemius	Rats
Piedade et al. (2008) [17]		
Turner et al. (2012) [2]	Quadriceps	Dogs

experimental model to study skeletal muscle laceration in rats, using transection of the gastrocnemius muscle, is described here.

2 Materials

1. Male 180–200 g Wistar rats.
2. Antibiotic: gentamicin.
3. Anesthetics: ketamine and xylazine.
4. Digital caliper rule (Digimed; Sao Paulo, Brazil).
5. Suture: nylon thread 3-0.
6. Antiseptic: povidone iodine (PVPI).
7. Heating lamp.

3 Methods

3.1 Animals and Preoperation Assessment (See Notes 1 and 2)

1. Keep rats in a temperature-controlled room with food (standard diet for rodents) and water available ad libitum.
2. Perform all experimental procedures that are executed during the light period, including anesthesia and surgery.
3. Keep animals in individual cages under controlled environmental conditions (22 ± 1 °C, with light-dark periods of 12 h).
4. Distribute rats randomly into two groups: (1) control and (2) injured.

5. Twenty-four hours before surgery, administer a single i.p. dose of gentamicin (80 mg/kg) to prevent bacterial infections (*see* **Note 3**).

6. Anesthetize animals with an i.p. injection of ketamine (90 mg/kg) and xylazine (10 mg/kg) (*see* **Note 4**).

7. Monitor reflexes (tail pinch and palpebral) and respiratory rate before and during the surgery to verify the anesthetic depth.

3.2 Surgical Procedure of Muscle Laceration Protocol (See Note 5)

1. Maintain animals in individual cages as above full recovery from anesthesia in order to avoid hypothermia.

2. Following mechanical hair cutting of the right hind limb, dissect and retract skin and subcutaneous tissue to expose the gastrocnemius muscle (*see* **Note 6**).

3. Perform all measurements (gastrocnemius muscle transection) using the digital caliper rule.

4. Use the contralateral hind limb (left) to a sham procedure.

5. Sectioning other gastrocnemius muscle (laceration) at 75% of the width (approximately 1.0 cm), 60% of the length from the distal insertion (about 2.5 cm up from the calcaneus flexed at 90°), and 50% of the thickness (Fig. 1) (*see* **Note 7**).

6. Suture the skin with the nylon thread.

7. After surgery, perform local antisepsis using PVPI.

8. Following the procedure of muscle laceration, present food and water ad libitum for all animals.

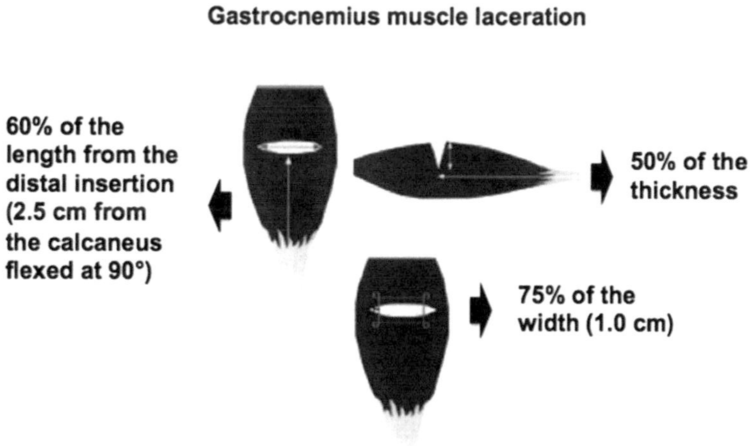

Gastrocnemius muscle laceration

60% of the length from the distal insertion (2.5 cm from the calcaneus flexed at 90°)

50% of the thickness

75% of the width (1.0 cm)

Fig. 1 Sectioning of the gastrocnemius muscle (laceration) is performed as follows: at 60% of the length from the distal insertion (2.5 cm from the calcaneus flexed at 90°), 75% of the width (1.0 cm), and 50% of the thickness. Adapted from Menetrey et al. [1] and presented in the Abreu PhD Thesis [20] and Abreu et al. [18]

**3.3 Postsurgery
Assessments**

1. Keep rats warm using a heating lamp (*see* **Note 8**).

2. Perform local antisepsis using PVPI, and keep animals in individual cages as above to avoid hypothermia.

3. Rats should be constantly evaluated until recovery (*see* **Note 9**).

---

4   Notes

1. Several studies evaluating muscle repair and regeneration in rodents with muscle injuries that mimic the damages observed in humans have been performed. The studies involving muscle damages that mimic the injuries in humans use diverse techniques that include the type of laceration (complete or partial transection) and animal models (Wistar rats, Sprague-Dawley rats, mice, dog) used. The variety of the models of injuries (type of muscle: gastrocnemius, soleus, tibialis anterior, quadriceps) and trauma makes difficult the comparison and reproduction of the human clinical conditions. We describe an experimental model of gastrocnemius muscle laceration in rats that allowed us to investigate the skeletal muscle recovery process [18].

2. Conduct all animal handling and surgical procedures according to the Guiding Principles for the Care and Use of Laboratory Animals.

3. In this phase, it is mandatory to ascertain the animal's physiological situations before an experiment (e.g., body condition score, respiratory rate, food and water intake, as well as defecation and urination).

4. Experimental procedures require anesthesia to obtain adequate immobilization and to reduce stress or pain.

5. The location of muscle laceration is standardized for all Wistar rats.

6. All the hair on the posterior portion of the selected calf is removed and the skin and subcutaneous tissue dissected in order to reach the gastrocnemius muscle and to permit suitable contact of the specific skeletal muscle.

7. In specific situations, bleeding can be controlled by compression.

8. Be sure to avoid burns.

9. Signs of discomfort and pain (e.g., redness and swelling at incision site, self-mutilation, aggressive behavior or weakness, cold or blue extremities, or hot or red extremities) should be dealt with immediately.

Acknowledgments

FAPESP, CNPq, CAPES, Dean's Office for Postgraduate Studies, and Research of the Cruzeiro do Sul University supported this study.

References

1. Menetrey J, Kasemkijwattana C, Fu FH, Moreland MS, Huard J (1999) Suturing versus immobilization of a muscle laceration. A morphological and functional study in a mouse model. Am J Sports Med 27:222–229

2. Turner NJ, Badylak JS, Weber DJ, Badylak SF (2012) Biologic Scaffold remodeling in a dog model of complex musculoskeletal injury. J Surg Res 176:490–502

3. Ghaly A, Marsh DR (2010) Aging-associated oxidative stress modulates the acute inflammatory response in skeletal muscle after contusion injury. Exp Gerontol 45:381–388

4. Menetrey J, Kasemkijwattana C, Day CS, Bosch P, Vogt M, Fu FH et al (2000) Growth factors improve muscle healing in vivo. J Bone Joint Surg Br 82:131–137

5. Minamoto VB, Grazziano CR, Salvini TF (1999) Effect of single and periodic contusion on the rat soleus muscle at different stages of regeneration. Anat Rec 254:281–287

6. Negishi S, Li Y, Usas A, Fu FH, Huard J (2005) The effect of relaxin treatment on skeletal muscle injuries. Am J Sports Med 33:1816–1824

7. Li Y, Huard J (2002) Differentiation of muscle-derived cells into myofibroblasts in injured skeletal muscle. Am J Pathol 161:895–907

8. Li Y, Foster W, Deasy BM, Chan Y, Prisk V, Tang Y et al (2004) Transforming growth factor beta1 induces the differentiation of myogenic cells into fibrotic cells in injured skeletal muscle: a key event in muscle fibrogenesis. Am J Pathol 164:1007–1019

9. Shen W, Li Y, Tang Y, Cummins J, Huard J (2005) NS-398, a cyclooxygenase-2-specific inhibitor, delays skeletal muscle healing by decreasing regeneration and promoting fibrosis. Am J Pathol 167:1105–1117

10. Kaar JL, Li Y, Blair HC, Asche G, Koepsel RR, Huard J et al (2008) Matrix metalloproteinase-1 treatment of muscle fibrosis. Acta Biomater 4:1411–1420

11. Chan YS, Hsu KY, Kuo CH, Lee SD, Chen SC, Chen WJ et al (2010) Using low-intensity pulsed ultrasound to improve muscle healing after laceration injury: an in vitro and in vivo study. Ultrasound Med Biol 36:743–751

12. Park JK, Ki MR, Lee EM, Kim AY, You SY, Han SY et al (2012) Losartan improves adipose tissue-derived stem cell niche by inhibiting transforming growth factor-beta and fibrosis in skeletal muscle injury. Cell Transplant 21:2407–2424

13. Hwang JH, Kim IG, Piao S, Jung AR, Lee JY, Park KD et al (2013) Combination therapy of human adipose-derived stem cells and basic fibroblast growth factor hydrogel in muscle regeneration. Biomaterials 34:6037–6045

14. Foster W, Li Y, Usas A, Somogyi G, Huard J (2003) Gamma interferon as an antifibrosis agent in skeletal muscle. J Orthop Res 21:798–804

15. Hwang JH, Ra YJ, Lee KM, Lee JY, Ghil SH (2006) Therapeutic effect of passive mobilization exercise on improvement of muscle regeneration and prevention of fibrosis after laceration injury of rat. Arch Phys Med Rehabil 87:20–26

16. Freitas LS, Freitas TP, Silveira PC, Rocha LG, Pinho RA, Streck EL (2007) Effect of therapeutic pulsed ultrasound on parameters of oxidative stress in skeletal muscle after injury. Cell Biol Int 31:482–488

17. Piedade MC, Galhardo MS, Battlehner CN, Ferreira MA, Caldini EG, de Toledo OM (2008) Effect of ultrasound therapy on the repair of gastrocnemius muscle injury in rats. Ultrasonics 48:403–411

18. Abreu P, Pinheiro CH, Vitzel KF, Vasconcelos DA, Torres RP, Fortes MS et al (2016) Contractile function recovery in severely injured gastrocnemius muscle of rats treated with either oleic or linoleic acid. Exp Physiol 101:1392–1405

19. Souza JD, Gottfried C (2013) Muscle injury: review of experimental models. J Electromyogr Kinesiol 23:1253–1260

20. Abreu P (2014) Effect of linoleic and oleic acids on regeneration of gastrocnemius muscle after laceration in rats. Ph.D. thesis. Instituto de Ciências Biomédicas, University of São Paulo, Brazil

Chapter 28

Neuropsychiatric Sequelae of Early Nutritional Modifications: A Beginner's Guide to Behavioral Analysis

Ann-Katrin Kraeuter, Paul C. Guest, and Zoltán Sarnyai

Abstract

Early parental nutritional interventions during prenatal development have been shown to result in neuropsychiatric sequelae in the adult offspring. In order to understand the impact of such nutritional interventions, the behavior of the animal has to be carefully analyzed. This chapter provides a step-by-step guide to conduct behavioral tests in adult mice for investigators without specific expertise or those without the equipment to carry out behavioral studies. We focus on tests tapping into the main behavioral abnormalities that correspond to mental illnesses. We describe the materials required and the detailed methods to conduct global assessment of parameters such as behavioral integrity and general well-being, psychomotor activity, social behavior, repetitive behavior, anxiety-like behavior, depression-like behavior, short-term spatial working memory, and spatial reference memory.

Key words Behavior, Mice, Nest building, Open field, Social behavior, Repetitive behavior, Elevated plus maze, Forced swim test, Y-maze, Working memory, Reference memory

1 Introduction

Developmental programming of adult noncommunicable diseases has been receiving increasing attention in recent years [1]. The impact of early fetal nutrition, including the lack of adequate supply of macro- and micronutrients, such as protein and vitamins (such as vitamin D), or the oversupply of certain type of nutrients, such as fat and carbohydrates, has been shown to be associated with the development of diseases in the adult offspring. For example, the link between parental nutrition and offspring metabolism [2] and specifically early protein malnutrition and the later development of type 2 diabetes has been well-established [3]. However, beyond the effects of early nutrition on the development of systemic, physical diseases, such as metabolic syndrome, type 2 diabetes, hypertension, and cancer, it is becoming increasingly clear that brain development and adult brain function are also critically influenced by such early stimuli [4–6]. For example, nutritional factors during

Paul C. Guest (ed.), *Investigations of Early Nutrition Effects on Long-Term Health: Methods and Applications*, Methods in Molecular Biology, vol. 1735, https://doi.org/10.1007/978-1-4939-7614-0_28, © Springer Science+Business Media, LLC 2018

pregnancy have been associated with the later development of anxiety, depression, schizophrenia, and autism spectrum disorders in the offspring [6, 7].

In order to better understand the impact of early nutrition on brain function, the study of the behavior of the offspring is inevitable as behavior is the main "biomarker" of brain function. Studies on parental nutrition, fetal, and early nutritional effects are often conducted by groups specialized in endocrinology, development, metabolism, or cardiovascular physiology, without specific behavioral expertise. The goal of this chapter is to provide a guide to basic behavioral analysis for researchers without specific expertise as well as an infrastructure for studying mouse behavior. We aimed to assemble a battery of behavioral tests that are relatively easy to carry out using the detailed notes provided in a laboratory with relatively minimal or no specific equipment. These tests will allow the investigators to obtain at least a preliminary answer as to which behavioral effects have occurred in their rodent models. These initial behavioral phenotyping studies can then lead to more specific and more labor- and resource-intensive analyses once a basic behavioral phenotype has been determined.

By considering the main mental health problems shown to be associated with general and specific early nutritional abnormalities, such as anxiety, depression, schizophrenia, cognitive problems, and autism spectrum disorders, we aimed to assemble a series of tests that can either be conducted individually or as a battery of tests to tap into behavioral endophenotypes associated with these conditions. Therefore, we describe tests to globally assess behavioral integrity and general well-being (nest building), psychomotor activity (open-field test), social behavior (social interaction, repetitive behavior, marble burying), anxiety-like behavior (open-field test and elevated plus maze), depression-like behavior (forced swim test), short-term spatial working memory (spontaneous alternation in Y-maze), and spatial reference memory (Y-maze) (Table 1). In order to help investigators with less expertise in analyzing mouse behavior, we provide a step-by-step instruction for each test based on the standard operating procedures developed in our laboratory. We also describe the materials necessary for the successful completion of these experiments.

Analyzing behavior with scientific rigor in mice is not different from other experimental methods in biomedical research in a sense that its successful completion requires the close adherence to established protocols. For the most stable results, mice need to be handled for at least a few days (3–5 min handling per day) in order to accustom them to the experimenter, who then will have to remain constant throughout the study. In addition, the temperature and lighting conditions of the facility need to be monitored and kept constant. If multiple behavioral tests are conducted on the same animals, these should be run in a way that the least stressful

Table 1
Summary of behavioral tests, brain functions, and associated brain areas

Behavioral function	Test	Brain region/system involved[a]	Relevance to human psychopathology[b]
Behavioral integrity and general well-being	Nest building	General (including hippocampus)	Various
Psychomotor activity	Open field	Mesolimbic/nigrostriatal dopamine systems	Schizophrenia, bipolar disorder, depression
Social behavior	Free social interaction	Various regions, including PFC, amygdala, VTA, nucleus accumbens, hypothalamus	Schizophrenia, autism spectrum disorders
Repetitive behavior	Marble burying	Cortico-basal ganglia-thalamic pathway	Obsessive-compulsive disorder, autism spectrum disorders
Fear/anxiety	Elevated plus maze	Amygdala	Generalized anxiety disorder
Learned helplessness	Forced swim	Amygdala, hippocampus	Major depression
Spatial working memory	Spontaneous alternation	PFC, hippocampus, septum, basal forebrain	Schizophrenia, various cognitive impairments, such as Alzheimer's disease
Spatial reference memory	Y-maze reference memory version	Hippocampus	Alzheimer's disease and other neurodegenerative disorders with hippocampal involvement

[a]Note that each of the behavioral functions results from coordinated activity of multiple brain areas. The areas listed here are the ones most often associated with the respective behavioral function in the literature, and this is for general guidance only

[b]Note that human psychopathologies are highly complex and cannot be modeled by single or even by multiple behavioral tests. The relevance of each of these behavioral tests to human psychopathology in this table is only approximate and for general guidance only

tests be carried out first followed by the more stressful ones. The tests described in this chapter can be carried out as a behavioral test battery. In this case, if time and experimental design are allowed (e.g., no acute intervention occurs), it is ideal to wait about 5–7 days between tests. However, we have conducted these tests in the form of a battery after acute, pharmacological, intervention in 4–5 consecutive days with good and reliable results. Behavior is inherently variable between animals, even if they are littermates from an inbred strain. According to our own experience and on the basis of power analysis, at least 12 mice per experimental group are desirable in order to achieve statistically significant results. If the quantification of behavior is carried out by manual scoring, the issue of potential bias also emerges. In order to minimize such bias and to increase scoring reliability, the tests need to run by an investigator blind to the experimental conditions, which is achieved by proper coding of the animals. The code should only be broken upon the completion of the statistical analysis. Finally, the experimenter has to be aware that mouse behavior is influenced by the time of the day. Mice are more active during the dark phase of their light-dark cycle. Therefore, mice tested early in the morning behave differently compared to mice tested late in the afternoon. To minimize the impact of the light-dark cycle, mice need to be tested at a consistent time during the day. If the testing takes 5–6 h a day, the experimental animals have to be randomized on the basis of their treatment conditions throughout the day of testing. Adherence to these guidelines can assure the reproducibility of behavioral analysis that could otherwise be highly variable.

2 Materials

2.1 Nest Building (See Note 1)

1. Nesting material – white compressed cotton (5 × 5 cm, mean weight 2.5 g) specifically made for mouse nest building in a research setting.
2. Camera.
3. Scoring criteria sheet.

2.2 Open Field (See Note 2)

1. Open-field box made of light-gray polyvinyl chloride (PVC) (420 × 420 × 420 mm) with a center zone in the middle of the box marked with permanent marker (205 × 205 mm).
2. TopScan Light® behavioral analysis software (Clever Sys Inc., Reston, VA, USA).
3. Webcam/camera.
4. Retort stand.
5. 70% ethanol in a spray bottle.
6. Two adjacent rooms (behavioral testing room and another area just outside the door for the computer and operator to sit) (*see* **Note 3**).

2.3 Social Interaction (See Note 4)

1. Open-field box as above.
2. Two rooms (as above).
3. Second scorer.
4. Two mice: target mouse and test mouse (gender age and size matched).

2.4 Marble Burying (See Note 5)

1. 15 marbles.
2. Camera.

3. Cage.

4. Sawdust.

2.5 Elevated Plus Maze (See Note 6)

1. Elevated plus maze made of light-gray PVC arms (360 × 60 mm) elevated off the ground (600 mm) (two of the arms are enclosed by walls 150 mm high and arranged opposite each other, and the two remaining open arms are also arranged opposite each other in a cross).

2. Two adjacent rooms as above.

2.6 Forced Swim (See Note 7)

1. 2 L beaker.

2. Warm water.

3. Thermometer.

4. Two rooms as above.

2.7 Spontaneous Alternation (See Note 8)

1. Light-gray PVC Y-shaped compartment (210 × 70 × 155 mm), with equal length arms.

2. Box big enough to place Y-maze into (to ensure animal cannot escape).

3. Two rooms as above.

2.8 Spatial Reference Memory in Y-Maze (See Note 9)

1. Light-gray PVC Y-shaped maze as above with equal length arms in a box to contain any animal that may jump out.

2. One divider.

3. Box big enough to place Y-maze into (to ensure animal cannot escape).

4. Three visual clues around the maze (can be pictures of circles and squares).

5. Two rooms as above.

3 Methods

3.1 Nest Building

1. Acclimatize mice at least 7 days prior to the experiment to enable all animals to familiarize themselves to the investigator's smell.

2. Singly house animals in the afternoon of the behavioral testing day.

3. To begin the experiment, provide each animal with nesting material (1 square) at 19:00 h.

4. Animals should not be disturbed during the dark phase while building the nest (as mice are nocturnal animals, most active behavior takes place at nighttime).

5. At 07:00 take the cage from the shelf to the workbench.

6. Take photos of the mouse experimental card first outlining the strain, sex, and identification (ID) number (do this for all tests).

7. Next take photographs of the nest first with and then without the mouse in the cage (place the ruler next to the nest for future reference).

8. Measure both the length and width of the nest.

9. Work methodically so that the first picture is always the experimental card and the next is the nest.

10. Label the photos when they are uploaded to the computer so that the experimental card and nest can be identified as coming from the same mouse.

11. Measure the height of each quarter of the nest and record it on the experimental card and in your lab book.

12. Score the nest and record data on the experimental card and in your lab book (*see* **Notes 10** and **11**).

3.2 Open-Field Test

1. Acclimatize mice as above.

2. On the morning of behavioral study, before transport to the behavioral testing room, weigh animal and record weight in lab book.

3. One hour before testing, place the animals in behavioral testing room to acclimatize to the room and reduce stress.

4. Place the video camera into the clamp of the retort stand.

5. Place the retort stand on a chair, so the camera can be positioned above the open-field box.

6. Connect the camera to computer via the USB cable (*see* **Note 12**).

7. Ensure that the open-field box is clean prior to testing.

8. Spray the open field with 70% ethanol in the spray bottle, and wipe down with paper towel.

9. Make sure the box is completely dry before beginning the experiment.

10. Set the timer for 15 min and leave it next to the computer.

11. Start recording.

12. For ease of identification, hold a piece of paper under the recording webcam with the strain, mouse ID, cage number, ethics number, and a short title of the experiment to be undertaken.

13. Remove animal from home cage, and place test animal into the center zone of the open-field box.

14. Leave behavioral testing room

15. Press start on the timer.

16. Leave the animal to explore the maze for 15 min (*see* **Note 13**).

17. Stop recording and save video recording (*see* **Note 14**).

18. Remove animal from the box and place it back into home cage (*see* **Notes 15** and **16**).

19. Clean all walls and the floor of the box with 70% ethanol and paper towel, before proceeding to the next animal.

20. Make sure the box is completely dry.

21. At completion of the experiment, clean all equipment and put away.

22. Return animals to the holding room.

3.3 Social Interaction

1. Acclimatize mice at least 7 days prior to the experiment as above.

2. On the morning of behavioral study, before transport to behavioral testing room, weigh animal and record weight in lab book as above.

3. One hour before testing, place the animals in behavioral testing room to acclimatize to the room and reduce stress as above.

4. Place the open-field box on the floor.

5. Place the video camera on a retort stand on a chair, so the camera can be positioned above the box.

6. Connect the camera to computer via the USB cable (*see* **Note 12**).

7. Ensure that the open-field box is clean prior to testing. Spray the OF with 70% ethanol in the spray bottle and wipe down with paper towel and make sure the box is completely dry before beginning the experiment.

8. Mark the target animal with a permanent marker on its tail.

9. Set the timer for 5 min, and leave the timer next to the computer.

10. Start recording.

11. Remove test mouse from the home cage, and place the test animal into the center zone of the box.

12. Place target animal into the center zone of the box.

13. Leave behavioral testing room.

14. Press start on the timer.

15. Leave the animals to explore the box for 5 min, and do not enter the room in this time.

16. Stop recording and save video recording (*see* **Notes 17** and **18**).

17. Remove animals (target and test) from open-field box, and place them back into home cage.

18. Clean all walls and the floor of the box before proceeding to the next animal.

19. At completion of the experiment, clean all equipment and put away.

20. Return animals to the holding room.

3.4 Marble Burying

1. Acclimatize mice as above.

2. Prepare a clean cage with 5 cm of sawdust.

3. Distribute 15 marbles evenly across two-thirds of the cage.

4. Place test animal facing the marbles in the one-third of the cage without marbles.

5. Leave behavioral testing room.

6. Let animal explore area for 30 min undisturbed.

7. Enter behavioral testing room and immediately remove animal from testing cage and place it into home cage.

8. Take photographs of the cage (Fig. 2).

9. Work methodically so that the first picture is always the experimental card and the next is the cage.

10. Label the photos when they are uploaded to the computer so that the experimental card and nest can be identified as coming from the same mouse.

11. Record in your lab book, how many marbles were completely burrowed, to 50% or not burrowed (*see* **Note 19**).

3.5 Elevated Plus Maze

1. Acclimatize the mice for at least 7 days prior to the experiment as above.

2. On the morning of behavioral study and before transport to behavioral testing room, weigh animal and record weight in lab book.

3. One hour before testing, place the animals in behavioral testing room to acclimatize to the room and reduce stress.

4. Place a video camera into a clamp of the retort stand on a chair, so the camera is positioned above the maze.

5. Connect the camera to computer via the USB cable (*see* **Note 12**).

6. Ensure that the maze is clean prior to testing by spraying with 70% ethanol and wiping down with paper towel (make sure the maze is completely dry before beginning the experiment).

7. Start recording setting a timer for 5 min.

8. For ease of identification, hold a piece of paper under the recording webcam with the mouse ID, cage number, ethics number, and a short title of the experiment to be undertaken as above.

9. Remove animal from home cage and place test animal in the intersection of the open and closed arms of the maze facing an enclosed arm.

10. Leave behavioral testing room.

11. Press start on the timer.

12. Leave the animal to explore the maze for 5 min, and do not enter the room in this time except in the case of the animal falling off the maze or seeming unwell.

13. Stop recording and save video recording (*see* **Note 20**).

14. Remove animal from the maze and place it back into home cage.

15. Clean and dry all walls and the floor of the maze before proceeding to the next animal as described above.

16. At completion of the experiment, clean all equipment and put away.

17. Return animals to the holding room.

3.6 Forced Swim

1. Acclimatize mice as above.

2. On the morning of behavioral study, before transport to behavioral testing room, weigh animal and record weight in lab book as above.

3. One hour before testing, place animals in behavioral testing room to acclimatize and reduce stress as above.

4. Place the video camera into the clamp of the retort stand.

5. Connect the camera to computer via the USB cable (*see* **Note 12**).

6. Ensure that the beakers are clean prior to testing (rinse with tap water and wipe with a paper towel to remove any dirt).

7. Fill the beaker to 1.4 L with warm tap water.

8. Record water temperature with the thermometer and maintain at 22–24 °C.

9. Place the beaker below the camera and check again that the computer screen can see the whole of the inside of the beaker.

10. Start recording.

11. Set the timer for 6 min.

12. Remove animal from home cage, and place it into the beaker by holding the mouse by the base of the tail.

13. Leave behavioral testing room.

14. Press start on the timer.

15. The animal is to be left undisturbed (no entry into testing room) to swim in the beaker for 6 min.

16. When the timer goes off, stop recording and remove the animal from the beaker.

17. Wipe off excess water with a tissue and place animal back into home cage.

18. Put four fresh tissues in the cage, and make sure no breeze from the air conditioning gets into the cage.

19. Go back to the computer and save video recording (*see* **Notes 21** and **22**).

20. Empty and rinse the beaker before proceeding to the next animal.

21. Refill with clean warm water and check the temperature as described above.

22. At completion of experiment, clean all equipment and pack away.

23. Place animals back into holding room.

3.7 Spontaneous Alternation

1. Acclimatize mice as above.

2. On the morning of behavioral study, before transport to behavioral testing room, weigh animal and record weight in lab book as above.

3. One hour before testing, place animals in behavioral testing room to acclimatize and reduce stress as above.

4. Place above the Y-maze a camera/webcam connected to a computer (*see* **Note 12**).

5. Ensure that the Y-maze is clean prior to testing as above.

6. Label arms A, B, and C (A will be the start arm).

7. Start recording.

8. For ease of identification, follow instructions above.

9. Remove animal from home cage, and place test animal into the distal part of the start arm, facing toward the center of the maze via holding the mouse by the base of the tail.

10. Leave behavioral testing room.

11. Leave animals undisturbed to explore the Y-maze for 8 min.

12. Stop recording and save video recording (*see* **Note 23**).

13. Remove animal from Y-maze and place it back into home cage.

14. Clean all walls and the floor of the Y-maze before proceeding to the next animal.

15. At completion of the experiment, clean all equipment and put away.

16. Place animals in their cages back into the holding room.

3.8   Spatial Reference Memory in Y-Maze

1. Acclimatize mice as above.

2. Weigh animal and record weight in lab book as above.

3. Place animals in testing room to acclimatize and reduce stress as above.

4. Clean the maze and box with 70% ethanol in the spray bottle, wipe, and dry.

5. Place the box with the maze on the floor.

6. Place the video camera into the clamp of the retort stand on a chair, so the camera can be positioned above the Y-maze.

7. Connect the camera to computer via the USB cable (*see* **Note 12**).

8. Label the arms A, B, and C (A will be the start arm).

9. Close off one of the arms with the divider (this is to be randomized and alternated between animals).

10. For ease of identification, follow instructions above.

11. Set the timer for 15 min.

12. Place the animal facing the center into one of the open arms (randomized and alternated between animals).

13. Leave behavioral testing room.

14. Press start on the timer.

15. Leave the animal to explore the maze for 15 min without disruption or entering the room.

16. Stop the recording and go back into the room to put test animal back into its home cage.

17. Remove divider.

18. Wait 1 h.

19. Start recording the second part of the experiment with no divider.

20. Set the timer for 5 min and leave this next to the computer.

21. Remove animal from home cage and place test animal into the distal part of the start arm (same arm you initially put the animal into), facing toward the center of the maze.

22. Leave behavioral testing room.

23. Press start on the timer.

24. Leave the animal to explore the maze for 5 min.

25. Stop recording and save video recording (*see* **Notes 24** and **25**).

26. Remove animal from Y-maze and place it back into home cage.

27. Clean all walls and the floor of the Y-maze before proceeding to the next animal.

28. At completion of the experiment, clean all equipment and put away.

29. Return animals to the holding room.

4   Notes

1. For the duration of all experiments, the handlers should not change perfumes, deodorants, etc. Within the facility, no culling or blood work should be undertaken 24 h prior to behavioral testing and on behavioral testing days as this will increase stress and alter behavioral outcomes. Throughout behavioral testing, ambient noise should be minimal within the facility. Nest building is a form of "luxury behavior" in mice. Mice build nests for comfort, for thermoregulation, and for nesting of the young [8, 9]. In order to build proper nests, mice have to be able to carry out an integrated series of complex behaviors that rely on the functional integrity of the sensory and motor systems of the animals. Therefore, this test can be used to assess general well-being and behavioral integrity. A normal mouse should be building a nest of a score of 4–5 according to Fig. 1. Nest building can vary between strains [10] and can also be altered by pharmacological intervention [11] and hippocampal lesions [12].

Fig. 1 Scoring guide for the nest building test. Score 0: untouched material. Score 1: nesting material is moved but scattered throughout the cage. Score 2: nest wall flat against the cage bedding. Score 3: the nest forms a cup but with less than 4 cm of wall height. Score 4: wall height 4–5 cm. Score 5: wall height greater than 5 cm, a perfectly formed nest

2. The open-field test analyzes locomotor activity, anxiety, and stereotypical behavior such as grooming and rearing [13]. Altered locomotor activity can be an indicator for altered neurological process reflecting abnormal brain development. This test may be used to assess general health and well-being of an animal in an objective manner. Animals that are unhealthy move less within the arena. A mouse, which is excessively stressed, will show less activity in the novel open field and increased stereotypical behavior [14, 15]. The open-field box can be subdivided into an exterior, peripheral, and a center zone. Mice prefer staying close to the walls and travel in the periphery. This behavior, called thigmotaxis, becomes especially pronounced in mice showing signs of anxiety-like behavior. A mouse, which is less anxious, will spend more time in the center [16, 17].

3. Two separate rooms are needed to ensure that during the test, the animal is undisturbed in the room and can be observed by the investigator on the computer.

4. The spontaneous social interaction test investigates how two mice, unfamiliar to one another, behave and interact in an open-field arena (40 × 40 cm). This test assesses social interest and sociability and is suitable for identifying ethologically relevant forms of social interactions between two unfamiliar mice. In the setting described here, we use the open-field apparatus, following a 15-min activity testing of the observed mouse, to allow it to freely interact with an unfamiliar, strain-, age-, sex-, and weight-matched target mouse.

5. Marble burying assesses repetitive behavior in mice [18]. Animals with increased repetitive behavior will bury more marbles compared to animals with less repetitive behavior (Fig. 2). Marble burying is increased in mouse models of autism/autism spectrum disorders and obsessive-compulsive behavior [19].

6. The elevated plus maze test is used to investigate an anxiety-like behavior [20] in mice and rats. This test is based on the natural behavior of rodents to avoid open and elevated places and on the conflict between such avoidance behavior and rodent innate curiosity to explore novel areas. The maze is elevated, which may increase fear-/anxiety-provoking effects of the open arm. A less anxious mouse will more frequently visit the open, more exposed arms of the maze. A mouse with elevated anxiety levels will be less likely to explore the open arms of the maze [20].

7. The forced swim test is a form of "learned helplessness," an endophenotypic feature of depression-like behavior in mice [21]. This test has also been used to test the efficacy of antidepressants [21]. The FST is based on the natural behavior of

A   B

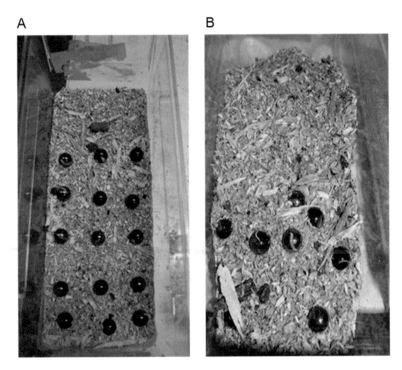

Fig. 2 Marble burying test. (**a**) Setting up the 15 marbles in the mouse cage. (**b**) Fully (8 marbles, not visible as covered by sawdust) and partially buried marbles along with non-buried ones, at the end of the marble burying session

the mouse to wanting to escape from water and to trying to stay afloat by swimming. A "depressed" mouse will give up swimming earlier than a "normal" mouse. "Depressed" mice will float more and earlier than "normal" mice.

8. The spontaneous alternation measured in a Y-maze is used to assess short-term, spatial working memory in mice [22, 23]. In this test mice are allowed to explore all three arms of the maze freely during testing. This exploration is driven by the mouse innate drive to explore previously unvisited, novel areas. Therefore, a mouse with intact spatial working memory will remember the arms previously visited and will show a tendency to enter a less recently visited arm. The number of arm entries and the number of triads are recorded in order to calculate the percentage of alternation. Many parts of the brain are involved in this task, including the hippocampus, septum, basal forebrain, and prefrontal cortex.

9. This test is used to assess spatial reference memory in mice [24–26]. In this test animals will be placed into a Y-shaped maze, with one arm closed off during training. The closed arm will be called the novel arm. After an inter-trial interval of 1 h, the animal should remember which arm of the maze it has not been in previously. Mice are generally curious and therefore

should investigate the novel arm longer and more often. If a mouse has a deficit in reference memory, one would expect that mouse cannot distinguish between the novel and other arms (does not spend more time in novel arm) [24].

10. Scoring criteria (Fig. 1)

0 = untouched material.

1 = moved but scattered throughout the cage.

2 = wall was flat against the cage bedding.

3 = still a cup but less than 4 cm.

4 = wall height 4–5 cm.

5 = wall height greater than 5 cm.

11. The finding of experimental animals with decreased nest building ability indicates reduced well-being, abnormal sensorimotor integrity, and general impairments in behavior.

12. Ensure that all tested areas are seen on the computer screen.

13. Do not enter the room in this time.

14. Label video file with strain, mouse ID, cage number, and a short title of the experiment being undertaken.

15. Multiple behavioral software are available (e.g., TopScan) for the analysis of OF videos. The total distance traveled, bouts into the center zone or exterior, time spent in the center zone or exterior, as well as velocity in both areas can be conveniently measured by such sofware. Stereotypical behavior is hand scored and double scored by an individual blinded scorer with an inter-rater reliability of at least 80%.

16. Experimental animals entering the center more often and with decreased velocity than control animals are less anxious [16, 17]. Altered total distance traveled within the arena may indicate altered neurological processes or motor impairment as well as excessive anxiety-like behavior (fewer entries into and time spent in the center area).

17. Social interaction is hand scored and double scored by an individual scorer blinded for the experimental condition with an inter-rater reliability of at least 80%.

18. Ways to score the behaviors are given here:

Facial sniffing – oral to oral contact between target and test mouse with clear separation of bodies. Score when (a) target and test mouse faces (head) are touching or 1 cm apart; (b) to count as a separate bout, test mouse has to move its head at least 2 cm away from the target; and (c) when target mouse is sniffing the test mouse face without being precipitated, no contact is awarded.

Anogenital sniffing – test mouse snout is in contact with the target mouse anogenital area. A score is given when the test mouse sniffs area behind the hip or underneath the hip. New contact is defined when the test and the target animals are more than 2 cm apart and the interaction is re-initiated. If the test mouse continously chases the target mouse or falls behind by more than 2 cm a new contact is not scored.

Pursuit – active purposeful movement toward target mouse exceeding approximately 10 cm resulting in either contact (oral to oral/oral to base) or close proximity (2 cm). Score when (a) test mouse approaches target from a distance more than 10 cm away and arrives either a 2 cm away or touches target mouse.

Active avoidance – active purposeful movement away from target exceeding approximately 3 cm. Score when (a) target mouse moves toward test mouse and physically touches or is within 2 cm and test mouse moves away by 3 cm, or if the test mouse chases the target mouse or falls behind (more than 2 cm), it is not considered a new contact. Time spent social is the sum of interfacial and anogenital sniffing and scored using a stop watch and calculated using the formula below:

$$\%\text{Time spent social} = \left(\frac{\text{Time spent social (s)}}{300 \text{ s}}\right) \times 100$$

Experimental animals with decreases in interfacial, anogenital sniffing, and pursuit and increases in active avoidance compared to control animals are considered to be less social. Increased interfacial, anogenital sniffing, and pursuit and decreased active avoidance indicate a more social animal.

19. Experimental animals with more marbles buried show increased repetitive behavior [18, 19].

20. Multiple behavioral software are available (e.g. TopScan from Clever Sys or EthoVision) for the analysis. There are multiple measures to analysis including total distance traveled, bouts into the center zone or exterior, time spent in the center zone or exterior, as well as velocity in both areas. Experimental animals with less anxiety will explore the open arms more often than control animals, whereas more anxious experimental animals will spent more time in the closed arms.

21. The forced swim test is hand scored and double scored by an individual unaware of the experimental conditions with an inter-rater reliability of at least 80%. Measures include time to first immobility, time spent, and bouts in immobility and mobility. Decreased mobility time or increased immobility time, compared to controls, shows "learned helplessness" or

depression-like behavior. Experimental animals showing shorter time to first immobility are also considered as showing a depression-like phenotype [21].

22. Monitor the animal closely on computer screen. Do not leave the screen unattended for any reason as you must always have sight of the animal. If the mouse sinks, the behavioral test has to be stopped immediately and the animal removed. Wipe the mouse with a tissue to remove excess water, and place it back into its home cage. Place four fresh tissues in the cage. Ensure the there is no breeze from air conditioning getting into the cage.

23. Continuous spontaneous alternation of Y-maze analysis included the analysis of all arm entries and alternations through hand scoring. An arm entry is defined as the entry of all four paws into one arm (A, B, or C). An alternation is defined as an entry into all three arms consecutively. To identify spontaneous alternation, an overlapping technique was used. For example, "A-C-B-C-A," this pattern consists of two alternations. The percent (%) alternation was established through the following calculation:

$$\%\text{Alternation} = \left(\frac{\text{Number of alternations}}{\text{Total number of arm entries} - 2}\right) \times 100$$

24. Experimental animals with reduced or increased total arm entries compared to control may indicate altered locomotor activity, which should be taken into consideration when interpreting percent alternation. Decreased percent alternation indicates impairment in spatial working memory; however reduced arm entries may influence alternation percentage [22, 23]. Increased percent alternation indicates an improvement in working memory.

25. Analyze the total distance traveled throughout the testing into all arms to establish locomotor activity. Identify start, novel, and other arms and record the number of entries into the arms as well as time spent in each arm. A mouse with intact spatial reference memory should enter significantly more often and should spend more time in the novel arm compared to the other arm. Mice tend to linger in the start arm a bit longer; therefore start arm is usually not considered as a base of comparison. Mice with impaired spatial reference memory will not be able to identify the novel arm as they do not remember which arm was visited previously. Therefore, an impaired memory is signified by the lack of difference between activity in the novel and other arms.

References

1. Langley-Evans SC (2015) Nutrition in early life and the programming of adult disease: a review. J Hum Nutr Diet 28(Suppl 1):1–14

2. Rando OJ, Simmons RA (2015) I'm eating for two: parental dietary effects on offspring metabolism. Cell 161:93–105

3. Ong TP, Ozanne SE (2015) Developmental programming of type 2 diabetes: early nutrition and epigenetic mechanisms. Curr Opin Clin Nutr Metab Care 18:354–360

4. Baskin R, Hill B, Jacka FN, O'Neil A, Skouteris H (2015) The association between diet quality and mental health during the perinatal period. A systematic review. Appetite 91:41–47

5. Emmett PM, Jones LR, Golding J (2015) Pregnancy diet and associated outcomes in the Avon longitudinal study of parents and children. Nutr Rev 73(Suppl 3):154–174

6. Faa G, Manchia M, Pintus R, Gerosa C, Marcialis MA, Fanos V (2016) Fetal programming of neuropsychiatric disorders. Birth Defects Res C Embryo Today 108:207–223

7. Morgese MG, Trabace L (2016) Maternal malnutrition in the etiopathogenesis of psychiatric diseases: role of polyunsaturated fatty acids. Brain Sci 6(3): pii: E24. doi:https://doi.org/10.3390/brainsci6030024

8. Lynch CB (1994) Evolutionary inferences from genetic analyses of cold adaptation in laboratory and wild populations of the house mouse. In: Boake CRB (ed) Quantitative genetic studies of behavioral evolution, 2nd edn. University Of Chicago Press, Chicago. ISBN-10: 0226062163

9. Deacon RM (2006) Assessing nest building in mice. Nat Protoc 1:1117–1119

10. Goto T, Okayama T, Toyoda A (2015) Strain differences in temporal changes of nesting behaviors in C57BL/6N, DBA/2N, and their F1 hybrid mice assessed by a three-dimensional monitoring system. Behav Process 119:86–92

11. Chiu HY, Chan MH, Lee MY, Chen ST, Zhan ZY, Chen HH (2014) Long-lasting alterations in 5-HT2A receptor after a binge regimen of methamphetamine in mice. Int J Neuropsychopharmacol 17:1647–1658

12. Deacon RM, Croucher A, Rawlins JN (2002) Hippocampal cytotoxic lesion effects on species-typical behaviours in mice. Behav Brain Res 132:203–213

13. Prut L, Belzung C (2003) The open field as a paradigm to measure the effects of drugs on anxiety-like behaviors: a review. Eur J Pharmacol 463:3–33

14. Kalueff AV, Tuohimaa P (2004) Grooming analysis algorithm for neurobehavioural stress research. Brain Res Brain Res Protoc 13:151–158

15. Dunn AJ, Berridge CW, Lai YI, Yachabach TL (1987) CRF-induced excessive grooming behavior in rats and mice. Peptides 8:841–844

16. Crawley JN (1985) Exploratory behavior models of anxiety in mice. Neurosci Biobehav Rev 9:37–44

17. Bale TL, Contarino A, Smith GW, Chan R, Gold LH, Sawchenko PE et al (2000) Mice deficient for corticotropin-releasing hormone receptor-2 display anxiety-like behaviour and are hypersensitive to stress. Nat Genet 24:410–414

18. Thomas A, Burant A, Bui N, Graham D, Yuva-Paylor LA, Paylor R (2009) Marble burying reflects a repetitive and perseverative behavior more than novelty-induced anxiety. Psychopharmacology 204:361–373

19. Malkova NV, CZ Y, Hsiao EY, Moore MJ, Patterson PH (2012) Maternal immune activation yields offspring displaying mouse versions of the three core symptoms of autism. Brain Behav Immun 26:607–616

20. File SE (1987) The contribution of behavioural studies to the neuropharmacology of anxiety. Neuropharmacology 26:877–886

21. Porsolt RD, Bertin A, Jalfre M (1978) "Behavioural despair" in rats and mice: strain differences and the effects of imipramine. Eur J Pharmacol 51:291–294

22. Swonger AK, Rech RH (1972) Serotonergic and cholinergic involvement in habituation of activity and spontaneous alternation of rats in a Y maze. J Comp Physiol Psychol 81:509–522

23. Drew WG, Miller LL, Baugh EL (1073) Effects of delta9-THC, LSD-25 and scopolamine on continuous, spontaneous alternation in the Y-maze. Psychopharmacologia 32:171–182

24. Sarnyai Z, Sibille EL, Pavlides C, Fenster RJ, McEwen BS, Toth MC (2000) Impaired hippocampal-dependent learning and functional abnormalities in the hippocampus in mice lacking serotonin(1A) receptors. Proc Natl Acad Sci U S A 97:14731–14736

25. Conrad D, Lupien SJ, Thanasoulis LC, McEwen BS (1997) The effects of type I and type II corticosteroid receptor agonists on exploratory behavior and spatial memory in the Y-maze. Brain Res 759:76–83

26. Conrad CD, Galea LA, Kuroda Y, McEwen BS (1996) Behav Neurosci 110:1321–1334

Chapter 29

Hyperlocomotion Test for Assessing Behavioral Disorders

Dan Ma and Paul C. Guest

Abstract

Under- or overfeeding during pregnancy can lead to behavioral deficits in the offspring in later life. Here, we present a protocol for setting up and carrying out the hyperlocomotion test for assessing behavioral symptoms such as psychosis or mania. As an example, we use the acute rat phencyclidine-injection model which exhibits hyperlocomotion and stereotypic behaviors, resembling the positive symptoms of schizophrenia.

Key words Nutritional programming, Behavioral deficits, Psychosis, PCP, Hyperlocomotion test

1 Introduction

Nutritional insults in early life caused by factors such as a low protein or high fat maternal diet can lead to a number of disorders in the offspring in later life [1]. These include increased incidence of metabolic disorders [2–4], cardiovascular conditions [5, 6], cancer [7, 8], and behavioral disorders [9–11]. For epidemiological examples of the latter, an increased incidence of schizophrenia has been observed in offsprings who were conceived during the Dutch Hunger Winter of 1944–1945 [12] and the Chinese Famine of the 1959–1961 [13]. Furthermore, emotional stressors and other traumas occurring during pregnancy, such as experiencing the death of a relative or loved one, earthquakes, floods, and other disasters, can lead to increased incidence of behavioral disorders in the offspring [14]. Again, this is believed to occur through altered programming of the brain and peripheral physiologies [15].

Animal models are used routinely in the study of psychiatric disorders in order to gain a better understanding of the underlying disease mechanisms. One of the most widely used models for schizophrenia is the acute phencyclidine (PCP) rat model. PCP is an antagonist of the N-methyl-D-aspartate (NMDA) receptor which induces behavioral changes that mimic some of symptoms of acute psychosis [16–18]. Furthermore, these effects can be

Paul C. Guest (ed.), *Investigations of Early Nutrition Effects on Long-Term Health: Methods and Applications*, Methods in Molecular Biology, vol. 1735, https://doi.org/10.1007/978-1-4939-7614-0_29, © Springer Science+Business Media, LLC 2018

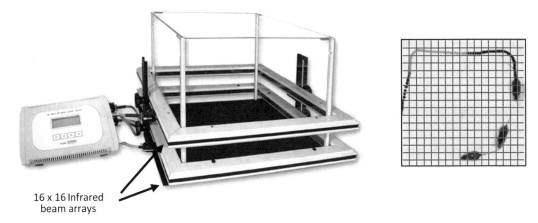

16 x 16 Infrared
beam arrays

Fig. 1 IR Actimeter System, Panlab, for measurement of locomotion and stereotypic behavior in rodents using a 16 × 16 grid of infrared beams

attenuated or blocked by co-administration of antipsychotic drugs [19].

Here we describe a protocol for production of the acute PCP rat model with a main focus on testing the resulting hyperlocomotion and stereotypical behaviors using the IR Actimeter System [20]. This device contains a 16 × 16 matrix of infrared beams for analyzing horizontal and vertical motion (Fig. 1). In a typical experiment, each beam break is recorded as one count and significant hyperlocomotion occurs 20–90 min after PCP injection [18]. Measurement of stereotypical behaviors can be achieved using a rating system that ranges from 0 (inactivity) to 5 (dyskinetic limb flexion and extension, head and neck movements, with gagging and weaving) [21]. This test may have applications in the study of anxiety, traumatic stress, schizophrenia, and other behavioral conditions that occur through antenatal nutritional programming deficits in rodents.

2 Materials

1. 280–300 g male Sprague-Dawley rats, housed in groups of 4 at 21 °C (±1 °C) using a 12 h/12 h light/dark cycle (lights on at 08:00) (*see* **Note 1**).

2. IR Actimeter System (Panlab; Barcelona, Spain) (Fig. 1) (*see* **Note 2**).

3. Vehicle: sterile 0.9% NaCl in water.

4. PCP: phencyclidine hydrochloride in the above vehicle (*see* **Note 3**).

3 Methods

1. Handle rats once a day for 1 week prior to the experimental day to aid their adjustment to the environment (*see* **Note 4**).

2. Beginning 48 h before the experiment, habituate the rats for 30 min each 24 h in the locomotion test box (*see* **Note 4**).

3. After a final 30 min habituation on the experimental day, inject the rats subcutaneously with vehicle or 5 mg/kg PCP, and place the rats back into the box (*see* **Note 5**).

4. Record locomotion activity at 10 min intervals for a total of 90 min using an observer who is blinded to the treatment group (*see* **Note 6**).

5. Record stereotypical behavior over the same period at 10 min intervals using an observer who is blinded to treatment group (*see* **Note 7**).

6. Use two-way ANOVA and post hoc tests to determine significant differences ($P < 0.05$) across the test groups (*see* **Notes 8** and **9**).

7. Express the data as mean ± SEM.

4 Notes

1. Food and water should be made available to the rats ad libitum. All experiments should be conducted during the light cycle and planned and performed in complete compliance with the appropriate regulations, such as the UK Animals Scientific Procedures Act of 1986 and the ethical policies of the Home Office.

2. The IR Actimeter System consists of a $45 \times 45 \times 35$ cm plastic box containing two infrared beam frames. These frames result in a three-dimensional 16×16 infrared beam matrix which allows analysis of both horizontal and vertical motions.

3. The PCP should be dissolved in saline at a concentration of 3.3 mg/mL, aliquoted, and stored at $-20\,^\circ$C for no more than 1 month. These conditions can be varied, but the user should ensure that no buffer incompatibility or other issues exist.

4. This should be performed by the person designated to carry out the locomotion test. We found that such habituation periods help to deliver more consistent behavioral responses across the different animals, compared to studies which do not use habituation. This may also be applicable across a range of behavioral tests.

5. To reduce the stress induced in any handling procedures, use gentle grabbing and holding techniques. In addition, the injection technique used should be the result of considerable experience and not attempted for the first time. Finally, attempts should be made to minimize noise with no additional strong lighting, as these conditions can also affect behavior.

6. Each beam break is recorded and differences in locomotion between PCP and control rats occurs usually between 20 and 90 min after the injection [18].

7. Measurement of stereotypical behavior can be performed using a variety of rating systems, such as the system described by Sturgeon et al. [21]. This uses a range of 0–5 which corresponds to inactivity and dyskinetic extension and flexion of limbs, head, and neck, gagging, and weaving, respectively. In general, rats injected with PCP lead to ratings of 2–4, and vehicle-treated rats have scores of 0–1 [18].

8. We used eight animals per group to help achieve statistical robustness. Achieving the right number can be difficult as one must attempt to use the minimum number of animals in all such experimentation studies.

9. If molecular studies are to be performed as in the case of biomarker analyses, a parallel group of animals should be used which have not been subjected to behavioral testing. This is because the testing can also induce molecular changes. In the parallel animals, cull and collect trunk blood or other tissues 30 min after injection as described [21].

References

1. Tarry-Adkins JL, Ozanne SE (2016) Nutrition in early life and age-associated diseases. Ageing Res Rev 39:96–105. https://doi.org/10.1016/j.arr.2016.08.003. Sep 1. pii: S1568-1637(16)30179-9. [Epub ahead of print]

2. Ong TP, Ozanne SE (2015) Developmental programming of type 2 diabetes: early nutrition and epigenetic mechanisms. Curr Opin Clin Nutr Metab Care 18:354–360

3. Jahan-Mihan A, Rodriguez J, Christie C, Sadeghi M, Zerbe T (2015) The role of maternal dietary proteins in development of metabolic syndrome in offspring. Forum Nutr 7:9185–9217

4. Arentson-Lantz EJ, Zou M, Teegarden D, Buhman KK, Donkin SS (2016) Maternal high fructose and low protein consumption during pregnancy and lactation share some but not all effects on early-life growth and metabolic programming of rat offspring. Nutr Res 36:937–946

5. Watkins AJ, Sinclair KD (2014) Paternal low protein diet affects adult offspring cardiovascular and metabolic function in mice. Am J Physiol Heart Circ Physiol 306:H1444–H1452. https://doi.org/10.1152/ajpheart.00981.2013

6. Raipuria M, Hardy GO, Bahari H, Morris MJ (2015) Maternal obesity regulates gene expression in the hearts of offspring. Nutr Metab Cardiovasc Dis 25:881–888

7. de Oliveira Andrade F, Fontelles CC, Rosim MP, de Oliveira TF, de Melo Loureiro AP, Mancini-Filho J et al (2014) Exposure to lard-based high-fat diet during fetal and lactation periods modifies breast cancer susceptibility in adulthood in rats. J Nutr Biochem 25:613–622

8. Govindarajah V, Leung YK, Ying J, Gear R, Bornschein RL, Medvedovic M et al (2016) In utero exposure of rats to high-fat diets perturbs gene expression profiles and cancer

susceptibility of prepubertal mammary glands. J Nutr Biochem 29:73–82

9. Palmer AA, Printz DJ, Butler PD, Dulawa SC, Printz MP (2004) Prenatal protein deprivation in rats induces changes in prepulse inhibition and NMDA receptor binding. Brain Res 996:193–201

10. Wu T, Deng S, Li WG, Yu Y, Li F, Mao M (2013) Maternal obesity caused by overnutrition exposure leads to reversal learning deficits and striatal disturbance in rats. PLoS One 8 (11):e78876. https://doi.org/10.1371/journal.pone.0078876

11. Bolton JL, Bilbo SD (2014) Developmental programming of brain and behavior by perinatal diet: focus on inflammatory mechanisms. Dialogues Clin Neurosci 16:307–320

12. Hoek HW, Brown AS, Susser E (1998) The Dutch famine and schizophrenia spectrum disorders. Soc Psychiatry Psychiatr Epidemiol 33:373–379

13. St Clair D, Xu M, Wang P, Yu Y, Fang Y, Zhang F et al (2005) Rates of adult schizophrenia following prenatal exposure to the Chinese famine of 1959–1961. JAMA 294:557–562

14. Guest FL, Martins-de Souza D, Rahmoune H, Bahn S, Guest PC (2013) The effects of stress on hypothalamic-pituitary-adrenal (HPA) axis function in subjects with schizophrenia. Rev Psiquiatr Clín 40:20–27

15. Guest PC (2017) Biomarkers and mental illness: it's not all in the mind, 1st edn. Copernicus, Cham. 2017 edition (14 Jan. 2017). ISBN-10: 3319460870

16. Allen RM, Young SJ (1978) Phencyclidine-induced psychosis. Am J Psychiatry 135:1081–1084

17. Adell A, Jimenez-Sanchez L, Lopez-Gil X, Romon T (2012) Is the acute NMDA receptor hypofunction a valid model of schizophrenia? Schizophr Bull 38:9–14

18. Ernst A, Ma D, Garcia-Perez I, Tsang TM, Kluge W, Schwarz E et al (2012) Molecular validation of the acute phencyclidine rat model for schizophrenia: identification of translational changes in energy metabolism and neurotransmission. J Proteome Res 11:3704–3714

19. Egerton A, Reid L, McGregor S, Cochran SM, Morris BJ, Pratt JA (2008) Subchronic and chronic PCP treatment produces temporally distinct deficits in attentional set shifting and prepulse inhibition in rats. Psychopharmacology (Berl) 198:37–49

20. http://www.harvardapparatus.com/compact-ir-actimeter.html

21. Sturgeon RD, Fessler RG, Meltzer HY (1979) Behavioral rating scales for assessing phencyclidine-induced locomotor activity, stereotyped behavior and ataxia in rats. Eur J Pharmacol 59:169–179

Chapter 30

2D-DIGE Analysis of Eye Lens Proteins as a Measure of Cataract Formation

Paul C. Guest

Abstract

This chapter describes the basics of two-dimensional difference gel electrophoresis (2D-DIGE) for multiplex analysis of two distinct proteomes. The example given describes the analysis of male and female rat lens soluble proteins labeled with fluorescent Cy3 and Cy5 dyes in comparison to a pooled standard labeled with Cy2. After labeling the proteomes are mixed together and electrophoresed on the same 2D gels. Scanning the gels at wavelengths specific for each dye allows direct overlay the two different proteomes. Differences in abundance of specific protein spots can be determined through comparison to the pooled standard.

Key words Metabolic disorders, Diabetes, Eye lens, Proteomics, 2D-DIGE, CyDyes

1 Introduction

The transparency of the eye lens is conveyed by precise and tight packing of the crystallin proteins into glass-like micro layers [1]. These proteins constitute around 80% of the soluble protein mass of the lens and are present in three basic forms termed α, β, and γ. The crystallin proteins are normally stable and do not turn over as the cells which produce them lose their nuclei during their differentiation to fibrous cells [2]. Therefore, any event altering the structure or the levels of specific crystallins, such as oxidation, could lead to a disruption of the lens architecture which, in turn, could alter the optical density and increase light scattering. This normally results in opacification of the lens as a cataract, which is the most common cause of visual loss in humans [3]. Several factors are associated with increased risk of cataract, including age, gender, and diseases such as type 2 diabetes [4, 5].

Two-dimensional (2D) gel electrophoresis studies have been carried out the lens from several species including humans and rodents [6–8]. Most of these studies have focused predominantly on the abundant crystallin proteins without considering the potential importance of the less abundant noncrystallin proteins. The

Paul C. Guest (ed.), *Investigations of Early Nutrition Effects on Long-Term Health: Methods and Applications*, Methods in Molecular Biology, vol. 1735, https://doi.org/10.1007/978-1-4939-7614-0_30, © Springer Science+Business Media, LLC 2018

latter are not easily detectable on 2D gels as these have normally been optimized for resolution of the crystallins. However, the noncrystallin proteins also have important roles in lens clarity and function. For example, the studies have shown that increased oxidation occurs in the aging eye, which can result in increased cross-linking of the crystallins [9, 10]. This suggests that perturbations in the activity of redox enzymes such as glutathione reductase and aldose reductase are likely to be altered during cataractogenesis [11].

This study describes a protocol for mapping of the noncrystallin proteins in rat lens using standard 2D gel electrophoresis. The conditions were optimized to allow resolution of the high molecular weight protein spots. The spots were excised from gels and the proteins identified by matrix-assisted laser desorption/ionization-time-of-flight (MALDI-TOF) mass fingerprinting. For presentation purposes, the identities of two of these proteins are given (αA crystallin dimer and aldose reductase), and we used 2D difference gel electrophoresis (DIGE) [12–14] to determine the relative levels of these proteins between female and male rats. This was carried out to gain further insight into the observation of significant sex differences in the propensity of rats and humans to develop age-related cataract [15, 16].

In the 2D-DIGE method, up to three protein extracts can be compared on a single gel by covalently labeling the proteins prior to electrophoresis with spectrally resolvable fluorescent dyes (Cy2, Cy3, and Cy5) [13, 14]. After electrophoresis, imaging of the gel at the wavelengths specific for each dye yields separate images for each proteome, and these can be overlaid for display of any protein spots that are present at different levels (Fig. 1). These proteins can then be identified by MALDI-TOF mass fingerprinting or other mass spectrometry methods [17, 18].

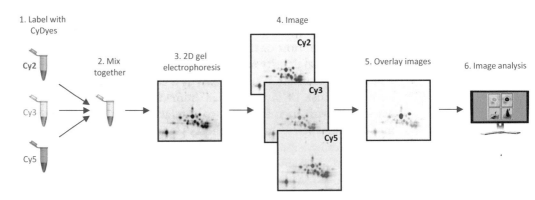

Fig. 1 Experimental flow of 2D-DIGE procedure incorporating the use of an internal standard

2 Materials

2.1 Tissue Extraction

1. Lenses from 20-week-old male and female Sprague-Dawley rats (*see* **Note 1**).

2. Extraction buffer: 30 mM Tris (pH 8.0) containing Complete Protease Inhibitors (Roche Diagnostics).

3. Soniprep 150 and microprobe (MSE UK Ltd.; London, UK) (*see* **Note 2**).

4. TL30 rotor (Beckman; Palo Alto, CA, USA) (*see* **Note 3**).

2.2 CyDye Labeling

1. 25 nmol Cy2™, 25 nmol Cy3™, and 25 nmol Cy5™ DIGE Fluor minimal dyes (GE Healthcare).

2. Anhydrous dimethyl formamide (DMF).

3. 10 mM lysine (*see* **Note 4**).

2.3 2D Gel Electrophoresis

1. 24 cm IPG strips (pH 3–10) (*see* **Note 5**).

2. Strip rehydration solution: 30 mM Tris (pH 8.0), 7 M urea, 2 M thiourea, 4% CHAPS, 2% dithiothreitol (DTT), 2% pH 3–10 IPG buffer (*see* **Note 6**).

3. Reducing equilibration buffer: 50 mM Tris (pH 6.8), 6 M urea, 30% glycerol, 2% sodium dodecyl sulfate (SDS), 1% DTT, 0.01% bromophenol blue (*see* **Note 7**).

4. Alkylating equilibration buffer: 50 mM Tris–Cl (pH 6.8), 6 M urea, 30% (v/v) glycerol, 2% (w/v) SDS, 2% iodoacetamide (IAA), 0.01% bromophenol blue (*see* **Note 8**).

5. 24 cm IPGphor ceramic strip holders (GE Healthcare) or similar.

6. First dimension isoelectric focusing (IEF): Ettan™ IPGphor™ 3 isoelectric focusing system (GE Healthcare) or similar.

7. Resolving gel buffer: 12.5% acrylamide, 0.33 N,N-'-methylenebisacrylamide, 0.375 M Tris–HCl (pH 8.8), 0.1% SDS, 0.1% ammonium persulfate, 0.025–0.09%tetramethylethylenediamine (TEMED) (*see* **Note 9**).

8. Second-dimension electrophoresis buffer: 25 mM Tris–HCl, 192 mM glycine, 0.1% SDS (pH 8.3).

9. 0.5% (w/v) agarose in SDS-PAGE electrode buffer.

10. Second-dimension, Ettan DALT6 Electrophoresis System (GE Healthcare) or similar.

11. Typhoon FLA 9500 Imager (GE Healthcare).

12. DeCyder™ 2D Software (GE Healthcare).

2.4 MALDI-TOF Mass Fingerprinting

1. SYPRO® Ruby Protein Gel Stain (Sigma-Aldrich).
2. ProPic™ protein spot picking robot (PerkinElmer Life Sciences).
3. ProGestTM digest robot (PerkinElmer Life Sciences).
4. α-Cyano-4-hydroxycinnamic acid matrix (CHCA) (Sigma-Aldrich).
5. MALDI-TOF target plates (PE Biosystems).
6. Voyager-DE STR Biospectrometry Workstation (PE Biosystems).
7. Protein Prospector MS-Fit (UCSF; CA, USA).

3 Methods

3.1 Lens Homogenization

1. Suspend lenses in extraction buffer at a concentration of 100 mg/mL.
2. Sonicate using an amplitude of 15 mm for 30 s.
3. Centrifuge at $100,000 \times g$ for 30 min at 4 °C using the TL30 rotor.
4. Collect the supernatants and proceed immediately to electrophoresis or store at −80 °C.

3.2 Protein Extraction and 2D-DIGE Analysis

1. Dilute each CyDye with 25 µL DMF immediately before labeling reaction to give 1 nmol stock solutions (*see* **Note 10**).
2. Add 0.4 µL Cy2 to 50 µg of pooled standard representing equal aliquots of all lens extracts (Table 1) (*see* **Note 11**).

Table 1
Experimental design for 2D-DIGE comparison of male and female rat lens soluble protein extracts. Three male and three female lens extracts and the pooled standard (equal aliquots of all male and female lens extracts) were labeled with the indicated CyDye, mixed together as shown, and electrophoresed on 2D gels

| Gel number | CyDye – sample combinations | | |
	Cy2	Cy3	Cy5
Gel 1	Pooled standard	Male 1	Female 1
Gel 2	Pooled standard	Male 2	Female 2
Gel 3	Pooled standard	Male 3	Female 3
Gel 4	Pooled standard	Female 1	Male 1
Gel 5	Pooled standard	Female 2	Male 2
Gel 6	Pooled standard	Female 3	Male 3

3. Add 0.4 µL Cy3 to 50 µg male lens extract.

4. Add 0.4 µL Cy5 to 50 µg of female lens extract (*see* **Note 12**).

5. Incubate samples 30 min on ice in the dark (*see* **Note 13**).

6. Add 1 µL 10 mM lysine and incubate 10 min on ice in the dark.

7. Top up the volume of each sample to 150 µL with rehydration solution.

8. Combine the three samples together as shown in Table 1 (*see* **Note 12**).

9. Add 450 µL sample mixture to IPG strips and hydrate 12 h at 20 °C in an IPGphor strip holder

10. Focus the IPG strip contents on the IPGphor system 1 h at 200 V, 1 h at 500 V, 1 h at 1000 V, and 8 h at 8000 at 20 °C applying a maximum current setting of 50 mA/strip (*see* **Note 14**).

11. After completion immerse the IPG strip for 10 min in 100 mL reducing equilibration buffer.

12. Remove this buffer and immerse the strip for 10 min in 100 mL alkylating equilibration buffer.

13. Prepare resolving acrylamide gel solution as required (*see* **Note 9**).

14. Add the APS and TEMED reagents to the solution last, and pour the mixture immediately between assembled low-fluorescence glass plates, leaving a space of approximately 3 cm from the top of the plates (*see* **Notes 15** and **16**).

15. Layer butanol on top of the gel to achieve a flat surface when it polymerizes.

16. After polymerization, rinse the butanol of the gel surface with water, and apply the equilibrated IPG strip so it rests immediately on top of the gel.

17. Seal the strip in pace with 0.5% agarose in second-dimension electrophoresis buffer on top (*see* **Note 17**).

18. Carry out second-dimension electrophoresis for 1 h at 60 V and then set at 30 µA/gel.

19. Stop electrophoresis when the dye front reaches the end of the gel.

20. Scan the gels between the glass plates using the Typhoon FLA 9500 Imager (*see* **Note 18**).

21. Export the images as tagged image file format (TIFF) files.

22. Analyze the images using the DeCyder Batch Processor and Differential In-Gel Analysis (DIA) software tools.

23. Compare the protein spot volumes on the Cy2-labeled pooled standard image with matching spots on the Cy3- or Cy5-labeled sample images (the male and female lens extracts).

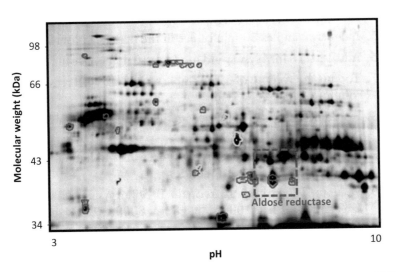
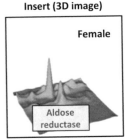
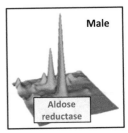

Fig. 2 2D-DIGE analysis showing increased levels of aldose reductase (+84%) in male versus female rat lenses. Only the high molecular weight region of the gel is shown. The insert shows a 3D view of the increased volume of a spot identified as aldose reductase (identified by MALDI-TOF mass fingerprinting as described below)

24. Match the images from each gel with the Biological Variation Analysis (BVA) software using the pooled standard image to normalize each protein spot.

25. Identify protein spots with differences in spot volume (abundance) between the male and female lenses using the software (Fig. 2) (*see* **Note 19**).

3.3 MALDI-TOF Mass Spectrometry

1. Stain gels with SYPRO Ruby according to the manufacturer's instructions (Fig. 3).

2. Excise spots using a sterile pipette tips or a spot picking robot if available (*see* **Note 20**).

3. Digest the excised gel spots in situ with sequencing grade porcine trypsin using a standard method [19] or a digest robot if available (*see* **Note 21**).

4. Mix 1 μL extracted digests with 0.5 μL 10 mg/mL CHCA on MALDI-TOF target plates.

5. Determine peptide masses using the Voyager-DE STR Biospectrometry Workstation or similar as described [20] (*see* **Note 22**).

6. Identify proteins by searching the UniProt databases using MS-Fit.

7. Carry out searches using a mass window of 10,000–200,000 Da of rodent and human sequences.

Normal exposure

Over-exposed upper gel region

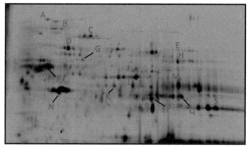

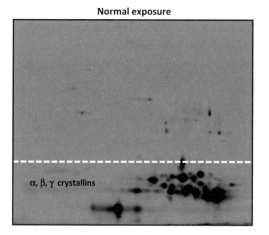

α, β, γ crystallins

Fig. 3 Identification of high molecular weight proteins in rat lens. (**Left**) 2D gel electrophoresis image of rat lens soluble protein extract (10 mg loading) showing mostly low molecular weight crystallin proteins using SYPRO Ruby fluorescent staining. (**Right**) 2D gel electrophoresis images of rat lens soluble proteins (100 mg loading) showing high molecular weight protein region. Predominant protein spots were identified by MALDI-TOF mass spectrometry

8. The search parameters should allow for carboxyamidomethylation of cysteine, oxidation of methionine, and modification of glutamine to pyroglutamic acid.

9. Accept positive identifications based on at least six matching peptide masses or greater than 30% peptide coverage of the theoretical sequences (Table 2).

4   Notes

1. Ensure that all procedures and approvals regarding the study of animals are in order in compliance with the Home Office Guidance on the Operation of the UK Animal Scientific Procedures Act of 1986 or similar.

2. Other sonicators can be used but the settings should be determined experimentally. A microprobe is favorable as this can reach to the bottom of a standard 1.5 mL-capacity microcentrifuge tube.

3. Other centrifuge rotors can be used but the swinging bucket variety is preferable.

4. Lysine is added as a quenching reagent after the labeling procedure.

5. IPG strip pH ranges should be selected to maximize visualization of the targeted proteins. Strips in the pH range 3–10 strips will provide resolution of acidic, neutral, and basic proteins.

Table 2
Identification of soluble lens proteins by MALDI-TOF mass spectrometry.
The spot identifiers correspond to those indicated on Fig. 2

Spot	UniProt code	Protein
A	Q13813	Spectrin alpha chain
B	P46462	Transitional ER ATPase
C	P13120	Gelsolin
D	P08109	Heat shock cognate 71 kDa
E	P40142	Transketolase
F	O88342	WD-repeat protein 1
G	P80316	T-complex protein 1 epsilon subunit
H	P11981	Pyruvate kinase M2 isozyme
I	P31000	Vimentin
J	P04764	Alpha enolase
K	P02523	βB1 crystallin (dimer)
L	P02523	βB1 crystallin (dimer)
M	P02523	βB1 crystallin (dimer)
N	P02570	Beta-actin
O	P02490	αA crystallin (dimer)
P	P02490	αA crystallin (dimer)
Q	P09117	Fructose-bisphosphate aldolase C
R	P07943	Aldose reductase
S	P04797	Glyceraldehyde 3-P dehydrogenase

6. The IPG buffer should match the pH range of the strips used for isoelectric focusing.

7. The DTT in this solution reduces disulfide bonds in proteins to yield free sulfhydryl groups. The dye is added to visualize the migration of the dye front upon second-dimension electrophoresis. However, this can be added at a later stage prior to second-dimension electrophoresis.

8. The IAA in this solution alkylates the free sulfhydryl groups to avoid reformation of disulfide bonds.

9. The acrylamide and N,N'-methylenebisacrylamide concentrations should be chosen based on the desired molecular weight range separation. The main rule to keep in mind is that low percentages of acrylamide help to resolve high molecular weight proteins and high percentages resolve lower molecular

weight proteins. Here a concentration has been chosen to allow resolution of proteins in the range of 10–200 kDa. Ammonium persulfate and TEMED should be added just before use as these reagents initiate the gel polymerization process.

10. DMF should be used as soon as possible after opening. We have noticed decreased labeling efficiency with DMF solutions over 1-month-old.

11. The use of an internal standard helps to minimize false positives and false negatives since it can serve as a control for each protein spot on all gels in the analysis. The standard is usually made by combining equal volumes of each extract.

12. The CyDyes may show preferential labeling of some proteins. This can be accounted for by reversing the dye/extract combinations as we have done here (Table 1).

13. The CyDyes are light sensitive and may suffer from photobleaching in excessive light.

14. We used a step voltage gradient as we and others have found that this helps to avoid horizontal streaking of protein spots on 2D gels.

15. This allows space to apply the equilibrated IPG strips.

16. These gels should be poured between low-fluorescence glass plates to reduce background fluorescence that can affect the imaging stage. Note that the Ettan DALT II system allows simultaneous running of multiple plates so that all second-dimension gels can be run under approximately the same conditions. This allows for better matching of the gel images in subsequent stages compared to gels run separately in different tanks.

17. Ensure that there are no air bubbles trapped between the strips and the gel tops to avoid distortion in the protein spot patterns.

18. The gels can be imaged after electrophoresis without disassembly of the low-fluorescence glass plates. This helps to maintain uniformity of the gels for improved imaging and subsequent matching. In addition, the gels can be scanned for different lengths of time to maximize detection of protein spots present at both high and low levels. This was required in the current study to detect high molecular weight noncrystallin proteins.

19. Within the same gel, matching is achieved automatically by direct overlay of the spots of the three images. Matching is achieved across different gels by the user carrying out land marking. This allows the Biological Variation Analysis function of the DeCyder software to warp and match the remaining

spots. The example shows higher levels of aldose reductase in male compared to female rat lenses [20] (Fig. 2 and Table 2).

20. We use the ProPic robot from PerkinElmer Life Sciences.

21. We use the ProGest robot from PerkinElmer Life Sciences.

22. Protein identification can be achieved using many kinds of protein-based mass spectrometry platforms. Here we have used peptide mass fingerprinting, and internal mass calibration was achieved using porcine trypsin autolysis digestion products.

References

1. Bloemendal H, de Jong W, Jaenicke R, Lubsen NH, Slingsby C, Tardieu A (2004) Ageing and vision: structure, stability and function of lens crystallins. Prog Biophys Mol Biol 86:407–485

2. Wistow G (2012) The human crystallin gene families. Hum Genomics 6:26. https://doi.org/10.1186/1479-7364-6-26

3. Hejtmancik JF, Kantorow M (2004) Molecular genetics of age-related cataract. Exp Eye Res 79:3–9

4. Beebe DC, Holekamp NM, Shui YB (2010) Oxidative damage and the prevention of age-related cataracts. Ophthalmic Res 44:155–165

5. Obrosova IG, Chung SS, Kador PF (2010) Diabetic cataracts: mechanisms and management. Diabetes Metab Res Rev 26:172–180

6. Shearer TR, Shih M, Mizuno T, David LL (1996) Crystallins from rat lens are especially susceptible to calpain-induced light scattering compared to other species. Curr Eye Res 15:860–868

7. Harrington V, Srivastava OP, Kirk M (2007) Proteomic analysis of water insoluble proteins from normal and cataractous human lenses. Mol Vis 13:1680–1694

8. Su SP, Song X, Xavier D, Aquilina JA (2015) Age-related cleavages of crystallins in human lens cortical fiber cells generate a plethora of endogenous peptides and high molecular weight complexes. Proteins 83:1878–1886

9. Babizhayev MA (2011) Mitochondria induce oxidative stress, generation of reactive oxygen species and redox state unbalance of the eye lens leading to human cataract formation: disruption of redox lens organization by phospholipid hydroperoxides as a common basis for cataract disease. Cell Biochem Funct 29:183–206

10. Babizhayev MA, Yegorov YE (2016) Reactive oxygen species and the aging eye: specific role of metabolically active mitochondria in maintaining lens function and in the initiation of the oxidation-induced maturity onset cataract--a novel platform of mitochondria-targeted antioxidants with broad therapeutic potential for redox regulation and detoxification of oxidants in eye diseases. Am J Ther 23:e98–117. https://doi.org/10.1097/MJT.0b013e3181ea31ff

11. Su S, Leng F, Guan L, Zhang L, Ge J, Wang C et al (2014) Differential proteomic analyses of cataracts from rat models of type 1 and 2 diabetes. Invest Ophthalmol Vis Sci 55:7848–7861

12. Unlü M, Morgan ME, Minden JS (1997) Difference gel electrophoresis: a single gel method for detecting changes in protein extracts. Electrophoresis 18:2071–2077

13. Knowles MR, Cervino S, Skynner HA, Hunt SP, de Felipe C, Salim K et al (2003) Multiplex proteomic analysis by two-dimensional differential in-gel electrophoresis. Proteomics 3:1162–1171

14. Aquino A, Guest PC, Martins-de-Souza D (2017) Simultaneous two-dimensional difference gel electrophoresis (2D-DIGE) analysis of two distinct proteomes. Methods Mol Biol 1546:205–212

15. Hiller R, Sperduto RD, Ederer F (1986) Epidemiologic associations with nuclear, cortical, and posterior subcapsular cataracts. Am J Epidemiol 124:916–925

16. Hales AM, Chamberlain CG, Murphy CR, McAvoy JW (1997) Estrogen protects lenses against cataract induced by transforming growth factor-beta (TGFbeta). J Exp Med 185:273–280

17. Intelicato-Young J, Fox A (2013) Mass spectrometry and tandem mass spectrometry characterization of protein patterns, protein markers and whole proteomes for pathogenic bacteria. J Microbiol Methods 92:381–386

18. Souza GH, Guest PC, Martins-de-Souza D (2017) LC-MSE, multiplex MS/MS, ion mobility, and label-free quantitation in clinical proteomics. Methods Mol Biol 1546:57–73

19. Shevchenko A, Tomas H, Havlis J, Olsen JV, Mann M (1996) In-gel digestion for mass spectrometric characterization of proteins and proteomes. Nat Protoc 1:2856–2860

20. Guest PC, Skynner HA, Salim K, Tattersall FD, Knowles MR, Atack JR (2006) Detection of gender differences in rat lens proteins using 2-D-DIGE. Proteomics 6:667–676

Chapter 31

Mass Spectrometry Profiling of Pituitary Glands

Divya Krishnamurthy, Hassan Rahmoune, and Paul C. Guest

Abstract

Many chronic diseases are associated with hypothalamic-pituitary-adrenal axis dysfunction. Therefore, proteomic profiling of the pituitary gland has potential to uncover new information on the underlying pathways affected in these conditions. This could lead to identification of new biomarkers or drug targets for development of novel therapeutics. Here we present a protocol for preparation of pituitary protein extracts and analysis of the major hormones and accessory proteins using liquid chromatography tandem mass spectrometry (LC-MS/MS). The same methods can be applied in the study of other tissues of the diffuse neuroendocrine system.

Key words Neuroendocrine disorders, Metabolic disorders, Pituitary, Hormone, Accessory protein, Proteome, LC-MS/MS

1 Introduction

The pituitary gland is located at the base of the brain [1]. The anterior lobe of this gland houses cell groups which produce pro-hormones such as prolactin, pro-opiomelanocortin (POMC) and growth hormone in response to physiological needs. The posterior lobe mostly contains neural projections from the hypothalamus and generates the prohormones oxytocin-neurophysin 1 and vasopressin-neurophysin 2-copeptin. The secretion of the mature forms of these hormones into the bloodstream is regulated by feedback and feed-forward mechanisms in the process of maintaining physiological homeostasis [2].

Pituitary abnormalities have been implicated in various pathological conditions, including psychiatric diseases [3–5], metabolic disorders [6, 7], immune diseases [8], and Cushing's syndrome [9]. This is not surprising as the hypothalamic-pituitary-adrenal (HPA) axis is involved in maintaining homeostasis of multiple neuroendocrine and physiological systems of the body. One of the products of HPA axis stimulation is cortisol, which is released from the adrenal cortex in response circulating adrenocorticotrophic

Paul C. Guest (ed.), *Investigations of Early Nutrition Effects on Long-Term Health: Methods and Applications*, Methods in Molecular Biology, vol. 1735, https://doi.org/10.1007/978-1-4939-7614-0_31, © Springer Science+Business Media, LLC 2018

hormone (ACTH), originating from the pituitary [10]. Cortisol levels have been proposed as a mechanism for programming fetal brain development. A study measured maternal plasma cortisol concentrations in mothers at 19 and 31 gestational weeks and then assessed brain development using structural magnetic resonance imaging and cognitive functions (Wechsler Intelligence Scale for Children IV and Expressive Vocabulary Test, Second Edition) when the children were 6–9 years old [11]. This showed that higher maternal cortisol concentrations during the third trimester were correlated with higher cortical thickness and enhanced cognitive performance in the children. However, other studies have proposed that dysregulation or pharmacological manipulation of maternal cortisol plasma concentrations can have negative consequences on the offspring in later life [12]. In animal models, glucocorticoid exposure leads to transcriptomic and epigenomic changes that can influence HPA function, growth, and behavior, and this can be inherited over several generations through both paternal and maternal transmissions [13]. Along these lines, maternal obesity and metabolic disorders increase the risk of the offspring developing behavioral disorders through effects on placental and neuroendocrine function [14].

The various anterior pituitary cell types (corticotropes, gonadotropes, lactotropes, somatotropes, and thyrotropes) and the neuronal projections in the posterior pituitary contain all of the major secreted hormones (Fig. 1). In addition, many of these cells contain other proproteins including members of the chromogranin

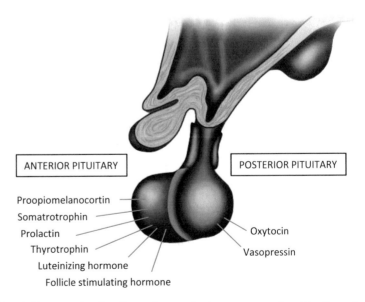

Fig. 1 Diagram showing the various major hormones produced in the anterior and posterior pituitary

and secretogranin family [15] as well as secretory granule factors involved in prohormone processing and secretion, such as the endoproteases prohormone convertase 1 (PCSK1) [16] and PCSK2 [17], the exopeptidase carboxypeptidase H (CPE) [18], and the carboxy-terminal amidating enzyme peptidyl-glycine alpha-amidating monooxygenase (PAM) [18, 19].

Given its vital role in maintaining physiological homeostasis, further characterization of the human pituitary proteome may provide new insights into the underlying causes and effects of many diseases such as type 2 diabetes, obesity, insulin resistance, and psychiatric disorders. Furthermore, the fact that the hormones and many of the accessory proteins are co-released into the bloodstream under specific physiological conditions suggests that they may also be used as blood-based biomarkers in clinical studies [20]. Here, we present a protocol for preparation and extraction of proteins from human postmortem pituitary glands for proteomic analysis using liquid chromatography tandem mass spectrometry (LC-MS^E). It was of particular interest to confirm identification of the major secreted pituitary hormones, other proproteins, and the proprotein-converting enzymes, given their potential utility as blood-based biomarkers in clinical studies.

2 Materials

1. Store postmortem pituitary glands or tissue of choice in liquid nitrogen (*see* **Note 1**).

2. Alumina mortar and pestle (Sigma-Aldrich).

3. Homogenization buffer A: 50 mM Tris–HCl (pH 8.0), EDTA-free protease inhibitor cocktail (Merck Calbiochem) (*see* **Note 2**).

4. Homogenization buffer B: 8 M urea, 50 mM sodium bicarbonate (pH 8.0) (*see* **Note 3**).

5. Sonifier device with microprobe (*see* **Note 4**).

6. BCA (bicinchoninic acid) Protein Assay Kit (*see* **Note 5**).

7. 50 mM NaHCO$_3$ (pH 8.0).

8. Reduction buffer: 5 mM dithiothreitol (DTT).

9. Alkylation buffer: 10 mM iodoacetic acid (IAA).

10. 25 fmol/μL yeast enolase (digested with trypsin) (Waters Corporation) (*see* **Note 6**).

11. Sequencing Grade Modified Trypsin (Promega).

12. Ultra Performance 10k psi nanoACQUITY liquid chromatography system (Waters Corporation).

13. 7 cm nanoelectrospray ionization emitter (10 mm tip) (New Objective; Woburn, MA, USA).

14. Q-Tof Premier Mass Spectrometer (Waters Corporation) (*see* **Note 7**).

15. Glufibrinopeptide B LockSpray calibration standard (Sigma-Aldrich).

16. ProteinLynx Global Server (PLGS) version 2.4 (Waters Corporation).

17. Rosetta Biosoftware Elucidator version 3.3 (Seattle, WA, USA) (*see* **Note 8**).

18. UniProt database [21].

3 Methods

1. Remove frozen intact pituitaries from liquid nitrogen storage and do not allow thawing.

2. Grind each pituitary to a powder on dry ice using a mortar and pestle on dry ice (*see* **Note 9**).

3. Aliquot and store powdered pituitary at −80 °C as required or proceed immediately to the next step.

4. Sonicate 100 mg tissue powder in 500 µL homogenization buffer A by gradually increasing power over 15 s to approximately 0.25 of the maximum power level (*see* **Note 10**).

5. Centrifuge 20 min at 13,000 × g at 4 °C.

6. Retain the supernatant and designate this as the soluble fraction.

7. Suspend the pellet in homogenization buffer B and repeat the sonication step.

8. Centrifuge 20 min at 13,000 × g at room temperature.

9. Retain the supernatant and designate this as the insoluble fraction (*see* **Note 11**).

10. Determine the protein concentrations of both fractions using the BCA kit.

11. Add 50 mM $NaHCO_3$ (pH 8.0) to each sample to give a final 100 µg protein in 100 µL buffer.

12. Incubate samples 30 min in reduction buffer at 60 °C (*see* **Note 12**).

13. Incubate samples 30 min in alkylation buffer at 37 °C in the dark (*see* **Note 13**).

14. Add trypsin at a 1:50 (trypsin/protein) ratio.

15. Incubate for 16 h at 37 °C.

16. Add a further trypsin aliquot after 16 h as above and continue the incubation for a further 2.5 h (*see* **Note 14**).

17. Stop digestions by adding 1.67 μL 10 M HCl.

18. Spike each sample with yeast enolase digest.

19. Desalt the samples with the trap column on the nanoACQUITY system using 97% buffer A/3% buffer B

20. Apply the samples to the BEH nanocolumn at 0.3 μL/min coupled via the nanoESI emitter to the mass spectrometer.

21. Using this system, the conditions should be initial 3% B; 3–30% B over 90 min; 30–90% B over 25 min; 90–95% B over 5 min; constant 95% B for 5 min; and 95–3% over 1 min.

22. Infuse Glufibrinopeptide B using the LockSpray and scan every 30 s for instrument calibration.

23. Acquire data as described previously [15] in MS^E mode (*see* **Note 15**).

24. Process the MS1 data using a combination of the PLGS and Elucidator softwares for alignment of time and m/z dimensions.

25. Extract the aligned peaks and integrate time, m/z, and peak volumes for determination of abundance levels if required (*see* **Note 16**).

26. Search the UniProt database with this data using the ion accounting algorithm as described previously [22] (*see* **Note 17**).

27. Set fixed modifications as carbamidomethylation of cysteine.

28. Set variable modifications as oxidation of methionine.

29. Set phosphorylation as serine, threonine, or tyrosine.

30. Set criteria for peptide identification as >fragment ions/peptide, > fragment ions/protein, > peptides/protein, detection in 2/3 replicates, and detection in 60% of samples.

31. Compile a list of all identified prohormones, proproteins, and converting enzymes using the above parameters (Table 1) (*see* **Note 18**).

4 Notes

1. The first step is to make sure that all of the necessary documentation is in place from the appropriate institutional ethical review boards regarding the storage and investigation of human tissues. In addition, appropriate data protection procedures should be in place.

Table 1
List of identified prohormones, proproteins, and converting enzymes

Protein name	Soluble (% sequence coverage)	Insoluble (% sequence coverage)
Prohormones		
Chorionic somatomammotropin	64	51
Follitropin β-chain	17	0
Galanin	43	12
Glycoprotein hormones α-chain	32	22
Growth hormone variant	71	66
Lutropin β-chain	70	79
Oxytocin-neurophysin 1	70	70
Prolactin	74	79
Pro-opiomelanocortin	76	83
Somatotropin (growth hormone)	89	76
Thyrotropin β-chain	63	98
Vasopressin-neurophysin 2-copeptin	70	69
Other proproteins		
Chromogranin A	80	56
Neuroendocrine protein 7B2	36	9
Neurosecretory protein VGF	11	30
ProSAAS	57	76
Secretogranin-1	73	71
Secretogranin-2	80	76
Secretogranin-3	19	4
Secretagogin	17	28
Converting enzymes		
Carboxypeptidase H	69	69
Prohormone convertase 1	9	16

2. This buffer can be used to extract mainly soluble proteins from cells and tissues.

3. This buffer should solubilize the integral membrane, membrane-attached, and high molecular weight proteins.

4. Many instruments can be used here. We used a Branson Sonifier 150 (Thistle Scientific; Glasgow, UK).

5. Other kits can be used but the investigator should make sure that all of their buffers and other reagents used in the experiment prior to protein determination are compatible with those used in the kit.

6. Trypsin-digested yeast enolase is used here as a standard to monitor the chromatography stage of the mass spectrometry procedure.

7. There are many mass spectrometers that can be used here. However, the experimental steps and conditions have been developed for use with the Q-Tof Premier.

8. The Rosetta Elucidator software can be used for alignment of peptide precursor ions in time and mass/charge (m/z) dimensions [23]. In this study, the use of this software led to higher percentages of protein coverage during the identification stage compared to other studies that used the PLGS system alone.

9. Whole tissue was used, as opposed to dissected tissue, to circumvent heterogeneity issues (the pituitary gland is comprised of distinct lobes and cell types).

10. Of course this is instrument dependent and should be determined experimentally. The main cautions are to avoid overheating of the sample and/or frothing.

11. Do not freeze or refrigerate solutions containing as this could lead to precipitation. This would make resuspension of the proteins difficult after thawing.

12. The presence of DTT reduces any disulfide bonds in proteins and converts them to free sulfhydryl groups.

13. The IAA reagent alkylates the free sulfhydryl groups to help minimize reformation of disulfide bonds and other modifications of proteins during subsequent steps.

14. The additional digestion period with trypsin should help to digest most proteins more completely. This will lead to increased sequence coverage and thereby improve protein identifications.

15. In the MS^E method, the energy is cycled rapidly from low to high to low in the collision cell. Intact peptide ions are detected and measured during the low-energy stage, whereas fragment ions are detected and measured during the high-energy stage. All fragment ions and the corresponding precursor peptide ions can be matched based on retention time, mass accuracy, and other physical properties [23]. We used a low collision energy of 5 eV and a ramped energy of 17–40 eV for the high-energy phase, over a full cycle time of 1.3 s.

16. All abundance measurements should be obtained after normalization. In this experiment, we used the total ion current.

17. To help reduce false identifications, the data were also used to search a randomized database with a maximum false discovery rate set at 4%.

18. In this analysis, we identified all of the major pituitary prohormones and other proproteins in both soluble and insoluble extracts. However, two of the major conversion enzymes, PCSK2 and PAM, were not identified. This was most likely due to the lower abundance of these proteins. Identification of these enzymes may be achieved after an enrichment step.

References

1. Kannan CR (2013) The pituitary gland: vol 1 (clinical surveys in endocrinology). Springer, Berlin. Softcover reprint of the original 1st ed. 1987 edition (4 Oct. 2013). ISBN-10: 1461290325

2. Papadimitriou A, Priftis KN (2009) Regulation of the hypothalamic- pituitary-adrenal axis. Neuroimmodulation 16:265–271

3. Takahashi T, Suzuki M, Velakoulis D, Lorenzetti V, Soulsby B, Zhou SY et al (2009) Increased pituitary volume in schizophrenia spectrum disorders. Schizophr Res 108:114–421

4. Krishnamurthy D, Harris LW, Levin Y, Koutroukides TA, Rahmoune H, Pietsch S et al (2013) Metabolic, hormonal and stress-related molecular changes in post-mortem pituitary glands from schizophrenia subjects. World J Biol Psychiatry 14:478–489

5. Naughton M, Dinan TG, Scott LV (2014) Corticotropin-releasing hormone and the hypothalamic-pituitary-adrenal axis in psychiatric disease. Handb Clin Neurol 124:69–91

6. Roelfsema F, Pereira AM, Veldhuis JD (2014) Impact of adiposity and fat distribution on the dynamics of adrenocorticotropin and cortisol rhythms. Curr Obes Rep 3:387–395

7. Joseph JJ, Golden SH (2016) Cortisol dysregulation: the bidirectional link between stress, depression, and type 2 diabetes mellitus. Ann N Y Acad Sci 1391(1):20–34. https://doi.org/10.1111/nyas.13217. Oct 17. [Epub ahead of print]

8. Straub RH, Buttgereit F, Cutolo M (2011) Alterations of the hypothalamic-pituitary-adrenal axis in systemic immune diseases - a role for misguided energy regulation. Clin Exp Rheumatol 29(5 Suppl 68):S23–S31

9. Pivonello R, De Martino MC, De Leo M, Simeoli C, Colao A (2016) Cushing's disease: the burden of illness. Endocrine 56(1):10–18. May 17. [Epub ahead of print]

10. Geer EB (2016) The hypothalamic pituitary adrenal axis in health and disease: cushing's syndrome and beyond, 1st edn. Springer, New York. 2017 edition (22 Dec. 2016). ISBN-10: 3319459481

11. Davis EP, Head K, Buss C, Sandman CA (2017) Prenatal maternal cortisol concentrations predict neurodevelopment in middle childhood. Psychoneuroendocrinology 75:56–63

12. Wood CE, Keller-Wood M (2016) The critical importance of the fetal hypothalamus-pituitary-adrenal axis. F1000Res 5.: pii: F1000 Faculty Rev-115. eCollection 2016. 10.12688/f1000research.7224.1

13. Constantinof A, Moisiadis VG, SG M (2016) Programming of stress pathways: a transgenerational perspective. J Steroid Biochem Mol Biol 160:175–180

14. Sullivan EL, Riper KM, Lockard R, Valleau JC (2015) Maternal high-fat diet programming of the neuroendocrine system and behavior. Horm Behav 76:153–161

15. Krishnamurthy D, Levin Y, Harris LW, Umrania Y, Bahn S, Guest PC (2011) Analysis of the human pituitary proteome by data independent label-free liquid chromatography tandem mass spectrometry. Proteomics 11:495–500

16. Seidah NG, Marcinkiewicz M, Benjannet S, Gaspar L, Beaubien G, Mattei MG et al (1991) Cloning and primary sequence of a mouse candidate prohormoneconvertase PC1 homologous to PC2, Furin, and Kex2: distinct chromosomal localization and messenger RNA

distribution in brain and pituitary compared to PC2. Mol Endocrinol 5:111–122

17. Smeekens SP, Avruch AS, LaMendola J, Chan SJ, Steiner DF (1991) Identification of a cDNA encoding a second putative prohormonecon-vertase related to PC2 in AtT20 cells and islets of Langerhans. Proc Natl Acad Sci U S A 88:340–344

18. Fricker LD, Evans CJ, Esch FS, Herbert E (1986) Cloning and sequence analysis of cDNA for bovine carboxypeptidase E. Nature 323:461–454

19. Glauder J, Ragg H, Rauch J, Engels JW (1990) Human peptidylglycine alpha-amidating monooxygenase: cDNA, cloning and functional expression of a truncated form in COS cells. Biochem Biophys Res Commun 169:551–558

20. Mains RE, Eipper BA (1984) Secretion and regulation of two biosynthetic enzyme activities, peptidyl-glycine alpha-amidating mono-oxygenase and a carboxypeptidase, by mouse pituitary corticotropic tumor cells. Endocrinology 115:1683–1690

21. http://www.uniprot.org/

22. Li GZ, Vissers JP, Silva JC, Golick D, Gorenstein MV, Geromanos SJ (2009) Database searching and accounting of multiplexed precursor and product ion spectra from the data independent analysis of simple and complex peptide mixtures. Proteomics 9:1696–1719

23. Neubert H, Bonnert TP, Rumpel K, Hunt BT, Henle ES, James IT (2008) Label-free detection of differential protein expression by LC/MALDI mass spectrometry. J Proteome Res 7:2270–2279

Chapter 32

Multiplex Immunoassay Profiling of Hormones Involved in Metabolic Regulation

Laurie Stephen and Paul C. Guest

Abstract

Multiplex immunoassays are used for rapid profiling of biomarker proteins and small molecules in biological fluids. The advantages over single immunoassays include lower sample consumption, cost, and labor. This chapter details a protocol to develop a 5-plex assay for glucagon-like peptide 1, growth hormone, insulin, leptin, and thyroid-stimulating hormone on the Luminex® platform. The results of the analysis of insulin in normal control subjects are given due to the important role of this hormone in nutritional programming diseases.

Key words Nutritional programming, Disease, Multiplex assay, Antibody, Hormone, Insulin, Biomarker, Luminex assay

1 Introduction

Studies over the last 30 years have focused on the effect of maternal malnutrition as a risk factor for development of metabolic diseases in the offspring in later life [1, 2]. The mechanism appears to involve epigenetic modifications resulting in perturbations of various hormonal and metabolic pathways along with disturbances in anti-oxidant capacity [3, 4]. Investigations have focused on unraveling the mechanisms involved in attempts to identify novel therapeutics biomarkers which could be used for early disease detection and treatment monitoring [5–8], as well as new therapeutic targets [9–11].

Blood serum or plasma contains thousands of circulating proteins including hormones such as insulin which are likely to be involved in metabolic diseases brought about by developmental programming deficits. However, most of these molecules are present only at very low concentration, which requires that any measurement system employed should have high sensitivity, such as those based on antibody detection [12]. Several studies have now been carried out which have shown that multiplex immunoassay

Paul C. Guest (ed.), *Investigations of Early Nutrition Effects on Long-Term Health: Methods and Applications*, Methods in Molecular Biology, vol. 1735, https://doi.org/10.1007/978-1-4939-7614-0_32, © Springer Science+Business Media, LLC 2018

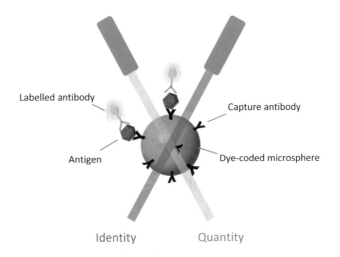

Fig. 1 Schematic diagram of a multiplex immunoassay microsphere with conjugated capture antibody, target analyte, detection antibody, and identification and quantitation lasers within the Luminex instrument

platforms such as the Luminex system can be used to measure both abundant circulating proteins such as clotting factors and transport proteins as low abundance molecules such as hormones and cytokines [13–15]. These platforms have the added benefits of maximizing the amount of information that can be obtained from a single sample, with reduced sample and reagent usage, and at a lower cost, compared to the traditional single target assays.

In the multiplex system, a sample is added to dye-coded microspheres containing specific capture antibodies, leading to binding of the target proteins (Fig. 1). Next, a second antibody containing a biotinylated label is added which also binds to the target in a "sandwich" configuration. After this, the mixtures are streamed through the Luminex instrument which employs lasers for identification of the antibody-microsphere conjugates and quantitative analysis of the bound proteins. This chapter describes the development of a 5-plex immunoassay which targets the hormones glucagon-like peptide 1, growth hormone, insulin, leptin, and thyroid-stimulating hormone. Data are presented on the insulin levels in serum samples from 27 young, healthy individuals as an example.

2 Materials

2.1 Samples

1. Serum samples collected from 27 control individuals matched for age and body mass index (Table 1) (*see* **Note 1**).

Table 1
Demographic information associated with serum samples used in the study

	Samples
Male/female	16/11
Age (years)	23.9 ± 2.9
Body mass index (kg/m^2)	22.5 ± 2.9

2.2 Microsphere Conjugation

1. Magnetic separation instrument (*see* **Note 2**).
2. 1–4 mL 12.5 × 10^6 magnetic microspheres/mL.
3. 125 μg/mL monoclonal antibodies (*see* **Note 3**).
4. N-hydroxysulfosuccinimide (Sulfo-NHS).
5. N-(3-dimethylaminopropyl)-N′-ethylcarbodiimide (EDC).
6. Activation solution: 100 mM monosodium phosphate (pH 6.0).
7. Coupling solution: 0.05 M 2-morpholino-ethane-sulfonic acid mono-hydrate (MES) (pH 5.0).
8. Blocking/storage solution: 10 mM N monosodium phosphate (pH 7.4), 150 mM sodium chloride, 0.02% Tween 20, 0.1% bovine serum albumin (BSA), and 0.05% sodium azide (PBS-TBN).

2.3 Detection Antibody

1. Antibodies targeting different epitopes in the same proteins as above (*see* **Note 3**).
2. Sulfo-NHS-LC biotin (Thermo Fisher Scientific).
3. Dialysis solution: PBS.

2.4 Multiplex Development

1. Assay solution: PBS, 1% BSA.
2. Wash solution: PBS, 0.02% Tween-20.
3. 100 μg/mL streptavidin, R-phycoerythrin (SAPE).
4. 96-well micro-titer plate.
5. Recombinant protein standards.
6. Blocking solution (*see* **Note 4**).

3 Methods

3.1 Bead Conjugation

1. Stand vials containing 1–4 mL of stock microspheres on the magnetic separation device for 2 min to allow microspheres to settle completely (*see* **Note 5**).

2. Take care not to unsettle the microspheres and remove the overlying solution.

3. Add 0.5 mL activation solution, vortex, and sonicate the beads to suspend them.

4. Place the suspension in the magnetic separator for 30–60 s to settle the spheres.

5. Remove the supernatant and resuspend in 0.4 mL activation solution.

6. Add activation solution to the sulfo-NHS solution to give a final concentration of 50 mg/mL.

7. Add 50 μL of the above solution to the tube containing the microspheres in activation solution and mix by vortexing.

8. Add activation buffer to EDC to give a final concentration of 10 mg/mL.

9. Add 50 μL of this solution to the tube containing the microspheres, activation solution, and sulfo-NHS and mix by vortexing.

10. Incubate 20 min in the dark using gentle rotation at room temperature (see **Note 6**).

11. Place the tube in the magnetic separator for 30–60 s, remove the supernatant and add 0.5 mL coupling solution.

12. Repeat this wash step and then leave the beads suspended in 0.45 mL coupling solution.

13. Add 0.2 mL antibody to the activated microspheres and mix by vortexing (see **Note 7**).

14. Incubate for 2 h while rotating at room temperature in the dark.

15. Place tube in magnetic separator for 30–60 s, remove supernatant and add 1 mL blocking/storage solution.

16. Resuspend and incubate 30 min while rotating at room temperature in the dark.

17. Place tube in magnetic separator for 30–60 s.

18. Remove the supernatant and wash twice as described above using 0.25 mL blocking/storage solution.

19. Resuspend in 0.25 mL blocking/storage solution.

20. Count the microspheres using a hemocytometer.

21. Adjust the concentration as required to 50×10^6 microspheres/mL and store at 4 °C.

3.2 Biotinylation

1. Prepare 10 mM biotin reagent fresh before use.

2. Add 10 mM biotin to antibody solutions at a 20:1 biotin/antibody ratio (see **Note 8**).

3. Incubate on ice for 2 h (*see* **Note 9**).

4. Remove surplus biotin reagent by dialysis using three exchanges of PBS.

5. Add BSA to 1% final concentration as well as a preservative for long term stability during storage.

3.3 Assay

1. Combine 5 μL each microsphere solution into one tube to give a final volume of 1.4 mL in assay solution (*see* **Note 10**).

2. Add 0.2 μg each recombinant protein to a final volume of 0.2 mL assay solution and create seven tenfold serial dilutions (*see* **Note 11**).

3. Combine 5 μg each biotinylated antibody solution to give a final 5 mL in assay solution (*see* **Note 12**).

4. Dilute serum and plasma samples 1:10 in assay solution (*see* **Note 13**).

5. Add 30 μL diluted sample (or standard solutions) to designated wells of the micro-titer plate.

6. Add 10 μL blocking solution.

7. Add 10 μL multiplex microsphere solution.

8. Incubate for 1 h at room temperature with gentle shaking.

9. Wash three times with 100 μL wash solution.

10. Add 40 μL multiplex detection antibody solution to each well.

11. Incubate for 1 h as above.

12. Add 20 μL SAPE to each sample.

13. Incubate for 30 min as above (*see* **Note 14**).

14. Wash three times with 100 μL wash solution.

15. Add 100 μL assay solution.

16. Incubate for 5 min as above.

17. Analyze on the Luminex 100 Analyzer.

3.4 Data Analysis

1. Carry out data analyses and determine the levels of each hormone in each sample (*see* **Notes 15** and **16**).

2. Identify significant differences ($p < 0.05$) between experimental and control samples using Student's *t*-test for each hormone measurement (Fig. 2) (*see* **Note 17**).

3.5 Storage and Use

1. Store all reagents separately in assay buffer using the volumes listed in protocol above.

2. Store beads, SAPE, and assay buffer at 4 °C.

3. Store all other components at −80 °C.

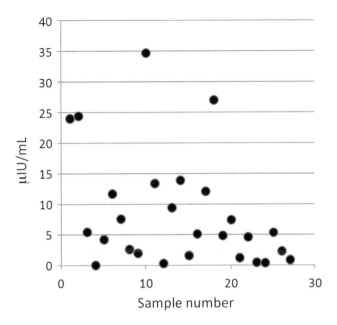

Fig. 2 Insulin levels (mlU/mL) in 27 control individuals as measured using the multiplex immunoassay

4. Make quality controls (QCs) by spiking recombinant protein standards into serum or plasma samples and store at $-80\ ^{\circ}C$.

5. Monitor assay performance regularly using the QCs.

4 Notes

1. The protocols of the study were approved by the ethical committees from the universities of Cologne, Muenster, and Magdeburg, Germany, from where the samples were collected as part of larger study [16].

2. A plate washer and other related instruments can also be used. This is simply a device to collect the beads for buffer exchange.

3. For each target protein, one antibody is used for capture and another for detection. These should recognize distinct epitopes on the protein to avoid steric hindrance and to provide maximum selectivity.

4. Many blocking reagents can be used here but we use Tru-Block (Meridian Life Sciences; Memphis, TN, USA).

5. Prior to coupling, the antibody solutions should be free of any other amines which will compete for binding. If the antibodies are in Tris-based buffers, these should be buffer exchanged (such as by dialysis) into a PBS-based solution. Likewise, antibodies in solutions containing stabilizing proteins such as BSA

or gelatin are purified using Protein A or Protein G followed by dialysis into PBS as above.

6. Prepare the antibodies designated for coupling during this step.

7. Note that each 1 mL microsphere is sufficient for approximately 40 μL plates.

8. The amount of biotin required for each 1 mg/mL antibody solution can be calculated using the formula

   1 mL IgG × 1 mg/mL × 20 mmol biotin/1 mmol IgG × 1 mmol IgG/150,000 mg IgG × 1000 μL/mL
   In this case, this equals to 0.133 mmol biotin (or 13 μL of the 10 mM biotin solution).

9. The incubation can also be carried out at room temperature for 30 min.

10. This can be used as a multiplex microsphere pool.

11. This generates an 8-point standard curve.

12. This can be used as a multiplex detection antibody pool.

13. The dilution is based on the expected concentrations of the target analytes in serum or plasma. Low abundance proteins such as the hormones targeted in this study require minimal dilutions (1:5–1:10) whereas abundant proteins such as the apolipoproteins will require high dilutions (1:1000000).

14. The SAPE concentration will vary in proportion to the number of analytes in the multiplex.

15. Optimize the readings by varying sample dilutions as required. As with most assays, the most accurate readings are obtained if the sample readings are derived from the linear region of the standard curve.

16. Assess cross-reactivity within the multiplex by comparison of the results with the corresponding results of single assays as required.

17. Here we have analyzed the readings of the insulin levels in samples from 27 control individuals as an example.

References

1. Carolan-Olah M, Duarte-Gardea M, Lechuga J (2015) A critical review: early life nutrition and prenatal programming for adult disease. J Clin Nurs 24:3716–3729

2. Lopes GA, Ribeiro VL, Barbisan LF, Marchesan Rodrigues MA (2016) Fetal developmental programing: insights from human studies and experimental models. J Matern Fetal Neonatal Med 23:1–7

3. Lee HS (2015) Impact of maternal diet on the epigenome during in utero life and the developmental programming of diseases in childhood and adulthood. Nutrients 7:9492–9507

4. Tarry-Adkins JL, Ozanne SE (2016) Nutrition in early life and age-associated diseases. Ageing Res Rev 39:96–105. pii: S1568-1637(16)30179-9. [Epub ahead of print]. https://doi.org/10.1016/j.arr.2016.08.003

5. Camm EJ, Martin-Gronert MS, Wright NL, Hansell JA, Ozanne SE, Giussani DA (2011) Prenatal hypoxia independent of undernutrition promotes molecular markers of insulin resistance in adult offspring. FASEB J 25:420–427

6. Martínez JA, Cordero P, Campión J, Milagro FI (2012) Interplay of early-life nutritional programming on obesity, inflammation and epigenetic outcomes. Proc Nutr Soc 71:276–283

7. Ortiz-Espejo M, Pérez-Navero JL, Olza J, Muñoz-Villanueva MC, Aguilera CM, Gil-Campos M (2013) Changes in plasma adipokines in prepubertal children with a history of extrauterine growth restriction. Nutrition 29:1321–1325

8. Tan HC, Roberts J, Catov J, Krishnamurthy R, Shypailo R, Bacha F (2015) Mother's pre-pregnancy BMI is an important determinant of adverse cardiometabolic risk in childhood. Pediatr Diabetes 16:419–426

9. Dixon JB (2009) Obesity and diabetes: the impact of bariatric surgery on type-2 diabetes. World J Surg 33:2014–2021

10. Khavandi K, Brownrigg J, Hankir M, Sood H, Younis N, Worth J et al (2014) Interrupting the natural history of diabetes mellitus: lifestyle, pharmacological and surgical strategies targeting disease progression. Curr Vasc Pharmacol 12:155–167

11. Allison BJ, Kaandorp JJ, Kane AD, Camm EJ, Lusby C, Cross CM et al (2016) Divergence of mechanistic pathways mediating cardiovascular aging and developmental programming of cardiovascular disease. FASEB J 30:1968–1975

12. Fulton RJ, McDade RL, Smith PL, Kienker LJ, Kettman JR Jr (1997) Advanced multiplexed analysis with the FlowMetrix system. Clin Chem 43:1749–1756

13. Bastarache JA, Koyama T, Wickersham NE, Mitchell DB, Mernaugh RL, Ware LB (2011) Accuracy and reproducibility of a multiplex immunoassay platform: a validation study. J Immunol Methods 367:33–39

14. Baker HN, Murphy R, Lopez E, Garcia C (2012) Conversion of a capture ELISA to a Luminex xMAP assay using a multiplex antibody screening method. J Vis Exp 65.): pii: 4084. https://doi.org/10.3791/4084

15. Stephen L (2017) Multiplex immunoassay profiling. Methods Mol Biol 1546:169–176

16. Guest PC, Wang L, Harris LW, Burling K, Levin Y, Ernst A et al (2010) Increased levels of circulating insulin-related peptides in first-onset, antipsychotic naïve schizophrenia patients. Mol Psychiatry 15:118–119

Chapter 33

Time-Resolved Fluorescence Assays for Quantification of Insulin Precursors in Plasma and Serum

Kevin Taylor, Ian Halsall, Paul C. Guest, and Keith Burling

Abstract

In metabolic diseases such as obesity and type 2 diabetes mellitus, the conversion of proinsulin to mature insulin can be impaired. This could mean that insulin molecules with lower activity toward the insulin receptor can be released under conditions of high metabolic demand, resulting in an inadequate glucoregulatory response. The chapter describes a fluorescent monoclonal antibody-based protocol for measurement of human proinsulin and the proinsulin conversion intermediates (split proinsulins). An example assay is presented using serum from non-diabetic, normal body mass index individuals.

Key words Metabolic disease, Obesity, Diabetes, Proinsulin conversion, Split proinsulin, Two-site immunoassay

1 Introduction

The radioimmunoassay for insulin was first established in 1959 as a sensitive way of measuring insulin in human plasma or serum in large numbers of samples [1]. However, the subsequent discovery that proinsulin-like molecules exist in human plasma meant that the existing assays for bioactive insulin may lack specificity, particularly since proinsulin and insulin have similar reactivity in most existing insulin assays [2–5]. This can be a problem in analyses of diabetic subjects since proinsulin-like molecules are present at higher concentrations compared with the levels found in normoglycemic control individuals [6]. Furthermore, a number of investigations have now shown that individuals with metabolic diseases, or those who are members of certain at-risk populations, have a higher proportion of proinsulin-like molecules in their circulation [7–9]. Other studies have shown that maternal diabetes [10] or a high fat maternal diet [11] can lead to higher levels of proinsulin and split proinsulin in the offspring later in life. Conversely, a randomized double-blind placebo-controlled trial showed that patients with type 2 diabetes who received a mixed meal and

Paul C. Guest (ed.), *Investigations of Early Nutrition Effects on Long-Term Health: Methods and Applications*, Methods in Molecular Biology, vol. 1735, https://doi.org/10.1007/978-1-4939-7614-0_33, © Springer Science+Business Media, LLC 2018

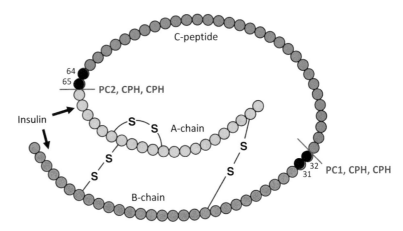

Fig. 1 Structure of human proinsulin showing the cleavage sites and processing enzymes. Amino acids 31 and 32 consist of arginine-arginine, and amino acids 64 and 65 are lysine-arginine. PC1 = prohormone convertase 1, PC2 = prohormone convertase 2, CPH = carboxypeptidase H

pioglitazone treatment had a significant reduction in fasting levels of proinsulin and split proinsulin compared to controls [12]. Measurement of these biomarkers has also been used to assist the diagnosis of insulinoma as this condition can be manifested by an increased proportion of precursors in relation to insulin in many patients [13]. The insulin precursors have also been measured to identify and characterize patients with defects in proinsulin processing [14]. Taken together, these findings indicate that these molecules can be important as biomarkers in the diagnosis and treatment of metabolic and other diseases.

In order to develop specific assays for proinsulin and split proinsulin, it is necessary understand the process of how mature insulin is generated in pancreatic beta cells (Fig. 1). This occurs by limited proteolysis in the insulin secretory granules through the action of two Ca^{2+}-dependent endoproteases (designated as prohormone convertase 1 and prohormone convertase 2, PC1 and PC2), which cleave the prohormone on the C-terminal side of arginine[31]-arginine[32] and lysine[64]-arginine[65] [15–17], and the Zn^{2+}-dependent carboxypeptidase H (CPH), which removes the exposed basic residues [18]. All of these enzymes are broadly distributed in neuroendocrine cells, where they are involved in posttranslational processing of other prohormones and proneuropeptides [19, 20].

Widespread introduction of time-resolved fluorescence assay technology using europium-labeled antibodies in the 1990s [21] offered the opportunity to modify the previously described immunoradiometric assays for insulin precursors [22] into a time-resolved immunofluorometric assay format. Time-resolved fluorescence assays benefit from increased reagent stability, safer analysis

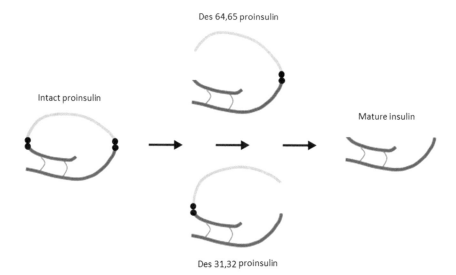

Fig. 2 Proinsulin, split proinsulins, and insulin structures. The split proinsulins are comprised of des 31,32-proinsulin (amino acids 31 and 32 removed) and des 64,65-proinsulin (amino acids 64 and 65 removed)

conditions, less laborious assay procedures, reduced analysis times, and greatly extended analytical range.

Here we describe the use fluorometric assays for intact proinsulin, total proinsulin (intact proinsulin and split proinsulins), and insulin (Fig. 2) using monoclonal antibodies developed by researchers in the Department of Clinical Biochemistry at the University of Cambridge in the United Kingdom (Fig. 3) [22].

2   Materials

2.1   Antibodies

1. Monoclonal antibodies 3B1 and A6 in cell culture medium [22].
2. CPT-3F11 C-peptide antibody [21] (Novo Nordisk; Bagsvaerd, Denmark) (*see* **Note 1**).
3. Protein G matrix (Pharmacia).
4. Protein G50 column (Pharmacia).
5. Antibody exchange buffer: 50 mM bicarbonate (pH 8.5).
6. Antibody dilution buffer: 50 mM bicarbonate (pH 9.2).
7. Assay buffer: 5% BSA in phosphate buffered saline (ph 7.4) (PBS) (containing 0.5% normal mouse serum).

2.2   Europium Labeling

1. Europium labeling kit containing Europium chelate of N1-(p-isothiocyanatobenzyl)-diethylenetriamine-N^1, N^2, N^2, N^3-tetraacetic acid (Wallac; Milton Keynes, UK).
2. Sepharose 6B column (Pharmacia).
3. Elution buffer: 50 mM Tris (pH 8.0), 0.9% NaCl, 0.5% sodium azide.

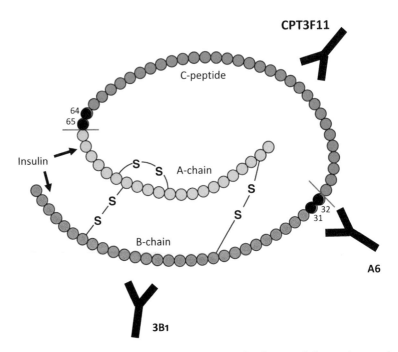

Fig. 3 Structure of proinsulin showing the antibodies used for capture and detection of proinsulin and des 31,32 proinsulin

2.3 Proinsulin-Related Standards

1. Intact proinsulin (National Institute of Biological Standards; 1st International Reference Reagent: 84/611; Potters Bar, UK).

2. Des 31,32 split proinsulin (Lilly Research Laboratories; Indianapolis, IN, USA).

3. Standard dilution buffer: 100 mM phosphate (ph 7.4), 5% w/v bovine serum albumin (BSA), 0.9% NaCl.

4. Quality controls: spiked proinsulin and des 31,32 split proinsulin expired blood transfusion human plasma known to be negative for HIV and hepatitis B virus.

2.4 Other Materials

1. Plate washer (Wallac).

2. Nunc MaxiSorp microtiter plates (Wallac).

3. DELFIA buffer (Wallac).

4. Wash buffer concentrate (Wallac).

5. Enhancement solution (Wallac).

6. DELFIA 1296-001 plate shaker (Wallac).

7. Multicalc data handling package (Wallac).

3 Methods

3.1 Preparation of Antibody Microtiter Plates

1. Prepare antibody coated microtiter plates for the intact proinsulin and 32,33 split proinsulin assays by diluting purified 3B1 antibody in 100 mM bicarbonate buffer (pH 9.2) to give a final antibody concentration of 1 μg/mL.

2. Add 200 μL diluted antibody solution to each well.

3. Leave plates overnight at 4 °C to allow passive adsorption of the antibody to the surface of the wells (*see* **Note 2**).

3.2 Preparation of Standards and Quality Controls

1. Prepare aliquots of a 400 pM working proinsulin standard of each proinsulin molecule and store at −20 °C until needed.

2. Dilute the working standards serially in PBS/BSA to give additional standards of concentrations of 100, 25, 6.25, and 1.56 pM.

3. Prepare control pools representing three concentrations in expired blood appropriate to the analytical range of the assay for each analyte.

3.3 Intact Proinsulin Assay

1. Wash plates coated with 3B1 antibody four times with wash solution using the plate washer (*see* **Note3**).

2. Block residual absorptive sites on the plates with 300 μL 5% BSA in PBS for 1 h per well.

3. Wash three times with wash solution.

4. Add 50 μL of samples, intact proinsulin standards and controls into appropriate wells, in duplicate.

5. Add 100 μL assay buffer (5% BSA in PBS with 0.5% v/v normal mouse serum) to each well.

6. Shake plates 20 min on the plate shaker and incubate overnight at 4 °C to allow binding of proinsulin and 32,33 split proinsulin to the solid phase antibody (*see* **Note 4**).

7. Wash plates four times with wash solution.

8. Add 200 μL of europium-labeled A6 antibody to each well (*see* **Note 5**).

9. Incubate 2.5 h at room temperature on a plate shaker.

10. Wash plates six times to remove excess labeled antibody and add 200 μL enhancement solution to each well.

11. Incubate on the plate shaker 5 min and leave to stand 10 min.

12. Read on the DELFIA 1232 fluorimeter.

13. Construct proinsulin standard curves and calculate unknown values using the Multicalc data handling package (Table 1) (*see* **Note 6**).

Table 1
Cross-reactivity of insulin-like molecules in the intact proinsulin assay

31,32 Split proinsulin	65,66 Split proinsulin	Insulin
0% at 400 pM	127% at 500 pM	0% at 2500 pM

Table 2
Cross-reactivity of insulin-like molecules in the des 31,32 proinsulin assay

Intact proinsulin	65,66 Split proinsulin	Insulin
87% at 400 pM	135% at 500 pM	0% at 2500 pM

3.4 Split Proinsulin Assay

1. Wash plates coated with 3B1 antibody four times with wash solution using the plate washer.
2. Block residual absorptive sites on the plates with 300 μL 5% BSA in PBS for 1 h per well.
3. Wash three times with wash solution.
4. Add 50 μL of samples, des 31,32 proinsulin standards and controls into appropriate wells, in duplicate.
5. Add 100 μL assay buffer (5% BSA in PBS with 0.5% v/v normal mouse serum) to each well.
6. Shake plates 20 min on the plate shaker and incubate overnight at 4 °C to allow binding of proinsulin and 32,33 split proinsulin to the solid phase antibody (*see* **Note 4**).
7. Wash plates four times with wash solution.
8. Add 200 μL 7 μg/mL europium-labeled CPT-3F11 antibody in DELFIA buffer to each well (*see* **Note 7**).
9. Incubate 2.5 h at room temperature on a plate shaker.
10. Wash plates six times to remove excess labeled antibody and add 200 μL enhancement solution to each well.
11. Incubate on the plate shaker 5 min and leave to stand 10 min.
12. Read on the DELFIA 1232 fluorimeter.
13. Construct des 31,32 proinsulin standard curves and calculate unknown values using the Multicalc data handling package (Table 2) (*see* **Notes 8–10**).

4   Notes

1. The Novo Nordisk anti C-peptide antibody (PEP-001) used previously for the 32,33 split proinsulin assay [22] exhibited poor binding characteristics when labeled with europium.

2. The 3B1 antibody is known to bind insulin with high affinity. The antibody concentration described for plate coating was targeted at 100% recovery of the analyte in the presence of 2500 pM insulin. For patient sera with an insulin concentration greater than this, pre-dilution of the sample with charcoal stripped human serum (Sigma) or equivalent is required. Extension of the plate coating time beyond 24 h was found to reduce the insulin binding capacity of 3B1 coated plates resulting in reduced recovery of analyte in the presence of insulin.

3. All subsequent washing steps use this instrument.

4. Both molecules contain the 3B1 epitope and will therefore be captured by the 3B1 monoclonal antibody (Fig. 3).

5. This should only recognize intact proinsulin as it requires an intact B chain/C-peptide junction between amino acids 30 and 31 (Fig. 3).

6. The intra-assay coefficients of variation for the intact proinsulin assay were 4.4% at 4.9 pM, 2.6% at 17 pM, and 3.0% at 85 pM ($n = 28$ for each level). Inter-assay precision was calculated from results of quality control materials representing three concentration levels that were assayed with each batch of samples. The inter-assay coefficients of variation for this assay were 10% at a mean concentration of 4.4 pM, 5.8% at a mean concentration of 23.5 pM, and 5.4% at a mean concentration of 87 pM ($n = 50$ for each level). The limit of detection (calculated as the mean + 3 standard deviations from 20 replicate analyses of a zero standard) within a batch was 0.8 pM.

7. This antibody should recognize intact proinsulin and split proinsulin as it recognizes an epitope within the C-peptide region (Fig. 3).

8. The intra-assay coefficients of variation for the des 31,32 proinsulin assay were 3.9% at 9.1 pM, 4.0% at 39 pM, and 3.0% at 85 pM ($n = 28$ for each level). Inter-assay precision was calculated from results of quality control materials representing three concentration levels that were assayed with each batch of samples. The inter-assay coefficients of variation for this assay were 10% at a mean concentration of 4.4 pM, 5.8% at a mean concentration of 23.5 pM, and 2.8% at a mean concentration of 101 pM ($n = 22$ for each level). The limit of detection (calculated as the mean $\pm$ 3 standard deviations from

20 replicate analyses of a zero standard) within a batch was 0.7 pM. To obtain a specific measure of 32,33 split proinsulin, it is necessary to take account of the intact proinsulin present in the specimen.

9. The demonstration of a high degree of cross reactivity from 65,66 split proinsulin in the intact proinsulin and 32,33 split proinsulin assays does not present difficulties as this moiety is not detectable in the plasma of fasted subjects [22]. The 32,33 split proinsulin assay exhibits a high level of cross reactivity with the other precursors and is therefore more correctly considered as a total proinsulins assay. Nevertheless, an accurate estimate of the 32,33 split proinsulin concentration can be made by using this assay in combination with the intact proinsulin assay and correcting for the presence of the measured intact proinsulin. The configuration of the assays described is such that insulin may compete for binding sites on the 3B1 solid phase capture antibody. It is essential that adequate recovery of the analyte is achieved in the presence of physiological concentrations of insulin. Insufficient binding capacity may lead to reduced recovery of analyte in spiked samples and an underestimation of analyte concentration in patient samples. Typically, we aim to achieve around 100% recovery in the presence of 2500 pM insulin. This should be adequate for all but the severest cases of insulin resistance even in stimulated (e.g., glucose tolerance test) samples. It is essential that recovery in the presence of insulin is assessed with each new batch of purified coating antibody.

10. Conversion of the assay to AutoDELFIA format greatly increases assay throughput. Three plates of each analyte (105 samples in duplicate) can easily be processed each day using this approach. Analysis of samples from large population studies revealed that some individuals have significantly high levels of heterophilic antibodies that swamp the blocking agents which are present in DELFIA multibuffer. These samples can be treated to remove the interfering antibodies and reanalyzed. In addition, in-house studies showed that intact proinsulin and 32,33 proinsulin are less prone to degradation than insulin and c-peptide. Hemolyzed samples do not give false results, and plasma and serum samples can undergo at least five freeze-thaw cycles without loss of reactivity.

References

1. Yalow RS, Berson SA (1959) Assay of plasma insulin in human subjects by immunological methods. Nature (London) 184:1648–1649
2. Steiner DF, Oyer PE (1967) The biosynthesis of insulin and a probable precursor of insulin by a human islet cell adenoma. Proc Natl Acad Sci U S A 57:473–480
3. Heding LG (1972) Determination of total serum insulin (IRI) in insulin-treated diabetic patients. Diabetologia 8:260–266

4. Roth J, Gorden P, Pastan I (1968) "Big insulin": a new component of plasma insulin detected by immunoassay. Proc Natl Acad Sci U S A 61:138–144

5. Given BD, Cohen RM, Shoelson SE, Frank BH, Rubenstein AH, Tager HS (1985) Biochemical and clinical implications of proinsulin conversion intermediates. J Clin Invest 76:1398–1405

6. Mako ME, Starr JI, Rubenstein AH (1977) Circulating proinsulin in patients with maturity onset diabetes. Am J Med 63:865–869

7. Kruszynska YT, Harry DS, Mohamed-Ali V, Home PD, Yudkin JS, McIntyre N (1995) The contribution of proinsulin and des-31,32 proinsulin to the hyperinsulinemia of diabetic and nondiabetic cirrhotic patients. Metabolism 44:254–260

8. Gelding SV, Andres C, Niththyananthan R, Gray IP, Mather H, Johnston DG (1995) Increased secretion of 32,33 split proinsulin after intravenous glucose in glucose-tolerant first-degree relatives of patients with non-insulin dependent diabetes of European, but not Asian, origin. Clin Endocrinol (Oxf) 42:255–264

9. Nagi DK, Knowler WC, Mohamed-Ali V, Bennett PH, Yudkin JS (1998) Intact proinsulin, des 31,32proinsulin, and specific insulin concentrations among nondiabetic and diabetic subjects in populations at varying risk of type 2 diabetes. Diabetes Care 21:127–133

10. Cooper MB, Al Majali K, Bailey CJ, Betteridge DJ (2008) Reduced postprandial proinsulinaemia and 32-33 split proinsulinaemia after a mixed meal in type 2 diabetic patients following sensitization to insulin with pioglitazone. Clin Endocrinol (Oxf) 68:738–746

11. Aaltonen J, Ojala T, Laitinen K, Poussa T, Ozanne S, Isolauri E (2011) Impact of maternal diet during pregnancy and breastfeeding on infant metabolic programming: a prospective randomized controlled study. Eur J Clin Nutr 65:10–19

12. Lindsay RS, Walker JD, Halsall I, Hales CN, Calder AA, Hamilton BA et al (2003) Scottish multicentre study of diabetes in pregnancy. Insulin and insulin propeptides at birth in offspring of diabetic mothers. J Clin Endocrinol Metab 88:1664–1671

13. Kao PC, Taylor RL, Service FJ (1994) Proinsulin by immunochemiluminometric assay for the diagnosis of insulinoma. J Clin Endocrinol Metab 78:1048–1051

14. O'Rahilly S, Gray H, Humphries PJ, Krook A, Polonsky K, White A et al (1995) Impaired processing of prohormones associated with abnormalities of glucose homeostasis and adrenal function. N Engl J Med 333:1386–1390

15. Davidson HW, Rhodes CJ, Hutton JC (1988) Intraorganellar calcium and pH control proinsulin cleavage in the pancreatic beta cell via two distinct site-specific endopeptidases. Nature (London) 333:93–96

16. Bailyes EM, Shennan KIJ, Seal AJ, Smeekens SP, Steiner DF, Hutton JC et al (1992) A member of the eukaryotic subtilisin family (PC3) has the enzymic properties of the type 1 proinsulin-converting endopeptidase. Biochem J 285:391–394

17. Bennett DL, Bailyes EM, Nielsen E, Guest PC, Rutherford NG, Arden SD et al (1992) Identification of the type 2 proinsulin processing endopeptidase as PC2, a member of the eukaryote subtilisin family. J Biol Chem 267:15229–15236

18. Davidson HW, Hutton JC (1987) The insulin-secretory-granule carboxypeptidase H. Purification and demonstration of involvement in proinsulin processing. Biochem J 245:575–582

19. Halban PA, Irminger J-C (1994) Sorting and processing of secretory proteins. Biochem J 299:1–18

20. Fricker LD (1991) Peptide processing exopeptidases: amino- and carboxypeptidases involved with peptide biosynthesis. In: Fricker LD (ed) Peptide biosynthesis and processing. CRC Press, Boca Raton, FL, pp 199–230. ISBN: 9780849388521

21. Hemmila I, Dakubu S, Mukhala VM, Siitari H, Lovgren T (1984) Europium as a label in time resolved immunofluorometric assays. Anal Biochem 137:335–343

22. Sobey WJ, Beer SF, Carrington CA, Clark PMS, Frank BH, Gray IP et al (1989) Sensitive and specific two site immunoradiometric assays for human insulin, proinsulin, 65-66 split and 32-33 split proinsulin. Biochem J 260:535–541

Chapter 34

Identification of Neural Stem Cell Biomarkers by Isobaric Tagging for Relative and Absolute Quantitation (iTRAQ) Mass Spectrometry

Paul C. Guest

Abstract

This chapter describes a proteomic analysis of neural progenitor cells using isobaric tagging for relative and absolute quantitation (iTRAQ) mass spectrometry. A detailed procedure is described for the isolation, proliferation, and differentiation of these cells, including a comparative iTRAQ mass spectrometry analysis of the precursor and differentiated states. In total, there were changes in the levels of 55 proteins, many of which are not resolved easily by other proteomic methods. Therefore, this method should be useful for the identification of important regulatory molecules in the study of other precursor cells involved in neuronal or metabolic regulation in nutritional programming diseases.

Key words Nutritional programming, Stem cells, Neural precursor cells, Growth factors, Proteomics, iTRAQ, Matrix-assisted laser desorption/ionization-time-of-flight mass spectrometry

1 Introduction

Nutritional insults in early life caused by factors such as a low-protein maternal diet can lead to diabetes in the offspring by impairing the regenerative capacity of pancreatic β-cells [1]. One study showed that high-fat diet-induced maternal obesity restricts proliferation of fetal hematopoietic progenitor cells while promoting differentiation [2]. It is likely that such effects extend to other organ systems of the body that exhibit active progenitor cell populations throughout life, including the brain. Within the brain, the best known regenerative regions are the subventricular zone and the hippocampal dentate gyrus [3]. Depletion of progenitor cells in these regions could have negative effects on regulation of vital body systems such as metabolic control, memory formation, cognition and behavior.

Neural progenitor cells are multipotent proliferating cells, which can differentiate into neurons, astrocytes and oligodendrocytes

Paul C. Guest (ed.), *Investigations of Early Nutrition Effects on Long-Term Health: Methods and Applications*, Methods in Molecular Biology, vol. 1735, https://doi.org/10.1007/978-1-4939-7614-0_34, © Springer Science+Business Media, LLC 2018

through signals in the central nervous system [4]. This is controlled tightly by endogenous growth factors, such as fibroblast growth factor (FGF) and epidermal growth factor (EGF) [5, 6]. Neural progenitor cells can be isolated from the subventricular zone of adult mouse brains and cultured in the presence of FGF or EGF to form neurospheres [7]. After sufficient numbers of neurospheres have been produced, withdrawal of these growth factors induces the progenitor cells to differentiate, providing an in vitro model of neurogenesis [8]. A complete understanding of the molecular cascades involved in this process is required to identify the pathways disrupted in disorders caused by nutritional programming deficits. This would be aided by technologies capable of multiplex analysis, given the limited quantities available of endogenous stem cell populations.

In comparison to traditional mass spectrometry techniques, the multiplexing potential of the isobaric tagging for relative and absolute quantitation (iTRAQ) mass spectrometry allows simultaneous analysis of multiple biological samples. Multiplexing helps to decrease the inherent variations in chromatographic analysis between the samples and requires fewer technical replicates to obtain improved accuracy in all measurements [9–11]. The approach involves derivatization of peptides from protein tryptic digests using four isobaric mass tags, which give identical chromatographic and mass fingerprinting properties to the differentially labeled peptides (for workflow, *see* Fig. 1). However, fragmentation of the labeled peptides by collision-induced dissociation in tandem mass spectrometry experiments produces mass spectra which contain fragment ions for the peptide amino acid sequences and four low-mass ions derived from the different mass tags [12]. The relative intensities of the mass tags correspond to the relative levels of the labeled peptides and are therefore representative of how the

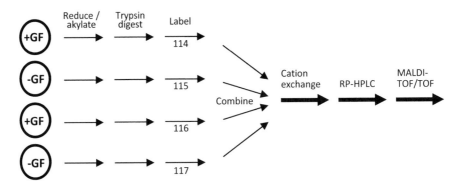

Fig. 1 Depiction of workflow for the shotgun isobaric tagging mass spectrometry method using the iTRAQ reagents. Equivalent amounts of whole cell lysates from neurospheres incubated in the presence (+GF) or absence (−GF) of EGF/FGF2 were subjected to parallel workflows using the standard protocol for reduction, alkylation, trypsinization, and derivatization with iTRAQ reagents 114, 115, 116, and 117. The labeled peptides were combined and subjected to cation exchange chromatography and HPLC prior to MALDI-TOF/TOF for identification and quantification

parent protein is affected in different physiological states. This chapter characterizes the proteomic changes induced by differentiation of adult subventricular zone progenitor cells using the iTRAQ mass spectrometry method.

2 Materials

2.1 Tissue Culture

1. Mouse subventricular zone cells (*see* **Note 1**).
2. Propagation medium: Dulbecco's Modified Eagle's Medium (DMEM)/F12 (1:1), penicillin/streptomycin/fungicide, B27, 20 ng/mL EGF, 20 ng/mL FGF-2 (*see* **Note 2**).
3. Differentiation medium: DMEM/F12 (1:1), penicillin/streptomycin/fungicide, B27 (*see* **Note 3**).
4. Wash medium: phosphate-buffered saline (PBS), pH 7.4.

2.2 Protein Extraction

1. Extraction solution: 500 mM triethylammonium bicarbonate (pH 8.5), 0.05% SDS, 0.1% Triton X-100 (Sigma-Aldrich) (*see* **Note 4**).
2. EDTA-free protease inhibitors (Sigma-Aldrich) (*see* **Note 5**).
3. Soniprep 150 with microprobe (MSE Ltd.; London, UK) (*see* **Note 6**).
4. Bicinchoninic acid (BCA) protein reagent (Pierce; Rockland, IL, USA).

2.3 iTRAQ Peptide Labeling

1. Reducing solution: 5 mM Tris(2-carboxyethyl) phosphine (TCEP) (*see* **Note 7**).
2. Alkylating solution: 10 mM methyl methanethiosulfonate (MMTS) (*see* **Note 7**).
3. iTRAQ reagent 4-plex kit (Applied Biosystems).
4. Stop solution: 1% formic acid.

2.4 Cation Exchange Chromatography (See Note 8)

1. Buffer A: 10 mM KH_2PO_4 (pH 3), 25% (v/v) acetonitrile.
2. Buffer B: 10 mM KH_2PO_4 (pH 3), 25% (v/v) acetonitrile, 1 M KCl.
3. 100 × 4.6 mm, 5 μm; 200 Å PolyC Polysulfoethyl A column (PolyLC Inc.; Columbia, MD, USA) (*see* **Note 9**).
4. PF2D system (Beckman Coulter Inc.) (*see* **Note 9**).

2.5 Reverse Phase High-Performance Liquid Chromatography (HPLC) (See Note 8)

1. HPLC buffer A: 2% (v/v) acetonitrile, 0.1% (v/v) trifluoroacetic acid (TFA).
2. HPLC buffer B: 85% (v/v) acetonitrile, 5% (v/v) isopropanol, 0.1% (v/v) TFA.

3. 0.3 × 5.0 mm, 3 μm, 100 Å C18 PepMap 100 trap column (Thermo Fisher Scientific) (*see* **Note 8**).

4. Famos Micro Autosampler and Switchos Micro Column Switching Module (Thermo Fisher Scientific) (*see* **Note 8**).

2.6 Matrix-Assisted Laser Desorption/Ionization-Time-of-Flight (MALDI-TOF) Mass Spectrometry (MS) and Tandem MS (MS/MS) Analysis (See Note 10)

1. Matrix: α-cyano-4-hydroxycinnamic acid.

2. ProBot MALDI microfraction collector (Thermo Fisher Scientific) (*see* **Note 11**).

3. MALDI target plate (Applied Biosystems).

4. 4700 Proteomics Analyzer (Applied Biosystems).

2.7 Data Analysis

1. 4000 Explorer™ software (V2; Applied Biosystems).

2. GPS Explorer™ and GPS Explorer™ (V3.5) search tool (Applied Biosystems).

3. UniProt database.

4. Microsoft Excel.

3 Methods

3.1 Culturing and Differentiation of Neural Precursor Cells

1. Plate cells in propagation medium (*see* **Note 12**).

2. Passage neurospheres using Accutase to produce single-cell suspensions every 7 days in propagation medium.

3. Enrich neural progenitor cells using a preplating protocol [13, 14], and use those from passages 7 to 10 to provide sufficient material for iTRAQ mass spectrometry analysis.

4. Split cells into two equal portions of 2×10^8 cells, and culture for 24 h in either propagation or differentiation medium to produce progenitor and differentiated cells, respectively (*see* **Note 3**).

5. Centrifuge cells 1 min at $400 \times g$ and discard supernatant.

6. Gently suspend cells in wash solution; repeat Subheading 3.1, **step 6**, three times; and leave the pellets on ice.

3.2 Protein Extraction

1. Suspend cell pellets at 100 mg/mL (tissue wet weight) in extraction buffer at 4 °C.

2. Sonicate 30 s at an amplitude of 15 μm (*see* **Note 13**).

3. Dilute samples in 200 μL extraction buffer containing EDTA-free protease inhibitors.

4. Leave samples 10 min on ice, centrifuge 10 min at $14,000 \times g$ at 4 °C.

5. Retain the supernatants.

6. Determine protein concentrations and store samples at −80 °C.

3.3 iTRAQ Labeling Procedure

1. Treat 100 µg protein from each sample with 5 mM TCEP for 1 h at 60 °C (*see* **Note 14**).

2. Incubate with 10 mM MMTS 10 min at room temperature (*see* **Note 15**).

3. Add 10 µg porcine trypsin, and incubate 16 h at 37 °C according to the iTRAQ Reagent Kit instructions.

4. Incubate the extracts from the progenitor cells with the 114 and 116 iTRAQ reagents and the extracts from the differentiated cells with the 115 and 117 reagents for 1 h at room temperature (*see* **Note 16**).

5. Add stop solution to a final concentration of 1%.

6. Combine all four reactions into a single tube and dry in a vacuum centrifuge (*see* **Note 17**).

3.4 Cation Exchange Chromatography

1. Suspend the iTRAQ tag-labeled peptide mixture in 4 mL of buffer A.

2. Load onto the PolyC column using the PF2D system.

3. Wash the column isocratically 30 min in the same buffer at a flow rate of 0.3 mL/min.

4. Elute peptides at a flow rate of 1 mL/min using 0–5% buffer B over 2 min, 5–10% B over 4 min, 10–22.5% B over 7 min, 22.5–50% B over 5 min, constant at 50% B for 13 min, 50–100% B over 6 min, constant at 100% B for 12 min, and 100–0.5% B over 1 min.

5. Collect fractions over 1 min intervals and dry in a vacuum centrifuge.

6. Suspend dried-labeled peptides in 250 µL of HPLC buffer A.

3.5 MALDI-TOF MS Survey Scan

1. Mix 0.45 nL each faction with 1 nL matrix using the ProBot microfraction collector, and spot onto a 192-well MALDI target plate according to the manufacturer's instructions.

2. Acquire mass spectrometry (MS) survey scan spectra of each fraction using the 4700 Proteomics Analyzer in positive ion reflectron mode.

3. Select the fractions containing the most intense peaks with a signal to noise ratio greater than 60 for HPLC fractionation (corresponding to fractions 39–51; Fig. 2) and tandem mass spectrometry (MS/MS) analysis (*see* **Note 18**).

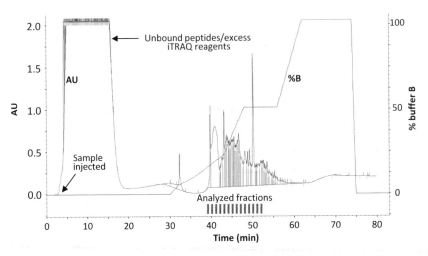

Fig. 2 Chromatogram showing cation exchange separation of iTRAQ-labeled peptides. Cation exchange separation of the combined iTRAQ-labeled peptides showing the relative absorbance units (AU), percentage buffer B (%B), and fractions subjected to further RP-HPLC separation

3.6 Reverse Phase HPLC and MALDI-TOF MS/MS Analysis

1. Auto-inject 50 μL of selected samples in buffer A onto the C18 PepMap 100 trap column using the Autosampler and Switching Module.

2. Elute peptides at a flow rate of 0.5 μL/min using 0–5% HPLC buffer B over 15 min, 5–35% B over 50 min, 35–65% B over 15 min, 65–95% B over 5 min, constant at 95% B for 10 min, and then 95–5% B over 5 min.

3. Mix 0.45 nL each faction with 1 nL matrix using the ProBot microfraction collector, and spot onto a 192-well MALDI target plate according to the manufacturer's instructions.

4. Acquire MS/MS spectra of each fraction using the 4700 Proteomics Analyzer in positive ion reflectron mode.

3.7 Data Analysis

1. Process MS and MS/MS spectra using the 4000 Explorer software with Gaussian smoothing at a filter width of seven points.

2. Set MS and MS/MS peak threshold detection using signal to noise ratios of 30 and 20, respectively.

3. Generate monoisotopic peak lists using the GPS Explorer™ and submit to the GPS Explorer™ (V3.5) search tool for Mascot-based identifications using the Swiss-Prot database with the following parameters: 0 or 1 missed trypsin cleavage, MMTS modification of N-terminal amino acids and lysine residues.

4. For all searches, set precursor ion tolerances at 50 ppm and fragment ion mass tolerances at 0.6 Da.

5. Extract peak areas for the iTRAQ 114, 115, 116, and 117 mass tags using the 4700 Explorer™ and match in an Excel data-sheet to the identified peptides retrieved from the MS/MS summary table in GPS Explorer (*see* **Note 16**).

6. Identify proteins on the basis of having at least 1 peptide with an individual ion score above the 99% confidence threshold ($P < 0.01$).

7. Group according to protein identity for calculation of average ratios and standard deviation (*see* **Note 19**).

8. Identify proteins which are differentially expressed (Fig. 3, Table 1) (*see* **Note 20**).

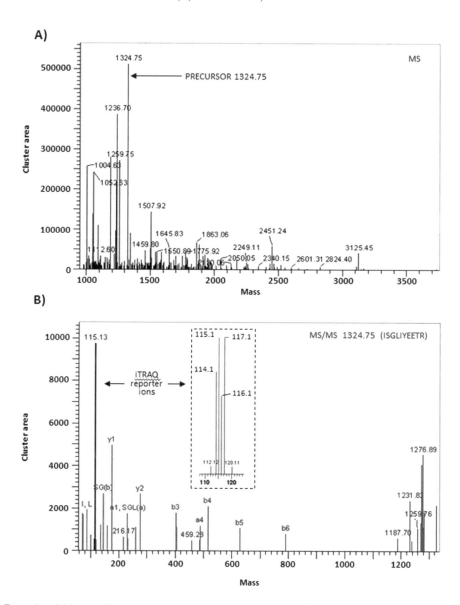

Fig. 3 Example of histone H4 sequencing and quantitation using iTRAQ mass spectrometry. (**A**) Example MALDI-TOF MS spectra indicating a 1324.748 m/z precursor ion identified as HIST1H4 (histone H4). (**B**) Example tandem mass (MS/MS) spectra of the 1324.788 m/z precursor ion (ISGLIYEETR) identified as HIST1H4 and expanded view of the low-m/z end of the MS/MS spectrum in showing relative abundances of the signature iTRAQ ions at 114.1, 115.1, 116.1, and 117.1 m/z

Table 1
Quantitation of 5 identified iTRAQ-labeled HISTH4 peptides following differentiation of neural progenitor cells by withdrawal of growth factors (GF)

Sequence	115/114 (−GF/+GF)	116/114 (+GF/+GF)	117/114 (−GF/+GF)
ISGLIYEETR	2.01	0.75	1.76
VFLENVIR	1.73	1.25	1.72
DAVTYTEHAK	1.72	1.08	1.80
TVTAMDVVYALK	1.61	0.87	1.34
KTVTAMDVVYALK	1.62	0.85	1.68
Calculations			
Mean ± SD	1.74 ± 0.16	0.96 ± 0.20	1.66 ± 0.19
Protein (5 peptides combined)	**1.83 ± 0.43**		

4 Notes

1. Mouse subventricular zone cells were dissected from adult mice and enzymatically and mechanically dissociated as described previously [8].

2. The B27 medium is used to support long-term growth of neuronal cells in culture.

3. Removal of growth factors stops the proliferation of neural progenitor cells and causes them to differentiate into neurons and glial cells.

4. A pH of 8.5 is optimal for peptide labeling.

5. Protease inhibitors should be used in this step to prevent degradation of proteins caused by proteases in the sample.

6. The microprobe allows easy disruption of samples in a 1.5 mL or 0.5 mL capacity microcentrifuge tube.

7. This solution should be made immediately prior to use.

8. Reagents should be analytical grade, solvents should be HPLC or LC-MS grade, and all solutions should be prepared with ultrapure water.

9. Other similar systems can be used but will most likely require different operating conditions than those presented here.

10. The current MALDI-TOF system was chosen for ease of operation. Other mass spectrometry devices can be used, such as electrospray ionization devices although these will require different operating conditions.

11. This instrument was chosen due to its accuracy and speed in spotting low volumes. However, spotting by hand onto the micro-target plate is possible.

12. Under these conditions, the cells grow as neurospheres with core material being more stemlike and the outer material representing more differentiated cells. Therefore, the serial passage procedure ensures a higher proportion of stem cells.

13. Conditions will vary depending on the sonication device used. Attempts should be made to minimize frothing as this can lead to insufficient homogenization. The best results can be achieved by introducing the probe just beneath the surface of the buffer and then gradually increasing the power to the desired amplitude.

14. Other reducing agents can be used here such as dithiothreitol (DTT).

15. Other alkylating agents can be used such as iodoacetamide (IAA).

16. This experimental design allows the following progenitor cell (PC) and differentiated cell (DC) comparisons to be made, which includes two like:like comparisons as an internal controls:

PC1:PC2 (114:116)
PC1:DC1 (114:115)
PC1:DC2 (114:117)
PC2:DC1 (116:115)
PC2:DC2 (116:117)
DC1:DC2 (115:117)

17. Due to the differential labeling, multiplexing is possible. Note that the peptides are easier to resuspend if they are not completely dry.

18. A high signal to noise ratio improves the chances of obtaining quantitatively accurate results.

19. In this experiment, proteins were selected which showed differences greater than 20%. In addition, the experimental design resulted in each run incorporating two plus/minus growth factor comparison for each peptide and each run was repeated on three separate occasions using different batches of neurospheres.

20. Only the data for histone H4 are shown. For further information, *see* Salim et al. [13].

References

1. Hill DJ (2011) Nutritional programming of pancreatic β-cell plasticity. World J Diabetes 2:119–126

2. Kamimae-Lanning AN, Krasnow SM, Goloviznina NA, Zhu X, Roth-Carter QR, Levasseur PR, Jeng S et al (2014) Maternal high-fat diet and obesity compromise fetal hematopoiesis. Mol Metab 4:25–38

3. Apple DM, Fonseca RS, Kokovay E (2017) The role of adult neurogenesis in psychiatric and cognitive disorders. Brain Res 1655:270–276

4. McKay RD (1997) Stem cells in the central nervous system. Science 276:66–71

5. Vescovi AL, Reynolds BA, Fraser DD, Weiss S (1993) bFGF regulates the proliferative fate of unipotent (neuronal) and bipotent (neuronal/astroglial) EGF-generated CNS progenitor cells. Neuron 11:951–966

6. Zigova T, Pencea V, Wiegand SJ, Luskin MB (1998) Intraventricular administration of BDNF increases the number of newly generated neurons in the adult olfactory bulb. Mol Cell Neurosci 11:234–245

7. Gottlieb DI (2002) Large-scale sources of neural stem cells. Annu Rev Neurosci 25:381–407

8. Johansson CB, Momma S, Clarke DL, Risling M, Lendahl U, Frisen J (1999) Identification of a neural stem cell in the adult mammalian central nervous system. Cell 96:25–34

9. Ross PL, Huang YN, Marchese JN, Williamson B, Parker K, Hattan S et al (2004) Multiplexed protein quantitation in Saccharomyces cerevisiae using amine-reactive isobaric tagging reagents. Mol Cell Proteomics 3:1154–1169

10. Nogueira FC, Domont GB (2014) Survey of shotgun proteomics. Methods Mol Biol 1156:3–23

11. Rauniyar N, Yates JR 3rd (2014) Isobaric labeling-based relative quantification in shotgun proteomics. J Proteome Res 13:5293–5309

12. Núñez EV, Domont GB, Nogueira FC (2017) iTRAQ-based shotgun proteomics approach for relative protein quantification. Methods Mol Biol 1546:267–274

13. Salim K, Kehoe L, Minkoff MS, Bilsland JG, Munoz-Sanjuan I, Guest PC (2006) Identification of differentiating neural progenitor cell markers using shotgun isobaric tagging mass spectrometry. Stem Cells Dev 15:461–470

14. Salim K, Guest PC, Skynner HA, Bilsland JG, Bonnert TP, McAllister G et al (2007) Identification of proteomic changes during differentiation of adult mouse subventricular zone progenitor cells. Stem Cells Dev 16:143–165

Chapter 35

Lab-on-a-Chip Device for Rapid Measurement of Vitamin D Levels

Harald Peter, Nikitas Bistolas, Soeren Schumacher, Cecilia Laurisch, Paul C. Guest, Ulrich Höller, and Frank F. Bier

Abstract

Lab-on-a-chip assays allow rapid analysis of one or more molecular analytes on an automated user-friendly platform. Here we describe a fully automated assay and readout for measurement of vitamin D levels in less than 15 min using the Fraunhofer in vitro diagnostics platform. Vitamin D (25-hydroxyvitamin D_3 [25 (OH)D_3]) dilution series in buffer were successfully tested down to 2 ng/mL. This could be applied in the future as an inexpensive point-of-care analysis for patients suffering from a variety of conditions marked by vitamin D deficiencies.

Key words Nutritional programming, Nutrient deficiency, Metabolic disease, Vitamins, Vitamin D, Lab-on-a-chip

1 Introduction

Vitamin D deficiency has a high prevalence across all ethnicities and age groups and occurs in a high proportion of the world's population [1]. It has a well-known role in calcium homeostasis, and its deficiency is now known to be a risk factor for several chronic diseases including obesity, dyslipidemia, cardiovascular disease, and type 2 diabetes mellitus [2–4]. In addition, it has been linked to neurological conditions such as schizophrenia [5], major depression [6], and Alzheimer's disease [7]. Maternal vitamin D deficiency has also been associated with pregnancy complications including preterm birth and intrauterine growth restriction, which may be due to alterations in placental function [8].

From these findings it can be seen that vitamin D testing is important. Various methods are now currently in use for this, including liquid chromatography combined with tandem mass spectrometry (LC-MS/MS) [9], enzyme-linked immunosorbent assay (ELISA) [10], and chemiluminescence microparticle

Paul C. Guest (ed.), *Investigations of Early Nutrition Effects on Long-Term Health: Methods and Applications*, Methods in Molecular Biology, vol. 1735, https://doi.org/10.1007/978-1-4939-7614-0_35, © Springer Science+Business Media, LLC 2018

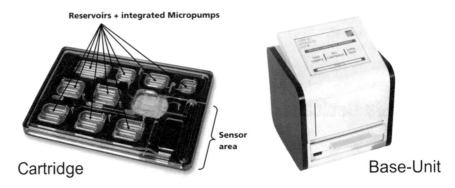

Fig. 1 Fraunhofer ivD-platform, allowing a multiplex immunoassay analysis and readout within 12 min. (Left) Lab-on-a-chip cartridge (size: 60 × 40 mm) with nine reservoirs and integrated micropumps, microfluidic channels, thermal control elements, electronics, and a sensor area for microarrays with up to 400 spots. (Right) Base unit (size: 14 × 14 × 14 cm) for control, readout, and data analysis of the lab-on-a-chip cartridge. Results can either be transferred to a computer or analyzed directly and presented on the display

immunoassay (CMIA) [11]. Although these methods are reliable and sensitive, they require considerable expertise in order to obtain reliable results. In addition, they are relatively expensive and normally require several hours to days for generation of the results.

Here, we describe the use of a rapid, automated, and integrated lab-on-a-chip-based system, developed by several Fraunhofer Institutes [12], that helps to overcome the above limitations. This platform can be used to run single or multiple immunoassays to generate a biomarker "score" [13]. The in vitro diagnostic (ivD) platform is comprised of a microfluidic cartridge and a base unit. The cartridge contains all of the essential elements for attachment of the relevant antibodies, including reservoirs for all reagents, integrated pumping mechanisms, and an optical transducer for integrated sensing (Fig. 1) [13–16]. The base unit houses the required electronic systems to control the cartridge, as well as an optical system for signal transduction and a touch screen for controlling the assay and viewing the result output. This platform delivers results in less than 15 min, with similar performance to the standard methods outlined above. We recently described the use of this platform for analysis of C-reactive protein in psychiatric disorders [17].

In this chapter, we present an automated microarray-based competitive immunoassay for the detection of 25-hydroxyvitamin D_3 levels using the Fraunhofer ivD-platform. Vitamin D is biologically inactive. The molecule undergoes hydroxylation in the liver and is metabolized into 25-hydroxyvitamin D [25(OH)D] and afterward in the kidney into 1,25-hydroxyvitamin D [1,25 $(OH)_2D$], the active metabolite. Both are of clinical relevance. We developed an assay that detects 25-hydroxyvitamin D_3 [25(OH) D_3], the major circulating form of vitamin D_3 (Fig. 2) [18, 19].

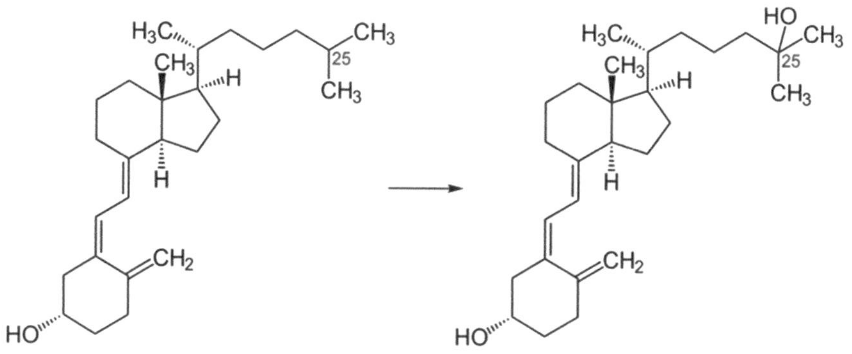

Fig. 2 Hydroxylation of vitamin D$_3$ to calcidiol (25-hydroxyvitamin D$_3$). Calcidiol is a prehormone, which is produced in the liver by hydroxylation of vitamin D$_3$. It can be used to measure a patient's vitamin D status

In plasma, 20–50 ng/mL vitamin D$_3$ is considered adequate for healthy people, and a concentration lower than 12 ng/mL indicates vitamin D deficiency [20]. Both ranges are detectable with our immunoassay.

Since vitamin D is a small molecule, we developed a test using a competitive immunoassay. In the first step, a specific monoclonal antibody against 25-hydroxyvitamin D$_3$ is immobilized onto cyclic olefin polymer (COP) slides using a noncontact piezo spotter (Fig. 3). Subsequently, the sample/antigen solution containing 25-hydroxyvitamin D$_3$ is incubated on the slide. After incubation the sensor field is rinsed with washing buffer to remove nonspecifically bound molecules. In the next step, a tracer (biotinylated 25-hydroxyvitamin D$_3$) is added to the microarray, which competes with the analyte in binding to the antibody, followed by another washing step. Finally, the addition of a fluorescently labeled streptavidin takes place, which binds to the tracer. After the last washing step, the concentration of bound analyte can be measured by the quantitative determination of bound fluorescence on the readout unit, which is inversely proportional to the analyte concentration.

2   Materials

2.1   Microarray Fabrication

1. QuadriPERM square tissue culture dish (Sarstedt; Nümbrecht, Germany).

2. Microarray spotter: sciFLEXARRAYER S11 (Scienion AG; Berlin, Germany) (*see* **Note 1**).

3. Epoxysilane slides: 3D-epoxy polymer slides (PolyAn; Berlin, Germany).

4. Phosphate-buffered saline (PBS): 1.5 mM KH_2PO_4, 137 mM NaCl, 2.7 mM KCl, 8 mM Na_2HPO_4 (pH 7.4).

5. PBS-IT: PBS with 0.004% IGEPAL, 1% trehalose.

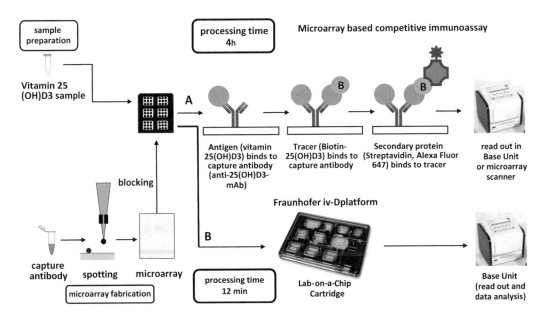

Fig. 3 Schematic of a microarray-based competitive immunoassay for the detection of vitamin 25(OH)D₃, including sample preparation and microarray fabrication. (**a**) Laboratory-based microarray protocol, using glass slides as support for the microarray, followed by several washing steps and a readout on a microarray scanner or the Fraunhofer ivD-platform base unit (4 h processing time). (**b**) Fully automated lab-on-a-chip-based competitive immunoassay allowing a detection of vitamin 25(OH)D₃ directly after 12 min. All reagents (tracer = biotin-labeled vitamin 25(OH)D₃, fluorescently labeled streptavidin) and washing solutions are pumped automatically over the microarray within the cartridge followed by a readout and data analysis within the base unit

6. Capture antibody: sheep antibody against 25-hydroxyvitamin D₃ [25(OH)D₃] (Bioventix; Farnham, Surrey, UK).

7. Internal spotting control antibody: Alexa Fluor 647 mouse anti-goat antibody (Molecular Probes; Eugene, OR, USA) (*see* **Note 2**).

8. Blocking buffer: 1% bovine serum albumin (BSA), 0.05% Tween 20 (*see* **Note 3**).

9. PBS-T: PBS, 0.05% Tween 20.

2.2 Competitive Immunoassay

1. QuadriPERM square tissue culture dish (Sarstedt).

2. ProPlate multi-array system (Grace Bio-Labs; Bend, OR, USA).

3. ProPlate adhesive seal strips (Grace Bio-Labs).

4. Antigen: 25-hydroxyvitamin D₃ [25(OH)D₃] (DSM Nutritional Products, Kaiseraugst, Switzerland).

5. Tracer: 25-hydroxyvitamin D₃, 3,3-Biotinylaminopropyl ether [Biotin-25(OH)D₃], Cat. No. H995815 (Toronto Research Chemicals Inc.; North York, ON, Canada).

6. Secondary protein: Alexa Fluor 647 streptavidin (Life Technologies).

7. Human serum (Fitzgerald Industries International; North Acton, MA, USA).

8. Displacement buffer: 220 mM acetate, 10% DMSO, 1% EtOH, 0.1% polidocanol, 0.1% ProClin 300 (Roche; Basel, Switzerland).

2.3 Microarray Scanning

1. Laser scanner (*see* **Note 4**).

2. Tecan Array-Pro Analyzer (Tecan Group) or alternative quantification software.

2.4 Automated Immunoassay, Washing, and Readout (Fraunhofer ivD-Platform)

1. Fraunhofer ivD cartridge (e.g., with the cardiac acute microarray) (Fraunhofer IZI-BB, Potsdam, Germany; BiFlow Systems, Chemnitz, Germany).

2. Fraunhofer ivD-platform (base unit).

2.5 Microarray Data Analysis

1. Spreadsheet software such as Microsoft Excel (Microsoft; Redmond, WA, USA).

3   Methods

3.1 Microarray Fabrication

1. Dilute capture antibody to 0.1 mg/mL in PBS-IT.

2. Dilute internal spotting control to 0.375 mg/mL in PBS-IT.

3. Pipette 30 μL spotting solutions into designated wells of a 384 well microtiter plate.

4. Use a noncontact spotter to print the probes onto epoxy-coated slides, setting the number of dots to three, the volume per dot to approximately 0.5 nL and the grid to 500 μm distance between the spots (*see* **Note 5**).

5. Shortly after the spotting process, the slides must be kept in a humid chamber with saturated NaCl at room temperature and protection from light (*see* **Note 6**).

6. To inactivate free epoxy groups, the slides need to be blocked directly before performing the immunoassay as indicated in the next seven steps (*see* **Note 7**).

7. Let the slide dry for 30–60 min at room temperature, protected from light and dust.

8. Immerse the slide in blocking solution for 30 s with fast up and down movements (*see* **Note 8**).

9. Incubate the slide in blocking solution for approximately 1 h at room temperature while shaking, protected from light.

10. Wash slides 10 s in PBS-T with fast up and down movements.

11. Afterwards wash the slides in PBS-T and PBS, each 5 min at room temperature while shaking, protected from light.

12. Wash slide 10 s in water with fast up and down movements.

13. Remove liquids carefully from the slide with a flow of nitrogen (*see* **Note 9**).

3.2 Competitive Immunoassay (See Note 10)

1. Prepare serial dilutions from the 25(OH)D$_3$ antigen ranging from 2000 to 4 ng/mL in a 1:2 series in blocking buffer, and mix each sample 1:1 with displacement buffer (*see* **Note 11**).

2. The following steps can be carried out to either measure an unknown sample or the samples for the calibration curve.

3. Assemble the ProPlate multi-array system on the slides.

4. Transfer 70 μL of each 25(OH)D$_3$ dilution or sample to be analyzed into the wells ensuring that no air bubbles form.

5. Seal the chamber with adhesive seal strips and incubate for 2 h at room temperature while shaking slightly (*see* **Note 12**).

6. Prepare 1 μg/mL tracer [Biotin-25(OH)D$_3$] and 10 μg/mL streptavidin (streptavidin, Alexa Fluor 647) in blocking solution.

7. Perform the following washing procedures between the incubation periods by removing the previous solution each time with a multichannel aspirator.

8. Wash the arrays three times 30 s with 250 μL PBS-T while shaking slightly.

9. Add 70 μL tracer into each well ensuring that no air bubbles form.

10. Seal the chamber with adhesive seal strips, and incubate 1 h at room temperature while shaking slightly.

11. Wash the arrays three times 30 s with 250 μL PBS-T while shaking slightly.

12. Add 70 μL streptavidin Alexa Fluor 647 (from **step 6**) into each well ensuring that no air bubbles form.

13. Seal the chamber with adhesive seal strips and incubate 1 h at room temperature while shaking slightly.

14. Wash arrays three times for 30 s with 200 μL PBS-T while shaking slightly.

15. Remove the ProPlate multi-array system from the slides, and wash directly 10 s in PBS and 5 s in water with fast up and down movements.

16. Remove liquids carefully from the slide with a flow of nitrogen without allowing droplets to dry on the surface.

17. For fluorescence image acquisition, image the slides using a laser scanner at 633 nm with the corresponding filter and an appropriate PMT/gain setting depending on the signal-to-noise ratio.

18. Quantify the fluorescent signals with the software provided by the scanner manufacturer.

3.3   Automated Immunoassay Procedure with the Fraunhofer ivD-Platform

1. Prepare the 25(OH)D$_3$ solution or sample to be analyzed as described above, and transfer it to the corresponding sample reservoir of the ivD cartridge (*see* **Note 13**).

2. Insert the cartridge into the ivD-platform base unit and start the immunoassay program.

3. The fluorescence image data can either be analyzed automatically within the base unit or exported for external analysis.

3.4   Data Analysis

1. After image acquisition and fluorescent signal quantification, subtract the local background of each spot from the raw spot intensity value, and calculate the mean net signal intensity (NI) and standard deviation (SD) of the replicates (Figs. 4 and 5).

2. Analyze the data depending on the individual scientific/clinical question.

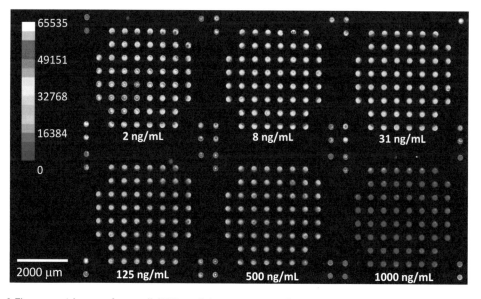

Fig. 4 Fluorescent image of a small (100 spot) immunoassay microarray, containing 47 25(OH)D$_3$ detection spots as well as positive, negative, and spotting control spots. Six array images taken from results obtained from samples containing six different concentrations of 25-OH-vitamin-D$_3$ shown as an example

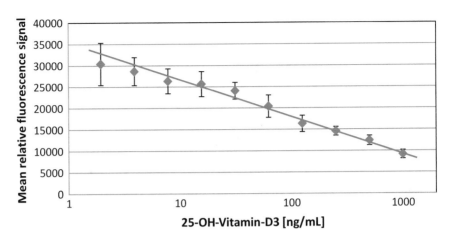

Fig. 5 Calibration curve of 25(OH)D$_3$ samples in buffer, allowing quantification of 25(OH)D$_3$ in a range from 2 to 1000 ng/mL

4   Notes

1. Other spotters can be utilized, but compatibility with the used materials and reagents is essential.

2. An alternative product for the internal spotting control could be goat anti-rabbit IgG (H+L), Alexa Fluor 647 (Life Technologies; Darmstadt, Germany).

3. As an alternative, 2.5% casein hydrolysate (biotin-free), 50 mM Tris base, 0.05% Tween 20, and 0.02% sodium azide can be used. All blocking reagents should be free of biotin.

4. We used the Tecan LS Reloaded (Tecan Group; Männedorf, Switzerland).

5. We use the sciFLEXARRAYER S11 for spotting and PolyAn polymer slides as the array surface. Using the indicated settings, the spots will have a diameter of about 150 μm.

6. At this stage, the slides should be processed within several days.

7. If not stated otherwise, use the quadriPERM dish for incubation.

8. We recommend using a Wafer Tweezer (Bernstein-Werkzeugfabrik Steinrücke GmbH; Remscheid, Germany).

9. Proteins can easily be removed or blown away if too much nitrogen pressure is applied without allowing droplets to dry on the surface.

10. If not stated otherwise, all solutions containing proteins should be kept on ice while working. In addition, to obtain reliable results when performing an immunoassay with an unknown

target antigen concentration, a freshly created calibration curve should be prepared to serve as a reference.

11. The concentrations of $25(OH)D_3$ can be customized by using a higher initial concentration and/or more or fewer dilution steps.

12. Make sure the slides are protected from light during all incubation periods as the fluorophores are light sensitive.

13. All other reservoirs are already prefilled with the necessary buffers and solutions. If an empty cartridge is being used, each reservoir can also be filled with custom buffers and reagents individually.

References

1. Holick MF, Chen TC (2008) Vitamin D deficiency: a worldwide problem with health consequences. Am J Clin Nutr 87:1080S–1086S

2. Rana S, Morya RK, Malik A, Bhadada SK, Sachdeva N, Sharma G (2016) A relationship between vitamin D, parathyroid hormone, calcium levels and lactose intolerance in type 2 diabetic patients and healthy subjects. Clin Chim Acta 462:174–177

3. Pannu PK, Calton EK, Soares MJ (2016) Calcium and vitamin D in obesity and related chronic disease. Adv Food Nutr Res 77:57–100

4. Milazzo V, De Metrio M, Cosentino N, Marenzi G, Tremoli E (2017) Vitamin D and acute myocardial infarction. World J Cardiol 9:14–20

5. Chiang M, Natarajan R, Fan X (2016) Vitamin D in schizophrenia: a clinical review. Evid Based Ment Health 19:6–9

6. Parker GB, Brotchie H, Graham RK (2017) Vitamin D and depression. J Affect Disord 208:56–61

7. Sommer I, Griebler U, Kien C, Auer S, Klerings I, Hammer R et al (2017) Vitamin D deficiency as a risk factor for dementia: a systematic review and meta-analysis. BMC Geriatr 17:16. https://doi.org/10.1186/s12877-016-0405-0

8. Yates N, Crew RC, Wyrwoll C (2017) Vitamin D deficiency and impaired placental function: potential regulation by glucocorticoids? Reproduction 153(5):R163–R171. https://doi.org/10.1530/REP-16-0647. Jan 30. pii: REP-16-0647

9. Kim HJ, Ji M, Song J, Moon HW, Hur M, Yun YM (2017) Clinical utility of measurement of vitamin D-binding protein and calculation of bioavailable vitamin D in assessment of vitamin D status. Ann Lab Med 37:34–38

10. Fattizzo B, Zaninoni A, Giannotta JA, Binda F, Cortelezzi A, Barcellini W (2016) Reduced 25-OH vitamin D in patients with autoimmune cytopenias, clinical correlations and literature review. Autoimmun Rev 15:770–775

11. Al-Haddad FA, Rajab MH, Al-Qallaf SM, Musaiger AO, Hart KH (2016) Assessment of vitamin D levels in newly diagnosed children with type 1 diabetes mellitus comparing two methods of measurement: a facility's experience in the middle eastern country of Bahrain. Diabetes Metab Syndr Obes 9:11–16

12. https://www.izi.fraunhofer.de/de/abteilung en/standort-potsdam/automatisierung/ivD-plattform-poc-technologien.html and www.ivd-platform.de

13. Schumacher S, Nestler J, Otto T, Wegener M, Ehrentreich-Förster E, Michel D et al (2012) Highly-integrated lab-on-chip system for point-of-care multiparameter analysis. Lab Chip 12:464–473

14. Schumacher S, Ludecke C, Ehrentreich-Förster E, Bier FF (2013) Platform technologies for molecular diagnostics near the patient's bedside. Adv Biochem Eng Biotechnol 133:75–87

15. Streit P, Nestler J, Shaporin A, Schulze R, Gessner T (2016) Thermal design of integrated heating for lab-on-a-chip systems. Proceedings of the 17th international conference on thermal, mechanical and multi-physics simulation and experiments in microelectronics and microsystems (EuroSimE), April 18–20. p. 1–6

16. Peter H, Wienke J, Bier FF (2017) Lab-on-a-chip multiplex assays. Methods Mol Biol 1546:283–294

17. Peter H, Wienke J, Guest PC, Bistolas N, Bier FF (2017) Lab-on-a-chip proteomic assays for psychiatric disorders. Proteomic methods in neuropsychiatric research. Adv Exp Med Biol 974:339–349

18. Christakos S, Ajibade DV, Dhawan P, Fechner AJ, Mady LJ (2010) Vitamin D: metabolism. Endocrinol Metab Clin North Am 39 (2):243–253

19. Dusso AS, Brown AJ, Slatopolsky E (2005) Vitamin D. Am J Physiol Renal Physiol 289: F8–F28. https://doi.org/10.1152/ajprenal.00336.2004

20. Vieth R (2006) What is the optimal vitamin D status for health? Prog Biophys Mol Biol 92:26–32

Chapter 36

Kidney Smartphone Diagnostics

P. R. Matías-García and J. L. Martinez-Hurtado

Abstract

Here we present a method for a mobile point-of-care (POC) testing of urinary albumin concentration, a biomarker of kidney damage and cardiovascular disease. The self-testing strips are meant to be interpreted by means of a smartphone application. The limits of detection range from 0.15 to 0.30 g/L urinary albumin, though results below 0.10 g/L are presented in a quantitative manner and estimates larger than this threshold are shown as categorical variables in a qualitative manner for increasing urinary albumin concentrations. Calibrated once under standard conditions, the app enables the user to capture problem samples and calculate the corresponding concentration. Negative and positive findings must be interpreted, taking into account the inherent limitations of the method, and professional health advice must be requested for diagnostic considerations. Acknowledgment of the association between early life nutrition and long-term renal health and the adoption of preventive strategies targeting high-risk groups is key for the reduction of the burden of chronic kidney disease on a global scale.

Key words Smartphone, Diagnostics, Mobile, Medical, Application, Quantitative assays, Kidney, CKD

1 Introduction

Chronic kidney disease (CKD) is a condition characterized by kidney damage in the forms of decreased renal function and increased urinary protein excretion (albuminuria) for at least 3 months [1]. In 2002, the National Kidney Foundation's Kidney Disease Outcomes Quality Initiative (NKF-KDOQI) proposed a model to outline the continuum of kidney damage progression in terms of renal function and estimated glomerular filtration rate (eGFR) [2]. However, the criteria and model for kidney disease in its original version were subject to debate in terms of coherence in risk estimation and diagnostic algorithms [3] so that after albuminuria was consistently identified as being associated with an increased risk of adverse outcomes independently from renal function estimates [4, 5], it was adapted to six eGFR and three albuminuria ordinal stages [6]. The aforementioned model identifies kidney damage as an early stage of CKD, which could potentially evolve

Paul C. Guest (ed.), *Investigations of Early Nutrition Effects on Long-Term Health: Methods and Applications*, Methods in Molecular Biology, vol. 1735, https://doi.org/10.1007/978-1-4939-7614-0_36, © Springer Science+Business Media, LLC 2018

to kidney failure. End-stage renal disease (ESRD), a subgroup of this latter stage, comprises patients whose survival is dependent on renal replacement therapy (RRT), either as dialysis or transplantation [2]. Deficient follow-up of early-stage patients can lead to progression to ESRD. It is in this context that the availability of tools for early detection and monitoring becomes increasingly important.

The prevalence of kidney disease in the general adult population in Europe, Asia, North America, and Australia has been reported to be within the 2.5–11.2% interval [7], while in high-risk populations, this figure rises to 50–60% [8]. Several studies have also shown a gender-specific disparity in estimated prevalence, which could be partially explained either by differences in age demographics for both groups or by inherent biological differences in glomerular physiology and hormonal pathways [7, 9]. The precision of early-stage CKD prevalence figures has been questioned because of the lack of standardized methods to calculate glomerular filtration rate, age-related decline in renal function, and disease heterogeneity in sampled populations [7, 8, 10]. In addition, there are issues related to underdiagnosis and low awareness in patients and healthcare providers [8, 11]. Nevertheless, evidence suggests that the prevalence of undetected CKD is as high in Europe as it is in the United States and is a rising concern [8]. Global trends show an association with an increase in incident diabetes and hypertension [5], and higher prevalence estimates were reported in countries with developed economies [12]. In terms of the epidemiological transition, population aging in developing countries will increase the absolute numbers of individuals prone to developing chronic conditions, doubling the burden on health systems dealing with both communicable and noncommunicable diseases [13].

The burden a disease imposes on societies can be further quantified in terms of its associated mortality rate and disability-adjusted life years (DALYs), a value that reflects the sum of total years of life lost (YLL) to premature mortality in prevalent cases and years lost to disability (YLD) in incident cases of the disease [14]. Data from the 2010 Global Burden of Disease study points out that CKD escalated from 27th to 18th place in the list of causes of age-standardized global deaths from 1990 to 2010. Such a change in ranking is only preceded by that of HIV/AIDS. CKD also rose from the 32nd to the 24th place in terms of causes of global YLL, being the third largest ranking ascent again after HIV and diabetes [10, 15]. Furthermore, kidney diseases ascended to the 19th place in the ranking for leading causes of DALY globally in 2015, accounting for 1.4% of the global DALY estimates [16]. Moreover, CKD is a currently acknowledged as a significant, independent risk factor for cardiovascular disease [5].

In view of the scaling prevalence of CKD, its uneven distribution, how little public health priority was it given until recent years (even considered to be a "silent" epidemic if compared to the equally prevalent diabetes [11]), its treatment costs, and associated mortality rates, CKD is now globally recognized as a public health preventable issue [17–19]. As such, the KDOQI proposed a conceptual model to guide public health efforts and healthcare interventions, in which primary prevention focuses on preventing risk factors of CKD from developing, while secondary and tertiary prevention measures are meant to slow progression and improve outcomes from complications in CKD patients in stages 1–4 and 5 (kidney failure), respectively [2, 20]. This fact evidences the need for tools to be used in kidney damage prevention strategies.

Kidney disease is a multifactorial pathology, comprised of an interplay of genetic and environmental risk factors that determines disease susceptibility, although the latter have the largest effects in non-monogenic chronic renal disease [19]. Diabetes, hypertension, obesity, infections, and autoimmune diseases are among the risk factors for CKD development. Of special relevance is the association between increasing obesity and renal disease, driven by diabetes and hypertension occurrence [2] but also by early life nutritional deficits.

Several animal experimental and human observational studies have found an association between high birth weight, exposure to maternal diabetes, and a rapid postnatal weight gain and adult obesity, diabetes, and CKD. Nutrition plays a key role in these associations as both an increased intrauterine nutrient availability followed by a deficient nutrition in early life and accelerated catch-up growth preceded by maternal malnutrition (as evidenced by low birth weight) have been associated with impaired adult renal function and hypertension [21, 22]. Moreover, maternal hyperglycemia and gestation length have a direct effect on fetal renal development and nephron number, and early life malnutrition has been observed to have a negative influence on structural physiological processes (such as nephrogenesis) and childhood kidney function [22]. Consumption of nutrient-rich food rather than energy-rich products should be promoted, so that exposure to risk factors for chronic disease during childhood is minimized.

Effective secondary prevention efforts should be focused on the early identification and treatment of individuals at greater risk of kidney disease [20], which is not an easy public health task due to the fact that CKD symptoms are not evident until higher stages of renal impairment are reached and once the window for preventing negative outcomes and delaying progress to renal failure is closed [23]. However, CKD screening in the general population is not recommended as a preventive measure as it does not meet the proposed criteria to assess economic viability and overall feasibility of screening programs [24]. Evidence from populations in which such a screening strategy was applied shows that this did not improve clinically relevant outcomes (kidney failure, cardiovascular events, and death), and false-positive results would impose stress and an unnecessary burden on health systems [8]. Furthermore, in settings where resources are limited, a greater public health benefit is expected when resources are allocated to preventing death and disability because of known (or prevalent) noncommunicable diseases (diabetes, hypertension, prevalent CKD), rather than detecting new (or incident) kidney cases [25]. While public health experts have emphasized the inadequacy of general population screening kidney disease in both high-income and low-income countries, the anticipated benefit of screening is greater in populations with a higher prevalence of kidney disease—termed elective screening or case finding, this strategy does meet many of the previously mentioned criteria for screening programs [25]. Individuals with diabetes, hypertension, cardiovascular disease, and autoimmune disease and family members of ESRD patients are considered to be at high risk of developing CKD and have been identified as a target for primary prevention strategies by means of urine protein analysis and serum creatinine screening [8, 20]. Furthermore, a recent statistical study consistently identified albumin (along with specific gravity, hemoglobin, diabetes mellitus, and hypertension) as a predictive attribute and as a cost-effective approach to detect CKD in a dataset of patients experiencing an early stage of this disease [26]. Therefore, awareness in population at high risk of developing CKD and affordable and accessible tools for screening and self-monitoring are key elements for improving prevention approaches.

Albumin-specific dipstick tests are widely available and used as a prescreening self-testing tool to guide professionals in the diagnosis of kidney disease. These tests use a pH-sensitive dye which shows color changes when in contact with negatively charged albumin, and, depending on its design, its specificity and sensitivity are quite high although these tests still face the problem of interobserver variability biasing the obtained results [27]. Therefore, it is paramount to have a quantitative or semiquantitative tool to minimize observer variability and fulfill the purpose of the existence of such tests. Even though albumin excretion is heavily dependent on exercise, diet, and even physical posture, spot morning urine testing to measure albuminuria is an acceptable mean of monitoring in cases in which the findings are confirmed from a 24-h urine sample collection [28].

Albuminuria's potential as biomarker has been suggested ever since it was first identified as an independent biomarker for renal function [29], and its association with CKD prognosis in all levels of eGFR and across different albuminuria measurement methods was confirmed as being consistent with data on 50 heterogeneous cohorts (divided into general population, high risk, and chronic kidney disease from Asia, Europe, North America, and Oceania) [3] and even in general populations with normal renal function [8]. The same data supported the current threshold values for albuminuria as a biomarker for kidney and cardiovascular disease [3, 5]. Therefore, self-testing of urine protein might be helpful in reducing the burden of disease in settings where its use would be recommended or at least support the monitoring of kidney disease progression. The trends and implications of self-testing in the provision and alleviation of the disease are still to be confirmed. However we believe providing the necessary tools to perform this in a global scale is a significant step forward.

Albumin in urine is a measure of increased glomerular permeability to macromolecules, and therefore it is an indicative of tissue damage and impaired endothelial function [27]. Urinary albumin can be detected once some degree of kidney injury is present, making it a late measure for its inclusion as primary prevention [30]. However, the same reason makes it an important biomarker in terms of secondary and tertiary prevention for high-risk populations or already diagnosed individuals [28]. The evidence indicates that even a minor increase confers increased risk of adverse kidney outcomes and cardiovascular disease [27, 28]. Additionally, results suggest that albuminuria can be considered as an indicator of healthcare quality and clinical follow-up [31]. Little information is available on its validity as an endpoint in epidemiological studies designed to test clinical therapies for renal disease progression, although methods such as those presented here pave the way for collecting more evidence [32]. Today, point-of-care diagnostics allows patients to obtain timely results without the need for expensive equipment and empowers them by progressively making them more aware and responsible for their own conditions [33]. Diagnostic technologies have come a long way from an early urine dip strip assay to detect proteinuria for diabetics and kidney disease patients (test dated in 1957) [33] to a time point in which the use of mobile phones as diagnostic tools is becoming increasingly feasible [34].

Here we demonstrate a protocol for evaluation of protein urine dipsticks as a possible global tool for CKD monitoring and prevention. The potential of this technology can be used to read other colorimetric test strips, photonic crystal arrays, or other hydrogel sensors for diagnostics, food, or other industrial applications [35–42]. This mobile application is designed for laboratory and clinical testing; even though it could serve help alleviate the burden of population monitoring, diagnostics and treatment are responsibility of the healthcare professionals and healthcare providers; mobile applications in the healthcare space must seek regulatory approval [43]; and its diagnostic performance should be evaluated and validated [28].

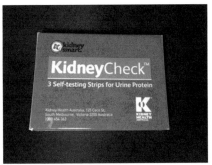

Fig. 1 Measurement and ideal positioning of smartphone and test strips

2   Materials

1. Urine samples collected from two healthy male volunteers aged 24 and 31 years old (*see* **Note 1**).

2. KidneyCheck™ self-testing strips (Siemens Healthcare Diagnostics Inc.; Erlangen, Germany) (*see* **Note 2**).

3. Fluorescent light source.

4. Colorimetrix v1.0 (29) (Fig. 1) smartphone app as described elsewhere [44], calibrated for a protein urine test with the KidneyCheck strips.

5. Phone model: iPhone 6S.

6. OS version: OS 10.2.1.

3   Methods (See Note 3)

3.1   Experimental Conditions

1. Ensure that environmental conditions are approximately 24 °C, 60% relative humidity.

3.2   Calibration

1. For calibration, record the standard test response chart from the manufacturer.

2. Open the Colorimetrix v1.0 (29) application.

3. In non-measuring mode open the camera capture to record the color information from the first concentration on reference chart.

4. Click on capture and save (*see* **Note 4**).

5. Input the associated concentration value for the recorded calibration point in the calibration curve (*see* **Note 5**).

6. Repeat Subheading 3.2, **steps 1–5** for as many calibration points as needed.

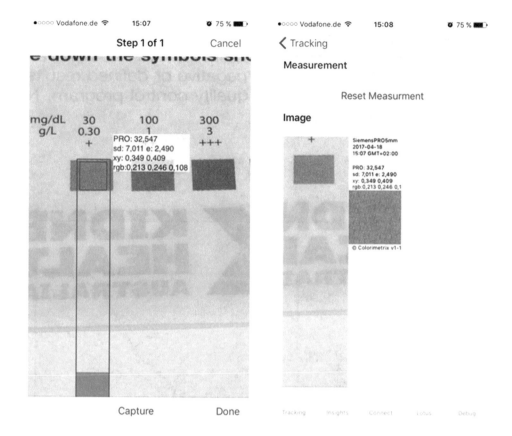

Fig. 2 Calibration steps illustrating the capture (**a**) and recording (**b**) of the calibration points. Alternatively, calibration can be performed with laboratory-grade solutions on test strips

7. Once calibration is completed, save all relevant inputs (analyte measured, strip type and name, and manufacturer) (Fig. 2) (*see* **Note 6**).

8. Request the app to show the newly recorded test calibration in measuring mode.

3.3 Measurement

1. In measuring mode, expose the test strips to problem samples, point the test area on the camera capture, and aim at the test area on the test strips.

2. Approximately 10 cm away from the strip, a zooming option is also available by pinching the screen with two fingers.

3. Freeze the camera feed by clicking capture.

4. Adjust the position of the strip capture frame by dragging the frame on the strip (the test capturing area can also be dragged to fit the test pad area).

5. Ensure that the test capture area is covered entirely by the test pad area.

Table 1
Calibration measurement results and corresponding variation as standard deviation

Conc. (mg/mL)	Meas. 1	Meas. 2	Meas. 3	Average	S.D.
0	2.53	3.75	5.64	3.97	1.57
10	13.95	13.14	13.91	13.67	0.46
30	30.41	34.45	47.51	37.46	8.94
100	118.86	134.79	177.91	143.85	30.55
300	1435.41	1783.99	2009.93	1743.11	289.43
2000	1057.5	1060.27	1014.74	1044.17	25.53

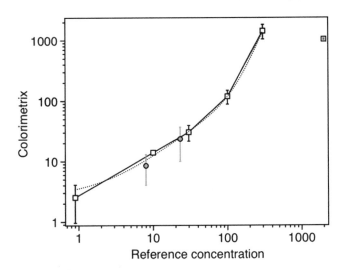

Fig. 3. Correlation with calibration curve values (white squares) and male volunteer values (green circles), axes values in mg/mL. Measurements are not effective for concentrations higher than 1000 mg/mL outlier shown to the right

6. Once in position click "done" (*see* **Note 7**).

7. Repeat as many times as necessary to capture as all problem data points needed.

3.4 Data Analysis

1. Evaluate the calibration measurement points to test the ability of the app to measure accurately all concentration ranges (Table 1) (*see* **Note 8**).

2. Plot the data from the test samples with the calibration curve in Fig. 3 (*see* **Notes 9** and **10**).

4 Notes

1. Follow waste disposal regulations for used solutions and materials. An artificial urine solution can also be used for calibration and to build problem samples [44, 45].

2. This protein urine test evaluates the excreted total protein in urine [46–48]. The strips are manufactured by Siemens Healthcare Diagnostics Inc. and are intended for self-testing by individuals or healthcare professionals. Total protein in urine is an indicative of premature kidney damage and can help assessing, predicting, and preventing kidney disease, thus acting upon dietary changes to control a potential chronic kidney disease problem. The KidneyCheck strips are firm plastic strips with a chemically treated pad for protein testing in urine. The reagents on the reaction pad are 0.3% w/w tetrabromophenol blue, 0.97% w/w buffer, and 2.4% w/w nonreactive ingredients. KidneyCheck is a registered trademark of Kidney Health Australia.

3. The methods consist of three parts: (a) calibration, one time setup of the mobile application under standard conditions; (b) measurement, scalable capturing of problem samples for population analysis; and (c) data analysis, to evaluate performance and set thresholds for a given application.

4. The application will store and process the relevant color information.

5. The value is stored in a calibration file that gathers all calibration points, and five points are recommended as a minimum for proper operation.

6. The calibration information is loaded into the application source code.

7. By clicking done, the smartphone application will save the corresponding values and calculate the corresponding concentration. The recorded values are saved in an internal database on the phone for immediate access and alternatively on a universally accessible, anonymized, and secured database in the cloud for further access.

8. As it can be noted, the variability of measurements above 100 mg/mL increases. Thus, the app is set to work on a threshold below 100 mg/mL in a quantitative manner and on a qualitative manner above that value. For example, a message can be displayed with the words "high" or "above threshold" still giving the user of the app useful information.

9. There is a deviation from a linear trend as the concentration increases above 100 mg/mL even though the standard deviation from the individual measurements is not high. There

seems to be a limitation on the green hues; further versions of the app will address for these and provide full accuracy across the entire range. Nevertheless, differences in both test samples were accurately captured by the mobile application.

10. Even though clinical and analytical studies provide accurate urine protein results for these types of tests, the color variation can be influenced by many factors, some of which are addressed in this protocol. The tests can detect 0.15–0.30 g/L of albumin as a trace result; however, because of the inherent variability of urines, the test area is more sensitive to albumin than to globulin, hemoglobin, B-J protein, and mucoprotein; thus a result does not rule out the presence of such proteins. Kidney Health Australia recommends to check after 12 months if the test result is negative (−) and contact health professionals to discuss results if the tests are positive (+, ++, +++, ++++). Negative test results are marked also as traces in the test instructions. + corresponds to 30 mg/mL, ++ to 100 mg/mL, +++ to 300 mg/mL, and ++++ as greater or equal to 2000 mg/mL. The range of detection of this test makes it a valuable tool to screen for albuminuria progression, since available epidemiological evidence suggests levels below the fixed albuminuria threshold may also require continuous monitoring [3].

Acknowledgment

The authors acknowledge Kidney Health Australia for providing the KidneyCheck test strips.

References

1. Bello A, Tonelli M, Jager K (2015) Epidemiology of kidney disease. In: Turner NN, Lameire N, Goldsmith DJ, Winearls CG, Himmelfarb J, Remuzzi G, Bennet WG, de Broe ME, Chapman JR, Covic A, Jha V, Sheerin N, Unwin R, Woolf A (eds) Oxford textbook of clinical nephrology, 4th edn. Oxford University Press, UK, pp 3–19. ISBN-10: 0199592543

2. Levey AS, Stevens LA, Coresh J (2009) Conceptual model of CKD: applications and implications. Am J Kidney Dis 53(3 Suppl 3): S4–16

3. Levey AS, de Jong PE, Coresh J, El Nahas M, Astor BC, Matsushita K et al (2011) The definition, classification, and prognosis of chronic kidney disease: a KDIGO controversies conference report. Kidney Int 80(1):17–28. https://doi.org/10.1038/ki.2010.483

4. Matsushita K, van der Velde M, Astor BC, Woodward M, Levey AS, de Jong PE et al (2010) Association of estimated glomerular filtration rate and albuminuria with all-cause and cardiovascular mortality in general population cohorts: a collaborative meta-analysis. Lancet 375(9731):2073–2081

5. Gansevoort RT, Correa-Rotter R, Hemmelgarn BR, Jafar TH, Heerspink HJ, Mann JF et al (2013) Chronic kidney disease and cardiovascular risk: epidemiology, mechanisms, and prevention. Lancet 382(9889):339–352

6. Group KDIGOKCW (2013) KDIGO 2012 clinical practice guideline for the evaluation and management of chronic kidney disease. Kidney Int Suppl 3. http://www.kdigo.org/clinical_practice_guidelines/pdf/CKD/KDIGO_2012_CKD_GL.pdf

7. Zhang QL, Rothenbacher D (2008) Prevalence of chronic kidney disease in population-based studies: systematic review. BMC Public Health 8:117. https://doi.org/10.1186/1471-2458-8-117

8. Lameire N, Jager K, Van Biesen W, de Bacquer D, Vanholder R (2005) Chronic kidney disease: a European perspective. Kidney Int Suppl (99):S30–S38. DOI: https://doi.org/10.1111/j.1523-1755.2005.09907.x

9. Silbiger SR, Neugarten J (2003) The role of gender in the progression of renal disease. Adv Ren Replace Ther 10:3–14

10. Jha V, Garcia-Garcia G, Iseki K, Li Z, Naicker S, Plattner B et al (2013) Chronic kidney disease: global dimension and perspectives. Lancet 382(9888):260–272

11. Zoccali C, Kramer A, Jager KJ (2010) Epidemiology of CKD in Europe: an uncertain scenario. Nephrol Dial Transplant 25:1731–1733

12. Hill NR, Fatoba ST, Oke JL, Hirst JA, O'Callaghan CA, Lasserson DS et al (2016) Global prevalence of chronic kidney disease - a systematic review and meta-analysis. PLoS One 11(7): e0158765. https://doi.org/10.1371/journal.pone.0158765

13. Correa-Rotter R, Naicker S, Katz IJ, Agarwal SK, Herrera Valdes R, Kaseje D et al (2004) Demographic and epidemiologic transition in the developing world: role of albuminuria in the early diagnosis and prevention of renal and cardiovascular disease. Kidney Int Suppl 92:S32–S37. https://doi.org/10.1111/j.1523-1755.2004.09208.x

14. The Global Burden of Disease: A Comprehensive Assessment of Mortality and Disability from Diseases, Injuries and Risk Factors in 1990 and Projected to 2020 (1996) Harvard School of Public Health on behalf of the World Health Organization and the World Bank, Cambridge. http://apps.who.int/iris/bitstream/10665/41864/1/0965546608_eng.pdf

15. Lozano R, Naghavi M, Foreman K, Lim S, Shibuya K, Aboyans V et al (2012) Global and regional mortality from 235 causes of death for 20 age groups in 1990 and 2010: a systematic analysis for the global burden of disease study 2010. Lancet 380(9859):2095–2128

16. World Health Organization W (2016) Global Health Estimates 2015: disease burden by cause, age, sex, by country and by region, 2000–2015 - global summary estimates. www.who.int/healthinfo/global_burden_disease/estimates/en/index2.html

17. Schoolwerth AC, Engelgau MM, Rufo KH, Vinicor F, Hostetter TH, Chianchiano D et al (2006) Chronic kidney disease: a public health problem that needs a public health action plan. Prev Chronic Dis 3(2):A57

18. Ayodele OE, Alebiosu CO (2010) Burden of chronic kidney disease: an international perspective. Adv Chronic Kidney Dis 17:215–224

19. Eckardt KU, Coresh J, Devuyst O, Johnson RJ, Kottgen A, Levey AS et al (2013) Evolving importance of kidney disease: from subspecialty to global health burden. Lancet 382(9887):158–169

20. Levey AS, Schoolwerth AC, Burrows NR, Williams DE, Stith KR, McClellan W (2009) Comprehensive public health strategies for preventing the development, progression, and complications of CKD: report of an expert panel convened by the Centers for Disease Control and Prevention. Am J Kidney Dis 53:522–535

21. Yim HE, Yoo KH (2015) Early life obesity and chronic kidney disease in later life. Pediatr Nephrol 30:1255–1263

22. Luyckx VA, Bertram JF, Brenner BM, Fall C, Hoy WE, Ozanne SE et al (2013) Effect of fetal and child health on kidney development and long-term risk of hypertension and kidney disease. Lancet 382(9888):273–283

23. Wouters OJ, O'Donoghue DJ, Ritchie J, Kanavos PG, Narva AS (2015) Early chronic kidney disease: diagnosis, management and models of care. Nat Rev Nephrol 11:491–502

24. Wilson JMG, Jungner G (1968) Principles and practice of screening for disease. Public health papers no. 34. World Health Organization, Geneva. http://apps.who.int/iris/bitstream/10665/37650/17/WHO_PHP_34.pdf

25. Remuzzi G, Benigni A, Finkelstein FO, Grunfeld JP, Joly D, Katz I et al (2013) Kidney failure: aims for the next 10 years and barriers to success. Lancet 382(9889):353–362

26. Salekin A, Stankovic J (2016) Detection of chronic kidney disease and selecting important predictive attributes. In: 2016 I.E. international conference on healthcare informatics (ICHI), 4–7 Oct 2016, pp 262–270. doi: https://doi.org/10.1109/ICHI.2016.36

27. Viswanathan G, Upadhyay A (2011) Assessment of proteinuria. Adv Chronic Kidney Dis 18:243–248

28. Polkinghorne KR (2006) Detection and measurement of urinary protein. Curr Opin Nephrol Hypertens 15:625–630

29. Verhave JC, Gansevoort RT, Hillege HL, Bakker SJ, De Zeeuw D, de Jong PE (2004) An elevated urinary albumin excretion predicts de novo development of renal function impairment in the general population. Kidney

Int Suppl (92):S18–S21. doi:https://doi.org/10.1111/j.1523-1755.2004.09205.x

30. Lopez-Giacoman S, Madero M (2015) Biomarkers in chronic kidney disease, from kidney function to kidney damage. World J Nephrol 4:57–73

31. Thorp ML, Smith DH, Johnson ES, Vupputuri S, Weiss JW, Petrik AF et al (2012) Proteinuria among patients with chronic kidney disease: a performance measure for improving patient outcomes. Jt Comm J Qual Patient Saf 38:277–282

32. Stoycheff N, Pandya K, Okparavero A, Schiff A, Levey AS, Greene T et al (2011) Early change in proteinuria as a surrogate outcome in kidney disease progression: a systematic review of previous analyses and creation of a patient-level pooled dataset. Nephrol Dial Transplant 26:848–857

33. Gubala V, Harris LF, Ricco AJ, Tan MX, Williams DE (2012) Point of care diagnostics: status and future. Anal Chem 84:487–515

34. Yetisen AK, Martinez-Hurtado JL, Garcia-Melendrez A, Vasconcellos FC, Lowe CR (2014) A smartphone algorithm with interphone repeatability for the analysis of colorimetric tests. Sensor Actuat B-Chem 196:156–160

35. Yetisen AK, Butt H, Volpatti LR, Pavlichenko I, Humar M, Kwok SJ et al (2016) Photonic hydrogel sensors. Biotechnol Adv 34:250–271. https://doi.org/10.1016/j.biotechadv.2015.10.005

36. Yetisen AK, Montelongo Y, Qasim MM, Butt H, Wilkinson TD, Monteiro MJ et al (2015) Photonic nanosensor for colorimetric detection of metal ions. Anal Chem 87:5101–5108

37. Yetisen AK, Montelongo Y, da Cruz Vasconcellos F, Martinez-Hurtado JL, Neupane S, Butt H et al (2014) Reusable, robust, and accurate laser-generated photonic nanosensor. Nano Lett 14:3587–3593

38. Hurtado JL, Lowe CR (2014) Ammonia-sensitive photonic structures fabricated in Nafion membranes by laser ablation. ACS Appl Mater Interfaces 6:8903–8908

39. Martinez-Hurtado JL, Davidson CA, Blyth J, Lowe CR (2010) Holographic detection of hydrocarbon gases and other volatile organic compounds. Langmuir 26:15694–15699

40. Martinez-Hurtado JL, Akram MS, Yetisen AK (2013) Iridescence in meat caused by surface gratings. Foods 2:499–506

41. Zawadzka M, Mikulchyk T, Cody D, Martin S, Yetisen AK, Martinez-Hurtado JL et al (2016) Photonic materials for holographic sensing. In: Serpe MJ, Kang Y, Zhang QM (eds) Photonic materials for sensing, biosensing and display devices. Springer International Publishing, Cham, pp 315–359. ISBN-10: 3319249886

42. Martinez Hurtado JL, Lowe CR (2015) An integrated photonic-diffusion model for holographic sensors in polymeric matrices. J Membr Sci 495:14–19. https://doi.org/10.1016/j.memsci.2015.07.064

43. Yetisen AK, Martinez-Hurtado JL, da Cruz Vasconcellos F, Simsekler MC, Akram MS, Lowe CR (2014) The regulation of mobile medical applications. Lab Chip 14:833–840

44. Martinez-Hurtado JL, Yetisen AK, Yun SH (2017) Multiplex smartphone diagnostics. Methods Mol Biol 1546:295–302

45. Free AH, Rupe CO, Metzler I (1957) Studies with a new colorimetric test for proteinuria. Clin Chem 3:716–727

46. Giordano AS, Allen N, Winstead M, Payton MA (1957) A new colorimetric test for albuminuria. Am J Med Technol 23:216–219

47. Longfield GM, Holland DE, Lake AJ, Knights EM Jr (1960) Comparison studies of simplified tests for glucosuria and proteinuria. Tech Bull Regist Med Technol 30:76–78

48. Pugia MJ, Lott JA, Profitt JA, Cast TK (1999) High-sensitivity dye binding assay for albumin in urine. J Clin Lab Anal 13:180–187

Chapter 37

A User-Friendly App for Blood Coagulation Disorders

Johannes Vegt and Paul C. Guest

Abstract

There is a strong association between a suboptimal maternal environment and increased risk of developing age-associated diseases such as type 2 diabetes, obesity, and cardiovascular disease in the offspring. Blood clotting time may be altered in all of these conditions, and it is also an important factor that requires monitoring in postoperative and cardiovascular disorder patients who are on coagulant medications. This chapter describes patient self-management of blood coagulation activity using a test strip device and the Coagu app. The app can also be used as a reminder of treatment times and for monitoring the effects of treatment over time.

Key words Proteomics, Blood coagulation, Clotting cascade, Anticoagulant test strips, Coagu app, Schizophrenia

1 Introduction

Obesity and poor nutrition during pregnancy can have long-term effects on the health of the offspring including risk of developing diseases such as cardiovascular disorders [1]. A recent meta-analysis found that the blood-based proteomic biomarker candidates most associated with cardiovascular disease include proteins associated with the blood coagulation cascade and wound healing [2]. Monitoring of the time it takes the blood to clot is also important for cardiovascular disorder and postoperative patients who have been prescribed anticoagulant therapies [3]. These medications are often used to minimize the risk of blood clots forming in the prevention of heart attacks, strokes, and blockages in blood vessels [4]. The use of home testing devices to measure the international normalized ratio (INR) of clotting time has been suggested as a means of improving patient self-management, as well as increasing the monitoring frequency and improving management and safety of long-term anticoagulation therapies [5].

Paul C. Guest (ed.), *Investigations of Early Nutrition Effects on Long-Term Health: Methods and Applications*, Methods in Molecular Biology, vol. 1735, https://doi.org/10.1007/978-1-4939-7614-0_37, © Springer Science+Business Media, LLC 2018

Patient self-monitoring is usually achieved using coagulation test strips combined with a user-friendly reader [6]. Based on the results of monitoring, an appropriate dose of an anticoagulant can then be self-administered to achieve the required target range. The Coagu app was developed by Appamedix in Berlin, Germany, as a means of improving the speed and accuracy of this process [7]. This app is based on a universal design, facilitating its use by people of all ages. It is now used by patients in more than 70 countries and has been acclaimed for its user friendliness by the International Design Centre Berlin and the International Funkausstellung (IFA) [8]. This chapter describes the basic use of the Coagu app.

2   Materials

1. Blood drop (*see* **Note 1**).
2. Clotting time test strips and meter (*see* **Note 1**).
3. Smartphone or similar device (*see* **Note 2**).
4. Coagu app (http://www.coagu.com/en/) (*see* **Note 3**).

3   Methods (Fig. 1)

3.1 INR Measurement

1. Insert the test strip into coagulation meter.
2. Insert a sterile lance into the lancet device (*see* **Note 1**).
3. Wash hands in soapy water and dry the designated fingertip thoroughly.
4. Pierce the tip of a finger and immediately apply the resulting blood drop onto the test strip.
5. After approximately 60 s, read the INR value which appears on the meter display (*see* **Note 4**).

3.2 App Usage: Determination of Anticoagulant and Dosage (See Note 5)

1. Open the Coagu app on your device.
2. For first use, configure the app according to specific needs (target INR range, drug names, and dose) (*see* **Note 6**).
3. Enter the measured value which is stored and displayed in the calendar and histogram of the app (Fig. 2) (*see* **Note 7**).
4. Determine what dosage you should take based on the INR reading (Figs. 1 and 3) (*see* **Note 8**).
5. The values are entered in the measurement history and can be visualized over a 6-month period in histogram format (Fig. 3) (*see* **Note 9**).

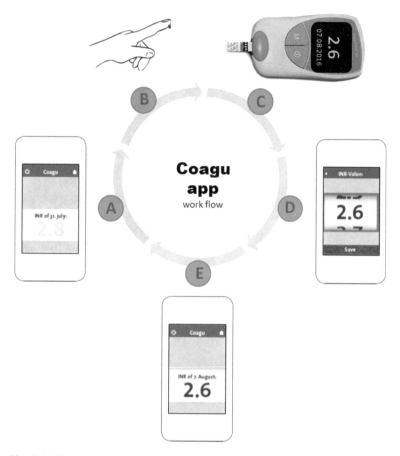

Fig. 1 (**a**) The readings of the last international normalized ratio (INR) measurement fade gradually over 7 days as a reminder to the patient to take the next measurement. (**b**) Gently pierce the tip of a finger to draw a small drop of blood. (**c**) Apply the blood drop immediately to the test strip contained in the measurement device, and the measured value appears on the meter display after 1 min (*see* **Note 3**). (**d**) Enter the INR value using a number picker and this is stored. (**e**) The value appears with the actual date on the start page of the app

6. The patient is reminded up to twice a day via a notification which occurs as an audible and visual signal with advise to set the time for taking their medication or some other action (Fig. 4) (*see* **Note 10**).

7. The patient may add a comment on the entry each day individually (Fig. 4) (*see* **Note 11**).

4   Notes

1. Sterile lances and lancet device for blood drop production comes with the CoaguChek instrument, but other similar devices can be used.

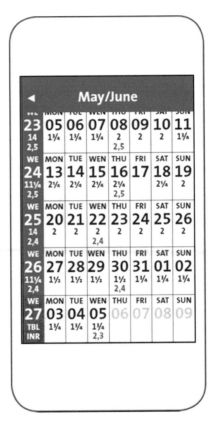

Fig. 2 Enter the measured value, which is stored and displayed on the calendar and histogram of the app. The histogram shows the measured INR values and medication use over time. Tendencies can be seen over a period of 6 months

2. Many devices can be used such as Apple iOS smartphones and tablets.

3. It should be noted that there is no connection between the producer of the CoaguChek measurement device (Hoffmann-La Roche AG) and Johannes Vegt (developer of the Coagu app).

4. Blood applied to the test strip mixes with the embedded ingredients to form a clot. The time required for clot formation is used to calculate the INR.

5. Data protection is important in the use and functioning of the app. All data should be managed by the user alone and not stored on an external server or as part of a cloud. In addition, the user should be aware that the Coagu app is not currently a medical product. For this, testing and approval by regulatory agencies in the targeted countries would be required, and acceptance as a medicinal product would be possible if the relevant health insurance companies accept liability.

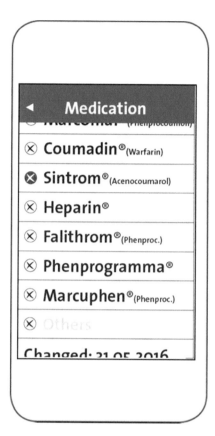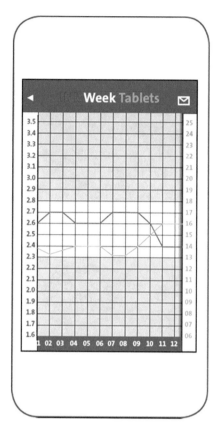

Fig. 3 Determine which medication you should take

6. The target range is the maximum and minimum INR values determined and related to each user by their doctor.

7. The average age of app users is approximately 60 years old. For this reason, large touch-sensitive surfaces were integrated into the app to allow for possible age-associated motor inaccuracies. Furthermore, larger font sizes are used at key points to allow for potential visual impairments.

8. If there is any uncertainty, one should consult their doctor.

9. The user can see any correlations between drug types, dosage, and INR reading to guide discussions with their doctor and help to determine any need for changes in medication and dosage, if required.

10. As an example, interruption or skipping the taking of the medication can be listed in the app.

11. Comments can be stored in the calendar and remain available in this format for 6 months. Using the app, the patient can also contact their physician so that their history could be updated, as required in cooperation with the health insurance agencies and higher health authorities.

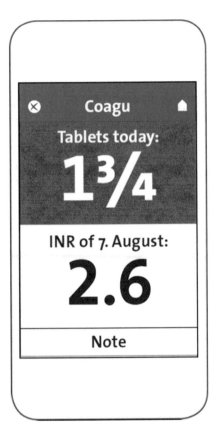

Fig. 4 Patients on anticoagulants should take their medication every day, and the app reminds them of the dose to be taken that day. After this, the patient confirms that they have taken the medication by pressing the display (this will be registered on the histogram and calendar)

References

1. Tarry-Adkins JL, Ozanne SE (2017) Nutrition in early life and age-associated diseases. Ageing Res Rev 39:96–105

2. Mokou M, Lygirou V, Vlahou A, Mischak H (2017) Proteomics in cardiovascular disease: recent progress and clinical implication and implementation. Expert Rev Proteomics 14:117–136

3. Vegt J (2017) Development of a user-friendly app for assisting anticoagulation treatment. Methods Mol Biol 1546:303–308

4. Chakrabarti R, Das SK (2007) Advances in antithrombotic agents. Cardiovasc Hematol Agents Med Chem 5:175–185

5. Pozzi M, Mitchell J, Henaine AM, Hanna N, Safi O, Henaine R (2016) International normalized ratio self-testing and self-management: improving patient outcomes. Vasc Health Risk Manag 12:387–392

6. van den Besselaar AM, Meeuwisse-Braun J, Schaefer-van Mansfeld H, van Rijn C, Witteveen E (2000) A comparison between capillary and venous blood international normalized ratio determinations in a portable prothrombin time device. Blood Coagul Fibrinolysis 11:559–562

7. http://www.coagu.com/en/

8. http://www.inr-austria.at/index.php?article_id=6

Chapter 38

Hormonal Smartphone Diagnostics

P. R. Matías-García, J. L. Martinez-Hurtado, A. Beckley, M. Schmidmayr, and V. Seifert-Klauss

Abstract

Mobile point-of-care diagnostics are paramount for the provision of healthcare. Hormonal diagnostics are powerful tools to monitor timely changes in human physiology. Hormone concentrations in serum directly correlate with urine excretions with minor time delays. Therefore, rapid tests for hormones in urine have been widely used for decades as means of early diagnostics, particularly in lateral flow immunoassay formats. However, the challenge of reading and interpreting these binary tests remains. Here we present a method for utilizing mobile technologies to quantitatively read and interpret hormonal test strips. The method demonstrates the detection of a urinary by-product of progesterone, pregnanediol glucuronide (PdG), and its relation to ovulation and the fertility cycle.

Key words Smartphone, Diagnostics, Mobile, Medical, Application, Quantitative assays, Hormones, Progesterone, PdG, Ovulation

1 Introduction

Hormones are recognized by specific receptors in the cells in which they are to exert an effect, depending on their secretion and circulation patterns. They play key roles in a variety of physiological processes, ranging from development and growth to metabolism regulation or sexual differentiation, and can act upon the same cells or tissues that secreted them or in neighboring ones. In terms of their chemical structure, three major groups can be distinguished: amino acid-, cholesterol-, and phospholipid-derived hormones [1]. Steroid hormones are synthesized in the adrenal cortex and gonadal tissues starting from cholesterol. Progesterone and estradiol play important roles in a range of different reproductive and nonreproductive physiological and metabolic processes [2]. Besides adrenal gland functions, these hormones are responsible for maintaining skeletal homeostasis by modulating apoptosis and differentiation in osteoblasts and cytokine production in bone marrow [3–6]. They are also involved in regulating both normal and

Paul C. Guest (ed.), *Investigations of Early Nutrition Effects on Long-Term Health: Methods and Applications*, Methods in Molecular Biology, vol. 1735, https://doi.org/10.1007/978-1-4939-7614-0_38, © Springer Science+Business Media, LLC 2018

disease-associated lung functions [7, 8], and there is evidence on a role in the immune responses to infections [9]. Other related neurosteroids are involved in regulating cognition, inflammation, mitochondrial function, neurogenesis and myelin synthesis [10–12], as well as injury and stress response within the central nervous system [13].

One of the most well-known roles of hormonal regulation in the human body is in reproductive functions. Hormonal regulation of female reproduction is controlled by the hypothalamic-pituitary-gonadal axis. Gonadotropin-releasing hormone (GnRH) is the primary factor released from the central nervous system, and this modulates the release of follicle-stimulating hormone (FSH) and luteinizing hormone (LH) from the anterior pituitary. As ovarian follicles develop and mature, they release estradiol (E2). The peak in E2 triggers a surge in LH, which allows the follicle to fully mature and be released from the ovary. After ovulation, the ruptured follicle becomes the corpus luteum and produces progesterone. Progesterone binds to endothelial cells within the uterus and prepares it for embryo implantation. If implantation occurs, cells within the embryo will produce human chorionic gonadotropin (hCG), which acts to maintain corpus luteum function until the placenta develops and takes over. Thus, progesterone and estradiol levels will remain high throughout pregnancy to help support fetal growth [14–17]. Therefore, FSH, LH, E2, P4, and hCG fluctuate as a function of ovulation and pregnancy status. This chapter focuses on the particular example of progesterone. Progesterone takes part in several metabolic processes, from gland remodeling during pregnancy prior to breastfeeding [18] to mediating tissue remodeling in the endometrium [15].

Progesterone synthesis in luteal cells depends on circulating cholesterol availability [16], which depends on its endogenous synthesis and dietary factors [1]. Progesterone levels may also depend on energy expenditure in relation to energy obtained from food intake, where its synthesis and secretion is reduced in settings with a negative energy balance (expenditure being larger than intake, as in exercising women) [19]. On the other hand, progesterone insufficiency has been identified in different conditions such as luteal phase deficiency [16]. Progesterone supplementation might be beneficial in treatment of women with this condition and has also been used in groups of women at increased risk of spontaneous preterm birth [20], as well as in women experiencing the menopausal transition to ameliorate their menstrual bleeding symptoms and in assisted reproduction [14].

Urinary pregnanediol glucuronide (PdG) appears to be a good clinical marker to monitor progesterone supplementation, even though it shows a larger variability when it is exogenously administered [21]. Methods, such as calendar calculations, examination of cervical mucus and a change in basal body temperature, as well as ovarian morphology changes, have been historically used to monitor and predict ovulation [22]. However, these methods monitor physiological signs which are initially triggered by hormonal changes. These cyclic fluctuations are evident when observing LH, estrogen, and progesterone serum levels, which are intimately related with urinary metabolites. Therefore these "side-effect" measurements are rendered secondary by the direct observation of hormonal fluctuations as indicative of ovarian activity and the menstrual cell cycle [23]. LH, FSH, estrone glucuronide (ElG), and PdG are four of such urinary metabolites [24] that have been largely studied because of their advantages for self-detection and point-of-care use [22]. Furthermore, there is evidence on the close relationship between urinary and ovarian excretion rates [15] and on the high correlation between urinary and serum reproductive hormones [25]. Therefore, they may be measured interchangeably for monitoring ovarian steroidogenesis [26].

Here we report a method for measuring urinary PdG. A threshold of 6.3 µmol/24 h urinary PdG was determined and validated in a large-scale, multinational study to be the most appropriate cutoff to assess the end of the fertility period in a menstrual cycle [15]. However, a more recent study has shown that, once a measurement has exceeded a threshold of 7 µmol/24 h, ovulation can be confirmed to have taken place and the cycle considered to be infertile from that point onward [27]. Considering that the life span of a human ovum does not exceed 24 h, there is evidence of urinary PdG consistently peaking a day (or more) after ovulation, thus making it a suitable marker for postovulatory infertility [28].

It is important to monitor urinary excretion of hormones for several reasons, the most important one being that women are frequently unfamiliar with the sequence and changes happening during their menstrual cycles, which is either a consequence of general misinformation or of many years of contraception methods altering their ovulatory cycles [28]. This, added to the expected normal inter- and intraindividual variability in menstrual cycle types and the continuum of ovarian function throughout the reproductive life [29], makes fertility prediction based solely on information from previous cycles information inaccurate [28, 30]. Ever since their introduction to the market, personal fertility monitors have generally been well received by the women interested in either contraception or assistance in achieving pregnancy. It is in these settings that a noninvasive approach to monitor progression of menstrual cycles and their corresponding hormonal fluctuations on a daily basis has become increasingly important. Moreover, a urinary hormonal assay provides the advantage of ease of sample collection and noninvasiveness, when compared to serum tests [15]. An example of such a device is the Ovarian Monitor, a validated test based on a homogeneous enzyme immunoassay [27].

As noted before, developing a simple yet competitive assay for the detection of urinary PdG is the next step that must be taken to bring this technology closer to women interested in monitoring their own ovarian function [31]. However, a thorough validation is needed to examine its suitability for self-testing and clinical validity of the information provided in terms of sensitivity and specificity and precision as compared to that of laboratory assays and experimental errors [32–34]. While hormone-monitoring devices have been made previously available in the market, they will greatly benefit from their use in combination with a smartphone app for conception/contraception, both as part of daily personal monitoring, and as a research tool in related scientific disciplines. Today, point-of-care diagnostics allow patients to obtain timely results without the need for expensive equipment and promote awareness and responsibility in patients [35]. Here we report a method for the combined used of such mobile technologies and analog hormonal measurements.

2   Materials

1. 50 mL artificial urine stock solution prepared as reported elsewhere [36] to match the concentrations in Table 1.

2. 5β-Pregnane-3α, 20α-diol glucuronide (Sigma-Aldrich) added to artificial urine in 1 mL vials to obtain the concentrations for two sets of PdG test strips (Table 2).

3. Morning urine samples collected from healthy 27-year-old female volunteer during different days of the menstrual cycle using the LH peak as reference (*see* **Note 1**).

4. Ovulation Double Check® "first-generation" test strips (MFB Fertility Inc.; Boulder, CO, USA) (*see* **Note 2**).

5. Fluorescent light source at standard laboratory conditions.

6. Colorimetrix v1.0(84) (Fig. 1) smartphone app as described elsewhere [37], calibrated for PdG tests with the "Ovulation Double Check" test strips.

7. Phone model: iPhone 6S.

8. OS version: OS 10.2.1.

Table 1
Artificial urine ingredients

Component	Concentration (mM)
Peptone L	1
Yeast extract	N/A
Lactic acid	1.1
Citric acid	2
Sodium bicarbonate	25
Urea	170
Uric acid	0.4
Creatinine	7
Calcium chloride. $2H_2O$	2.5
Sodium chloride	90
Iron(II) sulfate. $7H_2O$	0.005
Magnesium sulfate. $7H_2O$	2
Sodium sulfate. $10H_2O$	10
Potassium dihydrogen phosphate	7
Dipotassium hydrogen phosphate	7
Ammonium chloride	25
Distilled water to 1 L	
Hydrochloric acid to specific pH	
Sodium hydroxide to specific pH	

Table 2
PdG concentrations for first-generation PdG test strips

	PdG first-generation strips (μg/mL)
Threshold sensitivity:	4.8 μg/mL
Point 1	0
Point 2	0.5
Point 3	1
Point 4	2
Point 5	3
Point 6	4
Point 7	5
Point 8	10

3 Methods (See Note 3)

3.1 Experimental Conditions

1. Ensure that environmental conditions are approximately 24 °C, 60% relative humidity.

3.2 Calibration

1. Expose several PdG test strips to the stock solutions shown in Table 2 (*see* **Note 2**).

2. Record repetitions of several exposed test strips.

3. Open the Colorimetrix v1.0(84) application (Fig. 1).

4. In non-measuring mode, select PdG from the test selection menu.

5. Open the camera capture to record the color information from the first reference test strip.

6. Click on capture and save (*see* **Note 4**).

7. Input the associated concentration value for the recorded calibration point in the calibration curve (*see* **Note 5**).

8. Repeat the last three steps for as many calibration points as needed (Fig. 1).

9. Once the calibration is completed, save all relevant inputs (analyte measured, strip type and name, and manufacturer) (*see* **Note 6**).

10. Request the app to show the newly recorded test calibration in measuring mode.

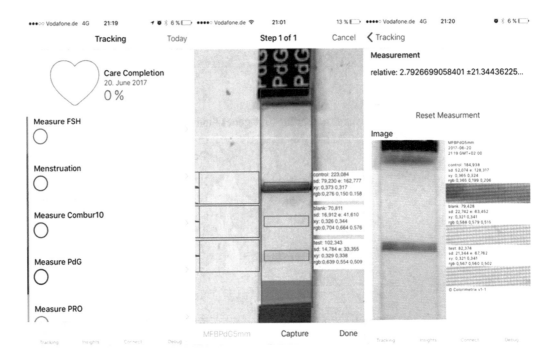

Fig. 1 App operation: (**a**) open test selection menu and select "Measure PdG"; (**b**) measurement screen, align strip to capturing areas; and **c** measurement screen storing the results

3.3 Measurement

1. In measuring mode, expose the test strips to test samples, point at the test area on the camera capture, and aim at the test area on the strips.

2. Approximately 10 cm away from the strip, a zooming option is available by pinching the screen with two fingers.

3. Freeze the camera feed by clicking capture.

4. The strip positioning can be adjusted by tap holding with one finger on the strip layout shown on the screen.

5. Ensure that the test capture area is covered entirely by the test pad area.

6. Once in position, click done (*see* **Note 7**).

7. Repeat as many times as necessary to capture all experimental data points needed (Fig. 1).

3.4 Data Analysis

1. Evaluate the calibration measurement points to test the ability of the app to measure accurately all concentration ranges (Table **2**, Fig. 2) (*see* **Note 8**).

2. Plot the data from the test samples with the calibration curve in Fig. 3 (*see* **Notes 9** and **10**).

3. Plot user data from daily measurements with LH as reference as in Fig. 4 (*see* **Notes 9** and **10**).

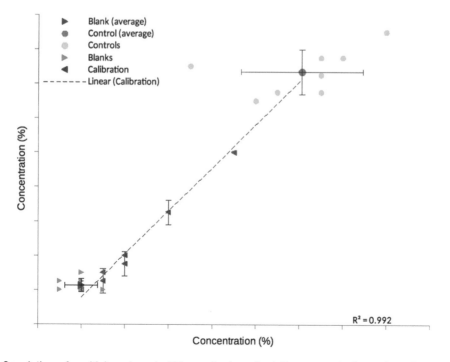

Fig. 2 Correlation of multiple values in XY coordinates of relative concentration values for calibration corresponding to the test lines: control, blank, and test area

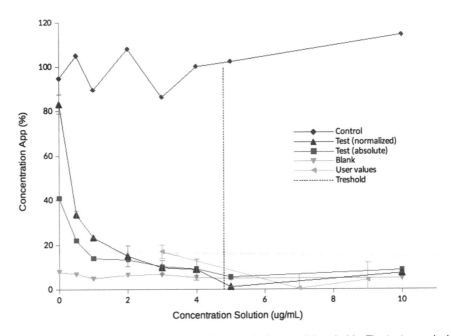

Fig. 3 Concentrations measured by the app and real concentrations and thresholds. The tests are designed to provide a positive value (when the test zone line is not noticeable) at 4.8 µg/mL, as specified by the manufacturer. There is an exponential decrease of the color as calibrated by the app. The dashed line shows the threshold. User values at different days during the cycle are inserted for reference (see Fig. 4)

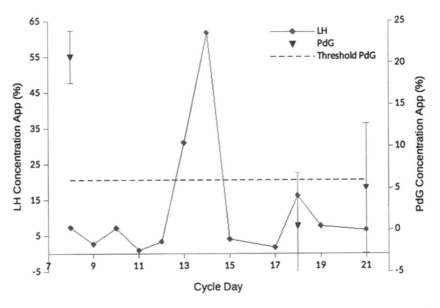

Fig. 4 User values measured along the cycle with mobile app. LH strips were used as reference to identify the time of ovulation. The concentration is shown in relative values from the maximum concentration. PdG is a competition assay, and the test line should decrease in coloration as the test becomes positive; the threshold line of the tests is shown in red

4   Notes

1. Follow waste disposal regulations for used solutions and materials. As suggested by the manufacturer of the PdG test strips, innocuous male urine can also be used as a base for calibration and to build experimental samples.

2. Ovulation Double Check tests are manufactured by MFB Fertility Inc. These tests measure PdG in urine which is the major progesterone metabolite in urine. The tests are lateral flow immunoassays in competitive format (i.e., a positive is shown by the absence of a colored line). A test and control line are visible when negative, and the color is provided by gold nanoparticle-bound antibody conjugates. To expose the test strips to stock solutions or morning urine, dip the exposure tip of the test strip into the liquid for 5 s, lay down on a flat nonabsorbent surface for 5 min, and read the result within 10 min.

3. The methods consist of three parts: (a) calibration, one time setup of the mobile application under standard conditions; (b) measurement, scalable capturing of problem samples for individuals and or laboratory solutions; and (c) data analysis, to evaluate performance and set thresholds.

4. The application will store and process the relevant color information.

5. The value is stored in a calibration file that gathers all calibration points. Five points are recommended as a minimum for proper operation.

6. The calibration information is loaded into the application source code.

7. By clicking done, the smartphone application will save the corresponding values and calculate the corresponding concentration. The recorded values are saved in an internal database on the phone for immediate access and alternatively on a universally accessible, anonymized, and secured database in the cloud for further access.

8. The app is set to measure all color information from the calibration points; these data are then loaded back into the app to activate it in measurement mode. The measurements of Fig. 2 demonstrate that control, blank, and test areas can be oriented in a linear fashion ideal for calibration. Not all tests give this linear trend. This is exclusive of lateral flow immunoassays.

9. Once the app is calibrated, a set of calibration strips is measured with the application. There is a rapid decay of signal as the hormonal concentration goes up. User values are included with given concentrations as indicated in the instructions chart by

the manufacturer for reference. The app gives low values with high error once the threshold is reached, meaning that the test is positive. A great advantage of measuring with the mobile application is that these tests can be interpreted in terms of concentrations beyond their binary functionality.

10. A user profile is recorded alongside LH measurements, to pinpoint the date of ovulation. PdG was measured at days −6, +4, and +7 from the ovulation date (estimated from the LH peak). As it can be observed, the test values go below the threshold indicating that ovulation has indeed happened and PdG has increased.

Acknowledgment

The authors acknowledge MFB Fertility Inc. for providing the PdG lateral flow test strips.

References

1. Nussey SS, Whitehead SA (2001) Endocrinology: an integrated approach, 1st edn. CRC Press, Boca Raton, FL. (14 May 2001). ISBN-10: 1859962521

2. Allen AM, McRae-Clark AL, Carlson S, Saladin ME, Gray KM, Wetherington CL et al (2016) Determining menstrual phase in human biobehavioral research: a review with recommendations. Exp Clin Psychopharmacol 24:1–11

3. Balasch J (2003) Sex steroids and bone: current perspectives. Hum Reprod Update 9:207–222

4. Seifert-Klauss V (2012) Progesteron und Knochen. Gynäkologische Endokrinologie 10 (1):37–44

5. Seifert-Klauss V, Schmidmayr M, Hobmaier E, Wimmer T (2012) Progesterone and bone: a closer link than previously realized. Climacteric 15(Suppl 1):26–31

6. Seifert-Klauss V, Prior JC (2010) Progesterone and bone: actions promoting bone health in women. J Osteoporos 2010:845180. https://doi.org/10.4061/2010/845180

7. Sathish V, Martin YN, Prakash YS (2015) Sex steroid signaling: implications for lung diseases. Pharmacol Ther 150:94–108

8. Gonzalez-Arenas A, Agramonte-Hevia J (2012) Sex steroid hormone effects in normal and pathologic conditions in lung physiology. Mini Rev Med Chem 12:1055–1062

9. Cabrera-Munoz E, Hernandez-Hernandez OT, Camacho-Arroyo I (2012) Role of estradiol and progesterone in HIV susceptibility and disease progression. Mini Rev Med Chem 12:1049–1054

10. Schumacher M, Mattern C, Ghoumari A, Oudinet JP, Liere P, Labombarda F et al (2014) Revisiting the roles of progesterone and allopregnanolone in the nervous system: resurgence of the progesterone receptors. Prog Neurobiol 113:6–39

11. Baulieu E, Schumacher M (2000) Progesterone as a neuroactive neurosteroid, with special reference to the effect of progesterone on myelination. Steroids 65:605–612

12. Wagner CK (2006) The many faces of progesterone: a role in adult and developing male brain. Front Neuroendocrinol 27:340–359

13. Brinton RD, Thompson RF, Foy MR, Baudry M, Wang J, Finch CE et al (2008) Progesterone receptors: form and function in brain. Front Neuroendocrinol 29:313–339

14. Filicori M (2015) Clinical roles and applications of progesterone in reproductive medicine: an overview. Acta Obstet Gynecol Scand 94(Suppl 161):3–7

15. Blackwell LF, Brown JB, Cooke D (1998) Definition of the potentially fertile period from urinary steroid excretion rates. Part II. A threshold value for pregnanediol glucuronide as a marker for the end of the potentially fertile period in the human menstrual cycle. Steroids 63:5–13

16. Mesen TB, Young SL (2015) Progesterone and the luteal phase a requisite to reproduction. Obstet Gynecol Clin North Am 42:135–151

17. Christensen A, Bentley GE, Cabrera R, Ortega HH, Perfito N, Wu TJ et al (2012) Hormonal regulation of female reproduction. Horm Metab Res 44:587–591

18. Macias H, Hinck L (2012) Mammary gland development. Wiley Interdisc Rev Dev Biol 1:533–557

19. Williams NI, Reed JL, Leidy HJ, Legro RS, De Souza MJ (2010) Estrogen and progesterone exposure is reduced in response to energy deficiency in women aged 25-40 years. Hum Reprod 25:2328–2339

20. Maggio L, Rouse DJ (2014) Progesterone. Clin Obstet Gynecol 57:547–556

21. Stanczyk FZ, Gentzschein E, Ary BA, Kojima T, Ziogas A, Lobo RA (1997) Urinary progesterone and pregnanediol. Use for monitoring progesterone treatment. J Reprod Med 42:216–222

22. Collins WP (1991) The evolution of reference methods to monitor ovulation. Am J Obstet Gynecol 165:1994–1996

23. Brown JB (2011) Types of ovarian activity in women and their significance: the continuum (a reinterpretation of early findings). Hum Reprod Update 17(2):141–158

24. Cekan SZ, Beksac MS, Wang E, Shi S, Masironi B, Landgren BM et al (1986) The prediction and/or detection of ovulation by means of urinary steroid assays. Contraception 33:327–345

25. Roos J, Johnson S, Weddell S, Godehardt E, Schiffner J, Freundl G et al (2015) Monitoring the menstrual cycle: comparison of urinary and serum reproductive hormones referenced to true ovulation. Eur J Contracept Reprod Health Care 20:438–450

26. Branch CM, Collins PO, Collins WP (1982) Ovulation prediction: changes in the concentrations of urinary estrone-3-glucuronide, estradiol-17 beta-glucuronide and estriol-16 alpha-glucuronide during conceptional cycles. J Steroid Biochem 16:345–347

27. Blackwell LF, Vigil P, Alliende ME, Brown S, Festin M, Cooke DG (2016) Monitoring of ovarian activity by measurement of urinary excretion rates using the ovarian monitor, part IV: the relationship of the pregnanediol glucuronide threshold to basal body temperature and cervical mucus as markers for the beginning of the post-ovulatory infertile period. Hum Reprod 31:445–453

28. Johnson S, Weddell S, Godbert S, Freundl G, Roos J, Gnoth C (2015) Development of the first urinary reproductive hormone ranges referenced to independently determined ovulation day. Clin Chem Lab Med 53:1099–1108

29. O'Connor KA, Ferrell R, Brindle E, Trumble B, Shofer J, Holman DJ et al (2009) Progesterone and ovulation across stages of the transition to menopause. Menopause 16:1178–1187

30. Blackwell LF, Vigil P, Cooke DG, d'Arcangues C, Brown JB (2013) Monitoring of ovarian activity by daily measurement of urinary excretion rates of oestrone glucuronide and pregnanediol glucuronide using the ovarian monitor, part III: variability of normal menstrual cycle profiles. Hum Reprod 28:3306–3315

31. Ecochard R, Leiva R, Bouchard T, Boehringer H, Direito A, Mariani A et al (2013) Use of urinary pregnanediol 3-glucuronide to confirm ovulation. Steroids 78:1035–1040

32. Blackwell LF, Vigil P, Gross B, d'Arcangues C, Cooke DG, Brown JB (2012) Monitoring of ovarian activity by measurement of urinary excretion rates of estrone glucuronide and pregnanediol glucuronide using the ovarian monitor, part II: reliability of home testing. Hum Reprod 27:550–557

33. Blackwell LF, Brown JB, Vigil P, Gross B, Sufi S, d'Arcangues C (2003) Hormonal monitoring of ovarian activity using the ovarian monitor, part I. Validation of home and laboratory results obtained during ovulatory cycles by comparison with radioimmunoassay. Steroids 68:465–476

34. Bouchard TP, Genuis SJ (2011) Personal fertility monitors for contraception. CMAJ 183:73–76

35. Yetisen AK, Martinez-Hurtado JL, da Cruz Vasconcellos F, Simsekler MC, Akram MS, Lowe CR (2014) The regulation of mobile medical applications. Lab Chip 14:833–840

36. Martinez-Hurtado JL, Yetisen AK, Yun SH (2017) Multiplex smartphone diagnostics. Methods Mol Biol 1546:295–302

37. Yetisen AK, Martinez-Hurtado JL, Garcia-Melendrez A, Vasconcellos FC, Lowe CR (2014) A smartphone algorithm with interphone repeatability for the analysis of colorimetric tests. Sensor Actuat B-Chem 196:156–160

INDEX

Printed in the United States
By Bookmasters